EVOLUTION IN THE MICROBIAL WORLD

Other Publications of the
*Society for General Microbiology**

THE JOURNAL OF GENERAL MICROBIOLOGY
THE JOURNAL OF GENERAL VIROLOGY

SYMPOSIA

1 THE NATURE OF THE BACTERIAL SURFACE
2 THE NATURE OF VIRUS MULTIPLICATION
3 ADAPTATION IN MICRO-ORGANISMS
4 AUTOTROPHIC MICRO-ORGANISMS
5 MECHANISMS OF MICROBIAL PATHOGENICITY
6 BACTERIAL ANATOMY
7 MICROBIAL ECOLOGY
8 THE STRATEGY OF CHEMOTHERAPY
9 VIRUS GROWTH AND VARIATION
10 MICROBIAL GENETICS
11 MICROBIAL REACTION TO ENVIRONMENT
12 MICROBIAL CLASSIFICATION
13 SYMBIOTIC ASSOCIATIONS
14 MICROBIAL BEHAVIOUR, 'IN VIVO' AND 'IN VITRO'
15 FUNCTION AND STRUCTURE IN MICRO-ORGANISMS
16 BIOCHEMICAL STUDIES OF ANTIMICROBIAL DRUGS
17 AIRBORNE MICROBES
18 THE MOLECULAR BIOLOGY OF VIRUSES
19 MICROBIAL GROWTH
20 ORGANIZATION AND CONTROL IN PROKARYOTIC AND EUKARYOTIC CELLS
21 MICROBES AND BIOLOGICAL PRODUCTIVITY
22 MICROBIAL PATHOGENICITY IN MAN AND ANIMALS
23 MICROBIAL DIFFERENTIATION

* Published by the Cambridge University Press, except for the first Symposium, which was published by Blackwell's Scientific Publications Limited.

EVOLUTION IN THE MICROBIAL WORLD

TWENTY-FOURTH SYMPOSIUM OF THE
SOCIETY FOR GENERAL MICROBIOLOGY
HELD AT
IMPERIAL COLLEGE LONDON
APRIL 1974

Published for the Society for General Microbiology

CAMBRIDGE UNIVERSITY PRESS

Published by the Syndics of the Cambridge University Press
Bentley House, 200 Euston Road, London NW1 2DB
American Branch: 32 East 57th Street, New York, N.Y.10022

Library of Congress Catalogue Card Number: 73–94353

ISBN: 0 521 20416 X

First published 1974

Printed in Great Britain
at the University Printing House, Cambridge
(Brooke Crutchley, University Printer)

CONTRIBUTORS

BAKER, J. R., MRC Biochemical Parasitology Unit, The Molteno Institute, Cambridge CB2 3EE.

CLARKE, PATRICIA H., Department of Biochemistry, University College, London WC1E 6BT.

DRAKE, J. W., Department of Microbiology, University of Illinois, Urbana, Illinois 61801, USA.

ELTON, R. A., MRC Virology Unit, Institute of Virology, University of Glasgow.

ESSER, K., Lehrstuhl Allgemeine Botanik, Ruhr-Universität, Bochum, Germany.

GABEL, N. W., Laboratory of Chemical Evolution, University of Maryland, College Park, Maryland 20742, USA.

HARTLEY, B. S., MRC Laboratory of Molecular Biology, Hills Road, Cambridge CB2 2QH.

JOKLIK, W. K., Department of Microbiology and Immunology, Duke University Medical Center, Durham, North Carolina 27710, USA.

KUBITSCHEK, H. E., Division of Biological and Medical Research, Argonne National Laboratory, Argonne, Illinois 60439, USA.

LEWIS, D. H., Department of Botany, University of Sheffield, Sheffield S10 2TN.

PECK, H. D., JR, Department of Biochemistry, University of Georgia, Athens, Georgia 30601, USA.

PONNAMPERUMA, C. A., Laboratory of Chemical Evolution, University of Maryland, College Park, Maryland 20742, USA.

POSTGATE, J. R., ARC Unit of Nitrogen Fixation, University of Sussex, Falmer, Brighton BN1 9QJ.

RICHMOND, M. H., Department of Bacteriology, University of Bristol, University Walk, Bristol BS8 1TH.

RUSSELL, G. J., MRC Virology Unit, Institute of Virology, University of Glasgow.

SKEHEL, J. J., National Institute for Medical Research, Mill Hill, London, NW7 1AA.

SNEATH, P. H. A., MRC Microbial Systematics Unit, University of Leicester, Leicester LE1 7RH.

STANIER, R. Y., Service de Physiologie Microbienne, Institut Pasteur, Paris 15e, France.

SUBAK-SHARPE, J. H., MRC Virology Unit, Institute of Virology, University of Glasgow.

WIEDEMAN, B., Department of Bacteriology, University of Bristol, University Walk, Bristol BS8 1TH.

CONTENTS

EDITORS' PREFACE

Studies on evolution are concerned with both the path taken and with the mechanisms of genetic variation and selection involved. Discussion of the path of evolution has in the past been based largely on comparative morphology and here, in spite of the recent application of electron microscopy, micro-organisms have little to offer when compared with plants and animals with their elaborate structural complexity and diversity. Now, however, when increasing weight is being given to comparative biochemistry and physiology, micro-organisms are of great value to the evolutionist, since the ease with which they can be grown in the laboratory has resulted in a knowledge of their biochemistry which rivals or surpasses that of higher animals or plants. Study of the mechanisms of evolution requires a knowledge of genetics, and the spectacular advances in microbial genetics during the last thirty years have made micro-organisms favourable material for such studies. Selection processes which have been the subject of many ingenious experiments and observations on higher animals have received little attention with micro-organisms, but with the use of the chemostat in which the operation of selection pressure can be observed over hundreds of generations this situation is being rectified. We therefore feel that the study of the evolution of micro-organisms is passing from a phase in which it was a negligible topic to one in which it will be of major importance and will make its own distinctive contribution to our understanding of evolution, and that this volume, which appears to be the first entirely devoted to evolution in the microbial world, comes at an appropriate time for summarizing recent developments and stimulating further work.

A few remarks on the logic behind the arrangement of chapters seem worthwhile. First, Sneath describes how recently developed objective methods of classification can be employed in phylogeny. Turning to the mechanisms of evolution, Drake discusses mutation rates, Richmond & Wiedeman and Esser recombination and Kubitschek selection. The evolution of nucleic acids is then dealt with by Subak-Sharpe, Elton & Russell and of proteins by Clarke and Hartley. After the evolution of molecules that of processes is considered: photosynthesis by Stanier, sulphur metabolism by Peck and nitrogen fixation by Postgate. Joklik and Skehel then discuss the evolution of viruses and Baker the evolution of one intensively studied parasitic genus, *Trypanosoma*. Lewis deals with the evolution of an ecological relationship, symbiosis, and finally Ponnamperuma & Gabel review recent experiments on the pre-biotic phase of evolution.

We thank the contributors and the staff of Cambridge University Press for their help.

Department of Biochemistry
Imperial College, London SW7 2AZ M. J. CARLILE

National Institute for Medical Research
Mill Hill, London NW7 1AA J. J. SKEHEL

PHYLOGENY OF MICRO-ORGANISMS

P. H. A. SNEATH

Medical Research Council, Microbial Systematics Unit,
University of Leicester, Leicester LE1 7RH

INTRODUCTION

Evolution is a process that we can observe only over the briefest time spans and in a minute fraction of living organisms, although we see its results over thousands of millions of years, and in every living creature we study. A major concern, therefore, is how we reconstruct the evolutionary past from the little we are able to examine.

It is convenient here to distinguish short-term from long-term evolution. We may hope to understand the processes operating in short-term evolution by observation of mutations and selection pressures. Much of this work is carried out in the framework of genetics, with predominantly experimental work, and a number of contributors to this symposium cover several aspects of it. In studying long-term evolution we have to reconstruct it from flimsy present-day evidence with little help from experiment or from biological principles other than the most general ones.

It is long-term evolution that I shall mainly discuss – the study of phylogeny. This will raise questions such as what data we have for reconstructing the past and how we use it to deduce a phylogeny. As we shall see, the very same questions frequently arise in short-term evolution.

Past attempts at microbial phylogeny

The history of phylogenetic studies in micro-organisms is strewn with wrecks of broken theories. Sometimes detailed phylogenies have been produced (for example for yeasts: Wickerham & Burton, 1961; Slodki, Wickerham & Cadmus, 1961), but it is now widely acknowledged that the evidence has been almost universally inadequate. Few attempts at phylogeny of micro-organisms have been made in recent years, and these have usually emphasized the problems in such work, stemming from the difficulty of deciding what properties of living micro-organisms are primitive (see Corliss, 1962, on protozoa and Klein & Cronquist, 1967, on thallophytes).

Bisset (1962), reviewing the main proposals on bacterial phylogeny, noted that the morphological scheme of Kluyver & van Niel (1936) cut across the physiological hypotheses of Lwoff (1944) and Knight (1945). The morphological scheme assumed that the most primitive morphology is that of a coccus; the physiological schemes postulate that the nutritionally less exacting organisms are the more primitive. Bisset's own suggestions were again different (Bisset, 1952, 1962). He considered that the most primitive bacteria are nutritionally inexacting spirilla with polar flagella, but that some forms are degenerate (see also Davis, 1964). Another thread dating at least to Orla-Jensen (1909), is that autotrophs are especially primitive, 'living fossils'. This was partly because their biochemistry was viewed as extraordinary, and also because it was thought the earliest forms of life must have possessed photosynthetic or chemoautotrophic powers because no other energy sources were then available. Recent theories of the origin of life (see Calvin, 1969; De Ley, 1968), that the earliest organisms were heterotrophs living in the 'prebiotic soup' of the primaeval oceans, have turned the physiological argument on its head. This has been reinforced by the growing feeling that autotrophs are not particularly remarkable after all (e.g. Lees, 1962; Kelly, 1971). There is at present little warrant to favour any particular one of these early ideas or to assume living bacteria are similar to ancient ones (De Ley, 1968); it must be admitted that development of the characters may have occurred independently in numerous lineages, and also reversal has probably occurred on many occasions.

The difficulty of unravelling this problem is partly due to our ignorance of what selection pressures operate in the major habitats over enormous periods of time. What is the difference, for example, between a Devonian swamp and one of today? Or between the ocean now and in early epochs? If evolution is dependent mainly on selection pressures, and if these have been the same for aeons, does this mean that present-day micro-organisms (saprophytes, at least) have not changed for long periods because they reached the limit of their adaptation many millions of years ago? De Ley (1968) discusses evidence that many bacterial species may have reached equilibrium in GC ratio, yet he notes other evidence that may imply radical changes in DNA sequence over the last million years. Yet other evidence from proteins and DNA pairing shows that homology can be recognized between very diverse bacteria. We are then faced with four possibilities: (*a*) bacteria have not altered greatly for long periods; (*b*) some parts of the genome are extraordinarily stable (conservative); (*c*) present-day bacteria are descended from a recent ancestor and have not diverged much; (*d*) recent transfer of genes accounts for the recognizable homologies, whereas other older genes are no longer recognizable as homologous at all. There is little critical evidence yet to support any one of these, and all may be partly true.

RELATIONSHIP AND CLASSIFICATION

Relationships that record the similarity between present properties of organisms, without reference to how they came to possess them, are phenetic. This is not synonymous with phenotypic relationship, as Starr & Heise (1971) have emphasized. Phenetic relationship includes genetic relationship in the sense of similarity between the present state of their genomes (Stanier, 1971), so that DNA pairing is phenetic but not phenotypic (Jones & Sneath, 1970).

Cladistic relationship describes the relationship by pathways of ancestry. It refers to how the characters of organisms arose, but not to their present properties. Cladistic and phenetic relationship are discordant if much convergent or parallel evolution has occurred, or if evolution rates have been very different.

Phenetic relationships are represented by phenograms, and cladistic ones by cladograms which are two types of dendrograms (tree-like diagrams). Fig. 1 illustrates a hypothetical example based on three positions in a protein sequence. It can be seen that organisms **b** and **d** have converged in the characters shown, and are identical in these. Consequently the phenogram and cladogram are discordant.

This figure also illustrates the difference between two components that add up to give phenetic similarity, patristic similarity and homoplastic similarity. The former is similarity in character states due to common ancestry; the latter is due to homoplasia (parallelism or convergence in evolution). Thus **a** and **c** are similar in one amino acid, glycine, because they have derived it from the common ancestor **g**, so this is patristic similarity; there is no homoplastic similarity. However, **a** and **d** are patristically similar in one character (glycine again), but have an additional homoplastic similarity in histidine, due to convergence; the observed phenetic similarity is the sum, giving similarity of two-thirds. Further distinctions can also be made (Sneath & Sokal, 1973) but these do not greatly affect our present discussion: the important point is that cladogeny is reconstructed from patristic similarity, which is why we wish to estimate it.

Fig. 1 emphasizes that without evidence on the characters of ancestors one cannot disentangle patristic and homoplastic similarity, and consequently phenetic and cladistic relations cannot be distinguished.

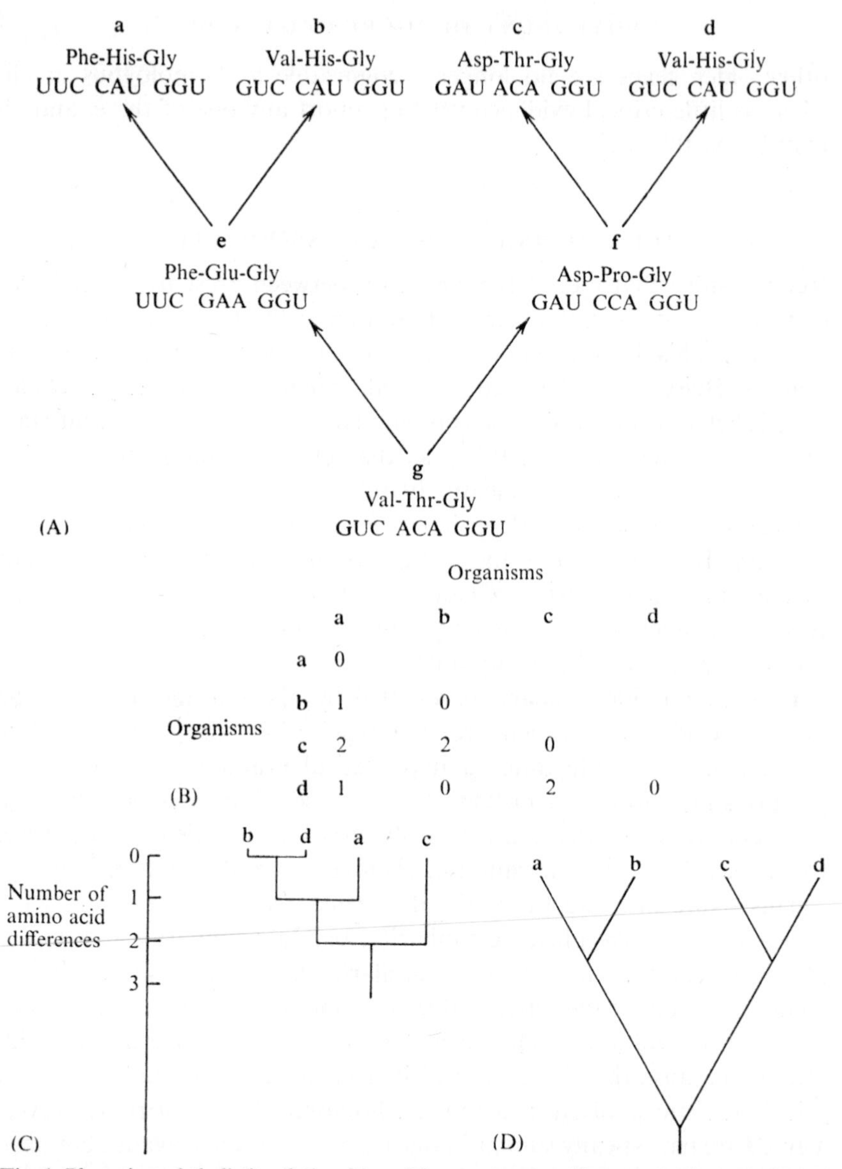

Fig. 1. Phenetic and cladistic relationships of four extant organisms, **a–d**. The hypothetical example is based on three positions in a protein sequence. The differences shown are the differences in amino acids. Similar, but slightly different, relationships would be shown by counting the number of differences in the nucleotide sequences instead. In (B) to (D) only the extant organisms have been considered: it is in theory possible to include ancestral forms if their characters are known, but this has not been done here in order to emphasize that this is seldom possible in practice.

(A) Phylogenetic tree (cladogram with characters added) of the four organisms and their ancestors **e**, **f** and **g**. The nucleotide codons are shown below the amino acids.

(B) Matrix of dissimilarities, consisting of the number of differences in amino acids between each pair of organisms.

(C) Phenogram derived by clustering the organisms on the basis of the dissimilarities. It has a defined scale, the phenetic dissimilarity between the organisms.

(D) Cladogram recording the genealogy alone. It shows the topology of branching, and has no defined scale unless one is added by using one of the numerical cladistic methods described in the text.

Indeed, even if fossils are available we are still heavily dependent on phenetic evidence. Very full arguments are given by Colless (1967, 1969*a*, *b*). Most work purporting to be phylogenetic is in reality phenetic.

Phenetic classifications are initially required to group organisms into taxa that are as homogeneous as possible, conventionally into species. The terms of Ravin (1963) are useful for distinguishing between different entities that are often called species without distinction. A taxospecies is a group of organisms (strains, isolates) that share a high proportion of similar properties, i.e. a tight phenetic cluster. A genospecies is a group that can exchange genes. A nomenspecies is a group bearing the same binominal name whatever its validity on other grounds. To this we may add a fourth term, of phylogenetic significance. This is a palaeospecies, used by geologists to mean a segment of an evolving lineage.

Much of our enquiry concerns species, and this is at two levels. Short-term evolutionary studies are particularly concerned with genetic relations within and between taxospecies, so that the distinction between taxospecies and genospecies is important. Long-term studies are primarily concerned with cladistic relations between taxospecies, though we shall note that gene exchange may be significant as well.

GENERAL CONCEPTS IN PHYLOGENETIC ANALYSIS

Cladistic reconstruction requires assumptions to be made about how evolution occurs. It is often thought that there are numerous evolutionary laws based on abundant evidence, but there are really few (Challinor, 1959) that can be relied upon in any given situation to decide, for example, whether convergence has occurred. The clearest discussion is by Hennig (1966), but few of his principles have scope in microbiology, except perhaps that of chorological progression; this postulates that a species exhibits primitive characters in the area of its origin, and it may have some application in studying the spread of disease or drug resistance.

The distinction between phenetic and phylogenetic methods lies less in the algorithms than in the assumptions underlying their use and the conclusions drawn from the results (Sneath & Sokal, 1973). The most important assumption is that of *parsimony of evolution*, a general concept dating from Darwin, and discussed in formal terms by Camin & Sokal (1965). Some such assumption underlies all phylogenetic methods, because without a general belief that evolution has followed the shortest pathways there is no constraint on the wildest of postulated pathways.

In its strictest form it requires one to draw the evolutionary pathways that imply the smallest total evolutionary change. It may be modified by subsidiary assumptions on whether evolutionary rates have been constant, whether there have been reversals in evolution, or whether convergence has occurred. These postulates can greatly affect the reconstruction.

It is doubtful if we can be sure about any of these in micro-organisms. Rapid or slow evolution, for example, can presumably occur in micro-organisms. Indeed, there is strong evidence for this from protein sequences (see below), but it does not greatly help us to recognize biochemical 'living fossils'. The biochemical genetics of the living fossil *Limulus*, the horseshoe crab, shows nothing out of the ordinary (Selander, Yang, Lewontin & Johnson, 1970).

One must distinguish between organisms whose properties have been observed, OTUs (operational taxonomic units), and those whose properties are to be reconstructed. The latter are *hypothetical taxonomic units* or HTUs (Farris, 1970). Their properties may include either character states or simply their ancestor–descendant (cladistic) relationships. In assembling the OTUs and HTUs into cladograms much use is made of graph theory (Fig. 2), in which the term graph is used in a special sense. A graph is a set of points (*nodes*) connected by links (*internodes*). The internodes may be given lengths (e.g. **a** has a given distance from **b**). Alternatively they may simply record a qualitative relation (e.g. **a** is an ancestor of **b**), in which event the topology alone is of interest. In either case it is usual to disregard the angles between the internodes, simply displaying them for greatest clarity. The nodes can be OTUs or HTUs.

If the nodes are all connected by some path the graph is *connected* (Fig. 2A, B). If there are the minimum number of connections (so that severing any internode leads to an unconnected graph) the graph is *minimally connected* (Fig. 2C). For many computational purposes only three internodes are allowed to meet at any node, but branching can also be polychotomous. A minimally connected graph is also called a *tree*. If the tree has no direction on its internodes it is an *unrooted tree*. Directed trees, or *rooted trees*, have one node V_0 as the root and directions shown to every other node (Fig. 2D).

Cladograms are rooted trees, whose root is the HTU representing the common ancestor, and the directionality leads to each tip (normally OTUs), via other nodes (normally HTUs, unless fossil OTUs are available).

The principle of evolutionary parsimony requires the discovery of the

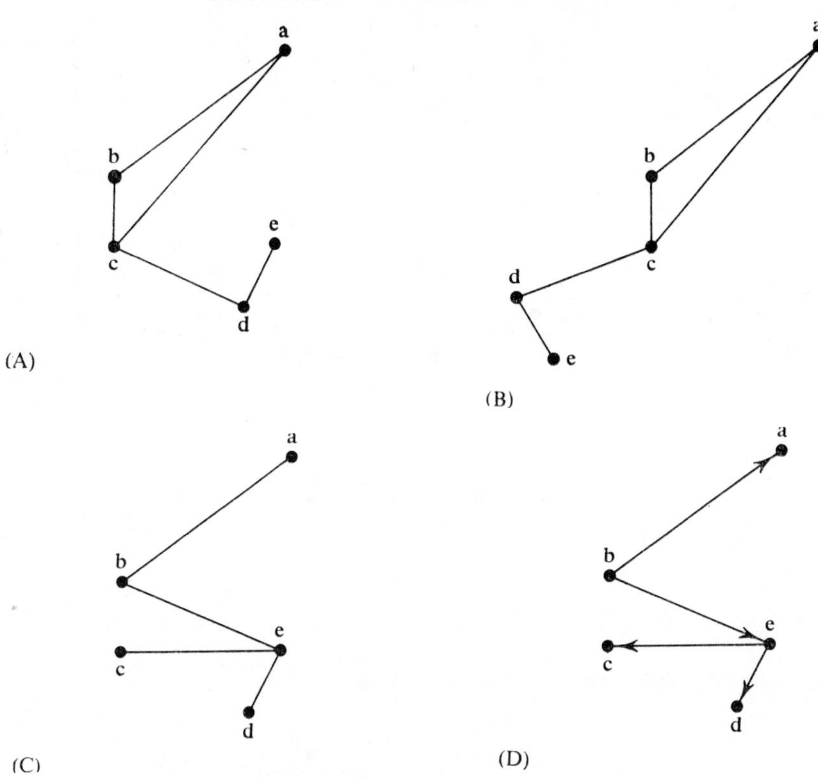

Fig. 2. Examples of different graphs (for details see text).
(A) Connected graph.
(B) Graph topologically equivalent to that in (A).
(C) Minimally connected graph or tree.
(D) Rooted tree: **b** is the root.

tree that has the minimum total length over the internodes, where these are measured in some unit of evolutionary change (e.g. number of mutations). These are *minimum length trees* (Fig. 3), of which there are two forms. The first is when no HTUs are permitted, a *shortest spanning tree* (Fig. 3B). Cladograms, however contain HTUs representing ancestors, and by introducing HTUs at appropriate positions the second form is produced, which is even shorter (Figs. 3C, 6C, 6D) and is called a *Steiner minimal tree* (Gilbert & Pollak, 1968). These then represent the most parsimonious cladograms, and cladistic reconstructions require the HTUs to be found: this is computationally difficult, so only approximations can usually be obtained. A minimal length tree may be unrooted if one cannot say which node is the common ancestor: most often a root is chosen by assuming constant evolution rates along the internodes.

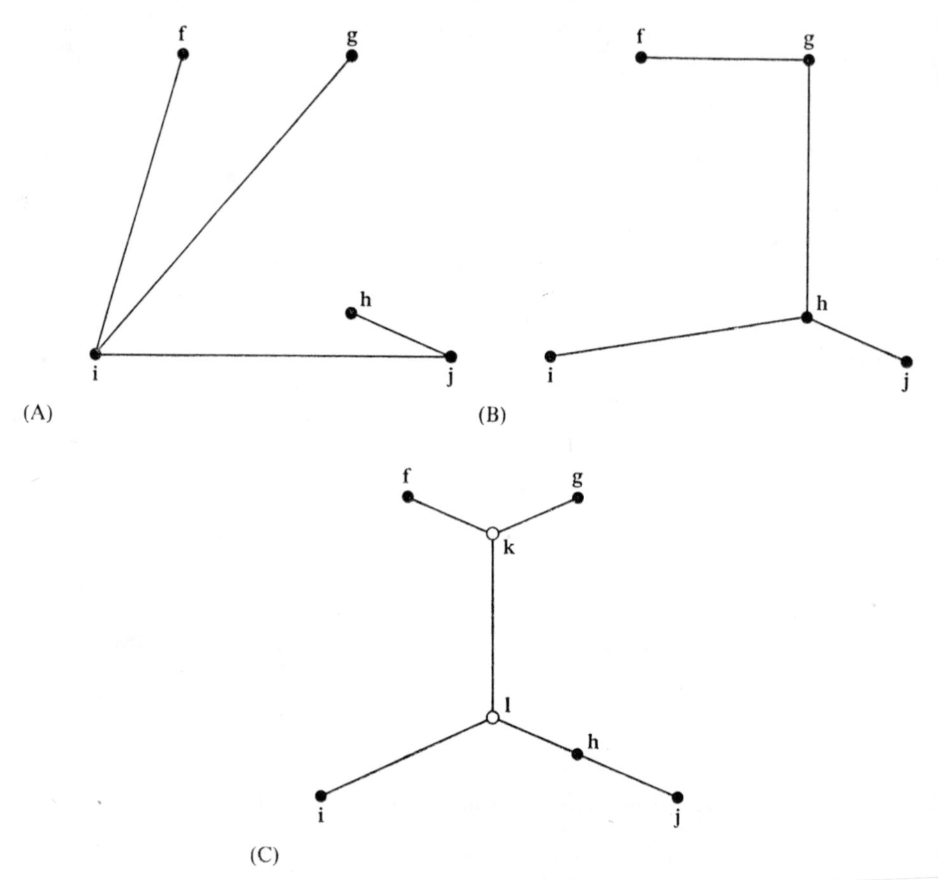

Fig. 3. Minimally connected and minimal length graphs.
(A) This is minimally connected but not of minimal length.
(B) Shortest spanning tree.
(C) Steiner minimal tree, obtained by introducing two new nodes, **k** and **l**, allowing a tree of somewhat shorter length (about 95 % of that in B).

Cladograms can express several different aspects of evolution, and it is becoming necessary to distinguish different forms, for which the following terms are suggested.

1. A *topocladogram* records only order of branching of the lineages.
2. A *chronocladogram* associates a time with each node.
3. An *orthocladogram* associates a minimum evolutionary distance with each internode.
4. Finally, we need to distinguish the last from the cladogram whose internodes represent the actual evolutionary change between nodes (not the minimal change, because the paths may not be straight), and *hodocladogram* (from Greek *hodos*, path) seems appropriate.

The difficulty of reaching firm cladistic conclusions is shown by the very large number of possible topocladograms for even a few OTUs. For rooted dichotomous trees there are $(2t-3)!/[2^{t-2}(t-2)!]$ alternatives with t OTUs – over 34 million for 10 OTUs (Cavalli-Sforza & Edwards, 1967). There are the same number of unrooted trees for $t+1$ OTUs. There are a great many cladograms that are of almost the same topology and almost of minimal length. At present, no practicable way is known of finding the correct one for most situations, so that approximate solutions are generally obtained. If fusions between lineages are permitted, the number of alternatives is greatly increased (see below).

NUMERICAL CLADISTICS

Numerical methods based on graph theory are now being used for cladistic reconstructions. They require the choice of a resemblance measure before the reconstruction can proceed.

Resemblance measures for phylogenetic studies

The most popular resemblance measures for cladistics are Manhattan distances, that reckon distances between points by adding displacements on each character axis, like the moves of a rook on a chess-board. They allow simpler computation of cladograms (and theoretically are better representations of successive mutations) than the more usual Euclidean distances that measure straight-line diagonals. They are available for morphological characters in micro-organisms, but unfortunately they are not suitable for protein or nucleic acid sequences; this is because their usefulness in cladistics depends on choosing the median character state for three OTUs, and there is no such median for the twenty amino acids or the four nucleotides, even though the number of differences is formally a Manhattan metric.

The commonest measure for protein sequences is the percentage of dissimilarities between amino acids when the two sequences are aligned to give the highest match (this match is taken to indicate the correct homology, often with allowance for gaps and insertions). Another measure, used chiefly by Fitch & Margoliash (1967, 1970) and their colleagues, is the minimal mutation distance. This is the sum over all comparable amino acid sites of the minimum number of changes in the nucleotide bases that will account for the amino acid differences. Thus to convert glutamic acid to proline requires at least two changes, from the codons GAA to CCA, or GAG to CCG. It leads to complications in reconstructing HTUs (Fitch, 1971*b*; Moore, Barnabas &

Goodman, 1973), and does not provide much more information than using amino acid differences (Sackin, 1971).

The rate of mutation at different sites in a protein is a complex subject (e.g. Kimura, 1969; Clark, 1970; Fitch, 1971*a*, 1972; Jukes & Holmquist, 1972) with disagreement on whether selectively neutral mutations are commonly preserved (e.g. Bradley, 1971; Barnabas, Goodman & Moore, 1971), so all sites are generally treated without distinction.

For nucleic acid pairing the usual measure is the per cent pairing compared with the homologous nucleic acid, but for serology there are a large variety of measures. Whatever measure is employed, it should bear a linear relation to evolutionary change, which few of them do unless transformed.

Transformations are needed for protein sequences because double and back mutations are not detected by simple frequency counts. Two sorts of correction have been proposed. The first is theoretical, for example that suggested by Margoliash & Smith (1965) in which the expected number of changes D' is $n \ln (n/(n-D))$, where there are D observed differences in n sites. Thus 50 % D corresponds to 69.3 % D'. The second, for example that used by Dayhoff and her colleagues, is empirical, based on the known frequency of point mutations between different amino acids (determined from numerous sequences); the corrected evolutionary change is expressed as *per cent accepted mutations* (PAMs). Thus 50 % D corresponds to 83 PAMs (Dayhoff, Eck & Park, 1972).

Other transformations are intended to express the probability of certain kinds of evolutionary change (Dayhoff *et al.* 1972). Thus an internode bearing rare changes (e.g. leucine to tyrosine) could be considered less likely than one bearing common changes (e.g. leucine to valine). Alternatively the rare changes could imply that the first internode represents a longer time span than the second. This might help to decide between cladograms that would otherwise be equally parsimonious.

DNA pairing also requires transformation, because it is not linearly related to evolutionary or phenetic distance (zero pairing implies indeterminately large distance). Though some transformations have been suggested (Hoyer, Bolton, McCarthy & Roberts, 1965; Sneath, 1971), study is needed on the most appropriate ones for cladistics.

Ultrametrics

An important observation by Kirsch (1969) and Jardine, van Rijsbergen & Jardine (1969) is that if evolution is constant and divergent the resulting phenetic relationships will fit a dendrogram perfectly. This is known as the ultrametric property. It is a consequence of the fact that all descendants on one branch from a common ancestor are equally related in time to all descendants on the other branch. Thus, all birds are equally related in time to all mammals by virtue of the common bird–mammal ancestor. Hence, with evolutionary divergence at constant rates the amount of phenetic difference will also be equal.

There are many exceptions (particularly in serology, where there is much current argument, e.g. Sarich & Wilson, 1967; Kirsch, 1969; Uzzell & Pilbeam, 1971) but the discrepancies are themselves of interest, as will be seen later. However, protein sequences often show good fit to ultrametrics, as is obvious on inspecting the tables in Dayhoff (1972). The importance of this is that when the ultrametric property holds, it allows easy cladistic reconstruction and also permits good estimation of geological times of divergence, once the resemblance scale has been calibrated against known geological events (e.g. Fitch & Margoliash, 1967).

Numerical cladistic methods

These methods fall into three classes. Those of the first operate only upon the resemblance matrix, without knowledge of the characters and character states. They attempt to reconstruct the cladogeny from a table of phenetic relations, for example DNA pairing. The second class requires knowledge of the characters and their states, for example protein sequences. The third class (e.g. the methods of Camin & Sokal, 1965, and Farris, Kluge & Eckardt, 1970) requires, in addition, knowledge of which character states are primitive; these methods are as yet of little use in microbiology because of the acute difficulty of deciding which are the primitive states.

Methods that operate only on a similarity matrix (most often converted into the more convenient form of phenetic distances) are restricted by assumptions on evolutionary rates – assumptions that will often be impossible to check. If the ultrametric property holds, however, it is reasonable to consider that a phenogram is a good representation of the correct topocladogram (and also of the correct chronocladogram and orthocladogram). If the ultrametric property holds only partly, then phenograms will only give correct topocladograms, and then only provided that branches are not very close and that rates are not too unequal with little convergence (Jardine *et al.* 1969; Colless, 1970). Moore (1971) shows that under plausible assumptions UPGMA clustering will in such situations give optimal estimates of cladistics. When evolution is divergent but rates are not constant the methods of Fitch & Margoliash (1967), Farris (1972), and Moore, Goodman &

Barnabas (1973) will give approximations to the most parsimonious cladogram.

Methods of the second class permit a limited amount of convergence or reversal of evolution, and do not require evolution rates to be constant. They are represented by Wagner trees (Farris, 1970), available for quantitative characters, and several methods for protein sequences that are more relevant to microbiology. They all add OTUs to a growing tree in such a way as to keep the length of the tree as short as possible, and finally allocate a root to the tree.

Protein sequence methods were introduced by Doolittle & Blombäck (1964) and developed by Eck & Dayhoff (1966): clear accounts are given by Dayhoff (1969, 1972) and Boulter *et al.* (1972). To start with, any three OTUs are joined by an HTU. The character states of the HTU are reconstructed by allocating (at each position in the sequence) the amino acid that is in the majority among the adjacent nodes. If all three are different this site is left blank in the HTU (Fig. 4). Another OTU is now fitted to each internode by constructing a temporary HTU; it is allocated to that internode that yields the minimum number of amino acid mutations (changes) along the tree. There are several rules to assist filling in the blanks by choosing ancestral states that minimize the mutations required, but a distinctive feature is that where a state is very uncertain the site is left blank rather than forcing a decision on slender grounds. In this way all OTUs are added. As in other methods, the solution may not be the very best possible, so that an

---

Legend for Fig. 4.

Fig. 4. The construction of HTUs by the method of Eck & Dayhoff for protein sequences. The HTUs are in boxes, and provisional HTUs (constructed for trial of the best position to attach a new OTU) are framed in dashed lines. Three sites in a protein are illustrated.

(A) Three OTUs, **a**, **b** and **c**, are linked to form a triad, by constructing an HTU from the majority amino acids on each branch. HTU 1 happens to be identical to **b** in this example: in practice there will usually be some differences if the sequence is a long one.

(B) A new OTU, **d**, is to be added. It is tried on each internode by constructing a provisional HTU, and counting the number of changes in the tree that are required if it is added to each internode. In this example **d** fits best to the internode between HTU 1 and **c**. This is because if it is fitted there the number of changes required is only increased by one. There are two changes required, plus one extra change that is unavoidable however the site marked ? is scored; however, two changes required by (A) can be subtracted, so only one extra is needed. Therefore a new HTU is confirmed and added here. If **d** is fitted upon the other internodes then a greater number of extra changes are needed; therefore these provisional HTUs are removed.

(C) The result of the fitting in (B). Note that the first site is left as unknown in HTU 2, because no one amino acid is in the majority on the surrounding nodes. After the tree is completed the average number of additional changes required by these gaps is calculated, and added to the others to give the total length of the tree.

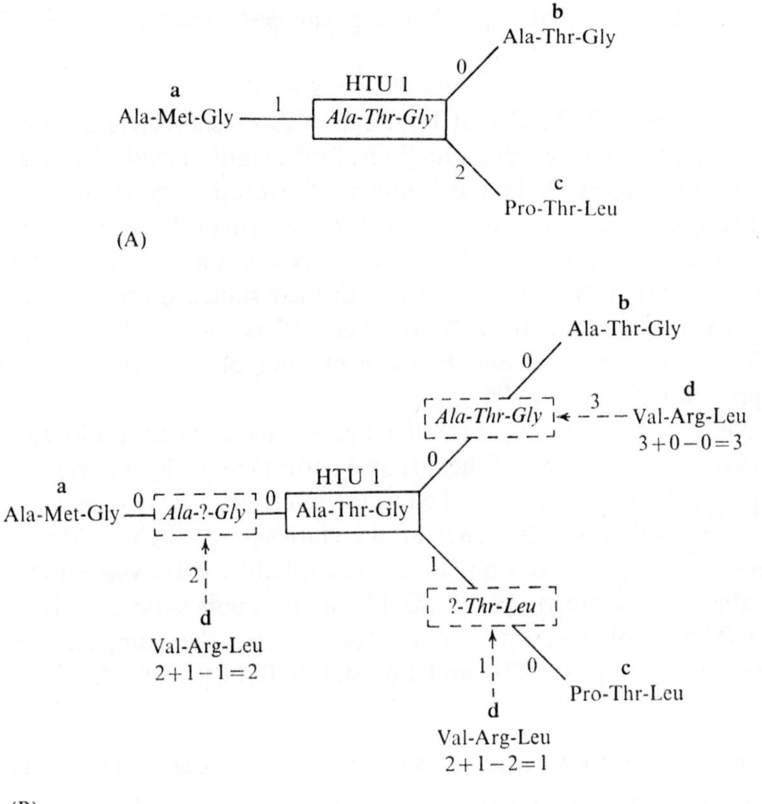

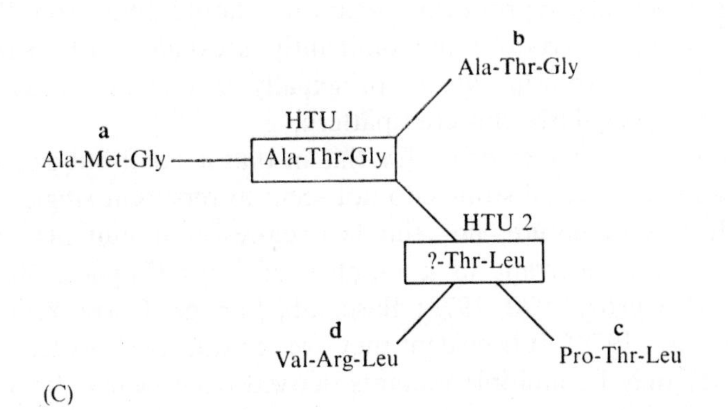

For legend to Fig. 4 see p. 12 opposite.

iterative procedure is employed to rearrange various branches in order to improve it. The root is finally chosen, usually so to make distances to the OTUs as equal as possible (e.g. smallest variance – Farris, 1972).

Errors and congruence

It is becoming realized that large numbers of characters are essential for accurate cladistic reconstructions. Fine details of cladograms depend on small differences in the number of evolutionary changes. These are subject to sampling error, which causes small departures from the ultrametric property. Evolutionary distances on cladograms should therefore properly be marked also with their standard errors, which are approximately given by $\sqrt{(D'/n)}$ where D' is the number of changes (after transformation) and n is the number of characters or sites in a protein (Dayhoff, 1972).

We also need to measure the agreement between cladograms or between a cladogram and the distance matrix from which it was derived. When real numbers are attached to the internodes the cophenetic correlation (Sokal & Rohlf, 1962) or the statistics of Fitch & Margoliash (1967) or Jardine & Sibson (1968) are available. A convenient measure of the ultrametric property would be the cophenetic correlation of a UPGMA phenogram. Topocladograms can be compared by the methods of Phipps (1971) and Bobisud & Bobisud (1972).

PHENETIC PATTERNS AND SHORT-TERM EVOLUTION

In the early days of numerical taxonomy it was noted that the phenetic patterns of sexually reproducing organisms should differ from those of organisms that reproduce predominantly asexually with occasional gene exchange (Sneath, 1962*a*), principally in that the latter would exhibit a less regularly clustered pattern.

There is now some evidence for this in that a small but appreciable proportion of bacterial strains do not seem to represent single samples from clusters in nature, nor simple drug-resistant mutants, but are isolated strains aberrant in a number of ways (Lapage, Bascomb, Willcox & Curtis, 1970, 1973; Bascomb, Lapage, Curtis & Willcox, 1973; Sneath, 1972). It is evident they are relevant to short-term evolution. They may be multiple mutants derived from nearby taxospecies, which have found some temporary niche, or they may instead have arisen as hybrids by one of the gene transfer mechanisms. We do not know their fate, whether they die out or hybridize yet further. Presumably most are poorly adapted and do not establish new taxospecies, but they may be the source of those taxospecies that are successful ones whenever new habitats arise.

We now are well aware that the widespread use of antibiotics is causing a new 'habitat' open to resistant forms (Meynell, Meynell & Datta, 1968). But there is one habitat that is relatively recent that may have become filled with progeny of aberrant forms, and that is the dairy industry, for until human stock-raising commenced some ten thousand years ago there was no dairy produce. Some milk or cheese taxospecies may be quite recent in origin. Man is continually creating new ecological niches, e.g. in large-scale storage of foodstuffs, so there are doubtless other examples.

Gene transfer in eukaryote micro-organisms follows broadly the usual pattern in higher organisms, with sexual reproduction and zygote formation, and presumably rare hybridization (though supplemented in fungi by mitotic recombination in heterokaryons). In bacteria, however, only a small part of the genome is transferred and it may be one reason for the relative ease with which wide taxonomic gaps can be crossed (Hedges, 1972). Similar phenomena are being reported in other groups (e.g. protozoa, see Ferone, O'Shea & Yoeli, 1970). The evidence on gene transfer in bacteria has been reviewed (Jones & Sneath, 1970), and numerous reports continue to appear of gene transfer between bacterial genera. There still seem to be two barriers across which no convincing transfer has been reported, that between undisputed Gram-negative and Gram-positive groups, and that between the Actinomycetales and other bacteria. But one must now accept that genes might be transferred in stepwise fashion through a long chain of genera, e.g. *Yersinia–Alcaligenes–Pseudomonas–Xanthomonas*. One limiting factor may be the need for integration into the bacterial chromosome at each step, but if the genes are carried on plasmids this may not be necessary (Falkow, 1965; Anderson, 1966). Geographic isolation plays much less part in restricting gene exchange in micro-organisms than in higher organisms: most groups are very widely distributed. Thus Ainsworth (1961) notes that about half the genera of fungi occur on more than one continent, and almost every bacterial genus must occur on every continent except Antarctica.

Extrachromosomal inheritance in bacteria has been well reviewed by Novick (1969). There is a growing view that because plasmids behave much like supernumerary chromosomes, the genome of bacteria can be considered as consisting of a number of replicons (Richmond, 1969). Some, the plasmids, are unstable and are continually merging and separating, whereas the chromosome is more stable and carries the genes most essential to cell division and energy supply (Richmond, 1970). Occasional recombination between chromosome and plasmids maintains a pool of changing plasmid genes (for a discussion on the origin of plasmids see Clowes, 1972; Richmond & Wiedeman, this vol.).

The phenetics of strains bearing plasmids has not yet been studied. An increasing number of genes are being found which can at least sometimes be plasmid-borne and are not responsible for the usual characters that geneticists investigate. These include genes controlling H_2S production (Stoleru, Gerbaud, Bouanchaud & Le Minor, 1972), mannitol and ribose fermentation in *Staphylococcus* (Schaefler, 1972) and perhaps aerial mycelium and fertility in *Streptomyces* (Okanishi, Ohta & Umezawa, 1970; Vivian, 1971; Chater & Hopwood, 1973). Richmond (1970) mentions other examples.

Grimont & Dulong de Rosnay (1972) raise the question of whether a phenetic subgroup of *Serratia* is associated with plasmid carriage. Problems have already risen for diagnostic bacteriology (e.g. lactose-positive typhoid bacilli, see Falkow, 1965). Because several per cent of the genome may be in plasmids, one might expect appreciable phenetic effects if they were looked for, paralleling those in higher organisms (Sneath & Sokal, 1973).

There may be some opportunity to study bacterial evolution over large time spans. Dr T. Cross (personal communication) has found viable spores of *Thermoactinomyces* in lake sediments 1000 to 1500 years old – a good deal older than the strains of *Bacillus* from soil reported to have survived about 300 years (Sneath, 1962*b*). Although, as Dr Cross points out, the effects of differential survival or loss of plasmids might be mistaken for an evolutionary change, yet detailed studies of single genes by serology or protein sequencing might give less ambiguous answers.

PHYLOGENETIC STUDIES OF MICRO-ORGANISMS

Data for phylogenetic studies

Many kinds of data can be used to study phylogeny. They include: (*a*) morphological characters, (*b*) physiological and biochemical characters, (*c*) gene linkage maps, (*d*) serological reactions, (*e*) nucleic acid pairing, (*f*) protein sequences, and (*g*) nucleic acid sequences. Two general points should be noted. Data from (*d*), (*f*) and (*g*) represent information from restricted portions of the genome, commonly single genes. Their evolutionary pattern usually reflects closely that of the whole genome, but exceptions are quite common. These exceptions, however, may themselves provide new insights into evolutionary history. Data from (*d*) and (*e*) are already in the form of phenetic similarities. It is not possible, therefore, to say what are the underlying character states, and difficult to know whether the antigens or nucleic acids are evolutionarily primitive or advanced. However, serology usually samples little of the genome, whereas nucleic acid pairing can, in principle, sample the entire genome.

Morphological, physiological and biochemical characters have been little used for numerical cladistics in micro-organisms. Unless large numbers of characters can be obtained they do not seem very promising. Gene linkage maps would certainly lend themselves to the kind of cladistic work that is done with banding patterns of dipteran chromosomes; they are, however, very laborious to compile, especially as one must be sure of the homologies between different genes of biosynthetic pathways. We have noted earlier that there are difficulties in obtaining suitable resemblance measures for serology. This leaves the last three classes of data, and because nucleic acid sequences are few, the main interest is in nucleic acid pairing and protein sequences. However, doublet frequencies of DNA can give much information about relationships between taxonomically distant bacteria (J. H. Subak-Sharpe, personal communication).

Most work has been done with DNA pairing, and De Ley discusses this at length in the evolutionary context. He has noted that several major groupings of bacteria could be outlined (De Ley, 1968) and that these are plausible phylogenetic groups. It has been pointed out above that DNA pairing is phenetic evidence, so that unless subjected to numerical cladistic analysis (e.g. Farris, 1972; Moore, Barnabas & Goodman, 1973) it does not give any specific information on cladogeny. When different fractions of DNA can be distinguished, this permits concepts such as that of a conserved core of genetic material resistant to evolutionary change (Dubnau, Smith, Morell & Marmur, 1965) or of ancestral portions of DNA (Park & De Ley, 1967), but interpretation depends on whether this material is evolutionarily conservative or has been transferred by recent gene exchange. Arguments based on the conservatism of functional classes of genes (e.g. ribosomal genes) are likely to prove as unsafe in micro-organisms as in higher organisms.

The most convincing cladogenies have so far been made by using protein sequences. Fitch & Margoliash (1970) list the reasons: the amount of information in a sequence is large; the evolutionary behaviour of proteins appears to follow fairly regular rules; conclusions are broadly consistent with geological evidence where it is available; it allows comparisons across the widest taxonomic gaps and the longest periods of time; homologies are usually unambiguous, or they at least provide a sequential framework in which one can hope to detect deletions in sequences and gene reduplications (as in the case of the globins). This last point is the weakest, because problems of homology continually crop up, and making allowance for insertions and deletions is apt to be arbitrary (for methods of study see, for example, Fitch, 1970; Needleman & Wunsch, 1970; Gibbs & McIntyre, 1970; Sackin, 1971). To these advantages we may add that protein sequences usually have reasonable ultrametric properties within any one family of proteins, although evolutionary rates for different families vary widely.

Recent work on microbial phylogeny

The most significant single fact to be accounted for is the wide gap between prokaryotes and eukaryotes. Their numerous differences are well discussed by Stanier (1970), Margulis (1970) and Flavell (1972). This gap is clearly shown by protein sequences wherever homologies are convincing. Unfortunately, this does not provide evidence on their origins, i.e. as to whether eukaryotes arose as prokaryotic symbionts (Margulis, 1970) or prokaryotes were derived from eukaryotes (Bisset, 1973*a*, *b*) or vice versa. These events are so far in the past that little convincing evidence is available for any theory, despite active palaeontological work on the early micro-organisms (e.g. Schopf, 1970).

Most evidence on micro-organisms is from cytochrome *c* sequences (Fig. 5). There is a good deal of uncertainty about the branching near the root of the cladogram, because several alternatives are almost equally parsimonious, and the inclusion of the prokaryote *Rhodospirillum* depends on accepting the rather tenuous homology with eukaryote sequences (Barker & Dayhoff, 1972). The homologies with other bacterial cytochromes are still poorer, so these are discussed separately. It can be seen that the number of PAM units from the root to any tip is about sixty-five, a fairly good ultrametric, so that the orthocladogram can be superimposed onto a time scale that covers the past 3000 crons (a cron is one million years – Huxley, 1957). The figure is thus also a chronocladogram, with minor discrepancies in constant evolution rates accommodated by bending the stems. The root (chosen to equalize the rates) at 2250 crons for the prokaryote–eukaryote divergence is consistent with geological and geochemical evidence. The protist *Crithidia* is, not unexpectedly, shown as a particularly early off-shoot of the eukaryote stem.

Most data for prokaryotes are on cytochrome *c*-551 for species of

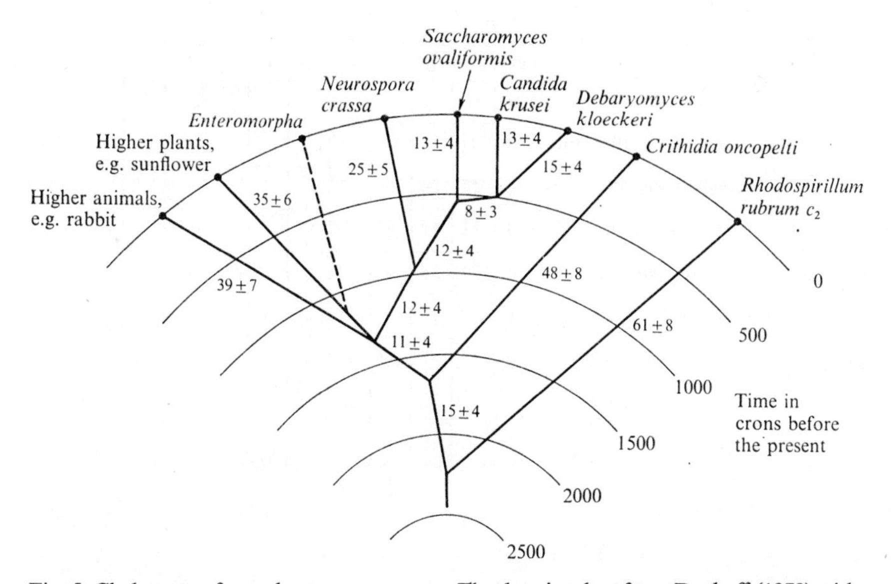

Fig. 5. Cladogram of cytochrome *c* sequences. The data is taken from Dayhoff (1972) with the addition of the sequence for *Crithidia oncopelti* (Pettigrew, 1972), using the method of Dayhoff *et al.* (1972). Brown *et al.* (1972) have positioned the alga *Enteromorpha* (Chlorophyta) near the base of the higher plant stem on unpublished data, but without exact positioning. The internodes are in PAM units with standard errors. The time scale is based on the estimate of Dayhoff *et al.* (1972).

Pseudomonas (Ambler & Wynn, 1973), with additional sequences kindly made available by Dr R. P. Ambler in advance of publication. I have chosen a manageable number for analysis by the cladistic method of Dayhoff and her colleagues, and the results are shown in Fig. 6, together with phenetic analyses for comparison.

The relationships, both phenetic and cladistic, of *Pseudomonas denitrificans* (though its taxonomic position is uncertain) and *Azotobacter vinelandii* are perhaps unexpected, but the other relationships are in good accord with general evidence given by Stanier, Palleroni & Doudoroff (1966) and Palleroni *et al.* (1970), including DNA pairing where available. The congruence between phenogram and cladogram is also noteworthy, although there are several alternative cladistic arrangements of **g** and **i** giving almost as good solutions. The tree shown in Fig. 6C is only slightly shorter than that implied by UPGMA (Fig. 6D). Unfortunately there is no way yet of estimating the time scale of these cladograms.

A limited amount of information from other proteins of micro-organisms is summarized by Dayhoff (1972). The ferredoxins show the expected close relationships between *Clostridium* species, with *Micro-*

Organism	OTU	a	b	c	d	e	f	g	h	i
P. aeruginosa P6009	a	0								
P. aeruginosa PR129B	b	1	0							
P. fluorescens 18	c	27	28	0						
P. fluorescens 50	d	26	27	3	0					
P. fluorescens 181	e	29	30	4	7	0				
P. stutzeri 221	f	26	27	23	26	24	0			
P. denitrificans 9496	g	35	35	39	37	40	32	0		
P. mendocina CH110	h	32	32	27	26	28	18	33	0	
A. vinlandii 8789	i	30	30	31	33	34	26	30	30	0

(A)

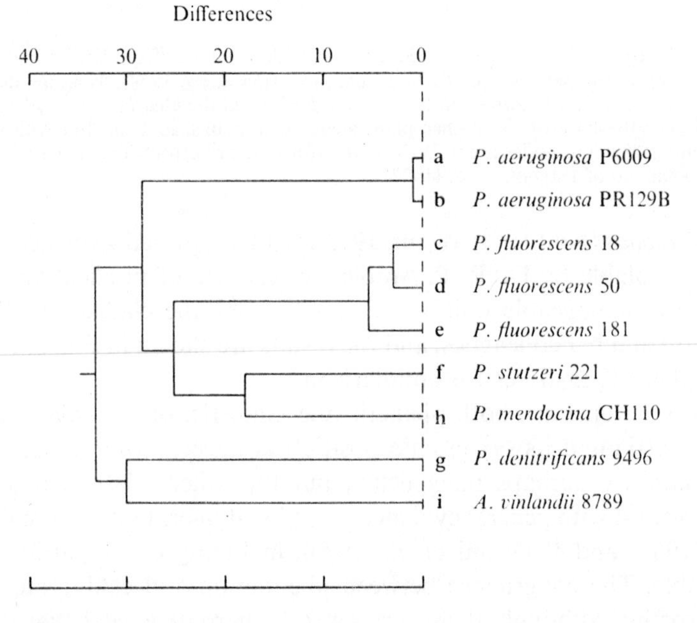

(B)

Fig. 6. Phenetic and cladistic analyses of cytochrome *c*-551 from *Pseudomonas* and *Azotobacter*. The sequences are from Ambler & Wynn (1973) and Ambler (personal communication). The strains are as follows:

OTU	Species	Strain
a	*Pseudomonas aeruginosa*	P 6009
b	*P. aeruginosa*	PR 129B (NCPPB 1224)
c	*P. fluorescens* biotype C	18 (ATCC 17400)
d	*P. fluorescens* biotype C	50 (ATCC 17427)

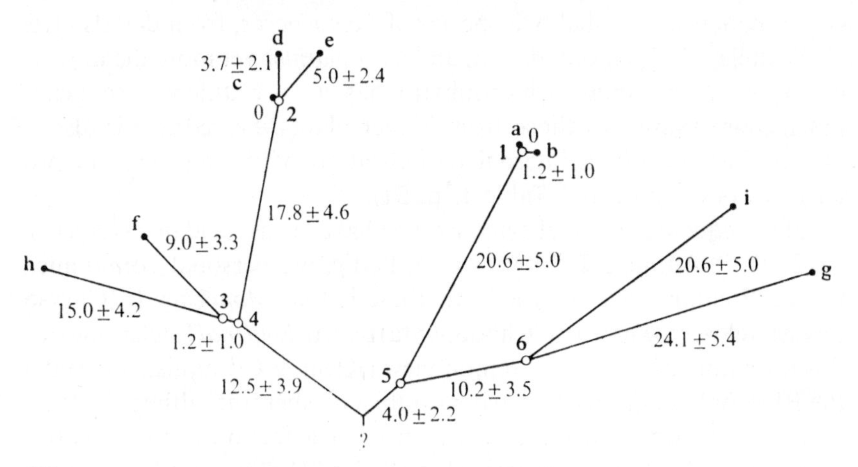

(C)

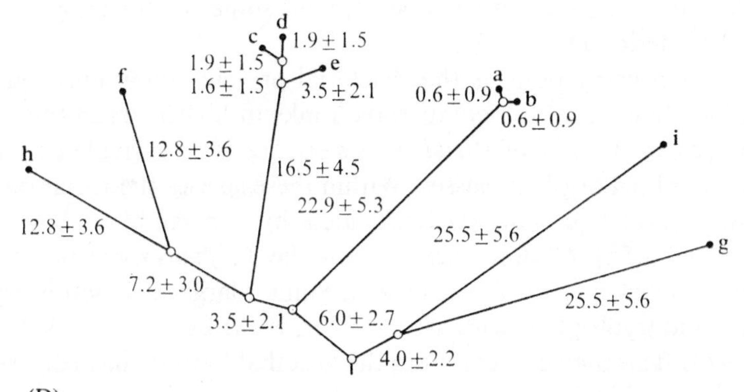

(D)

OTU	Species	Strain
e	*P. fluorescens* biotype C	181 (ATCC 11251)
f	*P. stutzeri*	221 (ATCC 17588)
g	*P. denitrificans*	NCIB 9496
h	*P. mendocina*	CH 110
i	*Azotobacter vinlandii*	NCIB 8789

The strains of *Pseudomonas* are named according to Stanier, Palleroni & Duodoroff (1966) and Palleroni *et al.* (1970), except for *P. denitrificans* which was not studied.

(A) Difference matrix between OTUs. The values shown are the number of differences in amino acids in the 82 positions of the sequence.

(B) UPGMA phenogram from the difference matrix.

(C) Orthocladogram derived by the Eck & Dayhoff method. The OTUs **a** to **i** are at the tips. The HTUs are labelled **1** to **6**, and two of them (**1** and **2**) are only formal representations, because they are identical to **a** and **c** respectively. The lengths of the internodes are in PAM units, with their standard errors appended. The root has been chosen to make the distances to the tips as equal as possible. The total length of the tree is 144.9 PAMs.

(D) A cladogram derived from the UPGMA phenogram (B) by assuming evolution was constant and divergent, but with lengths in PAM units. The total length is 146.8 PAMs.

It fits an ultrametric closely (cophenetic correlation of PAMs, 0.921; of *D*, 0.957).

coccus aerogenes (probably a species of *Peptococcus*, from details given by Whiteley, 1957) more distant, and *Chromatium* yet more distant; the eukaryote *Scendesmus* (Chlorophyta) has a very different sequence, much more similar to those from higher plants. Ferredoxin is also of interest because of an internal reduplication of the molecule in prokaryotes (see below and Table 1, p. 31).

A few sequences of algal cytochrome *f* have been elucidated (Laycock, 1972; R. P. Ambler, T. Meyer & G. Pettigrew, personal communications). Preliminary evidence from these is that *Monochrysis* (Chrysophyta) is closer to *Porphyra* (Rhodophyta) than to *Euglena* (Euglenophyta); this does not support the common view (Klein & Cronquist, 1967) that the Rhodophyta are particularly primitive or aberrant, although clearly more data is needed here. The sequences of a few other proteins (e.g. azurin) from bacteria, summarized in Dayhoff (1972) are still insufficient for cladistic analyses, but she does present some small cladogenies for certain bacteriophages and viruses.

One noteworthy point is that bacterial proteins show much greater variation than those for comparable ranks in higher organisms. Thus the cytochrome *c*-551 of *Pseudomonas* species is as variable as that of different animal or plant classes. Within these species the variation may also be large: the sequences provided by Dr Ambler showed an average of 8.5 % of differences within the *P. fluorescens* biotype C, although only 0.5 % for *P. aeruginosa*. Similar large variation is seen in azurins and tryptophan synthase (Ambler, 1968; Li, Drapeau & Yanofsky, 1973). This may give colour to the idea that bacteria have diversified primarily in biochemical modes as contrasted with the diversified morphology of animals and plants.

RETICULATE EVOLUTION

So far we have considered only cladogenies that branch with time, but fusion of lineages can also occur by hybridization. Fusion through allopolyploidy is believed to be common in plants. When there is much hybridization it leads to a reticulate pattern of evolution, and this sets severe problems for cladistic studies. Within sexually reproducing lineages there is, of course, extreme reticulation between the genomes of individuals.

The extent of gene transfer in bacteria has led Hedges (1972) to question whether it is possible to make cladogenies at all, except perhaps of single genes or proteins. He also notes what may be examples of gene transfer apart from drug resistance or laboratory experiments. Parti-

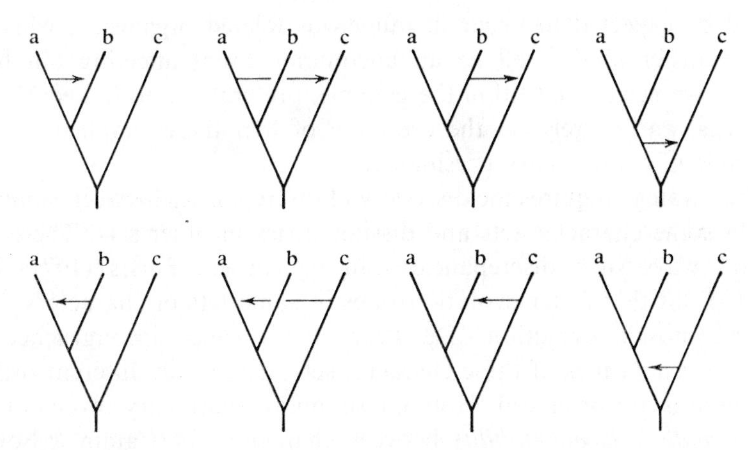

Fig. 7. Eight alternative cladograms produced by permitting a single transfer of genetic material between internodes of a branching cladogeny of three OTUs. If transfer is not directional, as in zygote formation, there are only four alternatives.

cularly noteworthy is the demonstration by Lüderitz, Staub & Westphal (1966) of the numerous parallels in somatic antigen side chains in *Escherichia* and *Salmonella*, suggesting there has been much gene transfer between these genera. How much reticulation occurs in viruses is an open question, but the wide host range of some arboviruses implies that if genetic exchange with host genome does occur, some viruses could be cladistically part insect, part bird and part mammal.

Reticulation raises two new problems: the great number of alternative cladogenies that must be distinguished, and how the effects of gene transfer can be distinguished from cladistic branching. The number of alternative cladograms grows extremely swiftly as the number of fusions increases (Sneath, 1971; Sneath & Sokal, 1973). If gene transfer occurs in a cladogeny of only three OTUs, there are eight possible transfers between internodes to distinguish (Fig. 7). And there are two other branching cladograms for three OTUs, giving twenty-four reticulate ones in all. Since there are almost $\frac{1}{2}t^2$ contemporaneous internodes in a cladogeny of t OTUs, the number of different patterns of a single gene transfer is about $t^4/4$ for each of the numerous rooted trees. For g transfers this approaches $(t^4/4)^g$ alternatives for each rooted tree, for example about 10^{14} for ten OTUs and two gene transfers.

One way of investigating this is to search systematically for *similarity within diversity* (Sneath, 1971), that is, for local similarity between the fine structure of single genes against the background of diversity of fine structure of other genes. Such similarity would either suggest recent gene transfer or extreme conservatism in evolution, but the latter would be expected to occur in numerous related organisms, whereas gene transfer might well be an uncommon event affecting but few. Clearly we need to partition the genome, preferably into linked blocks of genes, and to rely on the economy of hypotheses implied by the principle of evolutionary parsimony.

This strategy requires the discovery of discrepancies between similarities in some character sets and dissimilarities in other sets. There are several ways such discrepancies can be caused. Farris (1971) has pointed out that faster evolution rates in some sets of characters than others (mosaic evolution – De Beer, 1954) cause incongruence in phenetic similarities of these character sets; as a result different orthocladograms are produced. Homoplasia and evolutionary reversal produces *cladistic incompatibility* between character sets (Camin & Sokal, 1965), so that the sets imply different topocladograms.

Fig. 8 shows that gene transfer can mimic mosaic evolution if the rates of change differ in different internodes. In Fig. 8A is shown the effect on six characters forming a set *X* whose evolution is constant and divergent, and where one mutation takes place on each internode. The six characters of set *Y* show varying rates. The difference matrices in Fig. 8A show incongruence. They are, however, identical with those in B, where the same end-result has been given by constant divergent evolution with gene transfer. Evidently we cannot tell just from the extant OTUs which is the cause.

There are nevertheless ways in which we may hope to distinguish these phenomena. To obtain close mimicry one has to postulate specific changes in rate in different parts of the cladogram. It seems more likely that altered evolution rates would occur in one lineage (or in all lineages of a clade) for substantial periods if any major effect was to be produced. If so, most of the cladogram should obey the ultrametric property, and the exceptional portion should be distinguishable.

Legend for Fig. 8.

Fig. 8. Mosaic evolution and gene transfer.

(A) Three OTUs, **a**, **b** and **c** are shown with six two-state characters belonging to set *X* and six to set *Y*. The states of various ancestors are also shown, together with the matrices listing the number of differences between OTUs for the two sets of characters. It is seen that the comparisons **a b** and **b c** disagree in these matrices. The rates of evolution in *X* are constant, one change on each internode. The discrepancies are caused by the varying rates of change in *Y*, e.g. three changes immediately below **a**, but none just above the common ancestor.

(B) Gene transfer has caused the transfer of set *Y* from the ancestor of **c** to that of **b**, thus displacing the *Y* characters originally in the ancestor of **a** and **b**. Evolution is constant, one change on each internode. The difference matrices are identical with those in (A).

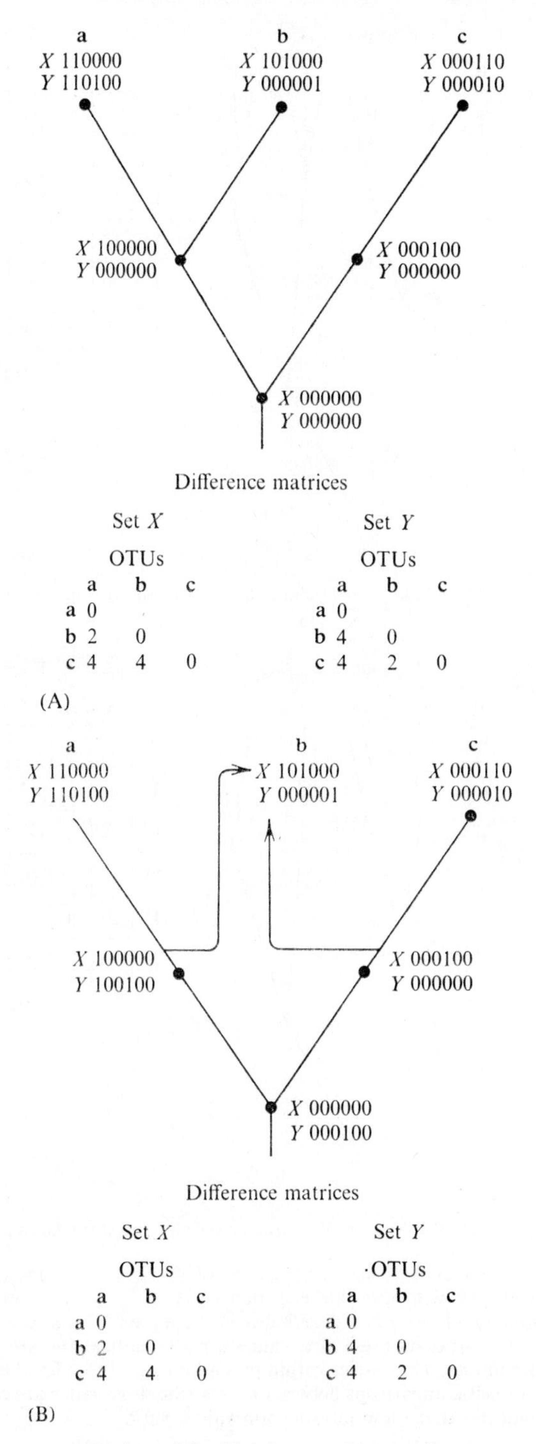

For legend for Fig. 8 see p. 24 opposite.

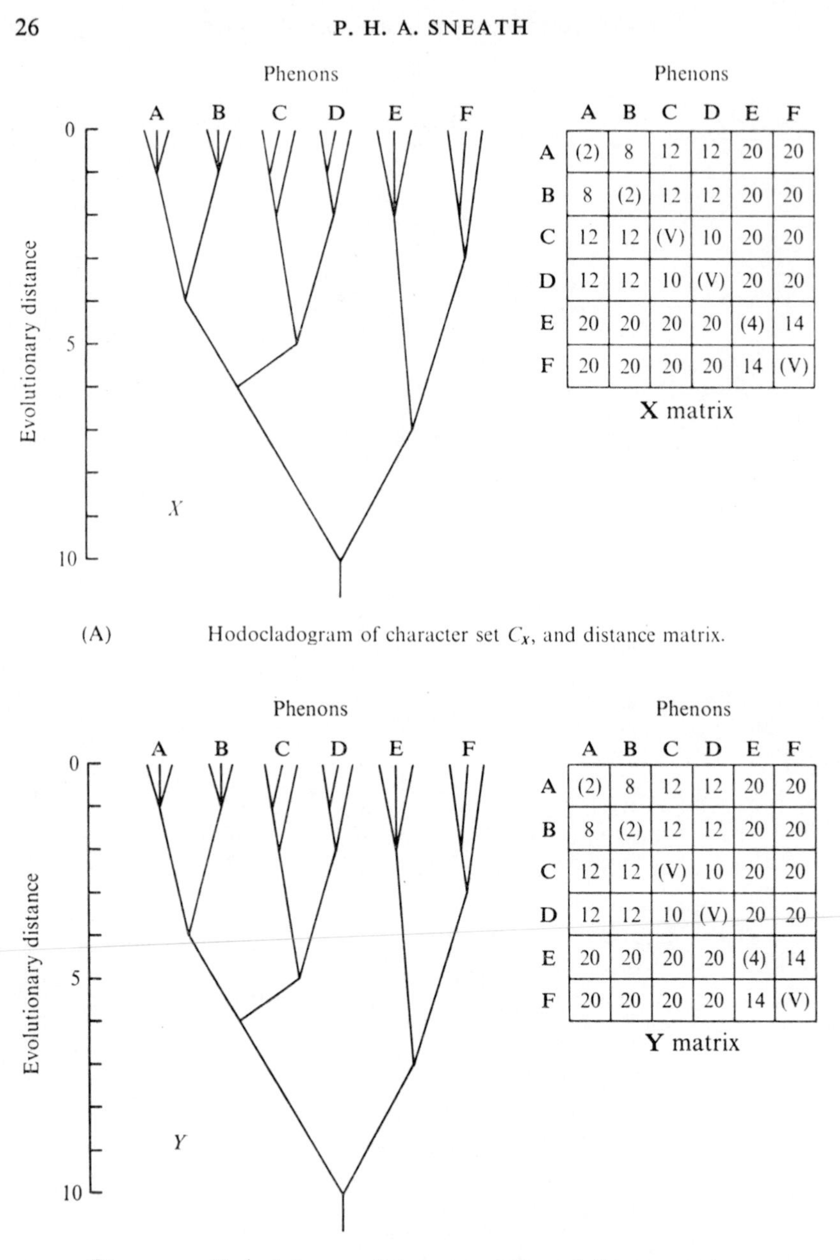

Phenons

	A	B	C	D	E	F
A	(2)	8	12	12	20	20
B	8	(2)	12	12	20	20
C	12	12	(V)	10	20	20
D	12	12	10	(V)	20	20
E	20	20	20	20	(4)	14
F	20	20	20	20	14	(V)

X matrix

(A) Hodocladogram of character set C_X, and distance matrix.

Phenons

	A	B	C	D	E	F
A	(2)	8	12	12	20	20
B	8	(2)	12	12	20	20
C	12	12	(V)	10	20	20
D	12	12	10	(V)	20	20
E	20	20	20	20	(4)	14
F	20	20	20	20	14	(V)

Y matrix

(B) Hodocladogram of character set C_Y, and distance matrix.

Fig. 9. Cladograms and scaled distance matrices of two sets of characters both obeying the ultrametric property, with the difference matrix. Each entry in the matrices **X** and **Y** represents nine individual distance values between the three OTUs of each of the phenons being compared. In most cases these nine values are all identical and are given as a single entry without parentheses. The entries within parentheses are those for the six intraphenon values that are not self-comparisons between OTUs (the three latter are necessarily zero). The symbol V indicates that these intraphenon values vary.

(C) If the **X** and **Y** matrices were previously scaled suitably, the **Y–X** matrix of differences consists entirely of zeros.

Phenons

	A	B	C	D	E	F
A	0	0	0	0	0	0
B	0	0	0	0	0	0
C	0	0	0	0	0	0
D	0	0	0	0	0	0
E	0	0	0	0	0	0
F	0	0	0	0	0	0
Mean	0	0	0	0	0	0
s.d.	0	0	0	0	0	0

Y–X matrix

For legend see p. 26 opposite.

Suppose we consider two sets of characters C_X and C_Y, and from each we derive distance (dissimilarity) matrices. The absolute values of distances may not be the same in the matrices because the numbers of characters, or the units of resemblance, or the underlying steady evolution rates, are not the same. We may, however, scale each matrix to equalize these (e.g. by dividing by the mean of all values of the appropriate matrix). This scaling is largely done when we calculate per cent differences in protein sequences, and is improved by transforming to units, like PAMs, that are additive over evolutionary change. Let the scaled matrices be **X** and **Y**. Although we may not have a true reference set let us treat **X** as the reference of comparison. Then if the ultrametric property holds completely the entries in **X** will be identical with the corresponding ones in **Y**, so that $\mathbf{Y}-\mathbf{X} = \mathbf{0}$. If a lineage or clade shows a distinctive phenomenon (such as faster evolution in C_Y) then some entries of $\mathbf{Y}-\mathbf{X}$ will not be zero. In fact there will be small non-zero values because of sampling error, but the magnitude of this will often be easy to estimate (see earlier). If the OTUs have been clustered and arranged in a suitable order it will assist the discovery of the effects described below. In Figs. 9–13 it is assumed that this has been done and that the resulting phenons are also clades.

Fig. 9 shows the result when the ultrametric property holds completely. In the succeeding figures only the cladograms and the $\mathbf{Y}-\mathbf{X}$ matrices are shown, because the **X** matrix is assumed to be the same in all. For simplicity, the separate entries for individual OTUs are replaced by one value for the phenon (as in most cases they are all in theory equal,

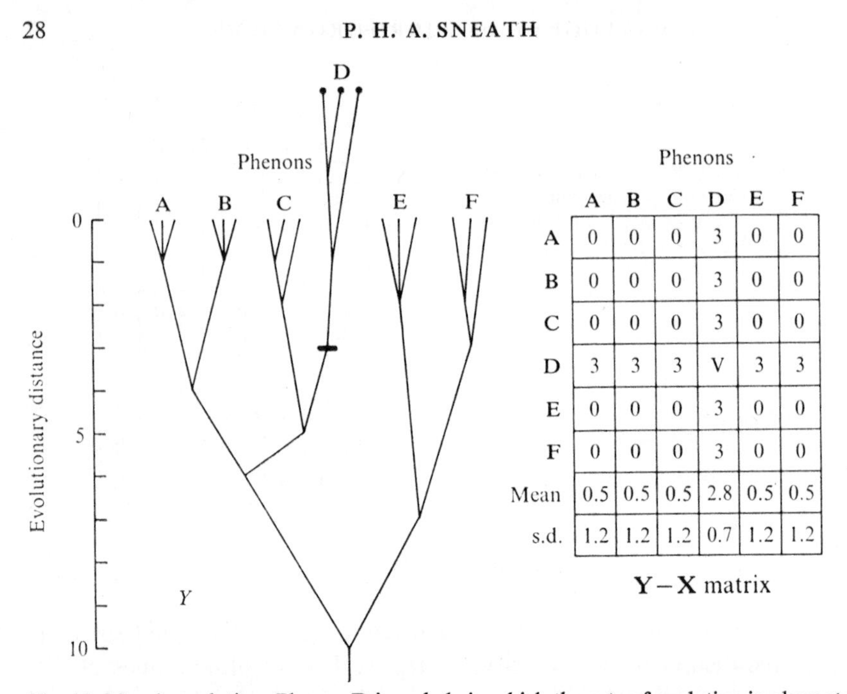

	A	B	C	D	E	F
A	0	0	0	3	0	0
B	0	0	0	3	0	0
C	0	0	0	3	0	0
D	3	3	3	V	3	3
E	0	0	0	3	0	0
F	0	0	0	3	0	0
Mean	0.5	0.5	0.5	2.8	0.5	0.5
s.d.	1.2	1.2	1.2	0.7	1.2	1.2

Fig. 10. Mosaic evolution. Phenon **D** is a clade in which the rate of evolution in character set C_Y became doubled at the point indicated by the cross-bar. The **Y**−**X** values between individual OTUs within the cell where **D** is compared with itself are theoretically $(R-1)$ times as large as the corresponding values in **X**, where R is the ratio of rates in C_Y to C_X. An approximate value for R can be obtained from the mean of column **D** by dividing this by $\frac{1}{2}X_{CD}$ (where this is half the X value for the nearest clade **C**, compared to **D**) and then adding 1. The means and standard deviations are approximate values only, because of the variable cells of the matrix.

apart from sampling error). Where they are individually important this is mentioned in the caption or text.

Mosaic evolution (Fig. 10) leads to a row and column **D**, that is consistently positive or else consistently negative, by about the same amount throughout (except within the clade). The standard deviation is therefore small. In this column even the most distant clades are affected. The difference in rate can, in theory, be estimated (see legend to Fig. 10) best from values between individual OTUs within the affected clade (these values are not zero).

Gene transfer (Fig. 11) gives a row and column that is positive in some cells and negative in others, and is zero in cells pertaining to distant clades unaffected by the transfer. The mean for the affected column is near zero, but the standard deviation is large. Within the affected clades the values for individual OTUs are zero (as in all clades in the figure).

Mistaken ancestral homology (Fig. 12) can occur if a gene redupli-

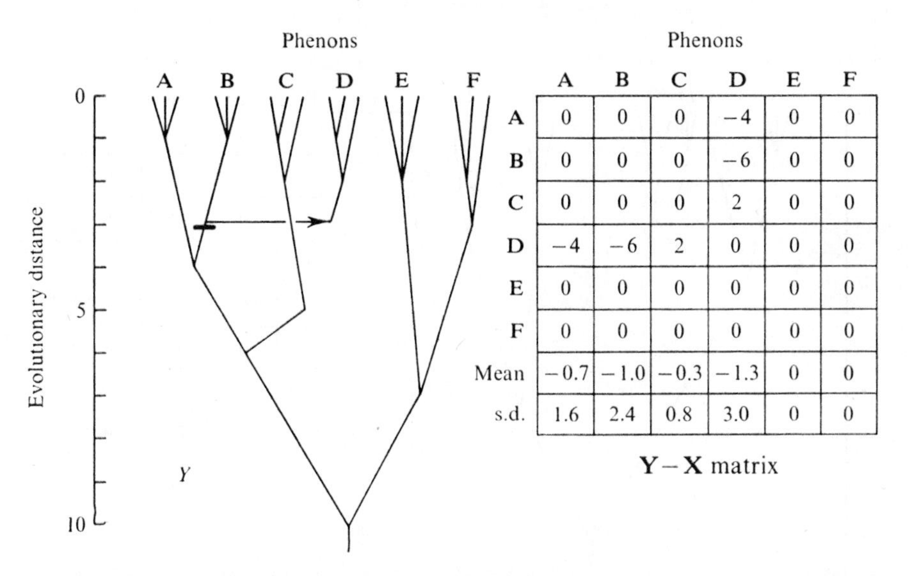

Phenons

	A	B	C	D	E	F
A	0	0	0	−4	0	0
B	0	0	0	−6	0	0
C	0	0	0	2	0	0
D	−4	−6	2	0	0	0
E	0	0	0	0	0	0
F	0	0	0	0	0	0
Mean	−0.7	−1.0	−0.3	−1.3	0	0
s.d.	1.6	2.4	0.8	3.0	0	0

Y−**X** matrix

Fig. 11. Gene transfer. The characters of C_Y in clade **D** have come from an ancestor of clade **B** at the point indicated by a cross-bar. The recipient clade shows several non-zero values, positive and negative. After excluding the column **D** the donor is indicated by the column with the most negative mean, **B**. The cell **B:D** indicates twice the distance to the transfer point, which was three units before the present.

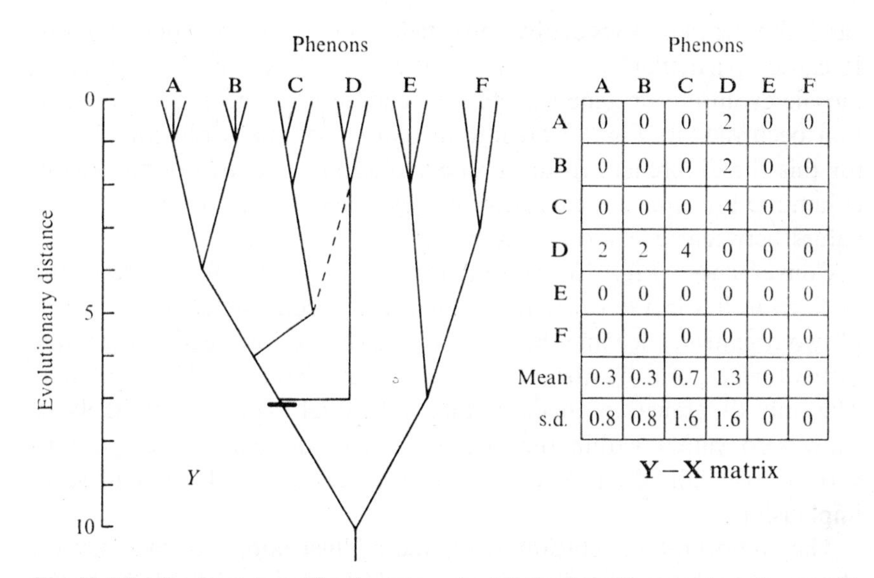

Phenons

	A	B	C	D	E	F
A	0	0	0	2	0	0
B	0	0	0	2	0	0
C	0	0	0	4	0	0
D	2	2	4	0	0	0
E	0	0	0	0	0	0
F	0	0	0	0	0	0
Mean	0.3	0.3	0.7	1.3	0	0
s.d.	0.8	0.8	1.6	1.6	0	0

Y−**X** matrix

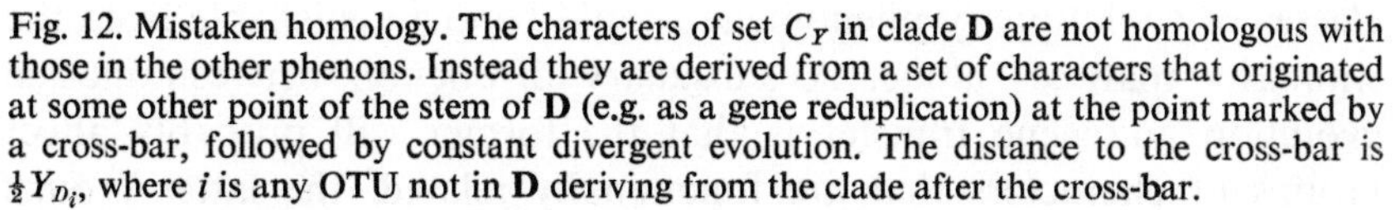

Fig. 12. Mistaken homology. The characters of set C_Y in clade **D** are not homologous with those in the other phenons. Instead they are derived from a set of characters that originated at some other point of the stem of **D** (e.g. as a gene reduplication) at the point marked by a cross-bar, followed by constant divergent evolution. The distance to the cross-bar is $\frac{1}{2}Y_{D_i}$, where i is any OTU not in **D** deriving from the clade after the cross-bar.

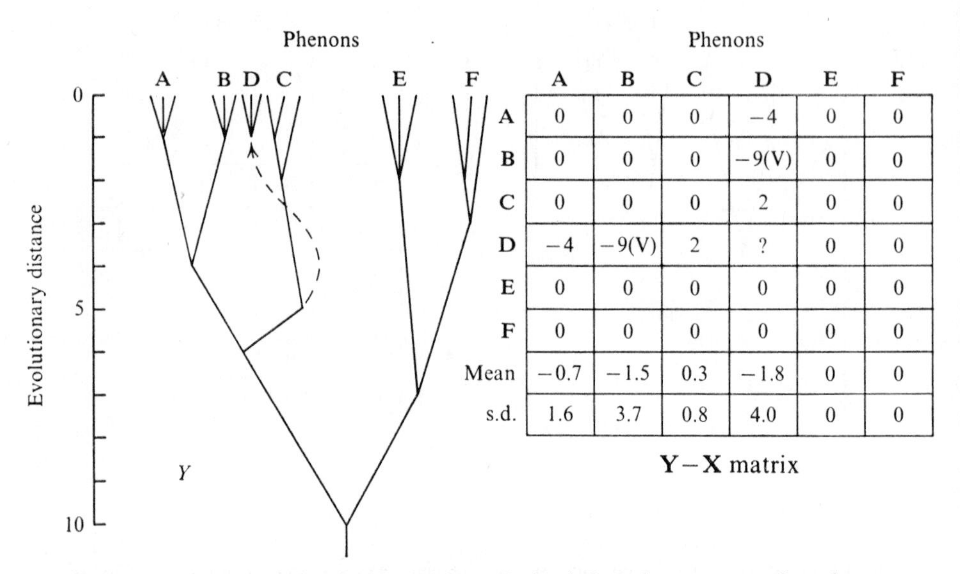

	A	B	C	D	E	F
A	0	0	0	−4	0	0
B	0	0	0	−9(V)	0	0
C	0	0	0	2	0	0
D	−4	−9(V)	2	?	0	0
E	0	0	0	0	0	0
F	0	0	0	0	0	0
Mean	−0.7	−1.5	0.3	−1.8	0	0
s.d.	1.6	3.7	0.8	4.0	0	0

Y−X matrix

Fig. 13. Convergence. The characters of set C_Y in clade **D** have converged so they are extremely similar to those of **B**. It is unlikely that, under the selection pressure causing convergence, individuals of **B** and **D** will be identical, so that the entries in the comparison of **B** to **D** will vary slightly, shown here by −9(V). The values within clade **D** are uncertain, depending on details of the convergence, but are probably small but variable.

cates and the two genes evolve separately, as in the mammalian globins. It causes particular difficulties because it can give results similar to mosaic evolution or gene transfer, depending just where the reduplication occurred. In Fig. 12 an intermediary position is illustrated, and for this the differences from mosaic evolution are two: not all cells in column **D** are equal, because two clades **E** and **F** are not affected; and values between OTUs of **D** are zero.

Convergence (Fig. 13) closely mimics gene transfer – indeed the distinction lies in the cause rather than the effect. The character states of two organisms become the same: this is caused by selection pressures in the former, and by abrupt transfer of genetic material in the latter. The only difference in effects is that individual OTUs are unlikely to show zero values within the affected converging clade, because convergence of individual OTUs on other individual OTUs would seem implausible.

The important conclusion from these illustrations is that several classes of evolutionary phenomena should be distinguishable from the similarity matrices alone. A subsidiary distinction between mosaic evolution and gene transfer is that the former will probably affect groups of characters that are functionally related, the latter groups

Table 1. **X**, **Y** *and* **Y** − **X** *matrices from the two halves of the reduplicated bacterial ferredoxin molecule, in PAM units (to nearest unit)*

X matrix

Species	OTUs	a	b	c	d	e
Clostridium pasteurianum	a	0	31	37	42	72
Clostridium butyricum	b	31	0	42	49	83
Clostridium acidi-urici	c	37	42	0	37	65
Micrococcus aerogenes	d	42	49	37	0	83
Chromatium	e	72	83	65	83	0

Y matrix

OTUs	a	b	c	d	e
a	0	7	26	37	121
b	7	0	32	37	121
c	26	32	0	49	121
d	37	37	49	0	121
e	121	121	121	121	0

Y − **X** matrix

OTUs	a	b	c	d	e
a	0	−24	−9	−5	49
b	−24	0	−10	−12	38
c	−9	−10	0	12	56
d	−5	−12	12	0	38
e	49	38	56	38	0
Mean	2.2	−1.6	9.8	6.6	36.2
s.d.	27.7	23.7	27.3	19.6	21.6

that are genetically linked, a point noted by Dubnau *et al.* (1965) in support of their views. Appropriate statistical significance tests remain to be developed.

The method is illustrated by Table 1, which shows the matrices for the two halves of bacterial ferredoxins, taken from Dayhoff (1972) with addition of values for *Chromatium* (but omitting an evident insertion). The mosaic incongruence due to the faster evolution of the second half in *Chromatium* is clearly seen in the **Y** − **X** matrix: the other values are all within two sampling standard errors (s.e. is about 13 units). The well-known differences in rates of globin evolution in mammals are also clearly shown by this technique, and correspond to the lengths on the cladograms in Dayhoff (1972). No evidence for gene transfer has been seen in the sequences studied so far, but it may be noted that the aberrant guinea pig insulin would fit a case of mosaic evolution *together* with mistaken homology; mistaken homology is implied because otherwise the increased rate would have to have started before the separation of birds and mammals.

Difference-matrix incongruences are of special consequence for DNA pairing, where the most powerful cladistic methods are not available. Poor pairing may be due to fairly even distribution of mismatches, suggesting that discrepancies between results of different techniques or DNA fractions are likely to be due to different evolution rates. The preponderance of evidence points this way (De Ley, 1968). Poor pairing would also occur if most of the DNA was very dissimilar, but a few genes were similar: this would be more in keeping with gene transfer. Park & De Ley (1967) were able to distinguish different fractions with distinctive properties, so that similarities for the separate fractions might, after linearizing transformation, be used for difference-matrix incongruence studies. Such studies could also be used to compare DNA pairing with other data, such as that from serology.

If numerous characters are available then one major problem would be to discover those character sets in which the characters are mutually compatible, with incompatibility between sets (if they could be found), so as to use these sets to discover cladistic phenomena of the kind just discussed. The number of different ways of choosing sets is extremely large. The methods of Le Quesne (1969, 1972) and Farris (1969) allow detection of incompatible pairs of characters. Le Quesne's method can be adapted to proteins: if $\mathbf{H}$ is a matrix of co-occurrences of amino acids in sites h and i, then if any off-diagonal element of $\mathbf{HH}'$ is greater than one the characters of h and i are incompatible. From these methods a 1, 0 $n \times n$ character compatibility matrix can be obtained, and complete linkage clustering will give the major compatible sets (though overlapping incompatibilities will prevent all characters from clustering).

Such sets can be used, as Le Quesne suggests, for cladistics, and M. J. Sackin (personal communication) notes that if there are $t-1$ or more wholly compatible characters one can readily construct a unique parsimonious rootless tree. Such sets would also be expected to give the maximum information in between-matrix incongruence analysis, as they would readily define different cladograms.

DISCUSSION AND CONCLUSIONS

Molecular detail is the greatest requisite for future phylogenetic work. This is also true for short-term evolution, because it seems unlikely that we will be able to get to grips with such problems as the origin and spread of drug resistance until we can obtain molecular detail of plasmid-borne genes. Even the severe problems of unravelling reticulate evolution may be feasible with abundant data. The development of electron microscopes capable of reading sequences directly from single molecules would revolutionize the field. It is also possible that polypeptides are sufficiently well preserved in some fossils to allow sequences to be obtained (Florkin, 1969).

The experience of constructing protein cladograms by hand will teach one that there are usually many cladogenies that are almost equally parsimonious, and estimation of parsimony is made difficult by complex rules for deciding the most probable amino acids in HTUs. One approach would be to replace the site by a vector of probabilities of each of the twenty amino acids, which is conceptually accurate and perhaps not too heavy on computation.

Yet even if we have the most parsimonious solution, we may find that its superiority over several other solutions is not statistically significant, so it would be unwise to think that it must, of necessity, be the correct solution. The philosophical attraction of the true cladogeny may blind us to the fact that all the cladogenies are only estimates of the correct one, just as phenograms are only estimates of phenetic structure. It would seem sensible to aim toward a situation where we could express the probability that a given solution was within a certain distance of the correct one, rather than just whether it was true or false.

Finally, to take stock of our position today, we may ask whether microbial phylogeny has in the past been mostly speculation with little of lasting value. I believe we must admit this, and look for new work based on molecular detail and numerical cladistic methods. We shall have to make some evolutionary assumptions, but at least they will be explicitly stated. At present we may conclude that the most reasonable view would be that molecular detail (and perhaps other characters) generally reflects divergent evolution, although with evolution rates that vary a good deal. That is, there has been little convergence as far as we can tell, and we would probably prefer to attribute apparent examples to gene transfer unless the evidence was overwhelming. Nevertheless, we must remember that assignment of ancestral homologies is critical to any interpretation, and this is a field where there are still many weaknesses.

But are we not in danger of falling into the same trap as our predecessors, of building phylogenies on insecure foundations? I believe that this danger recedes as we get better primary data and more sophisticated analytic methods. The advances should be real ones, with statistical support and additional corroboration from the uniform way evolution at the molecular level seems to occur in all kinds of organisms. It will need co-operation between many – biochemists, electron microscopists, systematists and computer scientists – but will offer large potential for a better understanding of the microbial world.

REFERENCES

Ainsworth, G. C. (1961). *Ainsworth & Bisby's Dictionary of the Fungi*, 5th edition. Kew, Surrey: Commonwealth Mycological Institute.

Ambler, R. P. (1968). Species differences in the amino acid sequences of bacterial proteins. In *Chemotaxonomy and Serotaxonomy*, ed. J. G. Hawkes, pp. 57–64. London: Academic Press.

Ambler, R. P. & Wynn, M. (1973). The amino acid sequences of cytochromes *c*-551 from three species of *Pseudomonas*. *Biochemical Journal*, **131**, 485–98.

Anderson, E. S. (1966). Possible importance of transfer factors in bacterial evolution. *Nature, London*, **209**, 637–8.

Barker, W. C. & Dayhoff, M. O. (1972). Detecting distant relationships: computer methods and results. In *Atlas of Protein Sequence and Structure* 1972, ed. M. O. Dayhoff, pp. 101–10. Washington, D.C.: National Biomedical Research Foundation.

Barnabas, J., Goodman, M. & Moore, G. W. (1971). Evolution of hemoglobin in primates and other therian mammals. *Comparative Biochemistry and Physiology*, **39**B, 455–82.

Bascomb, S., Lapage, S. P., Curtis, M. A. & Willcox, W. R. (1973). Identification of bacteria by computer: identification of reference strains. *Journal of General Microbiology*, **77**, 291–315.

Bisset, K. A. (1952). *Bacteria.* Edinburgh: Livingstone.

Bisset, K. A. (1962). The phylogenetic concept in bacterial taxonomy. In *Microbial Classification. Symposia of the Society for General Microbiology*, **12**, 361–73. Ed. G. C. Ainsworth & P. H. A. Sneath. London: Cambridge University Press.

Bisset, K. A. (1973*a*). Do bacteria have a nuclear membrane? *Nature, London*, **241**, 45.

Bisset, K. A. (1973*b*). This 'prokaryotic–eukaryotic' business. *New Scientist*, **57**, 296–8.

Bobisud, H. M. & Bobisud, L. E. (1972). A metric for classifications. *Taxon*, **21**, 607–13.

Boulter, D., Ramshaw, J. A. M., Thompson, E. W., Richardson, M. & Brown, R. H. (1972). A phylogeny of higher plants based on the amino acid sequences of cytochrome *c* and its biological implications. *Proceedings of the Royal Society, London*, B, **181**, 441–55.

Bradley, S. G. (1971). Phylogenetic relationships among actinomycetes. In *Recent Advances in Microbiology*, ed. A. Pérez-Miravete & D. Peláez, pp. 3–7. Mexico, D.F.: Asociacion Mexicana de Microbiologia.

Brown, R. H., Richardson, M., Boulter, D., Ramshaw, J. A. M. & Jefferies, R. P. S. (1972). The amino acid sequence of cytochrome *c* from *Helix aspersa* Müller (Garden snail). *Biochemical Journal*, **128**, 971–4.

Calvin, M. (1969). *Chemical Evolution.* New York: Oxford University Press.

Camin, J. H. & Sokal, R. R. (1965). A method for deducing branching sequences in phylogeny. *Evolution*, **19**, 311–26.

Cavalli-Sforza, L. L. & Edwards, A. W. F. (1967). Phylogenetic analysis: models and estimation procedures. *Evolution*, **21**, 550–70.

Challinor, J. (1959). Palaeontology and evolution. In *Darwin's Biological Work*, ed. P. R. Bell, pp. 50–100. London: Cambridge University Press.

Chater, K. F. & Hopwood, D. A. (1973). Differentiation in actinomycetes. In *Microbial Differentiation. Symposia of the Society for General Microbiology*, **23**, 143–60. Ed. J. M. Ashworth & J. E. Smith. London: Cambridge University Press.

Clark, B. (1970). Selective constraints on amino-acid substitutions during the evolution of proteins. *Nature, London*, **228**, 159–60.

Clowes, R. C. (1972). Molecular structure of bacterial plasmids. *Bacteriological Reviews*, **36**, 361–405.

Colless, D. H. (1967). The phylogenetic fallacy. *Systematic Zoology*, **16**, 289–95.

Colless, D. H. (1969*a*). The phylogenetic fallacy revisited. *Systematic Zoology*, **18**, 115–26.

Colless, D. H. (1969*b*). The interpretation of Hennig's 'Phylogenetic Systematics' – a reply to Dr Schlee. *Systematic Zoology*, **18**, 134–44.

Colless, D. H. (1970). The phenogram as an estimate of phylogeny. *Systematic Zoology*, **19**, 352–62.

Corliss, J. O. (1962). Taxonomic procedures in classification of protozoa. In *Microbial Classification. Symposia of the Society for General Microbiology*, **12**, 37–67. Ed. G. C. Ainsworth & P. H. A. Sneath. London: Cambridge University Press.

Davis, G. H. G. (1964). Notes on the phylogenetic background to lactobacillus taxonomy. *Journal of General Microbiology*, **34**, 177–84.

Dayhoff, M. O. (1969). Computer analysis of protein evolution. *Scientific American*, **221** (1), 86–95.

Dayhoff, M. O., ed. (1972). *Atlas of Protein Sequence and Structure* 1972, vol. 5. Washington, D.C.: National Biomedical Research Foundation.

Dayhoff, M. O., Eck, R. V. & Park, C. M. (1972). A model of evolutionary change in proteins. In *Atlas of Protein Sequence and Structure* 1972, vol. 5, ed. M. O. Dayhoff, pp. 89–99. Washington, D.C.: National Biomedical Research Foundation.

De Beer, G. R. (1954). *Archaeopteryx* and evolution. *Advancement of Science*, **11**, 160–70.

De Ley, J. (1968). Molecular biology and bacterial phylogeny. *Evolutionary Biology*, **2**, 103–56.

Doolittle, R. F. & Blombäck, B. (1964). Amino-acid sequence investigations of fibrinopeptides from various mammals: evolutionary implications. *Nature, London*, **202**, 147–52.

Dubnau, D., Smith, I., Morell, P. & Marmur, J. (1965). Gene conservation in *Bacillus* species. I. Conserved genetic and nucleic acid base sequence homologies. *Proceedings of the National Academy of Sciences, USA*, **54**, 491–8.

Eck, R. V. & Dayhoff, M. O. (1966). *Atlas of Protein Sequence and Structure* 1966. Silver Spring, Maryland: National Biomedical Research Foundation.

Falkow, S. (1965). Nucleic acids, genetic exchange and bacterial speciation. *American Journal of Medicine*, **39**, 753–65.

Farris, J. S. (1969). A successive approximations approach to character weighting. *Systematic Zoology*, **18**, 374–85.

Farris, J. S. (1970). Methods for computing Wagner trees. *Systematic Zoology*, **19**, 83–92.

Farris, J. S. (1971). The hypothesis of nonspecificity and taxonomic congruence. *Annual Review of Ecology and Systematics*, **2**, 277–302.

Farris, J. S. (1972). Estimating phylogenetic trees from distance matrices. *American Naturalist*, **106**, 645–68.

Farris, J. S., Kluge, A. G. & Eckardt, M. J. (1970). A numerical approach to phylogenetic systematics. *Systematic Zoology*, **19**, 172–89.

Ferone, R., O'Shea, M. & Yoeli, M. (1970). Altered dihydrofolate reductase associated with drug resistance transfer between rodent plasmodia. *Science*, **167**, 1263–4.

Fitch, W. M. (1970). Distinguishing homologous from analogous proteins. *Systematic Zoology*, **19**, 99–113.

Fitch, W. M. (1971*a*). Rate of change of concomitantly variable codons. *Journal of Molecular Evolution*, **1**, 84–96.

FITCH, W. M. (1971*b*). Toward defining the course of evolution: minimum change for a specific tree topology. *Systematic Zoology*, **20**, 406–16.

FITCH, W. M. (1972). Evolutionary variability in hemoglobins. In *Synthese, Struktur und Funktion des Hämoglobins*, ed. H. Martin & L. Nowicki, pp. 199–215. München: Lehmann.

FITCH, W. M. & MARGOLIASH, E. (1967). Construction of phylogenetic trees. *Science*, **155**, 279–84.

FITCH, W. M. & MARGOLIASH, E. (1970). The usefulness of amino acid and nucleotide sequences in evolutionary studies. *Evolutionary Biology*, **4**, 67–109.

FLAVELL, R. (1972). Mitochondria and chloroplasts as descendants of prokaryotes. *Biochemical Genetics*, **6**, 275–91.

FLORKIN, M. (1969). Fossil shell 'conchiolin' and other preserved biopolymers. In *Organic Geochemistry*, ed. G. Eglinton & M. T. J. Murphy, pp. 498–520. Berlin: Springer.

GIBBS, A. J. & MCINTYRE, G. A. (1970). The diagram, a method for comparing sequences. Its use with amino acid and nucleotide sequences. *European Journal of Biochemistry*, **16**, 1–11.

GILBERT, E. N. & POLLAK, H. O. (1968). Steiner minimal trees. *SIAM Journal of Applied Mathematics*, **16**, 1–29.

GRIMONT, P. A. D. & DULONG DE ROSNAY, H. L. C. (1972). Numerical study of 60 strains of *Serratia*. *Journal of General Microbiology*, **72**, 259–68.

HEDGES, R. W. (1972). The pattern of evolutionary change in bacteria. *Heredity*, **28**, 39–48.

HENNIG, W. (1966). *Phylogenetic Systematics*. Urbana, Illinois: University of Illinois Press.

HOYER, B. H., BOLTON, E. T., MCCARTHY, B. J. & ROBERTS, R. B. (1965). The evolution of polynucleotides. In *Evolving Genes and Proteins*, ed. V. Bryson & H. J. Vogel, pp. 581–90. New York: Academic Press.

HUXLEY, J. S. (1957). The three types of evolutionary process. *Nature, London*, **180**, 454–5.

JARDINE, N. & SIBSON, R. (1968). The construction of hierarchic and non-hierarchic classifications. *Computer Journal*, **11**, 177–84.

JARDINE, N., VAN RIJSBERGEN, C. J. & JARDINE, C. J. (1969). Evolutionary rates and the inference of evolutionary tree forms. *Nature, London*, **228**, 185.

JONES, D. & SNEATH, P. H. A. (1970). Genetic transfer and bacterial taxonomy. *Bacteriological Reviews*, **34**, 40–81.

JUKES, T. H. & HOLMQUIST, R. (1972). Estimation of evolutionary changes in certain homologous polypeptide chains. *Journal of Molecular Biology*, **64**, 163–79.

KELLY, D. P. (1971). Autotrophy: concepts of lithotrophic bacteria and their organic metabolism. *Annual Review of Microbiology*, **25**, 177–210.

KIMURA, M. (1969). The rate of molecular evolution considered from the standpoint of population genetics. *Proceedings of the National Academy of Sciences, USA*, **63**, 1181–8.

KIRSCH, J. A. W. (1969). Serological data and phylogenetic inference: the problem of rates of change. *Systematic Zoology*, **18**, 296–311.

KLEIN, R. M. & CRONQUIST, A. (1967). A consideration of the evolutionary and taxonomic significance of some biochemical, micromorphological, and physiological characters in the thallophytes. *Quarterly Review of Biology*, **42**, 105–296.

KLUYVER, A. J. & VAN NIEL, C. B. (1936). Prospects for a natural system of classification of bacteria. *Zentralblatt für Bakteriologie*, 2 Abt., **94**, 369–403.

KNIGHT, B. C. J. G. (1945). Growth factors in microbiology. Some wider aspects of nutritional studies with micro-organisms. *Vitamins and Hormones*, **3**, 105–228.

LAPAGE, S. P., BASCOMB, S., WILLCOX, W. R. & CURTIS, M. A. (1970). Computer identification of bacteria. In *Automation, Mechanization and Data Handling in Microbiology*, ed. A. Baillie & R. J. Gilbert, pp. 1–22. London: Academic Press.

LAPAGE, S. P., BASCOMB, S., WILLCOX, W. R. & CURTIS, M. A. (1973). Identification of bacteria by computer: general aspects and perspectives. *Journal of General Microbiology*, **77**, 273–90.

LAYCOCK, M. V. (1972). The amino acid sequence of cytochrome *c*-553 from the chrysophycean alga *Monochrysis lutheri*. *Canadian Journal of Biochemistry*, **50**, 1311–25.

LEES, H. (1962). The unremarkable autotrophs. *Lectures on Theoretical and Applied Aspects of Modern Microbiology*. University Park: University of Maryland Press.

LE QUESNE, W. J. (1969). A method of selection of characters in numerical taxonomy. *Systematic Zoology*, **18**, 201–5.

LE QUESNE, W. J. (1972). Further studies based on the uniquely derived character concept. *Systematic Zoology*, **21**, 281–8.

LI, S., DRAPEAU, G. R. & YANOFSKY, C. (1973). Amino terminal sequence of the tryptophan synthase alpha chain of *Serratia marcescens*. *Journal of Bacteriology*, **113**, 1507–8.

LÜDERITZ, O., STAUB, A. M. & WESTPHAL, O. (1966). Immunochemistry of O and R antigens of *Salmonella* and related Enterobacteriaceae. *Bacteriological Reviews*, **30**, 192–255.

LWOFF, A. (1944). *L'Évolution Physiologique. Études des Pertes de Fonctions chez les Microorganisms*. Paris: Hermann.

MARGOLIASH, E. & SMITH, E. L. (1965). Structural and functional aspects of cytochrome *c* in relation to evolution. In *Evolving Genes and Proteins*, ed. V. Bryson & H. J. Vogel, pp. 221–42. New York: Academic Press.

MARGULIS, L. (1970). *The Origin of Eucaryotic Cells*. New Haven: Yale University Press.

MEYNELL, E., MEYNELL, G. G. & DATTA, N. (1968). Phylogenetic relationships of drug-resistance factors and other transmissible bacterial plasmids. *Bacteriological Reviews*, **32**, 55–83.

MOORE, G. W. (1971). A mathematical model for the construction of cladograms. Ph.D. thesis, North Carolina State University.

MOORE, G. W., BARNABAS, J. & GOODMAN, M. (1973). A method for constructing maximum parsimony ancestral amino acid sequences on a given network. *Journal of Theoretical Biology*, **38**, 459–85.

MOORE, G. W., GOODMAN, M. & BARNABAS, J. (1973). An iterative approach from the standpoint of the additive hypothesis to the dendrogram problem posed by molecular data sets. *Journal of Theoretical Biology*, **38**, 423–57.

NEEDLEMAN, S. B. & WUNSCH, C. D. (1970). A general method applicable to the search for similarities in the amino acid sequences of two proteins. *Journal of Molecular Biology*, **48**, 443–53.

NOVICK, R. P. (1969). Extrachromosomal inheritance in bacteria. *Bacteriological Reviews*, **33**, 210–35.

OKANISHI, M., OHTA, T. & UMEZAWA, H. (1970). Possible control of formation of aerial mycelium and antibiotic production in *Streptomyces* by episomic factors. *Journal of Antibiotics*, **23**, 45–7.

ORLA-JENSEN, S. (1909). Die Hauptlinien des natürlichen Bakteriensystems. *Zentralblatt für Bakteriologie*, 2 Abt. **22**, 305–46.

PALLERONI, N. J., DOUDOROFF, M., STANIER, R. Y., SOLANES, R. E. & MANDEL, M. (1970). Taxonomy of the aerobic pseudomonads: the properties of the *Pseudomonas stutzeri* group. *Journal of General Microbiology*, **60**, 215–31.

PARK, I. W. & DE LEY, J. (1967). Ancestral remnants in deoxyribonucleic acid from *Pseudomonas* and *Xanthomonas*. *Antonie van Leeuwenhoek*, **33**, 1–16.

PETTIGREW, G. W. (1972). The amino acid sequence of a cytochrome *c* from a protozoan, *Crithidia oncopelti*. *FEBS Letters*, **22**, 64–6.

PHIPPS, J. B. (1971). Dendrogram topology. *Systematic Zoology*, **20**, 306–8.

RAVIN, A. W. (1963). Experimental approaches to the study of bacterial phylogeny. *American Naturalist*, **97**, 307–18.

RICHMOND, M. H. (1969). Extrachromosomal elements and the spread of antibiotic resistance in bacteria. *Biochemical Journal*, **113**, 225–34.

RICHMOND, M. H. (1970). Plasmids and chromosomes in prokaryotic cells. In *Organization and Control in Prokaryotic and Eukaryotic Cells. Symposia of the Society for General Microbiology*, **20**, 249–77. Ed. H. P. Charles & B. C. J. G. Knight. London: Cambridge University Press.

SACKIN, M. J. (1971). Crossassociation: a method of comparing protein sequences. *Biochemical Genetics*, **5**, 287–313.

SARICH, V. M. & WILSON, A. C. (1967). An immunological time scale for hominoid evolution. *Science*, **158**, 1200–3.

SCHAEFLER, S. (1972). Polyfunctional penicillinase plasmid in *Staphylococcus epidermidis*: bacteriophage restriction and modification mutants. *Journal of Bacteriology*, **112**, 697–706.

SCHOPF, J. W. (1970). Precambrian micro-organisms and evolutionary events prior to the origin of vascular plants. *Biological Reviews*, **45**, 319–52.

SELANDER, R. K., YANG, S. Y., LEWONTIN, R. C. & JOHNSON, W. E. (1970). Genetic variation in the horseshoe crab (*Limulus polyphemus*), a phylogenetic 'relic'. *Evolution*, **24**, 402–14.

SLODKI, M. E., WICKERHAM, L. J. & CADMUS, M. C. (1961). Phylogeny of phosphomannan-producing yeasts. II. Phosphomannan properties and taxonomic relationships. *Journal of Bacteriology*, **82**, 269–74.

SNEATH, P. H. A. (1962*a*). The construction of taxonomic groups. In *Microbial Classification. Symposia of the Society for General Microbiology*, **12**, 287–332. Ed. G. C. Ainsworth & P. H. A. Sneath. London: Cambridge University Press.

SNEATH, P. H. A. (1962*b*). Longevity of micro-organisms. *Nature, London*, **195**, 643–6.

SNEATH, P. H. A. (1971). Theoretical aspects of microbiological taxonomy. In *Recent Advances in Microbiology*, ed. A. Pérez-Miravete & D. Peláez, pp. 581–6. Mexico, D.F.: Asociacion Mexicana de Microbiologia.

SNEATH, P. H. A. (1972). Computer taxonomy. In *Methods in Microbiology*, ed. J. R. Norris & D. W. Ribbons, vol. 7A, pp. 29–98. London: Academic Press.

SNEATH, P. H. A. & SOKAL, R. R. (1973). *Numerical Taxonomy: The Principles and Practice of Numerical Classification.* San Francisco: W. H. Freeman.

SOKAL, R. R. & ROHLF, F. J. (1962). The comparison of dendrograms by objective methods. *Taxon*, **11**, 33–40.

STANIER, R. Y. (1970). Some aspects of the biology of cells and their possible evolutionary significance. In *Organization and Control in Prokaryotic and Eukaryotic Cells. Symposia of the Society for General Microbiology*, **20**, 1–38. Ed. H. P. Charles & P. H. A. Sneath. London: Cambridge University Press.

STANIER, R. Y. (1971). Toward an evolutionary taxonomy of the bacteria. In *Recent Advances in Microbiology*, ed. A. Pérez-Miravete & D. Peláez, pp. 595–604. Mexico, D. F.: Asociacion Mexicana de Microbiologia.

STANIER, R. Y., PALLERONI, N. J. & DOUDOROFF, M. (1966). The aerobic pseudomonads: a taxonomic study. *Journal of General Microbiology*, **43**, 159–271.

STARR, M. P. & HEISE, H. R. (1971). Comparison of the concepts of phenetic taxonomy and phylogenetic taxonomy. In *Recent Advances in Microbiology*, ed. A. Pérez-Miravete & D. Peláez, pp. 605–11. Mexico, D.F.: Asociacion Mexicana de Microbiologia.

STOLERU, G. H., GERBAUD, G. R., BOUANCHAUD, D. H. & LE MINOR, L. (1972). Etude d'un plasmide transférable déterminant la production d'H_2S et la résistance à la tétracycline chez 'Escherichia coli'. *Annales de l'Institut Pasteur*, **123**, 743–54.

UZZELL, T. & PILBEAM, D. (1971). Phyletic divergence dates of the hominoid primates: a comparison of fossil and molecular data. *Evolution*, **25**, 615–35.

VIVIAN, A. (1971). Genetic control of fertility in *Streptomyces coelicolor* A3(2): plasmid involvement in the interconversion of UF and IF strains. *Journal of General Microbiology*, **69**, 353–64.

WHITELEY, H. R. (1957). Fermentation of amino acids by *Micrococcus aerogenes*. *Journal of Bacteriology*, **74**, 324–36.

WICKERHAM, L. J. & BURTON, K. A. (1961). Phylogeny of phosphomannan-producing yeasts. I. The genera. *Journal of Bacteriology*, **82**, 265–8.

THE ROLE OF MUTATION IN MICROBIAL EVOLUTION

J. W. DRAKE

Department of Microbiology, University of Illinois, Urbana, Illinois 61801, *USA*

INTRODUCTION

The tenet that mutation generates the raw material of evolution is widely accepted by biologists, and is no longer sufficiently controversial to be particularly interesting in itself. Of much greater contemporary interest, however, is the extent to which mutation is continuously involved in microbial evolution, and the extent to which mutation rates themselves show evidence of having evolved.

MUTATIONAL ADAPTATION

Short-term adaptation is our best available model for the role of mutation in the broader evolutionary changes which produce strains so distinct as to constitute different species. While quantitative studies of base pair homology or amino acid sequence divergence no doubt produce valid indices of the extent of evolution, it is generally very difficult to determine whether strain differences arise solely by serial mutation, or also involve to some extent parallel mutations brought together by gene transfer mechanisms such as sexuality. However, in strictly controlled laboratory experiments designed to follow serial adaptations within a clone genomic mixing is generally absent, and whatever adaptation occurs is likely to be the result of mutation alone.

It should be stressed at this point that not all adaptations observed under carefully controlled conditions are necessarily mutational in origin. Phase variation in *Salmonella*, for instance, while fulfilling some of the more simplistic definitions of mutation, seems hardly likely to result from changes in DNA base sequences (Lederberg & Iino, 1956). Novick & Weiner (1957) simulated a mutation-like process by maintaining *Escherichia coli* in the presence of low levels of lactose, insufficient to activate the *lac* operon. By occasional random events, however, some cells became induced, thereby producing a permease which concentrated the lactose sufficiently within the cell to maintain the induced state. While such pseudomutational events are probably rare, it is advisable to consider their possible participation in any particular system.

Adaptation by strong selection for pre-existing or induced mutants is such a common laboratory procedure as not to require consideration here. A very elegant system of serial adaptation by mutation to a gradually changing biochemical environment is well documented by Clarke in this volume. While direct selection experiments of this type are typically aimed at a well-defined phenotype, an indirect type of adaptive process has been described in which the biochemical nature of the phenotypic change is poorly understood, but which illustrates important evolutionary principles. This process is periodic selection, considered by Kubitschek in this volume and described only briefly here.

When a bacterium such as *E. coli* is grown in a nearly constant environment such as a chemostat, selectively neutral spontaneous mutations such as resistance to phage T 5 accumulate. At intervals which are initially of the order of 100 generations most of the accumulated mutants suddenly disappear, and the build-up of new mutants begins again. The basis for this fluctuation is the periodic replacement of the population by one which is better adapted to chemostat growth: since the T 5-resistant mutants are still relatively rare at the time of the changeover, the better adapted variant will almost always be T 5-sensitive. The better adapted variants exhibit shorter generation times, and are genetically stable in the usual sense that they maintain their advantageous features over many generations. The classic examples of periodic selection are described by Novick & Szilard (1950), Atwood, Schneider & Ryan (1951), and Novick (1958).

Periodic selection is an interesting evolutionary model for several reasons. It clearly represents adaptation to a complex environmental change, since it would be naive to suppose that the chemostat does not differ in many important ways from the more typical niches occupied by *E. coli*. Furthermore, Atwood *et al.* (1951) observed that strains adapted to a chemically defined medium displayed a selective *disadvantage* compared to their predecessors when returned to nutrient broth: this adaptation therefore resembles the kind of divergence which is one of the most important early aspects of speciation. Periodic selection also appears to represent the result of serial multiple mutations, there being no evidence for gene transfer except to direct descendents. Finally, periodic selection offers a simple mechanism for the fixation of selectively neutral, or even slightly deleterious, mutations: the chromosomal composition of the variant defines that of the new strain until further mutational divergence occurs, and any coincidental mutation in this variant becomes fixed.

The potential evolutionary significance of periodic selection, which is in some ways akin to the local population extinctions which are reputed to occur frequently in higher organisms (MacArthur, 1972), makes it desirable to determine whether mutation as ordinarily understood is its cause. Recent experiments with mutator strains of *E. coli* indicate that this is in fact the case. When chemostats are seeded with mixtures of mutator and non-mutator strains, which are otherwise identical except for an indicator marker, the mutator strain almost invariably outgrows the non-mutator strain (Gibson, Scheppe & Cox, 1970; Nestman & Hill, 1973). In the case of the *mutT* mutator strain, for instance, this selective advantage is about 0.5 % at cell densities below 10^7 ml^{-1}, and rises to about 3 % at cell densities above 10^9 ml^{-1}. Under continuous growth conditions these are very substantial selective advantages.

If high mutation rates are so advantageous, why are they not more generally observed? It seems likely that high mutation rates become profitable only when mutational adaptation can occur at many loci. Under highly adapted conditions, on the other hand, the high mutation rate simply increases the mutational load, leading to extensive genomic wastage. In the wild, some balance may be struck between these two extremes. It seems most unlikely that rapid exponential growth is ever maintained for very long, and when growth is cyclically controlled by variable sources of nutrition, periodic mutational adaptation can be expected to occur. During a period of unlimited nutrient supply, for example, rapid-growth mutants will excel, but when the nutrient supply becomes limiting, different types of mutants, those best adapted to competition for a more limited food supply, will come to the fore. This type of environmental fluctuation is engendered by the microbe itself, and generates the selection patterns referred to by MacArthur (1972) as the 'r' and 'k' varieties.

ARE OPTIMAL MUTATION RATES FEASIBLE?

Several different types of selective forces probably affect average mutation rates, acting through mechanisms of group selection in which local populations, rather than individual organisms, are subjected to selection. Very high mutation rates, for example, produce correspondingly high frequencies of mutationally defective individuals, and thus constitute a selective disadvantage for well adapted populations in relatively stable environments. Very low mutation rates, on the other hand, will reduce a population's genetic variance, making successive adaptations to fluctuating environments less likely and increasing the probability of complete extinction of subpopulations. Another factor to consider is the physiological cost of maintaining low mutation rates. While the enzymatic machinery of DNA replication and repair might theoretically evolve towards continuously increasing accuracy, it cannot do so indefinitely. If a particular error-avoidance system is to be continuously improved, for example, the eventual result is likely to be a decreased rate of DNA synthesis, since the rates of activation and dissociation of substrates on the surface of an enzyme will tend to decrease as more and more atoms of the substrate molecules are bound. Decreased velocities might be circumvented, however, by greatly increasing the number of replication forks per genome, as observed, for instance, in eukaryotes. Alternatively, a rapid but less accurate process of DNA synthesis could be reinforced by a sophisticated set of post-replication repair systems. In either case, however, the organism commits a greater proportion of its genome and its energy expenditure to increased accuracy, and thus still pays an increased physiological price.

A very different kind of limit to arbitrarily low mutation rates may arise from an evolutionary commitment to a particular mechanism of error avoidance. In a highly evolved enzymatic apparatus, improvements directed towards a particular set of goals will tend to reach maxima which represent the best that can be done without major changes in amino acid composition. Even though small numbers of simultaneous amino acid substitutions can no longer produce significant improvements, more gross changes might still be effective, but are virtually excluded by the product of mutation rate and population size and by a lack of selectively neutral or advantageous intermediate states. This situation is analogous to that of physically metastable states in which the activation energy needed to reach a lower energy level is so great as to preclude a transition within a finite time.

Only a few attempts have been made to apply the above principles to estimating the feasibility and magnitude of optimum mutation rates. Kimura (1967), for instance, calculated optimum mutation rates so as to minimize mutational and substitutional genetic loads, obtaining values in the neighborhood of 6 % per genome per generation. This value, as we shall see, is approximately correct, since microbes exhibit mutation rates per genome per generation of the order of 0.1 %, while the corresponding rate for *Drosophila* is about 100 %. Levins (1967) introduced the factor of short-term environmental fluctuations into his calculations, with the expectation that a significantly higher optimum would result. This was in fact obtained, but the calculated rates, about 1 % per locus per generation, were three or four orders of magnitude higher than typically observed in nature.

A curious effect of sexuality on the optimum mutation rate has been claimed by Leigh (1970). While agreeing with Kimura that an optimal rate could be selected in asexual systems, Leigh concluded that relatively high rates of recombination tend to dissociate the genes controlling mutation rates from their target genes, with the result that selection in sexual systems could not produce optimal rates. Instead, mutation rates would approach physiologically limited minima. High mutation rates in sexual organisms could only be produced by poorly understood group selection mechanisms. Nevertheless, as we shall see, the mutation rate of a highly sexual organism such as *Drosophila* turns out to be very high.

One of the best indications that optimum mutation rates can indeed be selected would be to discover organisms with markedly changed mutation rates, such as the mutator mutant mentioned in the previous section. It was therefore gratifying, upon examining an organism (bacteriophage T 4) which ordinarily displays a rather high mutation rate per base pair replication, to observe that certain of its mutations exhibit *antimutator* properties; thus far, such mutations have been observed in the genes for DNA polymerase (Drake & Allen, 1968; Drake *et al.* 1969) and for gene 32 protein (Drake, 1973). These systems will be described more fully in a later section. The anticipated relationship between physiological cost and accuracy is also hinted at by recent in-vitro studies on mutator, normal and antimutator alleles of the DNA polymerase gene (Muzyczka, Poland & Bessman, 1972). One mechanism by which this polymerase maintains accuracy is to excise incorrectly inserted residues; and even correctly inserted residues are sometimes excised, at least *in vitro*. The rate of excision, however, is abnormally low in a mutator strain and abnormally high in an antimutator strain; in an extreme case, over 90 % of the residues polymerized *in vitro* by an antimutagenic DNA polymerase are excised.

COMPARATIVE RATES OF MUTATION

It is highly desirable, while considering the role of mutation in microbial evolution, to know the magnitudes of typical mutation rates. Despite the very large number of mutation rate determinations which have been performed in the past half-century, we are faced with a paucity of suitable comparative data. First of all, technically satisfactory measurements are available for only a few organisms. Second, we are primarily concerned with average rates, so that studies on reversion rates, which range from over 10^{-1} to less than 10^{-11} per generation, are not very useful. What are required instead are estimates of total forward mutation rates per genome. This means that we must have reliable forward mutation rates for at least several genes, together with good estimates of the sizes of these genes and of the genome itself.

It is also necessary to know what proportion of the genome actually comprises the mutational target, in the sense that mutations therein have biological significance. Very extensive genetic investigations of the enteric bacteria *E. coli* and *S. typhimurium*, for instance, and of their bacteriophages T 4 and lambda, have identified such large proportions of the genome that the projected mutational target size corresponds rather well to the total genome. This condition is much less well satisfied in eukaryotes, however, for at least two reasons. First, a variable but sometimes very large proportion of the genome exhibits base sequence redundancy (of which the high ploidy of many plants is an extension), and until we gain much deeper insights into the functional significance of these sequences, it remains unclear whether they should be included within the mutational target. Slave gene models (Callan, 1967), for instance, although increasingly untenable except for rather restricted portions of the genome, imply that a large fraction of the genome might be virtually indifferent to mutation. Second, although many mutation rates to particular phenotypes have been determined, the physical sizes of the corresponding nuclear DNA regions usually remain unknown, making extrapolation to genomic mutation rates very difficult. There is, fortunately, one exception. Mukai, Chigusa, Mettler & Crow (1972) have determined the mutation rate for spontaneous recessive mutations on chromosome II of *Drosophila melanogaster*, and when their value is scaled up to the total *diploid* genome, it approaches 100 %, that is, an average of one new (mildly deleterious) mutation per fly per generation. This value, however, is a minimum, and could be substantially higher if even less deleterious mutations could be detected, or if additional mutations were revealed under different scoring conditions.

A further general difficulty in determining mutation rates arises when one asks what proportion of mutations are actually scored under laboratory conditions. There are both theoretical and experimental reasons for suspecting that this proportion is frequently rather small. The many synonymous codons, for example, probably cause about 25 % of base pair substitutions to go undetected. Some additional

Table 1. *Comparative spontaneous mutation rates**

Organism	Base pairs per genome	Mutation rate per base pair replication	Mutation rate per genome per generation
Bacteriophage lambda	4.7×10^{4}	2.4×10^{-8}	0.001
Bacteriophage T 4	1.8×10^{5}	1.1×10^{-8}	0.002
Salmonella typhimurium	3.8×10^{6}	2.0×10^{-10}	0.001
Escherichia coli	3.8×10^{6}	4.0×10^{-10}	0.002
Neurospora crassa	4.5×10^{7}	5.8×10^{-11}	0.003
Drosophila melanogaster	4.0×10^{8}	8.4×10^{-11}	0.93†

* Most values are from Drake (1969, 1970) with small improvements. For instance, the data of Bujard, Mazaitis & Bautz (1970) were used to estimate the size of the T4 *rII* region, and the data of Woodward (1956) were used for *N. crassa* mutations to arginine and citrulline auxotrophy, assuming target sizes of eleven and four cistrons of 10^{3} base pairs each, respectively, and averaging with the previous values for the *ade-3* locus, assuming final population sizes of 10^{10} throughout.

† The *Drosophila* values are per diploid genome *per sexual generation*, i.e. per generation of flies, not per generation of cells.

fraction of base pair substitutions also go undetected because the resulting amino acid substitutions are tolerated under laboratory conditions. That such mutations do occur at typical rates, however, is demonstrated by the discovery of 'cryptic' mutations of bacteriophage T 4 (Koch & Drake, 1970): these mutations produce such small phenotypic effects that they are rarely observed under standard conditions, and very specialized procedures must be employed to demonstrate their existence. Most mutation rates are therefore underestimates. Fortunately, the extent of underestimation is limited. Several per cent of base pair substitutions generate chain-terminating mutations which are scored with nearly 100 % efficiency. Base pair addition and deletion mutations of the frameshift variety will also be scored with nearly 100 % efficiency, and frameshift mutations usually comprise at least 20 % of *detected* spontaneous mutations (Drake, 1970). It is therefore possible to calculate that observed mutation rates are unlikely to be less than about a quarter of total mutation rates.

Comparative spontaneous mutation rates were tabulated by Drake (1969, 1970), and an updated version of these values appears in Table 1. Mutation rates per genome per replication turn out to be remarkably uniform over a wide range of microbial genome sizes, a 1000-fold increase in genome size being compensated by a 1000-fold decrease in mutation rates per base pair. It would, however, be highly desirable to obtain values for organisms outside of this taxonomically limited field.

It seems most unlikely that the remarkably constant mutation rates illustrated in Table 1 are coincidental. Indeed, there is anecdotal evidence that even smaller genomes are correspondingly more mutable per base pair: reversion rates in the small RNA viruses are frequently very high, so that attempts to measure recombination in these organisms frequently fail because of the lack of sufficiently stable markers. Furthermore, there is little evidence that organisms listed in Table 1 differ greatly in mutation rate from other unlisted organisms: median reversion rates in many fungi (Esser & Kuenen, 1967) and bacteria are typically no higher or lower than those observed in *Neurospora crassa* and in *E. coli* or *S. typhimurium*. It appears, therefore, that the selective forces which determine microbial mutation rates are remarkably uniform in diverse species, and that it is the total genomic mutation rate which is selected.

A further remarkable aspect of Table 1 is the sharp reversal of trends between *Neurospora* and *Drosophila*: instead of exhibiting the same mutation rate per genome per sexual replication, *Drosophila* exhibits the same mutation rate per base pair per replication. Neither the ninefold increase in genome size nor the greatly increased number of DNA replications per (sexual) generation appears to have been compensated by decreased rates of mutation per base pair replication. This result suggests that decreases in the per-base-pair mutation rate may have reached some evolutionary limit at the approximate level of the fungi, and that per-genome mutation rates in the higher eukaryotes are limited only by the necessity of avoiding the 'mutational suicide' of the species.

THE GENETIC DETERMINATION OF MUTATION RATES

Mutation rates are determined at two very different levels: at localized sites within genes, and throughout the genome.

Control at the level of individual base pairs is implied by the great range of reversion rates of different mutations within the same gene, for instance the transition A:T → G:C; rates within a single gene can vary by two orders of magnitude. These differences are commonly attributed to the still mysterious effects of neighboring base pairs. This 'explanation' is reasonable in terms of the contemporary theory of the gene, and is supported by experimental evidence. R. E. Koch (1971), for instance, showed that an A:T → G:C base pair substitution increased the rate of the A:T → G:C transition at the adjacent position by about twenty-fold, and also increased the rate of the T:A → C:G transition two base pairs away by about four-fold. Even more dramatic effects have been observed for frameshift mutagenesis (Okada *et al.* 1972), where, in contrast to base pair substitution, something is known

Table 2. *Genetic determinants of mutation rates in bacteriophage T4*

Gene	Function	Effects on mutation rates	References
30	DNA ligase	Weak frameshift mutator alleles	1
32	DNA unwinding	Strong base pair substitution and frameshift mutator and antimutator alleles	2, 3, 4
42	dCMP hydroxymethylase	G:C→A:T mutator alleles	4, 5
43	DNA polymerase	Strong base pair substitution and frameshift mutator and antimutator alleles	2, 4, 6–9
44	DNA synthesis	Weak frameshift mutator alleles	2
46	Deoxyribonuclease	Weak frameshift mutator alleles	2
47	Deoxyribonuclease	Weak frameshift mutator alleles	2
62	DNA synthesis	Uncharacterized mutator allele	10
hm	Unknown	Weak base pair substitution mutator, promotes misrepair	4
px	Generalized repair	Part of 'misrepair' system	4
td	Thymidylate synthetase	A:T→G:C mutator allele	3, 11
v	Dimer excision	Promotes ultraviolet mutagenesis	12
y	Generalized repair	Part of 'misrepair' system	4

References are to mutation studies; further references about primary gene functions are located therein. 1: R. E. Koch & Drake (1973). 2: Bernstein (1971). 3: Bernstein, Bernstein, Mufti & Strom (1972). 4: Drake (1973 and unpublished results). 5: G. R. Greenberg (personal communication). 6: Speyer (1965). 7: Drake & Allen (1968). 8: Drake *et al.* (1969). 9: Drake & Greening (1970). 10: J. D. Karam (personal communication). 11: Smith, Green, Ripley & Drake (1973). 12: Drake (1966*a*).

about the specific role of neighboring bases, the rate of frameshift mutagenesis being a function of the amount of local base sequence redundancy. Altogether, the magnitude of neighboring base pair effects upon local mutation rates is sufficient to permit rather large differences between different genes – perhaps even a 100-fold variation – without placing undue restrictions upon base pair sequences.

Genome-wide determination of mutation rates is implied by the existence of mutator mutations. A considerable number of such mutations have been identified, but it is generally unknown, with the exception of certain bacteriophage T4 mutators, just what functions are involved. One major exception, however, is the error-prone, recombination-like, generalized repair system of *E. coli* (Witkin, 1969), for which a rather specific molecular model is available (Rupp, Wilde, Reno & Howard-Flanders, 1971), and of which an analogue also exists in bacteriophage T4 (Drake, 1973). This system, however, mainly controls induced mutation rates, and only slightly affects spontaneous mutation rates.

The thirteen genes of bacteriophage T4 which are known to affect mutation rates are listed in Table 2. Their functions are known specifically in six cases, and in a general sense in six more. In most of these examples, effects on mutation rates have been observed by constructing double mutants containing a *ts* (temperature-sensitive) allele of the gene in question plus a well characterized *rII* mutation, and comparing the reversion rate of the *rII* mutation in the *ts* and the ts^+ backgrounds. In some cases, however, *rII* mutations induced in the forward direction have been collected and subjected to reversion analysis with base analogues, hydroxylamine and proflavin (inducing A:T ↔ G:C, G:C → A:T, and frameshift mutations, respectively).

Consistently strong effects upon mutation rates, including both mutator and antimutator effects, have been observed with *ts* alleles of genes 43 and 32. Gene 43 encodes the T 4 DNA polymerase, which is continuously required for replication but is at most marginally implicated in repair synthesis, and which therefore appears to act mainly as a replicase. Recent studies *in vitro* suggest that the T 4 DNA polymerase maintains accuracy by two distinct mechanisms (Muzyczka *et al.* 1972; Hershfield and Nossal, 1972; Hershfield, 1973; Schnaar, Muzyczka & Bessman, 1973). First, incorrect bases may be discriminated against in the polymerization step itself. Second, incorrectly polymerized residues can be excised by a 3′ exonuclease activity which resides in the DNA polymerase molecule itself. These two mechanisms are enhanced in the antimutagenic DNA polymerases, and are impaired in the mutagenic DNA polymerases. While these two mechanisms were originally predicted as alternatives (Kornberg, 1969; Drake *et al.* 1969), that both appear to operate is an example of the principle that nature tends to employ all feasible means to solve evolutionary problems.

Gene 32 encodes a structural protein of DNA synthesis, which binds specifically to single-stranded regions of DNA and maintains them in the extended configuration. It also binds to the DNA polymerase itself. It is not clear as yet, however, whether the role of gene 32 in determining mutation rates depends upon its orienting effects upon the template and/or primer strands, or upon its interaction with the DNA polymerase molecule.

It should be stressed that, besides exhibiting both mutator and antimutator effects, *ts* alleles of genes 32 and 43 affect all types of point mutations. To a first approximation, the proportions of different kinds of spontaneous mutations do not vary markedly in phages, bacteria and fungi (Drake, 1970). At present, the major determinants of mutation rate are unknown except for phage T 4, but it will be very interesting to discover whether the broad specificities of genes 32 and 43 are also observed in the corresponding genes of higher forms. For instance, in the evolutionary process of reducing mutation rates per base pair, are only a few determinants with broad specificity uniformly improved, or are additional systems created, each with specific error-avoidance properties? A possible example of the latter is the *mutT* gene of *E. coli*, which acts concomitantly with DNA synthesis and which appears to specifically suppress the transversion A:T → C:G (Cox, 1970; Yanofsky, Cox & Horn, 1966).

Rather specific mutation rate increases are produced by mutations in the T 4 genes controlling pyrimidine metabolism. The *td* gene encodes the viral thymidylate synthetase, and defects in this gene enhance mutation rates at A:T base pairs, especially when other sources of thymine (such as the corresponding host enzyme and the breakdown of host DNA by viral deoxyribonucleases) are inhibited (Bernstein, Bernstein, Mufti & Strom, 1972; Smith, Green, Ripley & Drake, 1973). Phage T 4 employs 5-hydroxymethylcytosine (5-HMC) in its DNA instead of cytosine, and lesions in gene 42, which encodes dCMP hydroxymethylase, enhance mutation rates at G:C base pairs. There is some reason to question, however, whether simple 5-HMC deprivation is the sole cause of this mutator effect, since substantial mutagenesis occurs at temperatures which do not greatly reduce enzyme activity (Drake, 1973; R. G. Greenberg, personal communication). Furthermore, the gene 42 protein also participates in some additional function required for DNA synthesis, and is inserted into a very large molecular weight component (Chiu & Greenberg, 1968; R. G. Greenberg, personal communication). Although a kinase step falls between 5-HMCMP synthesis and DNA synthesis, it is intriguing to speculate that information flow may occur between the gene 42 protein and the DNA polymerase itself, in the sense that the DNA polymerase may direct the dCMP hydroxymethylase to produce residues upon demand.

The theory of frameshift mutagenesis assigns the final step in the process to DNA ligase (Streisinger *et al.* 1966). It was therefore somewhat surprising to observe that mutationally defective ligases produce, at most, rather small effects upon frameshift mutation rates (Bernstein, 1971; Koch & Drake, 1973). A reasonable conclusion from these results is that decisions of whether or not to accept the mispaired intermediates of frameshift mutagenesis are initially made by the DNA polymerase, and that the ligase thereafter performs blindly.

Like its host, bacteriophage T 4 encodes a generalized repair system which is considerably less accurate than is DNA replication itself. This 'misrepair' system is currently defined by the mutations *px*, *y*, and *1206* (Drake, 1973; Symonds, Heindl & White, 1973). These mutants produce four simultaneous effects: reduced burst size, reduced recombination rate, increased sensitivity to diverse inactivating agents, and (at least in the case of *px* and *y*) decreased sensitivity to mutagenesis by ultraviolet irradiation and methyl methanesulfonate. The mutation *hm* enhances misrepair and also spontaneous mutation, and may therefore also be a part of the same system (Drake, 1973). The *v* gene is involved in a different repair system, the excision of pyrimidine dimers, and when defective, shunts ultraviolet-induced lesions into the misrepair system (Drake, 1966*a*, 1973). As we shall see in the next section, the accumulation of lesions in resting genomes may be an important part of the natural mutation process, and these repair genes therefore probably play a substantial role in the mutational basis of evolution.

THE TIME/GENERATION PARADOX

Amino acid sequences are now available for several homologous proteins from diverse organisms, and correlations have emerged between frequencies of amino acid substitutions and time since phylogenetic divergence. The evolutionary rates of amino acid substitution are, for the most part, proportional to time, especially for putatively neutral mutations (Kimura, 1969; Fitch, 1972; Sarich & Wilson, 1973). In most laboratory experiments, however, mutation rates are proportional, not to time, but to generations. Assuming the correctness of the measurements, there appear to be only two ways to resolve this inconsistency. The first is to assume that mutation rates in nature are proportional to biological rather than to absolute time, but that the fraction of mutations which become fixed increases with increasing generation time by amounts just sufficient to offset the decreased number of generations per unit of time. The second is to assume that mutation rates in nature, unlike those in the laboratory, are proportional to time rather than to generations.

The first assumption becomes particularly difficult to maintain when one considers that generation times between the more rapidly multiplying microbes and the longer-lived mammals differ by factors as large as a million, while mutation rates per base pair differ by a factor of only about a thousand. Furthermore, it is not clear what forces could bring about the precise adjustments in rates of mutation fixation which are required by the first assumption.

The possibility that mutation rates in nature are proportional to time is more tenable. Despite the fact that most laboratory mutation rates are generation-dependent, some certainly are not. Thus, when

E. coli is maintained in the chemostat, mutant accumulation is generation-dependent when growth is limited by glucose, but becomes time-dependent when growth is limited by amino acids (Novick & Szilard, 1950; Kubitschek & Gustafson, 1964). The basis for this effect is a curious process of nuclear selection which causes the rate of DNA replication to become proportional to time rather than to generations (H. E. Kubitschek, personal communication). Possibly even more important, however, are the effects of microbial life-styles. Continuous growth will occur only rarely under natural conditions, the more typical mode of existence probably being similar to the 'feast or famine' regime postulated by A. L. Koch (1971). Under this regime, the typical genome is usually replicating slowly or not at all. Furthermore, since microbial subpopulations frequently become extinct, for instance with the deaths of hosts or with seasonal variations in free-living populations, new populations are likely to be initiated from organisms whose genomes have been dormant for long intervals. It is therefore reasonable to postulate that typical evolutionary lineages contain high proportions of 'resting' intervals during which organisms replicate their DNA slowly or not at all, but do accumulate mutations in a time-dependent manner. The question then arises of whether the non-dividing gene actually does mutate at a significant rate.

This was a vexing question for many years, and was only solved when fungal spores (Auerbach, 1959) and bacteriophage T 4 particles (Drake, 1966*b*; Drake & McGuire, 1967) were monitored over long intervals at temperatures ranging down to 0 °C. Such populations did in fact accumulate mutations, or at least DNA lesions which generated mutations when replication again ensued. The total rate of recessive lethals accumulated at 30 °C in *Neurospora* spores was about 0.33 % per week, and extrapolations from the T 4 data suggest a corresponding rate of about 0.26 % per week at 30 °C. Furthermore, calculations based on data for stored *Drosophila* sperm (Rinehart, 1969) suggest a total rate of about 0.25 % per week at 26 °C. The similarity of these values has been noted by Auerbach & Kilbey (1971), but it is probably fortuitous. First, the temperature-dependencies of the *Neurospora* and T 4 rates differed markedly. Second, progressively decreasing fractions of the genome were probably sampled for T 4, *Neurospora* and *Drosophila*: I estimate efficiencies of detection of accumulated mutants to have been about 20 % for T 4, about 2 % for *Neurospora*, and only about 0.2 % for *Drosophila*. In contrast to generation-dependent mutation rates, therefore, time-dependent mutation rates appear to be approximately proportional to genome size, and roughly constant per codon.

Table 3. *Reversion analysis responses* (rII → r+) *of time-dependent mutants of bacteriophage T 4*

Effects of mutagenic agents				
Base analogues	Proflavin	Hydroxyl-amine	No. and % of *rII* mutants*	Estimated spontaneous background†
+	−	+	34	5 (for first three rows)
+	−	−	9	
+	−	?	5	
−	+	−	29	32
−	−	−	23	

* From Drake & McGuire (1967).

† The estimated contribution from the spontaneous background was calculated from the typical proportion (70 %) of *rII* mutants among spontaneous *r* mutants, and the typical fraction (86.4 %) of spontaneous *rII* mutants refractory to base analogues (Freese, 1959). The estimated five background *rII* mutants reverted by base analogues might be distributed anywhere among the three classes defined by their hydroxylamine responses. From extensive unpublished observations, however, the great majority of the estimated 32 background *rII* mutants not reverted by base analogues are expected to be proflavin-revertible.

We have recently begun to study the molecular basis of replication-independent mutagenesis in bacteriophage T 4. Two observations laid the basis for this work. First, stocks of T 4 *rII* mutants were observed to accumulate revertants linearly with time over intervals of several hundred days, the mutation rate exhibiting a marked temperature dependence (Drake, 1966*b*). Only those mutants reverted, however, which could also respond to the cytosine-specific mutagen hydroxylamine. Second, mutants which had accumulated in the forward direction were subjected to reversion analysis (Drake & McGuire, 1967). These mutants are described in Table 3. When the spontaneous background is excluded, these mutants consist primarily of two types, those reverted by hydroxylamine and those reverted neither by base analogues nor by proflavin. It therefore seems very likely that the main pathway of mutagenesis at neutral pH consists of the transversion G:C → C:G, with a minority pathway by the transition G:C → A:T. Two additional observations support this conclusion. First, a substantial proportion of the transversion mutants would probably be unable to revert by a C:G → T:A transition, depending upon what amino acids happen to be acceptable at the mutated site. These mutants would be represented by the twenty-three unable to respond either to base analogues or to proflavin. Second, when mutants induced by a transversion mechanism revert by transitions, they should frequently produce pseudorevertants. When the revertant phenotypes of the thirty-three mutants revertible by hydroxylamine were examined, at least twenty of the mutants clearly generated pseudorevertants (revertants able to grow on K12(λ) cells, but producing *r*-like plaques on B cells).

We (R. H. Baltz and J. W. Drake, unpublished) have recently observed that replication-independent reversion rates are greatly increased at higher temperatures and lower pHs, and evidence is now accumulating for two distinct processes acting on G:C base pairs. One is predicted by the results of Shapiro & Klein (1966), and consists of the deamination of cytosine to produce uracil, which corresponds to the mutational pathway G:C $\rightarrow$ A:T. This pathway may be represented in Table 3 by the nine mutants reverted by base analogues but not by hydroxylamine (and perhaps also by the five mutants whose hydroxylamine responses could not be determined). This process is strongly proton-catalyzed, and shows a marked rate increase down to about pH 4.0, which agrees well with the estimated N3 pK_a of 5-HMC in DNA. The other process corresponds to the postulated G:C $\rightarrow$ C:G transversion. By examining the phenotypes of revertants of specific G:C mutants, it is possible to detect a second class of revertants whose appearance is much less enhanced at low pH. We are now exploring the possibility that these revertants may represent the transversion pathway, and the further possibility that guanine, rather than cytosine, is the primary mutational target.

The amino acid substitutions which reveal a time-dependent rate of evolution result, of course, from a variety of mutational pathways, and not just from base pair substitutions originating at G:C base pairs. Amino acid additions and deletions, which presumably arise by the same mechanism which generates frameshift mutations, are too few in number to provide reliable estimates of whether their evolutionary accumulation proceeds by time- or generation-dependent processes. We must, however, inquire about the source of time-dependent base pair substitutions originating at A:T base pairs. Such mutations might arise in two possible ways. First, DNA embedded in cytoplasm (or nucleoplasm) may accumulate lesions in A:T base pairs. Second, lesions may accumulate in resting DNA which trigger misrepair, from which a wide variety of mutations result (Drake, 1973). Only time will tell.

SUMMARY

Mutation plays a continuous and critical role in microbial evolution, both during relatively short-term adaptations and also during long intervals of time. Microbial organisms display a remarkably constant total mutation rate per genome per generation, despite 1000-fold variations in genome size. This observation, and the frequent occurrence of mutator and antimutator mutations, indicate that mutation rates are genetically determined at apparently optimal magnitudes. Several of the genetic determinants of mutation rate have been characterized in bacteriophage T 4. The most important are the DNA polymerase, the associated gene 32 (unwinding) protein, and an error-prone generalized repair system.

Although rates of mutation measured in the laboratory are usually proportional to numbers of generations or to rates of DNA replication, rates of evolution estimated from selectively nearly-neutral amino acid substitutions are proportional to time itself. This paradox can be resolved by postulating that typical long-term lineages contain high proportions of 'resting' intervals during which the organism does not replicate its DNA, but does accumulate mutations.

Previously unreported studies from this laboratory were supported by grant VC-5L from the American Cancer Society, grant GB-30604X from the National Science Foundation, and Public Health Service grant AI04886 from the National Institute of Allergy and Infectious Diseases.

REFERENCES

ATWOOD, K. C., SCHNEIDER, L. K. & RYAN, F. J. (1951). Periodic selection in *Escherichia coli*. *Proceedings of the National Academy of Sciences, USA*, **37**, 145–55.

AUERBACH, C. (1959). Spontaneous mutations in dry spores of *Neurospora crassa*. *Zeitschrift für Vererbungslehre*, **90**, 335–46.

AUERBACH, C. & KILBEY, B. J. (1971). Mutation in eukaryotes. *Annual Review of Genetics*, **5**, 163–218.

BERNSTEIN, H. (1971). Reversion of frameshift mutations stimulated by lesions in early function genes of bacteriophage T 4. *Journal of Virology*, **7**, 460–6.

BERNSTEIN, C., BERNSTEIN, H., MUFTI, S. & STROM, B. (1972). Stimulation of mutation in phage T 4 by lesions in gene 32 and by thymidine imbalance. *Mutation Research*, **16**, 113–19.

BUJARD, H., MAZAITIS, A. J. & BAUTZ, E. K. F. (1970). The size of the *r*II region of bacteriophage T 4. *Virology*, **42**, 717–23.

CALLAN, H. G. (1967). The organization of genetic units in chromosomes. *Journal of Cell Science*, **2**, 1–7.

CHIU, C.-S. & GREENBERG, G. R. (1968). Evidence for a possible direct role of dCMP hydroxymethylase in T 4 phage DNA synthesis. *Cold Spring Harbor Symposia on Quantitative Biology*, **33**, 351–9.

COX, E. C. (1970). Mutator gene action and the replication of bacteriophage λ DNA. *Journal of Molecular Biology*, **50**, 129–35.

DRAKE, J. W. (1966*a*). Ultraviolet mutagenesis in bacteriophage T 4. I. Irradiation of extracellular phage particles. *Journal of Bacteriology*, **91**, 1775–80.

DRAKE, J. W. (1966*b*). Spontaneous mutations accumulating in bacteriophage T 4 in the complete absence of DNA replication. *Proceedings of the National Academy of Sciences, USA*, **55**, 738–43.

DRAKE, J. W. (1969). Comparative rates of spontaneous mutation. *Nature, London*, **221**, 1132.

DRAKE, J. W. (1970). *The Molecular Basis of Mutation*. San Francisco: Holden-Day.

DRAKE, J. W. (1973). The genetic control of spontaneous and induced mutation rates in bacteriophage T 4. *Genetics*, **73** (Supplement), 45–64.

DRAKE, J. W. & ALLEN, E. F. (1968). Antimutagenic DNA polymerases of bacteriophage T 4. *Cold Spring Harbor Symposia on Quantitative Biology*, **33**, 339–44.

DRAKE, J. W., ALLEN, E. F., FORSBERG, S. A., PREPARATA, R.-M. & GREENING, E. O. (1969). Genetic control of mutation rates in bacteriophage T 4. *Nature, London*, **221**, 1128–32.

DRAKE, J. W. & GREENING, E. O. (1970). Suppression of chemical mutagenesis in bacteriophage T 4 by genetically modified DNA polymerases. *Proceedings of the National Academy of Sciences, USA*, **66**, 823–9.

DRAKE, J. W. & MCGUIRE, J. (1967). Characteristics of mutations appearing spontaneously in extracellular particles of bacteriophage T 4. *Genetics*, **55**, 387–98.

ESSER, K. & KUENEN, R. (1967). *Genetics of Fungi*. New York: Springer-Verlag.

FITCH, W. M. (1972). Does the fixation of neutral mutations form a significant part of observed evolution in proteins? *Brookhaven Symposia in Biology*, **23**, 186–216.

FREESE, E. (1959). On the molecular explanation of spontaneous and induced mutations. *Brookhaven Symposia in Biology*, **12**, 63–75.

GIBSON, T. C., SCHEPPE, M. L. & COX, E. C. (1970). Fitness of an *Escherichia coli* mutator gene. *Science*, **169**, 686–8.

HERSHFIELD, M. S. (1973). On the role of deoxyribonucleic acid polymerase in determining mutation rates. *Journal of Biological Chemistry*, **248**, 1417–23.

HERSHFIELD, M. S. & NOSSAL, H. G. (1972). Hydrolysis of template and newly synthesized deoxyribonucleic acid by the 3′ to 5′ exonuclease activity of the T 4 deoxyribonucleic acid polymerase. *Journal of Biological Chemistry*, **247**, 3393–404.

KIMURA, M. (1967). On the evolutionary adjustment of spontaneous mutation rates. *Genetical Research, Cambridge*, **9**, 23–34.

KIMURA, M. (1969). The rate of molecular evolution considered from the standpoint of population genetics. *Proceedings of the National Academy of Sciences, USA*, **63**, 1181–8.

KOCH, A. L. (1971). The adaptive responses of *Escherichia coli* to a feast and famine existence. *Advances in Microbial Physiology*, **6**, 147–217.

KOCH, R. E. (1971). The influence of neighboring base pairs upon base-pair substitution mutation rates. *Proceedings of the National Academy of Sciences, USA*, **68**, 773–6.

KOCH, R. E. & DRAKE, J. W. (1970). Cryptic mutants of bacteriophage T 4. *Genetics*, **65**, 379–90.

KOCH, R. E. & DRAKE, J. W. (1973). Ligase-defective bacteriophage T 4. I. Effects on mutation rates. *Journal of Virology*, **11**, 35–40.

KORNBERG, A. (1969). Active center of DNA polymerase. *Science*, **163**, 1410–18.

KUBITSCHEK, H. E. & GUSTAFSON, L. A. (1964). Mutation in continuous cultures. III. Mutational responses in *Escherichia coli*. *Journal of Bacteriology*, **88**, 1595–7.

LEDERBERG, J. & IINO, T. (1956). Phase variation in *Salmonella*. *Genetics*, **41**, 743–57.

LEIGH, E. G. (1970). Natural selection and mutability. *American Naturalist*, **104**, 301–9.

LEVINS, R. (1967). Theory of fitness in a heterogeneous environment. VI. The adaptive significance of mutation. *Genetics*, **56**, 163–78.

MacArthur, R. H. (1972). *Geographical Ecology*. New York: Harper & Row.

Mukai, T., Chigusa, S. I., Mettler, L. E. & Crow, J. F. (1972). Mutation rate and dominance of genes affecting viability in *Drosophila melanogaster*. *Genetics*, **72**, 335–55.

Muzyczka, N., Poland, R. L. & Bessman, M. J. (1972). Studies on the biochemical basis of spontaneous mutation. I. A comparison of the deoxyribonucleic acid polymerases of mutator, antimutator, and wild type strains of bacteriophage T 4. *Journal of Biological Chemistry*, **247**, 7116–22.

Nestman, E. R. & Hill, R. F. (1973). Population changes in continuously growing mutator cultures of *Escherichia coli*. *Genetics*, **73** (Supplement), 41–4.

Novick, A. (1958). Some chemical bases for evolution of micro-organisms. In *Perspectives in Marine Biology*, ed. A. A. Buzzati-Traverso, pp. 533–46. Berkeley: University of California Press.

Novick, A. & Szilard, L. (1950). Experiments with the chemostat on spontaneous mutations of bacteria. *Proceedings of the National Academy of Sciences, USA*, **36**, 708–19.

Novick, A. & Weiner, M. (1957). Enzyme induction as an all-or-none phenomenon. *Proceedings of the National Academy of Sciences, USA*, **43**, 553–66.

Okada, Y., Streisinger, G., Owen, J., Newton, J., Tsugita, A. & Inouye, M. (1972). Molecular basis of a mutational hot spot in the lysozyme gene of bacteriophage T 4. *Nature, London*, **236**, 338–41.

Rinehart, R. R. (1969). Spontaneous sex-linked recessive lethal frequencies from aged and non-aged spermatozoa of *Drosophila melanogaster*. *Mutation Research*, **7**, 417–23.

Rupp, W. D., Wilde, C. E., Reno, D. L. & Howard-Flanders, P. (1971). Exchanges between DNA strands in ultraviolet-irradiated *Escherichia coli*. *Journal of Molecular Biology*, **61**, 25–44.

Sarich, V. M. & Wilson, A. C. (1973). Generation time and genomic evolution in primates. *Science*, **179**, 1144–7.

Schnaar, R. L., Muzyczka, N. & Bessman, M. J. (1973). Utilization of aminopurine deoxynucleoside triphosphate by mutator, antimutator and wild type DNA polymerases of bacteriophage T 4. *Genetics*, **73** (Supplement), 137–40.

Shapiro, R. & Klein, R. S. (1966). The deamination of cytidine and cytosine by acidic buffer solutions. Mutagenic implications. *Biochemistry*, **5**, 2358–62.

Smith, M. D., Green, R. R., Ripley, L. S. & Drake, J. W. (1973). Thymineless mutagenesis in bacteriophage T 4. *Genetics*, **74**, 393–403.

Speyer, J. F. (1965). Mutagenic DNA polymerase. *Biochemical and Biophysical Research Communications*, **21**, 6–8.

Streisinger, G., Okada, Y., Emrich, J., Newton, J., Tsugita, A., Terzaghi, E. & Inouye, M. (1966). Frameshift mutations and the genetic code. *Cold Spring Harbor Symposia on Quantitative Biology*, **31**, 77–84.

Symonds, N., Heindl, H. & White, P. (1973). Radiation sensitive mutants of phage T 4. A comparative study. *Molecular and General Genetics*, **120**, 253–9.

Witkin, E. M. (1969). Ultraviolet-induced mutation and DNA repair. *Annual Review of Genetics*, **3**, 525–52.

Woodward, V. W. (1956). Mutation rates of several gene loci in *Neurospora*. *Proceedings of the National Academy of Sciences, USA*, **42**, 752–8.

Yanofsky, C., Cox, E. C. & Horn, V. (1966). The unusual mutagenic specificity of an *E. coli* mutator gene. *Proceedings of the National Academy of Sciences, USA*, **55**, 274–81.

PLASMIDS AND BACTERIAL EVOLUTION

M. H. RICHMOND AND B. WIEDEMAN*

Department of Bacteriology, University of Bristol, University Walk, Bristol BS8 1TH

Plasmids play a crucial part in the growth and survival of many bacterial populations. This was first realised from experiments on some species of enteric bacteria (Watanabe, 1963; Meynell, Meynell & Datta, 1968) and some strains of *Staphylococcus aureus* (Novick, 1967; Richmond, 1968*c*), but latterly similar structures have been detected in bacilli (Carlton & Helinski, 1969) as well as in a very wide range of Gram-negative species – members of the Pseudomonadales (Holloway, 1969; Sykes *et al.* 1972) as well as Eubacteriales (Meynell, 1973). Not that these elements are present in all strains at all times; certain isolates – and among them strains very valuable to molecular biologists – are found to be 'plasmid-free', but as time goes on it becomes more and more probable that these genetic elements are present in all bacterial species at some time or another and that they play an important role in their evolution.

It is not intended in this article to go over the evidence that bacterial plasmids are, indeed, important in evolution. A number of reviews in the last year or so have been into this fairly thoroughly (see, for example, CIBA Symposium, 1968; Richmond, 1968*c*; Jones & Sneath, 1970; New York Academy Symposium, 1971) and the widespread study of the role of R factors in the spread of antibiotic resistance through bacterial populations has given us the very vivid impression that in observing this phenomenon we are indeed watching evolution in action (see, for example, Anderson, 1968; New York Academy Symposium, 1971; Richmond, 1973). Nor will plasmid transfer be described at length, since this topic has been well reviewed recently (Curtiss, Charamella, Stallions & Mays, 1968; Clowes, 1972; Meynell, 1973). What is proposed is to take a more detailed look at bacterial extrachromosomal DNA to try to see how it is adapted to facilitate evolution. Nor will discussion be limited to plasmids alone. The similarities between these elements and some phages suggests that an understanding of the latter, particularly with respect to their recombination

* Present address: Klinikum der Johann Wolfgang Goethe-Universität, Zentrum der Hygiene, 6, Frankfurt a. Main, 70, Paul-Ehrlich Strasse, 40.

and integration properties, may throw some light on the sort of genetic flexibility that might be inherent but as yet not fully revealed in true plasmids.

PLASMID/CHROMOSOME INTERACTIONS

The role of 'homology'

The overall process of F integration to form the Hfr state, with its strong superficial, but rather misleading, analogies to the integration of the phage λ (Campbell, 1962), is familiar enough not to need restating in detail. The current view is that a given F factor may integrate at a number of distinct points on the *Escherichia coli* chromosome – two in the case of derivatives of F_B (Broda, 1967) and five in the case of F_J (Jacob & Wollman, 1961). Moreover the direction of insertion may vary from site to site. Thus, for example, J10 is inserted clockwise at about 87 minutes and J6 anticlockwise at 75 minutes on the *E. coli* chromosome (Hayes, 1968).

Why these particular points are accessible is not certain. Phage lambda integrates at a specific site between the *gal* and *bio* genes but there seems to be no homology between the lambda DNA itself and the chromosome at the point of integration (Davis & Parkinson, 1971); and this observation fits with the genetic tests (Signer, 1968). Similarly, other specialised transducing phages like ϕ80 (Signer, 1966), P2 (Calendar & Lindahl, 1969) and P22 (Smith & Stocker, 1962) seem to have specific sites of attachment but little homology with the chromosome.

Another example in which integration is likely to occur without there being extensive homology between the interacting DNAs is provided by the mutator phage μ1 (Taylor, 1963; Toussaint, 1969). This phage can integrate at a very large number of different chromosomal sites. Indeed it is this ability that gives the phage its name, since it may integrate into genes essential for bacterial growth and consequently produce widespread and serious mutational effects. Phage μ1 may even integrate into a λ genome that is itself integrated at $attB_\lambda$ (Toussaint, 1969). Such a wide range of integrative sites argues strongly that base sequence homology can play little part in the integration of this phage.

Against this background, the early papers on F integration speak of the need for 'limited homology' between the F plasmid and the bacterial chromosome (Scaife, 1966) but only recently has any direct evidence been provided that this homology may exist. Starlinger and his colleagues have investigated the properties of two DNA sequences

(IS1 and IS2) which have been detected both in the chromosome of *E. coli* K12 and in the F factor F_8 (Hirsch, Saedler & Starlinger, 1972; Starlinger & Saedler, 1972). IS1 contains 800 nucleotide pairs and is present at a level of about eight copies per cell, while IS2 is 1400 base pairs long and present as five copies per cell, both measurements being made on F^- cultures. The fact that DNA regions that can be shown to be homologous both by hybridisation on membranes and by examination as heteroduplexes in the electron microscope, exist both in chromosomal DNA and in F implies that some homology must exist between these two replicons; and this, in turn, suggests that the few copies of IS1 and IS2 present in strains of *Escherichia coli* may represent some, at least, of the regions whereby F is integrated into the chromosome (Starlinger & Saedler, 1972; Saedler & Heiss, 1973). So far the IS1 and IS2 DNA have only been mapped in the middle of the *gal* operon – a site for F integration not found in naturally occurring strains. However, the strains used for this purpose were laboratory isolates with somewhat bizarre properties when compared with naturally occurring *Escherichia coli* and, moreover, there is some suggestive evidence that IS1 and IS2 may integrate into a large number of distinct locations on the *Escherichia coli* chromosome. Starlinger's experiments may, therefore, have detected the presence of DNA that can provide the necessary base sequence homology to facilitate recombination between a plasmid and the bacterial chromosome, particularly when the host's recombination enzymes are used for the purpose (see later). This possibility is further discussed later (see p. 80).

Certain R factors (notably R1–19 and R100–1) can promote the transfer of chromosomal markers, although it is uncertain whether true integration is involved (Pearce & Meynell, 1968; Meynell, 1973). Certainly, these plasmids do not seem to be able to achieve an Hfr state of the type found with F; and the extent of their homology with the chromosome is unknown. Stable Hfr clones can, however, be obtained in ColB (Frédéricq, 1965) and ColV–K94 (Kahn, 1968) carrying lines, but once again the extent of sequence homology between the DNA of these plasmids and any point on the bacterial chromosome is unknown, even though ColV–K94 seems to have a strong preference for integration at one of three main sites (Kahn, 1969). Some ColI plasmids have properties similar to R1–19 and R100–1 rather than to other Col factors (Smith & Stocker, 1966); but once again the extent of homology with any part of the DNA is unknown.

In summary, therefore, the only plasmid where integration seems to involve sequence homology is F. In contrast, many of the phages that act as episomes do not seem to have homology with their point of insertion. It seems likely however that two broad classes of plasmid exist: those which share a common DNA sequence with part of the chromosome, and those that do not.

Before going on to discuss the nature of the enzymes that may catalyse interaction of plasmid DNA with other DNA in the cell, it is worth reviewing briefly what is known about the extent of nucleotide sequence similarity needed for DNA/DNA interaction and the extent to which some degree of mismatch impairs the process. Evidence on these points is not precise, particularly *in vivo*, but some idea can be gained (Franklin, 1971).

First the presence of two similar nucleotide sequences in the cell does not necessarily lead to recombination. Heteroduplex formation can occur between λ DNA and several fragments of the *Escherichia coli* chromosome (Bolton *et al.* 1964), yet recombination at these sites is rare (Gottesmann & Yarmolinsky, 1968). Similarly, penicillinase plasmids with a considerable length of sequence homology can coexist without recombination in *Staphylococcus aureus* (Novick & Richmond, 1965), and the same is true of R factors (Meynell, 1973). The base sequences of a number of tRNAs in *Escherichia coli* are very similar (Russell *et al.* 1970), yet these regions do not seem to lead to genetic instability through recombination. Similarly, the repetition of punctuation sequences in DNA might also be sites for recombination if nucleotide sequence in the presence of a functional RecA system was sufficient for recombination.

The in-vitro interaction of DNA strands can be examined by physicochemical techniques and the stability of a duplex can be stated in terms of the median temperature of denaturation of the duplex: T_m. At least twelve nucleotide pairs seem to be needed for a stable duplex to form *in vitro* and the T_m for such a duplex is reduced detectably by as little as 1 % difference between the interacting sequences (McCarthy & Church, 1970). This value of twelve nucleotides is particularly provocative since Thomas (1966) has calculated that this is the length of DNA whose sequence is unlikely to be repeated in a single replicon the size of the *Escherichia coli* chromosome. So all regulatory signals may have to be accommodated within a sequence of less than twelve nucleotide pairs. On the basis of this information, therefore, short common base sequences in two interacting pieces of DNA will probably not lead to recombination, and large common stretches need not do so. Whether interaction occurs seems to depend to a great extent on the nature of the enzymes available.

The role of enzymes in integration and excision

The similarity between F and λ integration tends to break down even further when one considers the enzymes involved in the two processes; and these differences also help to explain why base sequence homology may be needed for F integration but not for lambda and other similar phages. Lambda integration occurs by a crossing-over between two genetic regions *attB* in the chromosome and *attP* in the lambda genome, the process being catalysed by a phage specified 'integrase' enzyme (Signer, 1968). Presumably genetic homology between the two *att* regions is unnecessary because the integrase system is capable of recognising the two relevant base sequences. Whether this is so or not, lambda integration is a highly specialised process specific for a very restricted range of phages and for a single site on the chromosome. Two sets of observations tend to reinforce this view: first, lambda integration is unimpaired in *recA*$^-$ hosts, thus stressing the independence of the process from other recombination processes in the cell (Brooks & Clark 1967; Echols, 1971). Secondly, phage deletions that no longer specify the synthesis of an active 'integrase' enzyme greatly reduce the frequency with which λ can integrate. Furthermore, what integration remains in such circumstances occurs at sites other than *attB* and depends now on the presence of an unmutated *recA* gene in the host (Signer, 1968, 1969; Shimada, Weisberg & Gottesman, 1971).

Although their investigation is less far advanced, it seems certain that a number of other specialised transducing phages specify their own integration enzymes, since integration at the appropriate *attB* sites is unaffected by the presence of a *recA*$^-$ state in the host. Furthermore, some specialised transducing particles derived from such phages, such as ϕ80d*lac* for example, are found to have lost an 'integrase' function through formation of the specialised particle and the insertion of this recombinant phage genome occurs in the *lac* region of the *Escherichia coli* rather than at $attB_{\phi 80}$ (Signer, 1966).

Since our interest here is primarily in the way that plasmids may play their role in bacterial evolution, an important consideration must be the extent to which genetic information can pass from one replicon to another. Integration of phage λ at its chromosomal attachment site may be followed by excision, and sometimes when this occurs chromosomal DNA passes to the plasmid state as a recombinant with some of the phage DNA. A typical example, which has been known for some time, is the formation of specialised λd*gal* transducing particles, but similar specialised situations may be formed with several other phages

Table 1. *Effect of* recA *mutants on the ability of various plasmids to mobilise chromosomal markers in* Escherichia coli *K*12

Sex factor	Transfer frequency for pro$^+$ [strA]	
	Rec$^+$	RecA
F	3.5×10^{-5}	1.1×10^{-8}
ColV2	4.0×10^{-6}	4.0×10^{-9}
ColB1	9.1×10^{-7}	6.0×10^{-9}
ColVB	4.0×10^{-4}	4.8×10^{-9}
R1 *drd*19	2.0×10^{-4}	1.8×10^{-4}
R192 *drd*F7	4.0×10^{-6}	4.8×10^{-9}
ColIb *drd*	5.0×10^{-9}	6.0×10^{-9}
R64 *drd*-11	20.0×10^{-6}	4.2×10^{-9}
R114 *drd*3	3.0×10^{-7}	5.6×10^{-9}

Cross: *metB nalA* × *proA his trp strA*.
Data from Moody & Hayes (1972).

(for a general account of phage λ see Hershey, 1971). In the case of λ, the excision of the phage genome involves the *int* product, but also requires another phage specified product – the so-called excision enzyme(s). How it is that the *int* and *xis* gene products co-operate to excise the lambda genome from the chromosome, and in particular, that some chromosomal material is sometimes excised as well, is still quite unclear. Nevertheless, the overall process is well enough substantiated: excision of a phage genome may well carry chromosomal markers to the plasmid state. One example where mobilisation involves chromosomal antibiotic resistance has been described by Kondo & Mitsuhashi (1964). Phage P1 can mobilise the chloramphenicol acetyl transferase gene in *Escherichia coli* to produce a converting phage for the Cm marker.

In contrast to these specialised phages, F integration seems relatively unspecialised. Moody & Hayes (1972) have studied the mobilisation of chromosomal genes by F in both RecA$^-$ and Rec$^+$ donors and shown that a mutation in this region reduces the efficiency of chromosomal integration by at least 1000-fold (Table 1). On the other hand, mutations in *recB* or *recC* have little deleterious effect (Hall & Howard-Flanders, 1972). This argues that the *recA* gene product(s) play an inescapable part in the mobilisation of chromosomal markers by F, but that the products of *recB* and *recC* do not. Unexpectedly, however, the need for the *recA*$^+$ state in the donor is not absolute, since some mobilisation can take place when *recA* is inactive (Moody & Hayes, 1972). This argues that although substantially dependent on host-specified systems, F may specify products that will carry out the same function as the *recA* host's gene although much less effectively; and a similar conclusion has been reached by Wilkins (1969). In general, few R factors seem to be able to integrate and certainly no examples are known where the process leads to a true Hfr state (see above). On the other hand R100–1, like F, can achieve 'integrative suppression' (Nishimura, Caro, Berg & Hirota, 1971); that is, integration of the R or F factor can allow chromosomal replication to occur in a mutant where chromosomal initiation has been abolished. Although integration has not been formally proved in this series of experiments, it seems likely to have occurred since integrative suppression cannot occur in *recA*$^-$ hosts (Nishimura *et al.* 1971).

Certain R factors – and again notably R1 – can mobilise chromosomal markers and this process is unaffected by the presence of *recA*$^-$ in the donor (Moody & Hayes, 1972). Similarly, transfer of *pro* and *his* by ColIb*drd* is unaffected by the state of the *recA* gene in either the donor or the recipient, and it seems possible, therefore, that both these plasmids may have some plasmid-linked genes specifying enzymes that catalyse interaction with the chromosome. However, since neither of these plasmids can give rise to an Hfr state, neither of these enzyme systems seems adequate to achieve the steps necessary to promote transfer of the whole chromosome.

In contrast to this observation, *recA*$^-$ donors do substantially reduce chromosomal marker transfer by ColB, ColV and the R factors R64, R144 and R192 (Moody & Hayes, 1972). This suggests that the generalised recombination system of the host must be used for chromosomal mobilisation by these plasmids; but in this connection it is important to remember that I-like factors cannot form an Hfr state (Nishimura *et al.* 1971), nor can they achieve integrative suppression (Moody & Runge, 1972), unlike the F factor which otherwise has similar properties.

U.v. resistance

The exact relationship between the enzymes specified by the Rec genes of the host cell (some of which are nucleases; Barbour & Clark, 1970) and the enzymes which confer resistance to ionising radiation, is not fully understood at present. It is certain, however, that some plasmids confer resistance to u.v. irradiation and this function is due to a plasmid specified nuclease (Moody & Hayes, 1972). Whether these same enzymes can catalyse integration or facilitate any other type of interaction between plasmid and chromosome is unknown. Certainly plasmids that specify resistance to ionising radiation may give rise to chromosomal mutator effects; and this argues for some ability of these products to modify the host chromosome. Whether they can ever catalyse recombination is less clear.

In summary, therefore, three levels of plasmid interaction with the chromosome seem to have been detected so far. First, integration without the need for extensive sequence homology but with a number of highly specific enzymes to catalyse the process (e.g. phage λ); secondly, integration at a restricted number of sites using some sequence homology and the normal recombination processes of the cell (e.g. F); and thirdly a type of interaction which may lead to the mobilisation of chromosomal markers but which does not seem to lead to a true Hfr state. The details of this last type of mobilisation are less well understood but may involve interaction with a single strand of the chromosome or some other incomplete recombination process (Evenchik, Stacey & Hayes, 1969; Moody & Hayes, 1972). Mobilisation in this way is not impossible since only one DNA strand passes to the recipient on conjugation (Vapnek & Rupp, 1970, 1971). The extent of homology involved in this third type of interaction (if indeed it really occurs) is also completely unknown. All, however, have important evolutionary implications for the populations in which they are occurring.

PLASMID/PLASMID INTERACTIONS

Recombination between pairs of plasmids has been reported frequently in enteric bacteria (see reviews by Clowes, 1972, and Meynell, 1973), *Pseudomonas aeruginosa* (Holloway, Krishnapillai & Stanisich, 1971) and *Staphylococcus aureus* (Novick, 1969; Richmond & Johnston, 1969). Indeed, attempts to insert a plasmid into a cell already carrying an incompatible element leads either to the exclusion of the incomer, to replacement of the resident by the incomer or to recombination between the two plasmids; and an ability to select for a marker on each element usually yields a recombinant population (Novick, 1969). This type of recombination has been shown to occur readily between pairs of F-like R factors (Mitsuhashi *et al.* 1962; Watanabe *et al.* 1964; Meynell & Datta, 1968), and between segregants of the same R factor (Watanabe & Lyang, 1962; Nisioka *et al.* 1970). In the last example, the recombinant plasmid was found to have a molecular length equal to the sum of the parental elements.

Recombination between compatible R factors is less common (Falkow *et al.* 1971) but does occur. The recombinant of phage P22 and the R factor R222 has already been mentioned (Watanabe, Sakaizumi & Furuse, 1968) and recombination between RP4 and R64 (P and I class R factors respectively) also occurs (Datta *et al.* 1972). Some of the most complex recombinant plasmids have been produced by Frédéricq

(1969). These elements carry a wide range of markers derived from a number of compatible parental sources. For example, one recombinant that exists stably in *Escherichia coli* has the pattern Sm Cm Tc ColB *tonB cysB* ColV and contains material from three parental replicons. However, all the examples quoted up to this point arose in Rec$^+$ bacteria and it is impossible to decide whether the recombination was catalysed by specific plasmid products or by the host's generalized recombination system.

When one comes to consider the recombination of plasmids – whether compatible or not – in Rec$^-$ hosts the picture becomes much more confused. A plasmid-mediated Rec system has certainly been implicated in some types of plasmid interaction. Willetts (1971) has reported that recombination can occur between R100–1 and a transfer defective (*tra*$^-$) mutant of F*lac* in the RecA$^-$ host and recombinants between R222 and ColV2 also arise in the absence of any host RecA function (Moody, 1970).

There have been a number of reports that incompatible R factors can recombine in Rec$^-$ hosts. Takano (1966) showed recombination of two segregants of R222 in such strains and intracistronic recombination between two chloramphenicol-sensitive mutants derived from R222 has been claimed by Hashimoto, Iyobe & Mitsuhashi (1969), also in a Rec$^-$ host. However the Japanese experiments are hard to interpret. The authors never state the exact type of their Rec$^-$ mutants, and since plasmid/plasmid recombination can occur in *recB*$^-$ and *recC*$^-$ hosts (Foster & Howe, 1971), it is possible that these experiments were carried out in strains in which the RecA gene was intact. Moreover this view is supported by the experiments of Hoar (1970) who could show no recombination between derivatives of R100 in a *recA*$^-$ host.

Perhaps the most clear-cut study on recombination/complementation between plasmids in RecA$^+$ and RecA$^-$ hosts is that of Foster & Howe (1971). Chloramphenicol-sensitive alleles of the R factor R1 were isolated and shown to recombine to produce a chloramphenicol-resistant phenotype in *Escherichia coli* K12 Rec$^+$, and similar results were obtained with chloramphenicol sensitive derivatives of R100. Some of the Cm^s mutants of R100 recombined with R1 mutants in a similar way, but all the recombination between these two R factors ceased in a K12 *recA*$^-$ strain. On the other hand, recombination in *recB*$^-$ and *recC*$^-$ hosts was decreased only to about 10% of that found in the controls (Table 2). In practice some phenotypically chloramphenicol-resistant colonies were obtained in the *recA*$^-$ host but in all cases resistance could be shown to be the result of a chloramphenicol acetyl

Table 2. *Frequency of recombination between Cm-sensitive mutants of R100*

Host strain	Cm-resistant recombinants per donor cell: R100 Cm99 × R100 Cm84	R100 Cm127 × R100 Cm84
rec$^+$	3×10^{-5}	6×10^{-7}
recA	$< 5\times10^{-8}$ (< 0.002)	$< 1\times10^{-8}$ (< 0.01)
recB	2×10^{-6} (0.07)	2×10^{-8} (0.03)
recC	5×10^{-6} (0.2)	9×10^{-8} (0.1)

Ratios of recombination frequencies in *rec*$^-$ relative to *rec*$^+$ hosts are shown in parentheses.

Data from Foster & Howe (1971).

transferase enzyme that arose by intra-allelic complementation between the two R factors and not by recombination (Foster & Howe, 1971).

A rather different situation has been reported in a preliminary form by Moody (1970). The colicinogenic factor ColV2 can apparently recombine with both R1 and R15 in a *recA*$^-$ mutant of *Escherichia coli* K12. The observations of Foster & Howe mentioned above show that R1 seems to have no plasmid-linked recombination system, and it therefore seems probable that ColV2 may specify these enzymes because the ColV2/R1 recombinants arise in *recA*$^-$ lines at about 10 % of the frequency found in *recA*$^+$. In this connection it is worth remembering that ColV2 is one of the few plasmids other than F that can achieve a true Hfr state in *Escherichia coli* (Kahn, 1968).

Before leaving the question of the enzymes that may be involved in plasmid/plasmid interactions, it is important to mention the Red system in phage λ (Hershey, 1971). The *red* genes specify one or more enzymes that catalyse generalised recombination between lambda genomes. These enzymes can even compensate for *recB*$^-$ or *recC*$^-$ in *Escherichia coli* but cannot repair *recA*. Thus, episomes such as lambda can specify their own generalised recombination systems, and these can, under certain circumstances, allow recombination that does not include lambda as one of the interacting elements. Furthermore, similar phage-linked recombination genes are found in T4 (Bernstein, 1968) and in P22 (Botstein & Matz, 1970).

This survey of the literature shows that our knowledge of the types of system that can facilitate recombination between plasmids is still very incomplete. At the extremes one has two relatively clear-cut situations: first, the R factors R1 and R100 seem to have an absolute requirement for the host *recA* system – the cell's generalised recombination system. At the other extreme, phage λ specifies a function specifically adapted to catalyse recombination between replicons – the Red system. In between these extremes there seem to be a number of intermediate states. ColV2 is reported to have a plasmid-specified recombination system which can certainly catalyse interaction with the chromosome and maybe also with other plasmids. F seems to be able to integrate at a low rate in *recA*⁻ although it is not yet known whether recombination between F and other plasmids can occur under these circumstances. ColIb (Howarth, 1965) and certain other plasmids (Hoar, 1970) increase bacterial resistance to ultraviolet light, and this sometimes involves the synthesis of a specific DNA repair system (Moody & Hayes, 1972) which also has the effect of causing a plasmid-specified 'mutator' effect (Howarth, 1966). A number of plasmids can mobilise chromosomal markers without achieving an Hfr state. All these intermediate situations imply that a plasmid may specify enzymes that catalyse interaction between DNAs, and whether these effects lead to recombination or not they are all likely to influence the evolutionary potential of the clones in which they occur.

FRAGMENTATION AND INTERACTION OF NATURALLY OCCURRING PLASMIDS

Most of the work described up to this point has involved plasmids with relatively stable properties whose characters have been studied in well documented host strains, notably *Escherichia coli* K12. Unhappily most plasmids when newly isolated from clinical or other material are neither so stable nor so easy to handle. Furthermore, the bacteria in which they are first detected may not be as tractable as K12.

It is certain that many naturally occurring plasmids undergo spontaneous fragmentation and recombination, so that single bacteria can give rise to clones containing a wide range of derivative plasmids. For example Clowes and his colleagues (Nisioka *et al.* 1969, 1970; Clowes, 1972) have studied the segregation and recombination of R222 (alias R100 and NR1), a plasmid that confers resistance to sulphonamide, streptomycin, chloramphenicol and tetracycline. This plasmid exists stably as a single molecular species in *Escherichia coli* but gives rise to a range of derivative plasmids in *Proteus mirabilis*. Equilibrium centrifugation of DNA prepared from a culture derived ultimately from a single *Proteus* organism carrying R222 gives a complex pattern (Fig. 1). Analysis of the molecular species of DNA present, both by centrifugation and electron microscopy, shows the culture to contain at least six distinct molecular species (Table 3). On the basis of this and further genetic evidence, Clowes (1972) has suggested that these species

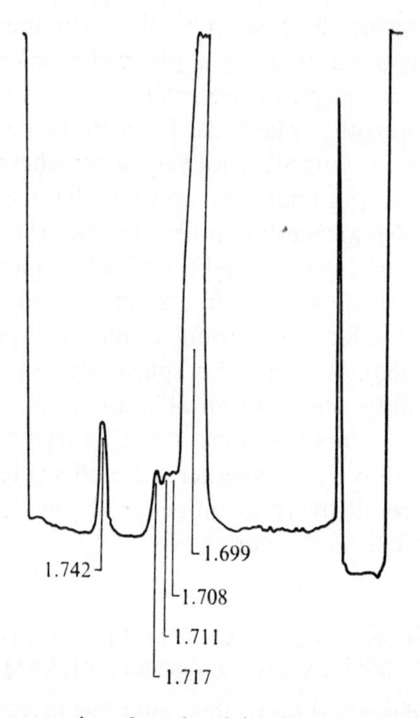

Fig. 1. Microdensitometer tracing of an ultraviolet absorption photograph after analytical centrifugation for 27 h at 44000 r.p.m. in a CsCl gradient of DNA from a *Proteus* R222+ strain. The major peak at 1.699 g cm^{-3} represents *Proteus* chromosomal DNA while the peak at 1.742 g cm^{-3} is reference DNA (*Bacillus subtilis* phage SPO1 DNA). Data from Nisioka, Mitani & Clowes (1969).

Table 3. *Contour length measurements and molecular weights of R222, its segregants and its recombinants in* Escherichia coli

Plasmid	Mean contour length (μm)	Mol. wt. (daltons × 10^{-6})
Parental		
R222/R4	33.6	70
Segregant		
R222/R3N	30.3	63
R222/R1A	22.3	46
R222/R1B	25.6	53
R222/R1C	25.7	53
R222/R1D	25.4	53
R222/R1E	25.8	53
Recombinant		
R1S × R3N	33.5	69
R1B × R3N	33.2	69

Data from Nisioka, Mitani & Clowes (1970). For the detailed genetic constitution of the plasmids, see Clowes (1972).

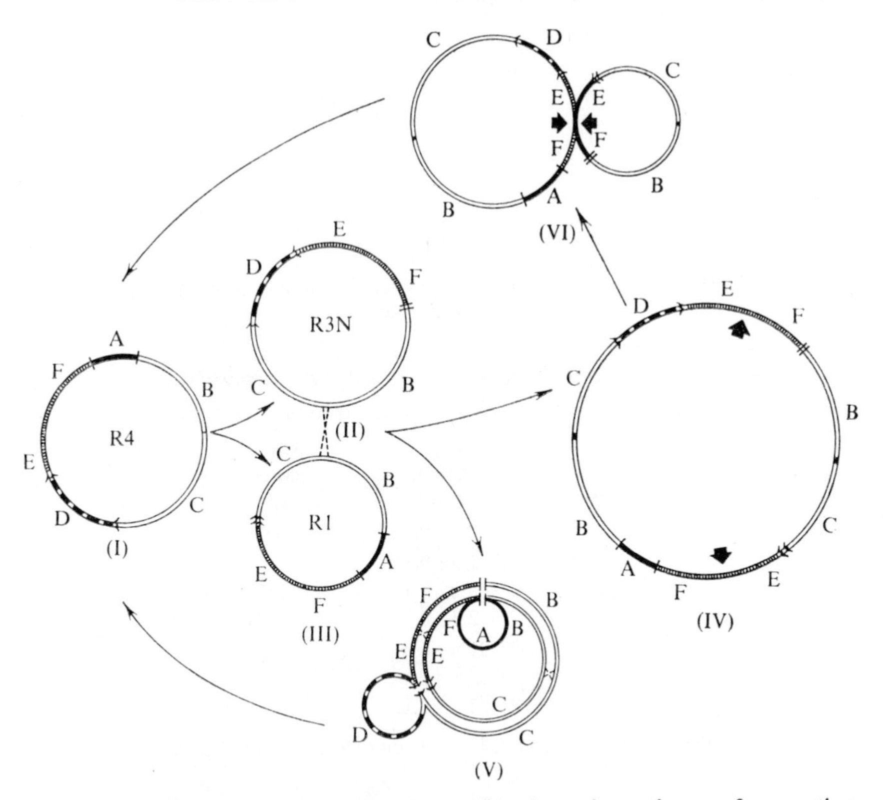

Fig. 2. Schematic diagram representing the possible alternative pathways of segregation and recombination of R222 in *Proteus* strains. Data from Clowes (1972).

arise by a primary segregation of a single large parental plasmid followed by recombination between the various fragments in various ways (Fig. 2). In this way a single clone derived from a single organism carrying a single molecular species of plasmid DNA may contain all the molecular reassortment implicit in the initial plasmid (see, for example, Richmond 1968*a*); and if additional recombinations can occur with other replicons in the cell, the evolutionary potential of the situation is enormous.

Another example of plasmid fragmentation has been described by Anderson (1968), Anderson & Lewis (1965*a*, *b*) and recently analysed at a molecular level by Clowes (1972). The original plasmid carried infectious resistance to streptomycin (S), ampicillin (A) and tetracycline (T) and was designated ΔAST. This plasmid segregated according to the pattern shown in Fig. 3 – the linkage between Δ and T being apparently indissoluble. Clowes has now analysed the segregant plasmids produced by these strains and their properties are shown in

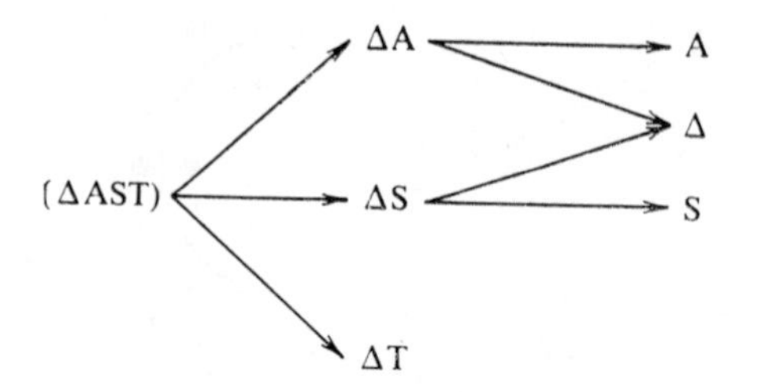

Fig. 3. Reassortment of markers between various plasmid conformations in the ΔAST system.

Table 4. *Contour length and molecular weights of a number of plasmids of the Δ Ap Sm Tc system*

Plasmid	Contour length (μm)	Mol. wt. (daltons $\times 10^{-6}$)
Δ	29.1 ± 0.8	60
S	2.9 ± 0.13	6
ΔS	29.3 ± 1.1	61
	2.7 ± 0.13	5.6
ΔA	29.5 ± 1.4	61
	2.7 ± 0.15	5.6
Δ-T	32.3 ± 1.1	67

Data from Anderson & Lewis (1965*b*) and Clowes (1972).

Table 4. In view of the transmissibility of the segregants one might expect recombinant elements to have arisen – for example ΔAS and ΔAT. However, genetic experiments suggest that such plasmids are rare in the culture harbouring this plasmid (Anderson, 1968) and the molecular experiments support this view (Clowes, 1972). This example therefore provides a pattern distinct from that described immediately before: fragmentation is common but recombination is rare (cf. Nisioka *et al.* 1970 with Anderson, 1968 and Clowes, 1972).

There is no evidence, as yet, of the role that the host's recombination system plays in plasmid fragmentations and recombinations of the type described in this section.

In an attempt to analyse such systems further, we in Bristol have studied the properties of a plasmid which carries the markers

Ap Tc Sm Cm ColI

together with two distinct transfer factors, called t_1 and t_2. The plasmid was originally detected in *Salmonella typhimurium* but it can exist in a wide range of enteric species, including Salmonellas and *Escherichia coli*. The existence of a single plasmid with the markers stated above

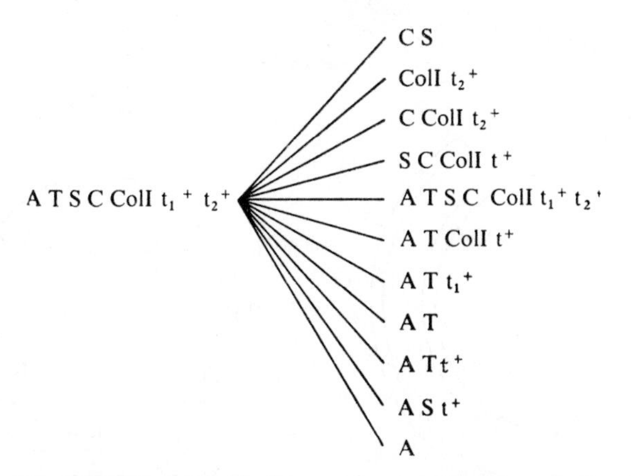

Fig. 4. Fragments of R1767 obtained after one-hour matings by selection for one drug resistance or colicin I production only.

can be deduced from the transfer properties of the plasmid (see below) and the presence of two transfer factors may be inferred from the compatibility relationships of the two most common fragments arising after transfer.

The exact molecular configuration of the markers in the donor has been examined both by conjugation and by transduction. The situation is highly complex and a full description of what occurs is not yet complete; nevertheless, certain types of plasmid interaction can be discerned. Moreover, the *recA* system of the host seems to play no part in them.

If the donor expressing all the markers is mated with a plasmid-less recipient *Escherichia coli*, and the transfer of a single marker selected, two main types of transcipient are found. With ampicillin selection the primary product is Ap TC t_1, while with chloramphenicol Cm Sm ColI t_2 is predominant. The same pattern was found when selection was for tetracycline or streptomycin resistance, respectively. Nor did the replacement of *Escherichia coli* by *Salmonella typhimurium* as recipient affect the pattern transferred.

In addition to these two main products, however, a variable number of other patterns were found with lower abundance (Fig. 4). Two are worth comment. Firstly, some transcipients carried all the markers originally present in the donor, although mating was only continued for one hour; secondly, certain transcipients only carried single markers (or sometimes pairs of markers) which were not transferable. The patterns obtained in these experiments are compatible with the map shown in Fig. 5.

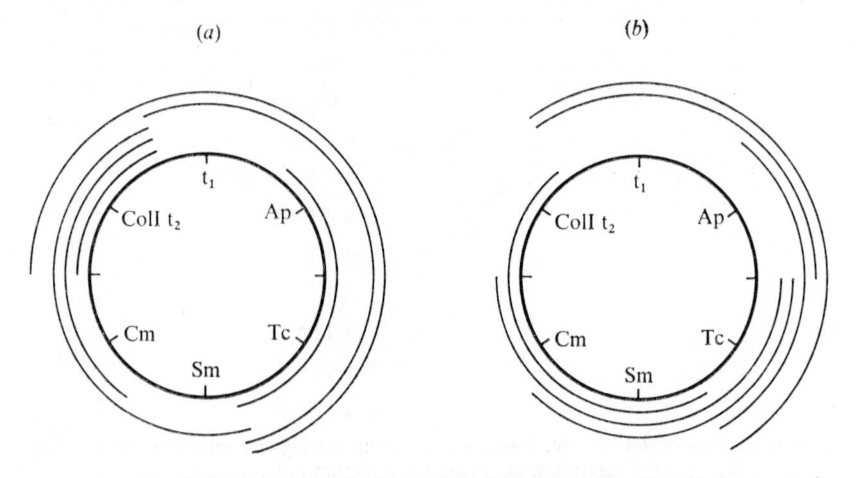

Fig. 5. Tentative linkage map of R1767 deduced (*a*) from the plasmid fragmentation patterns obtained in mating experiments, but neglecting the rare fragment ASt^+, and (*b*) from transduction patterns obtained with phage P1kc, again neglecting ASt^+.

Table 5. *The marker patterns obtained from* 50 *transcipient colonies obtained when the cross* Salmonella typhimurium *R*1767 (*Ap Tc Sm Cm ColI* t_1^+ t_2^+) × *E.* coli R^- *was selected on plates containing nalidixic acid and either ampicillin* (*Ap*) *or chloramphenicol* (*Cm*)

Pattern obtained by selection on CmNx				Pattern obtained by selection on ApNx		
Cm Sm/ coll t^+	Cm coll t^+	Ap Tc Cm Sm ColI t^+	Cm Sm	Ap Tc Col⁻ t^+	Ap t^+	Ap Tc Cm Sm ColI t^+
48/50	—	2/50	—	12/12	—	—
47/50	—	3/50	—	25/25	—	—
48/50	—	2/50	—	24/25	1/25	—
50/50	—	—	—	24/24	—	—
50/50	—	—	—	11/13	—	2/13
45/50	5/50	—	—	24/25	1/25	—
49/50	1/50	—	—	50/50	—	—
49/50	—	1/50	—	16/18	2/18	—
48/50	—	2/50	—	1/2	—	1/2
49/50	—	1/50	—	—	—	4/4
—	—	13/13	—	7/7	—	—
46/50	3/50	1/50	—	13/14	—	1/14
50/50	—	—	—	23/25	2/25	—
48/50	—	2/50	—	5/5	—	—
50/50	—	—	—	8/8	—	—
50/50	—	—	—	19/20	—	1/20
50/50	—	—	—	20/20	—	—
46/50	—	3/50	1/50	4/4	—	—
50/50	—	—	—	24/24	—	—
50/50	—	—	—	8/9	1/9	—

In order to get more information about the quantitative aspects of this experiment the donor culture expressing all the markers was plated on nutrient agar without selection and twenty individual colonies picked into nutrient broth. These separate clones were then mated with a plasmid-less *Escherichia coli* recipient and the progeny plated on ampicillin or chloramphenicol agar with nalidixic acid as a counter-selective agent in a mating experiment lasting one hour (Table 5). With chloramphenicol as selective agent, 8/20 donors gave transfer of Cm Sm ColI t_1 only. Eleven others gave transfer of this marker pattern together with another. In nine of these, the additional pattern was the complete pattern from the donor while in two the additional pattern was Cm ColI t_1 only. Perhaps the most interesting pattern transferred was found in only one case out of the twenty tested: this was the transfer of the whole pattern alone. This must be good evidence for the existence of a single plasmid in the donor comprising all the markers, since two separate plasmids in the donor culture would be extremely unlikely to give 100 % linkage on transfer in a cross of this type. Furthermore, the fact that such a pattern of events was found in as many as one colony of the twenty tested argues that the linkage of all markers cannot be uncommon in these donor populations.

Selection of this cross for transfer of ampicillin resistance gave a similar picture. A wide range of patterns was obtained; but once again 1/20 colonies transferred all the markers from the donor. These experiments argue, therefore, that the full pattern Ap Tc Sm Cm ColI t_1 t_2 exists in some organisms of the donor population, although the behaviour on transfer suggests that fragmentation to smaller linkage groups occurs freely. Moreover, they suggest that there is a continuous reassortment of markers between various plasmid configurations – perhaps as outlined in Fig. 6.

In order to gain further information about the plasmid configuration in the cultures expressing the pattern Ap Tc Sm Cm ColI t_1 t_2, a transductional analysis was carried out with phage P1. The information from this source confirmed the data obtained by conjugation. However, transduction of the complete marker pattern was never obtained, possibly because the molecular size of this element was greater than the amount of the DNA that can be accommodated on the head of phage P1 (Ikeda & Tomizawa, 1968).

Molecular studies on donor cultures carrying all the markers showed the bacteria to contain up to about 20 % of their DNA in a plasmid form. Moreover the molecular weight of the main components of this material was about 10×10^6 daltons. It seems probable therefore that

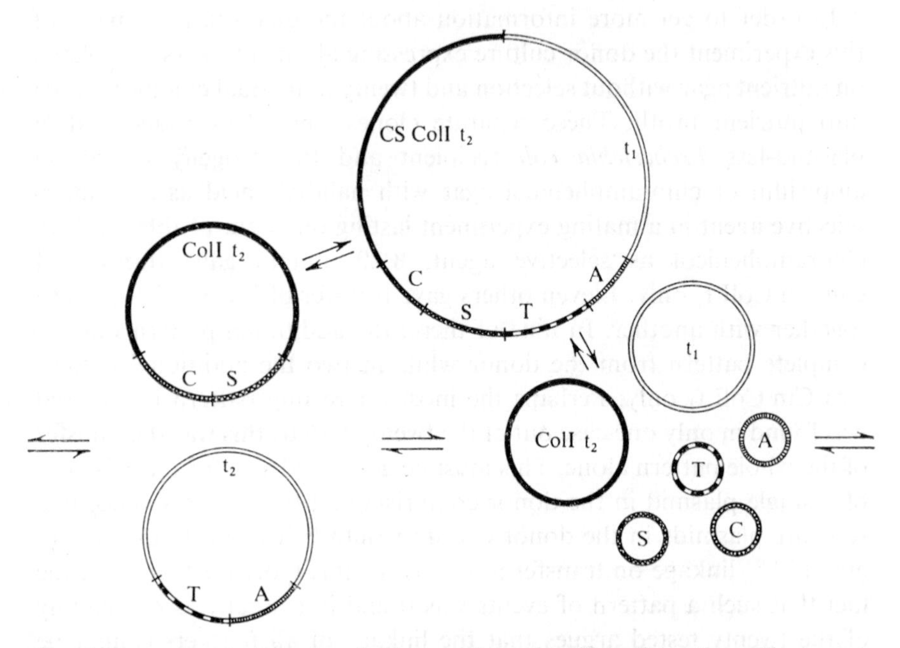

Fig. 6. Reassortment of markers between various plasmid conformations. The arrows pointing outwards indicate that there are more possibilities than shown here.

any large plasmids containing transfer factors, and replicating under stringent control (Clowes, 1972), will be obscured by the non-transmissible small fragments replicating in a relaxed manner. Elucidation of the nature and abundance of the individual replicons in these bacteria must await electron microscopy of the various types of DNA molecule present.

These experiments described here suggest strongly that there is a great deal of molecular assortment going on between a large number of different replicons – some transmissible and others not – in these strains. The question arises, therefore, as to whether this process requires a functional *recA* system at any stage, or whether it occurs under the control of a plasmid specified system. Accordingly, the donor expressing the pattern Ap Tc Sm Cm ColI t_1 t_2 was mated with a plasmid-less *Escherichia coli* which carried *recA13*. Very similar results to those found in *recA*$^+$ were obtained, even when a number of cycles of transfer and selection of single clones were carried out in the *recA*$^-$ strains. Thus, we can conclude that the molecular interactions of the various replicons to be found in these experiments may occur without the aid of a functional *recA* system, either in the donor or in the recipient.

PLASMID STUDIES IN MAN

Although carried out with plasmids originally isolated in the wild, the experiments reported up to this point were all carried out in the laboratory. The difficulty with any attempt to show plasmid recombination, either with the chromosome or with another plasmid, under natural conditions is that everything must be inferred from the characteristics of the elements known to be present initially and what appears subsequently. In many cases the first of these points is particularly difficult to establish. If, for example, one is concerned with possible plasmid recombination in bacteria inhabiting the gut of man, how can one be sure that a plasmid that has all the characteristics of a recombinant may not have been present as a rare member of the bacterial population from the outset?

There have been a number of attempts to show that plasmid recombination occurs in nature by comparing the antibiotic resistance markers on R factors present over a period during which some selective agent has been used. However, marker pattern alone is very dubious evidence for plasmid identity and hence for recombination. An R factor of molecular weight 60×10^6 daltons carries enough DNA for about 100 genes, and consequently the nature of four or five resistance determinants must inevitably give a poor indication of the nature of the structure as a whole.

At present, there are two possible ways of overcoming this difficulty. The first is to compare the plasmids isolated in the early and late stages of the observation period and compare their base sequence by DNA/DNA hybridisation. This approach gives a relatively crude measure of plasmid identity, but at least the measurement refers to the whole of the plasmid DNA, not just to the part that specifies the known gene products (Falkow, Hääpala & Silver, 1969). The other approach, and one which potentially has much greater resolving power, is to use the electron microscope to examine heteroduplexes obtained by annealing the DNA strands from the two plasmids to be compared. Such heteroduplexes have 'out-loops' of single-stranded DNA wherever the two elements are not homologous (Fig. 7) (Westmoreland, Szybalski & Ris, 1969); and this therefore gives a very sensitive measure of the distribution of sequence similarities and differences along the plasmid DNA.

One recent careful investigation of a change of antibiotic resistance pattern from Sm Tc to Sm Tc Ap Km during an epidemic of *Shigella flexneri* infection treated with kanamycin, showed that the 'recombinants' in this case carried two distinct plasmids (Sm Tc Ap + Km) rather than a single element that had acquired its Km marker by recombination (Farrer *et al.* 1972). However, the acquisition of additional markers in the wild is not always due to the uptake of a second compatible plasmid: recombination may also occur under natural conditions. Molecular studies on the nature of some R factors found in strains of *Klebsiella pneumoniae* and *Pseudomonas aeruginosa* isolated from burned patients (Lowbury *et al.* 1969) showed that a single plasmid with the marker pattern Ap Ne/Km Tc Cm fragmented to give a smaller element with the pattern Ap Ne/Km Tc on transfer from a Klebsiella strain to *P. aeruginosa* (Ingram, Richmond & Sykes, 1973; L. C. Ingram, unpublished observations). In these experiments the criteria for claiming a relationship between the two plasmids was the extent of DNA/DNA hybridisation pattern between them, as well as the nature of the gene products they specified. All of the smaller plasmid was represented in the larger, but only about 70 % of the larger in the smaller.

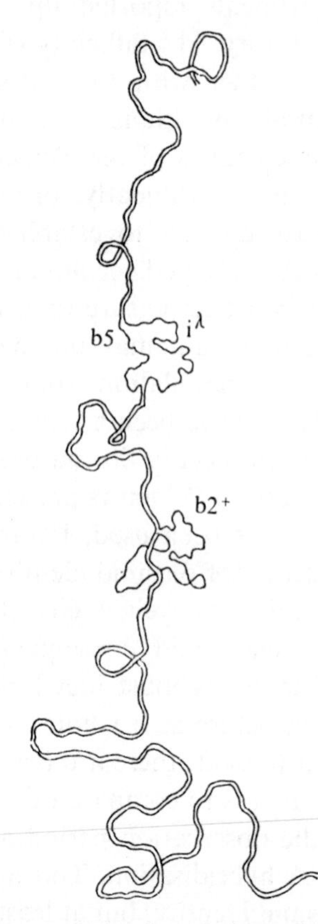

Fig. 7. A drawing of a heteroduplex between the left-hand strand of wild-type λ and the right-hand strand λ b2b5. Two regions are unpaired and have different structures. From Westmoreland, Szybalski & Ris (1969).

Plasmid recombination has also been shown in experiments carried out in volunteers to whom organisms carrying well characterised R factors were fed, followed by courses of antibiotics designed to select any recipients which had received the donor R factor by transfer or any strains harbouring plasmids derived from the donor by recombination. R-factor transfer and recombinant formation was then investigated by feeding a strain of *Escherichia coli* marked genetically for identification which carried an R factor of known properties. Transfer can be detected by identifying the donor plasmid either in the ingested potential recipient or in any endogenous coliforms other than the marked donor. Recombinants can be detected by showing that a substantial piece of the donor plasmid can be detected in another plasmid appearing in the second phase of the experiment.

In practice, no transfer could be detected in any experiment in the absence of antibiotic selection. However, treatment of the volunteer with antibiotics resulted in the emergence of a number of types of resistant organism, and among them some strains carrying plasmids which could be shown by molecular tests to be recombinant of the fed R factor and another element – presumably present at the onset of the experimental sequence. In one volunteer the fed R factor was R1 (which confers resistance to Ap Cm Km Sm Su) and transfer to a serotype O9 strain was found following treatment with ampicillin. However, this volunteer also carried a serotype O2 strain of *Escherichia coli* with the marker pattern Ap Cm Km Sm Su Tc after treatment whereas earlier the majority of type O2 strains in his alimentary tract apparently expressed the resistance pattern Ap Sm Tc. The type O2 Ap Cm Km Sm Su Tc-resistant organism was found to carry a single plasmid which had a molecular weight of about 90×10^6 daltons, about two thirds of which was derived from R1 and the remaining third from the Am Sm Tc plasmid present in the type O2 strain initially. So in this volunteer the feeding of an R factor resulted in some recombinant plasmids formed between R1 and a resident plasmid being selected by ampicillin treatment (Anderson, Gillespie & Richmond, 1973; Anderson, Ingram, Richmond & Wiedeman, 1973; L. C. Ingram, unpublished experiments).

PLASMIDS AND EVOLUTION: SOME SPECULATIONS

Evidence adduced elsewhere argues strongly that the transmissibility of plasmids, whether by conjugation or transduction, plays a crucial part in bacterial evolution because it facilitates the transfer of large blocks of genetic information between bacteria in a population (Anderson, 1968; Meynell *et al.* 1968; Richmond, 1968*a*, *b*; Jones & Sneath, 1970). But what about the compilation of the genetic information carried by the plasmids themselves? This is not immutable; and indeed rapid evolution of the marker patterns carried by plasmids may occur under natural conditions (see, for example, Anderson, 1968). Furthermore, chromosomal markers may be mobilised and this phenomenon, taken in conjunction with the other aspects of plasmid variation and transfer, adds up to the most potent system for the evolution of bacterial populations.

The evolutionary potential of a plasmid must depend to a great extent on its recombination properties and work that is emerging now seems to divide these elements into two broad classes with perhaps a third of even greater potential. First, there are the plasmids which have their own recombination systems specified entirely by genes that are themselves part of the plasmid. At its most sophisticated this system may specify both integration at a specific point in the host genome and also some generalised recombination between these plasmids themselves. Phage λ provides a typical example of this pattern, and this type of organisation may turn out to be much more common among bacterial plasmids than is realised at the moment. Paradoxically, perhaps, this type of system has a relatively limited evolutionary potential, if only because the interactions that occur are so site specific.

The second type of plasmid is typified by F. Such elements integrate at a site where there seems to be some sequence homology with another replicon by using the host's generalised recombination enzymes. For this to occur homology seems to be needed; and below we speculate how this may arise.

The third – and potentially most important – class of plasmid, whose presence is little more than a hunch at present, is a class of elements analogous to the mutator phage μ1. Such plasmids would be the most promiscuous of all and would be expected to integrate all over the genome. However, the majority of such interactions would probably be lethal, since they would occur in the middle of essential genes. Only those that occurred in non-essential areas, or precisely between cistrons, would survive. Thus the number of viable points of interaction over the genome would be limited and this would be reflected in a restricted number of integration sites. Furthermore, these interactions would not be confined to the chromosome. Saedler & Heiss (1973) have already reported that regions of homology to the insertion fragments IS1 and IS2 may be found both in the chromosome and other plasmids in *Escherichia coli* K12 and indeed IS1 and IS2 might themselves be fragments of a mutator-like particle acting in the manner envisaged here. It follows, therefore, that apart from its evolutionary potential in its own right, the activities of an element like phage μ1 may set up just the regions of homology between replicons needed to facilitate recombination of the type found with F; and the sites of such interaction would be restricted to those where integration would not be lethal.

In summary, therefore, the recombinational properties of plasmids are certain to be of the greatest importance as a source of bacterial variation, and these properties, taken in conjunction with their transfer ability, truly make them the crucible of bacterial evolution.

We are grateful to the Wellcome Foundation for the award of a Wellcome Anglo-German Exchange Fellowship tenable in the Department of Bacteriology in the University of Bristol to one of us (B. W.). The original work described here was supported partly by this grant and partly by the Programme Grant for the Molecular and Epidemiological Study of Bacterial Plasmids awarded to M. H. R. by the Medical Research Council.

REFERENCES

Anderson, E. S. (1968). The ecology of transferable drug-resistance in the enterobacteria. *Annual Review of Microbiology*, **22**, 131–80.

Anderson, E. S. & Lewis, M. J. (1965*a*). Drug resistance and its transfer in *Salmonella typhimurium*. *Nature, London*, **206**, 579–83.

Anderson, E. S. & Lewis, M. J. (1965*b*). Characterisation of a transfer factor associated with drug resistance in *Salmonella typhimurium*. *Nature, London*, **208**, 843–9.

Anderson, J. D., Gillespie, W. A. & Richmond, M. H. (1973). Chemotherapy and antibiotic resistance transfer between Enterobacteria in the human gastrointestinal tract. *Journal of Medical Microbiology*, **6**, 461–73.

Anderson, J. D., Ingram, L. C., Richmond, M. H. & Wiedeman, B. (1973). Studies on the nature of the plasmids arising from conjugation in the human gastrointestinal tract. *Journal of Medical Microbiology*, **6**, 475–86.

Barbour, S. D. & Clark, A. J. (1970). Biochemical and genetic studies on recombination proficiency in *E. coli*. I. Enzymatic activity associated with the *recB*$^+$ and *recC*$^+$ genes. *Proceedings of the National Academy of Sciences, USA*, **65**, 955–61.

Bernstein, H. (1968). Repair and recombination in phage T 4. I. Genes affecting recombination. *Cold Spring Harbor Symposia on Quantitative Biology*, **33**, 325–31.

Bolton, E. T., Britten, P. J., Byers, T. J., Cowie, D. B., Hoyer, B., Kato, Y., McCarthy, B., Miranda, M. & Roberts, R. B. (1964). Report on the Biophysics Section. *Carnegie Institution of Washington Yearbook*, **63**, 366–91.

Botstein, D. & Matz, M. J. (1970). A recombination function essential to the growth of bacteriophage P2. *Journal of Molecular Biology*, **54**, 417–25.

Broda, P. (1967). The formation of Hfr strains in *Escherichia coli* K12. *Genetical Research, Cambridge*, **9**, 35–47.

Brooks, K. & Clark, A. J. (1967). Behaviour of λ bacteriophage in a recombination deficient strain of *Escherichia coli*. *Journal of Virology*, **1**, 283–93.

Calendar, R. & Lindahl, G. (1969). Attachment of prophage P2: gene order at different host chromosomal sites. *Virology*, **39**, 867–81.

Campbell, A. M. (1962). Episomes. *Advances in Genetics*, **11**, 101–45.

Carlton, B. C. & Helinski, D. R. (1969). Heterogeneous circular DNA elements in vegetative culture of *Bacillus megaterium*. *Proceedings of the National Academy of Sciences, USA*, **64**, 592–9.

CIBA symposium (1968). CIBA Foundation Symposium, *Bacterial Episomes and Plasmids*, ed. G. E. W. Wolstenholme & M. O'Connor. London: S. & A. Churchill.

Clowes, R. C. (1972). Molecular structure of bacterial plasmids. *Bacteriological Reviews*, **36**, 361–405.

Curtiss, R., Jr, Charamella, L. J., Stallions, D. R. & Mays, J. A. (1968). Parental functions during conjugation in *E. coli* K12. *Bacteriological Reviews*, **32**, 320–48.

Datta, N., Hedges, R. W., Shaw, E. J., Sykes, R. B. & Richmond, M. H. (1972). Properties of an R factor isolated from *Pseudomonas aeruginosa*. *Journal of Bacteriology*, **108**, 1244–9.

DAVIS, R. W. & PARKINSON, J. S. (1971). Deletion mutants of bacteriophage lambda. III. Physical structure of *attφ*. *Journal of Molecular Biology*, **56**, 403–23.

ECHOLS, H. (1971). Lysogeny: viral repression and site-specific recombination. *Annual Review of Biochemistry*, **40**, 827–54.

EVENCHIK, Z., STACEY, K. A. & HAYES, W. (1969). Ultraviolet induction of chromosome transfer by autonomous sex factors in *E. coli*. *Journal of General Microbiology*, **56**, 1–14.

FALKOW, S., HÄÄPALA, K. K. & SILVER, R. P. (1969). Relationships between extrachromosomal elements. In CIBA Foundation Symposium, *Bacterial Episomes and Plasmids*, ed. G. E. W. Wolstenholme & M. O'Connor, pp. 136–62. London: J. & A. Churchill.

FALKOW, S., TOMPKINS, L. S., SILVER, R. P., GUERRY, P. & LE BLANC, D. J. (1971). The replication of R factor DNA in *E. coli* K12 following conjugation. *Annals of the New York Academy of Sciences*, **182**, 153–71.

FARRER, W. E., JR, EDSON, M., GUERRY, P., FALKOW, S., DRUSIN, L. M. & ROBERTS, R. B. (1972). Interbacterial transfer of R factors in the human intestine: *In vivo* acquisition of R factor-mediated kanamycin resistance by a multiresistant strain of *Shigella sonnei*. *Journal of Infectious Diseases*, **126**, 27–33.

FOSTER, T. & HOWE, T. G. B. (1971). Recombination and complementation between R factors in *Escherichia coli* K12. *Genetical Research, Cambridge*, **18**, 287–97.

FRANKLIN, N. C. (1971). Illegitimate recombination. In *The Bacteriophage Lambda*, ed. A. D. Hershey, pp. 175–94. Cold Spring Harbor Laboratory.

FRÉDÉRICQ, P. (1965). Genetics of colicinogenic factors. *Zentralblatt für Bakteriologie, Parasitenkunde, Infektionskrankheiten und Hygiene*, **196**, 142–51.

FRÉDÉRICQ, P. (1969). In CIBA Foundation Symposium: *Bacterial Episomes & Plasmids*, ed. G. E. W. Wolstenholme & M. O'Connor, pp. 163–78. London: J. & A. Churchill.

GOTTESMANN, M. E. & YARMOLINSKY, M. R. (1968). The integration and excision of the λ genome. *Cold Spring Harbor Symposia on Quantitative Biology*, **33**, 735–48.

HALL, J. D. & HOWARD-FLANDERS, P. (1972). Recombinant F factors from *Escherichia coli* K12 strains carrying *recB* or *recC*. *Journal of Bacteriology*, **110**, 578–84.

HASHIMOTO, H., IYOBE, S. & MITSUHASHI, S. (1969). Unstable mutants of R factor. *Japanese Journal of Microbiology*, **13**, 343–9.

HAYES, W. (1968). *The Genetics of Bacteria and Their Viruses*, 2nd edition. Oxford & Edinburgh: Blackwell.

HERSHEY, A. D. (1971). The bacteriophage lambda. *Cold Spring Harbor Laboratory*.

HIRSCH, H. J., SAEDLER, H. & STARLINGER, P. (1972). Insertion mutations in the control regions of the galactose operon of *Escherichia coli*. II. Physical characterisation of the mutations. *Molecular and General Genetics*, **115**, 226–35.

HOAR, D. I. (1970). Fertility regulation in F-like resistance transfer factors. *Journal of Bacteriology*, **101**, 916–20.

HOLLOWAY, B. W. (1969). Genetics of Pseudomonas. *Bacteriological Reviews*, **33**, 419–43.

HOLLOWAY, B. C., KRISHNAPILLAI, V. & STANISICH, V. (1971). Pseudomonas genetics. *Annual Review of Genetics*, **5**, 425–46.

HOWARTH, S. (1965). Resistance to the bactericidal effect of ultraviolet radiation conferred on enterobacteria by the colicine factor *colI*. *Journal of General Microbiology*, **40**, 43–55.

HOWARTH, S. (1966). Increase in the frequency of ultraviolet-induced mutation brought about by the colicine factor, *col*I in *Salmonella typhimurium*. *Mutation Research*, **3**, 129–34.

IKEDA, H. & TOMIZAWA, J. (1968). Prophage P1, an extrachromosomal replication unit. *Cold Spring Harbor Symposia on Quantitative Biology*, **33**, 791–8.

INGRAM, L. C., RICHMOND, M. H. & SYKES, R. B. (1973). The characteristics of the plasmids in a series of strains isolated from burned patients. *Antimicrobial Agents and Chemotherapy*, in press.

JACOB, F. & WOLLMAN, E. L. (1961). In *Sexuality and the Genetics of Bacteria*. New York: Academic Press.

JONES, D. & SNEATH, P. H. A. (1970). Genetic transfer and bacterial taxonomy. *Bacteriological Reviews*, **34**, 40–81.

KAHN, P. L. (1968). Isolation of high-frequency recombining strains from *Escherichia coli* containing the V colicinogenic factor. *Journal of Bacteriology*, **96**, 205–14.

KAHN, P. L. (1969). Evolution of a site of specific genetic homology on the chromosome of *Escherichia coli*. *Journal of Bacteriology*, **100**, 269–75.

KONDO, E. & MITSUHASHI, S. (1964). Drug resistance of enteric bacteria. IV. Active transducing bacteriophage P1*Cm* produced by the combination of R factor with bacteriophage P1. *Journal of Bacteriology*, **88**, 1266–76.

LOWBURY, E. J. L., KIDSON, A., LILLY, H. A., AYLIFFE, G. A. J. & JONES, R. J. (1969). Sensitivity of *Pseudomonas aeruginosa* to antibiotics: emergence of strains highly resistant to carbenicillin. *Lancet*, ii, 448–52.

MCCARTHY, B. J. & CHURCH, R. B. (1970). The specificity of molecular hybridisation reactions. *Annual Review of Biochemistry*, **39**, 131–50.

MEYNELL, E. & DATTA, N. (1968). In Ciba Foundation Symposium, *Bacterial Episomes and Plasmids*, ed. G. E. W. Wolstenholme & M. O'Connor, pp. 120–35. London: J. & A. Churchill.

MEYNELL, E., MEYNELL, G. G. & DATTA, N. (1968). Phylogenetic relationships of drug resistance factors and other transmissible bacterial plasmids. *Bacteriological Reviews*, **32**, 55–88.

MEYNELL, G. G. (1973). In *Bacterial Plasmids*. London: Macmillan.

MITSUHASHI, S., HARADA, K., HASHIMOTO, H., KAMEDA, M. & SUZUKI, M. (1962). Combination of two types of transmissible drug-resistance factors in a host bacterium. *Journal of Bacteriology*, **84**, 9–16.

MOODY, E. E. M. (1970). Recombination between R factors and a colicinogenic factor. *Microbial Genetics Bulletin*, **32**, 8.

MOODY, E. E. M. & HAYES, W. (1972). Chromosome transfer by autonomous transmissible plasmids: the role of the bacterial recombination (Rec) system. *Journal of Bacteriology*, **111**, 80–5.

MOODY, E. E. M. & RUNGE, R. (1972). The integration of autonomous transmissible plasmids into the chromosome of *E. coli* K12. *Genetical Research, Cambridge*, **19**, 181–90.

NEW YORK ACADEMY SYMPOSIUM (1971). *Annals of the New York Academy of Sciences*, vol. 182.

NISHIMURA, Y., CARO, L., BERG, C. M. & HIROTA, Y. (1971). Chromosome replication of *E. coli*. IV. Control of chromosome replication and cell division by an integrated episome. *Journal of Molecular Biology*, **55**, 441–56.

NISIOKA, T., MITANI, M. & CLOWES, R. C. (1969). Composite circular forms of R factor deoxyribonucleic acid molecules. *Journal of Bacteriology*, **97**, 376–85.

NISIOKA, T., MITANI, M. & CLOWES, R. C. (1970). Molecular recombination between R factor deoxyribonucleic acid molecules in *Escherichia coli* host cells. *Journal of Bacteriology*, **103**, 166–77.

Novick, R. P. (1967). Penicillinase plasmids of *Staphylococcus aureus*. *Federation Proceedings*, **26**, 29–38.

Novick, R. P. (1969). Extrachromosomal inheritance in bacteria. *Bacteriological Reviews*, **33**, 210–35.

Novick, R. P. & Richmond, M. H. (1965). Nature and interaction of the genetic elements governing penicillinase synthesis in *Staphylococcus aureus*. *Journal of Bacteriology*, **90**, 467–80.

Pearce, L. E. & Meynell, E. W. (1968). Mutation to high level streptomycin-resistance in R^+ bacteria. *Journal of General Microbiology*, **50**, 173–6.

Richmond, M. H. (1968*a*). Extrachromosomal elements and the spread of anti-biotic resistance in bacteria. *Biochemical Journal*, **113**, 225–34.

Richmond, M. H. (1968*b*). Plasmids and chromosomes in Prokaryotic cells. In *Organization and Control in Prokaryotic and Eukaryotic Cells, Symposia of the Society of General Microbiology*, **20**, 249–77. Ed. H. P. Charles & B. C. J. G. Knight. London: Cambridge University Press.

Richmond, M. H. (1968*c*). The plasmids of *Staphylococcus aureus* and their relation to other extrachromosomal elements in bacteria. *Advances in Microbial Physiology*, **2**, 43–88.

Richmond, M. H. (1973). Resistance factors and their ecological importance to bacteria and man. *Progress in Nucleic acid Research and Molecular Biology*, **13**, 191–244.

Richmond, M. H. & Johnston, J. H. (1969). Genetic Interactions of penicillinase plasmids in *Staphylococcus aureus*. In CIBA Foundation Symposium, *Bacterial Episomes and Plasmids*, ed. G. E. W. Wolstenholme & M. O'Connor, pp. 179–96. London: J. & A. Churchill.

Russell, R. L., Abelson, J. N., Lardy, A., Gefter, M. L., Brenner, S. & Smith, J. T. (1970). Duplicate genes for tyrosine transfer RNA in *Escherichia coli*. *Journal of Molecular Biology*, **47**, 1–13.

Saedler, H. & Heiss, B. (1973). Multiple copies of the insertion DNA sequences IS1 and IS2 in the chromosome of *Escherichia coli* K12. (In Press.)

Scaife, J. (1966). F-prime factor formation in *E. coli* K12. *Genetical Research, Cambridge*, **8**, 189–96.

Shimada, K., Weisberg, R. A. & Gottesman, M. E. (1971). Prophage lambda at unusual chromosomal sites. I. Location of the secondary attachment sites and the properties of the lysogen. *Journal of Molecular Biology*, **63**, 483–503.

Signer, E. R. (1966). Interaction of prophages at the att_{80} site with the chromosome of *Escherichia coli*. *Journal of Molecular Biology*, **15**, 243–55.

Signer, E. R. (1968). Lysogeny: the integration problem. *Annual Review of Microbiology*, **22**, 451–88.

Signer, E. R. (1969). Plasmid formation: a new mode of lysogeny by phage λ. *Nature, London*, **223**, 158–60.

Smith, S. M. & Stocker, B. A. D. (1962). Colicinogeny and recombination. *British Medical Bulletin*, **18**, 46–51.

Smith, S. M. & Stocker, B. A. D. (1966). Mapping of prophage P22 in *Salmonella typhimurium*. *Virology*, **28**, 413–19.

Starlinger, P. & Saedler, H. (1972). Insertion mutations in microorganisms. *Biochemie*, **54**, 177–85.

Sykes, R. B., Grinsted, J., Ingram, L. C., Saunders, J. R. & Richmond, M. H. (1972). In *Bacterial Plasmids and Antibiotic Resistance*, ed. V. Krcmery, L. Rosival & T. Watanabe, pp. 27–35. Berlin: Springer-Verlag.

Takano, T. (1966). Behaviour of some episomal elements in a recombination-deficient mutant of *Escherichia coli*. *Japanese Journal of Microbiology*, **10**, 201–10.

TAYLOR, A. L. (1963). Bacteriophage-induced mutation in *Escherichia coli. Proceedings of the National Academy of Sciences, USA*, **50**, 1043–51.

THOMAS, C. A., JR (1966). Recombination of DNA molecules. *Progress in Nucleic Acid Research and Molecular Biology*, **5**, 315–37.

TOUSSAINT, A. (1969). Insertion of phage Mu1 within prophage λ. *Molecular and General Genetics*, **116**, 89–92.

VAPNEK, D. & RUPP, W. D. (1970). Asymmetric segregation of the complementary strands during conjugation in *Escherichia coli. Journal of Molecular Biology*, **53**, 287–303.

VAPNEK, D. & RUPP, W. D. (1971). Identification of individual R-factor DNA strands and their replication during conjugation. *Journal of Molecular Biology*, **60**, 413–24.

WATANABE, T. (1963). Infective heredity of multiple drug resistance in bacteria. *Bacteriological Reviews*, **26**, 23–8.

WATANABE, T. & LYANG, K. W. (1962). Episome-mediated transfer of drug resistance in Enterobacteriaceae. V. Spontaneous segregation and recombination of resistance factors in *Salmonella typhimurium. Journal of Bacteriology*, **84**, 422–30.

WATANABE, T., NISHIDA, H., OGATA, C., ARAI, T. & SATO, S. (1964). Episome-mediated transfer of drug resistance in *Enterobacteriaceae*. VII. Two types of naturally-occurring R factors. *Journal of Bacteriology*, **88**, 716–26.

WATANABE, T., SAKAIZUMI, S. & FURUSE, C. (1968). Superinfection with R factors by transduction in *Escherichia coli* and *Salmonella typhimurium. Journal of Bacteriology*, **96**, 1796–802.

WESTMORELAND, B. D., SZYBALSKI, W. & RIS, H. (1969). Mapping of deletions and substitutions in heteroduplex DNA molecules of bacteriophage lambda by electron microscopy. *Science*, **163**, 1343–8.

WILKINS, B. M. (1969). Chromosome transfer from F-lac$^+$ strains of *Escherichia coli* K12 mutant at *recA*, *recB* or *recC. Journal of Bacteriology*, **98**, 599–604.

WILLETTS, N. S. (1971). Plasmid specificity of two proteins required for conjugation in *E. coli* K12. *Nature, London*, **230** 183–5.

BREEDING SYSTEMS AND EVOLUTION

K. ESSER
Lehrstuhl Allgemeine Botanik, Ruhr-Universität, Bochum, Germany

INTRODUCTION

Since the time of Darwin, evolution has never lost its attraction for scientists and for laymen. If the word evolution comes up it is associated by most people with pedigrees exhibiting the relationships and descent of various taxa of living beings. Such diagrams are based mainly on palaeontological studies. Since fossils are not equally common from all periods of the earth's history and not all tribes of living beings (e.g. bacteria and fungi) have structures suitable for preservation for long periods of time, the schemes of evolutionary relationships have numerous uncertainties and question marks. We probably will have to tolerate this, due to limited possibilities for reconstructing the long path of evolution. I am not sure whether much can be done at present to fill the gaps in evolutionary pedigrees with facts instead of speculation. However, there is another aspect of evolution which also comes to our attention. This is the question of the mechanisms which have caused the ramifications of and within the various branches of these pedigrees.

It is generally accepted that *mutation, recombination and selection are the main agencies of evolution.* In this sequence each phenomenon is a prerequisite for the next. Thus recombination cannot be effective without mutation and the efficiency of selection is tremendously enhanced when mutation is followed by recombination. Owing to this close connection between all three phenomena, any kind of genetic mechanism which will affect one of them will also concern the whole process of evolution. Mutation and selection are relatively inaccessible to genetic control, because mutation can occur spontaneously and is undirected, and selection depends mainly on environmental conditions. In contrast *recombination*, although also undirected and occurring spontaneously, *requires the bringing together of different genetic material which is achieved by sexual or parasexual processes.* Both of these latter phenomena are *controlled by* genetic factors which display their action in so-called *breeding systems.*

The progress in genetics and in molecular biology achieved within the last few decades has also unveiled the genetic determinants of the breeding systems. Therefore the knowledge of the action and the interaction of the breeding systems will contribute to a better understanding of recombination and therefore of evolution. A consideration from this point of view will stress the functional aspects of evolution. This shift from its classical descriptive treatment to some of its genetic causes and therefore to its physiological and biochemical consequences is evident from the other papers of this symposium. Thus, evolution has lost many of its mystifying attributes and becomes more and more understandable to us in terms of scientific parameters.

It is the aim of this paper, after briefly reviewing some facts about the various breeding systems, to direct attention to some generalities which may have been overlooked, partly due to the nomenclatorial confusion which still exists in this field of biology. The fungi, with their tremendous versatility in sexual and parasexual processes, will be given most attention. However, we hope that it will become evident that the systems and mechanisms, some of which were first detected in this taxon, are not restricted to fungi and are acting in all other living beings, including the prokaryotes. This might help to meet the criticism of Francis Crick (1970):

> There is also a major problem to which I believe biologists have given insufficient attention. All biologists essentially believe that evolution is driven by natural selection, but someone from the more exact sciences could well point out that it has yet to be established that the rate of evolution can be adequately explained by the processes which are familiar to us. It would not surprise me if nature has evolved rather special and ingenious mechanisms so that evolution can proceed at an extremely rapid rate – recombination is an obvious example...To solve this problem we may need a rather complete knowledge of many biological systems, both at the molecular level and at the ecological level and at all levels in between. For this reason I doubt if it will become a mature subject within the period we are discussing, although it would be surprising if there were not some initial attacks on it.

We do not intend in this review to go into experimental details. References for the original papers may be found in the following reviews, if not quoted in the text. General aspects on breeding systems and fungal genetics: Kniep (1928), Brieger (1930), Lewis (1954), Esser & Kuenen (1967), Fincham & Day (1971), Esser (1971); special aspects, fungi: Whitehouse (1949*a*, *b*), Burnett (1956), Papazian (1958), Raper & Esser (1964), Esser & Raper (1965), Davis (1966), Emerson (1966), Roper (1966), Horenstein & Cantino (1969), Raper & Flexer (1970), Koltin, Stamberg & Lemke (1972), Carlile (1973); higher plants: East (1940), Linskens & Kroh (1967), Arasu (1968), Townsend (1971); animals: Grell (1968), Esser & Blaich (1973).

DEFINITION AND DESCRIPTION OF VARIOUS BREEDING SYSTEMS

Breeding systems are responsible for the realization of plasmogamy and karyogamy, thus creating a basis for meiotic and somatic recombination. The genetic factors acting in the various breeding systems have to be clearly distinguished from those genetic traits which are responsible for morphological differentiation, e.g. formation of sex organs. Whereas the latter category are morphogenetic genes, the first class of genes have only a regulatory function. The completion of a life cycle depends on the action of both types of hereditary factors.

Since we have already described the various breeding systems elsewhere in detail (Esser, 1971) we will here confine ourselves to short definitions on the basis of Fig. 1. In this scheme we have used the fungi as an example, because within the fungi the greatest diversity of breeding systems occurs, including the phenomenon of heterokaryosis which is restricted to fungi and protozoa. The main breeding systems as presented in the large rectangle are also applicable to other eukaryotes.

Monoecism and dioecism

These depend on the capacity of an organism to contribute one or both nuclei to karyogamy. Thus a monoecious individual can both donate and receive a nucleus. An organism which has only one or the other potentiality is dioecious. If dioecism is correlated with differentiation into male and female individuals it is called morphological dioecism, but if the partners in a mating cannot be distinguished by morphological criteria we call the system physiological dioecism. In this case to neither of the partners can a male or female capacity be attributed; therefore they are called mating types and their physiological difference is usually indicated by the symbols + and −. Apart from the genes responsible for sexual morphogenesis there are no special genes determining monoecism. In contrast to higher organisms, dioecism in fungi is not controlled by sex chromosomes but by single genes.

Incompatibility

Incompatibility may be defined in a very general way as a genetically determined prevention of karyogamy which is not caused by sterility factors. Therefore, it concerns not only the sexual cycle but also the parasexual cycle as will be discussed below. On the basis of their genetic action two different systems are recognized: homogenic and heterogenic incompatibility, each of which can be caused by various genetic mechanisms.

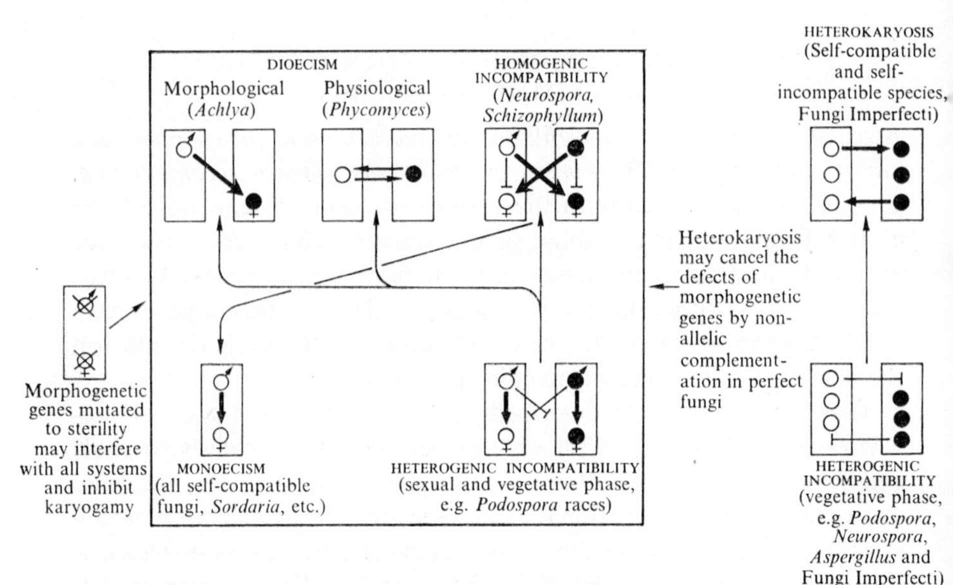

Fig. 1. Action and interaction of breeding systems in fungi. The large rectangle displays the main systems. Heterokaryosis occupies the right side of the figure. Representative organisms for each system are given in parentheses. The small rectangles represent single individuals (except for heterogenic incompatibility, where the rectangles represent single *Podospora* races). The male and female symbols represent nuclei of different sex. Differences in the genetic equipment of nuclei are characterized by white and black. In the case of physiological dioecism and heterokaryosis, where sexual differentiation of the nuclei cannot be proved, they are symbolized by white and black circles. The thick arrows indicate the direction of karyogamy or heterokaryotization respectively. The blocked arrows indicate that karyogamy or heterokaryotization is impossible. Interactions of the different systems are depicted by thin arrows. The top half of the diagram shows systems that increase outbreeding, the bottom half systems that decrease it. From Esser (1971).

Homogenic incompatibility

This is present when karyogamy is prevented by the presence of identical incompatibility factors in the two strains. So far as is known it concerns only the sexual cycle and occurs mainly in fungi and higher plants. In the most simple case two allelic genes, usually called + and −, control this mating system. Compatibility comes about only in the + × − combination, whereas combinations of like mating types (+ × + and − × −) do not lead to karyogamy. The general principle exemplified in this so-called bipolar mechanism is also the basis of more complicated mechanisms, in which the incompatibility factors are multiple alleles of one or more loci (p. 99) and of the interactions between the haploid pollen tubes and the diploid stylar tissue in higher plants.

It should be realized, that the term mating type is used both for strains which are physiological dioecious as well as for strains which are homogenically incompatible (Fig. 1). Whereas in the first case this expression indicates the lack of sexual differentiation, in the latter it denotes the inability of hermaphrodites to show self-fertilization.

Heterogenic incompatibility

This is present when karyogamy is prevented by the presence of non-identical incompatibility factors in the two strains. This system concerns both the sexual cycle and also the vegetative phase. It occurs between geographical races and is widely distributed within all living beings (Esser & Blaich, 1973).

In contrast to the many detailed genetic studies of the various mechanisms of the homogenic system, there are few thoroughly analysed examples of the heterogenic system. This lack of information results mainly from the difficulty of overcoming the failure of karyogamy by detours or experimental tricks. The minimal requirement for heterogenic incompatibility is a single allelic difference, whereas in most cases, especially in wild populations, more than one locus is involved, as will be shown later (p. 95).

Heterokaryosis

This is the association of genetically different nuclei in a common cytoplasm, and is widespread in the fungi. It is not a breeding system *per se*, like the ones previously discussed. However, by initiating via hyphal fusion plasmogamy in somatogenic fungi (e.g. higher Basidiomycetes), it indirectly controls the sexual cycle. The main function of heterokaryosis is evident in the Fungi Imperfecti, where hyphal fusion is a prerequisite for the parasexual cycle. Therefore it is not surprising that heterokaryosis is controlled either by the genes of the homogenic incompatibility system or by those of the heterogenic system, or by both.

GENERAL ASPECTS OF THE ACTION AND INTERACTION OF THE BREEDING SYSTEMS WITH RESPECT TO EVOLUTION

As already mentioned in the introduction, *breeding systems influence evolution* only *indirectly*, that is *via recombination*. The biological significance of recombination is the creation of new genomes by combining the hereditary traits of germ cells or vegetative nuclei in a new fashion. The efficiency of recombination increases with the heterogeneity of the genetic material to be mated, because the greater their genetic diversity the greater is the chance of forming new genomes having an advantage in selection. Thus the effect of recombination in true breeding individuals is almost zero and very slight in inbreeding organisms. This becomes especially evident in organisms with haploid life cycles, where the nuclei undergoing karyogamy all originate from the single nucleus of one spore and are identical as long as no spontaneous mutation occurs during their vegetative propagation. From this it follows that any mechanism which prevents or diminishes inbreeding enhances the efficiency of recombination. This is achieved in nature by the action and interaction of the various breeding systems, which may be deduced from Fig. 1 as follows:

1. *Inbreeding is decreased by dioecism.* Both mechanisms, morphological and physiological, do not allow self-fertilization. In both cases, two organisms which differ genetically by sex or mating factors (and hence probably in other ways) are required for karyogamy. The *same effect* is attained in many monoecious species by *homogenic incompatibility.* Despite the hermaphroditic character of each individual no self-fertilization, but only cross-fertilization, is possible. In this way the species obtains the same outbreeding efficiency as if it were dioecious. It now becomes understandable why homogenic incompatibility is mainly restricted to plants, because nature has evidently developed this mating system as a compensation for the evolutionary disadvantage of inbreeding caused by monoecism, which predominates in the plant kingdom in contrast to the animal kingdom where dioecism prevails. Owing to the limiting effect of dioecism and homogenic incompatibility on self-fertilization, wild strains or wild populations which are controlled by one of these systems are to a large extent heterogeneous, or in case of diploids, heterozygous.

2. *Inbreeding is increased* most drastically *by monoecism.* There is then not much chance of a quick distribution of genetic diversity within a species. This may result in the formation of local races which become more and more genetically diverse. Isolation is then strengthened *by heterogenic incompatibility*, which creates mating barriers between different races of a single species. When heterogenic incompatibility overlaps dioecism or homogenic incompatibility (see arrows in Fig. 1) the species concerned is split into small breeding units which are identical with geographical races.

In the thoroughly analyzed example of the Ascomycete *Podospora anserina* (p. 95) we showed that, as a result of heterogenic incompatibility, a sample of nineteen geographical races had only thirteen combinations (= 7.6 %) that exhibited an unrestricted exchange of genetic material, and that in seventy-five combinations (43.9 %) there was no exchange at all. In the remaining combinations there was only partial exchange, either due to barrage formations or reduction in fruiting body production.

From this it follows that recombination of the genetic information occurs no longer within the species as a whole, but is restricted to single races. Mutations with selective advantages occurring in one race are not transferred to other races. In this way the race and not the species becomes the unit of evolution. Thus, heterogenic incompatibility has to be acknowledged as a genetic isolation mechanism, which has an effect for speciation similar to geographical barriers.

The taxonomist may object that races which cannot be successfully crossed with each other are no longer races but different species. This may be true in many cases. However, in the examples we have analyzed and quoted elsewhere (Esser & Blaich, 1973) these races exhibit no morphological difference which might place them in different species according to the classical rules of taxonomy, and moreover where heterogenic incompatibility occurs within these species the A–B–C scheme is effective which enable us to distinguish between cross sterility and incompatibility, i.e. two races A and B which never cross with each other are fertile with a third race C and therefore the gross chromosomal divergencies which are responsible for sterility are not present.

There is another point which should not be overlooked in this connection. As we will see later (p. 96), in fungi heterogenic incompatibility always prevents heterokaryon formation. Owing to this effect, when present, accompanying genetic isolation, the exchange of harmful cytoplasmic components such as viruses or mutant-suppressive mitochondria is stopped (Caten, 1972). Thus the distribution of cell diseases is counteracted and the survival of healthy cells or tissues is favoured, which provides a selective advantage for strains showing heterogenic incompatibility.

3. The action of *heterokaryosis* is not as evident as that of the sexual systems occupying the central part of Fig. 1. However, it is obvious that an exchange of nuclei between fungal mycelia leading to heterokaryotic hyphae increases the chance of outbreeding, because in fungi all nuclei of a mycelium have the chance of participating in the sexual cycle. In perfect fungi the action of heterokaryosis as a control mechanism for breeding is not very important. In monoecious forms exhibiting self-fertilization it may increase the possibility of outbreeding, but in dioecious or homogenically incompatible species the outbreeding effect already occurs. In the Fungi Imperfecti, however, heterokaryosis

is the *only* system for promoting outbreeding, because there is no recombination except by the parasexual cycle. Heterokaryosis, the only way of bringing genetically different nuclei together, can be totally blocked by heterogenic incompatibility affecting the vegetative phase. Where no heterokaryon formation can occur, the genetic isolation of two imperfect strains is complete, but in perfect fungi failure to form vegetative heterokaryons does not prevent recombination via the sexual cycle. Thus the Fungi Imperfecti may have survived because of heterokaryosis, and heterogenic incompatibility may prevent the survival of some genotypes.

4. All these statements would be unrealistic if we overlook the existence of the *morphogenetic genes* (p. 89), which are responsible for sexual differentiation. It is well known that in nature mutations at these loci often occur which block either the female or the male capacity, or both. These mutated genes interfere with all the breeding systems responsible for the sexual cycle, and hence may cancel all the positive evolutionary effects brought about by the action of breeding systems.

Here, and this is unique for fungi, the importance of heterokaryosis again becomes obvious, because there are many examples (e.g. in the ascomycete *Sordaria*; Esser & Straub, 1958) showing that the negative effect of the mutated morphogenetic genes may be nullified by non-allelic complementation. In other haploids lacking the possibility of heterokaryosis there is no way of overcoming the effects of these 'sterility' genes. In diploids the effects of 'sterility' genes can be overcome if the sterility concerns only one sex and the appropriate partner is available, since if the 'sterility' genes are recessive, they will be ineffective in the heterozygote.

From the preceding considerations it should be evident that in nature the mating and propagation of a species or race is very seldom controlled by a single breeding system, but by the interaction of different systems. By the continuous interaction of all these systems the transfer of genetic information from individual to individual, from race to race, and sometimes from species to species, is controlled. Thus, via recombination, the production of ameliorated genomes which have a better chance of surviving the hard game of selection and hence strengthening existing branches or provoking new ramifications of the evolutionary tree, is favoured. In the next section we will present some examples in more detail to emphasize this contention. In this connection it needs to be stressed that the existence of heterogenic incompatibility often has been overlooked, because in general it is the aim of the geneticists to study true breeding strains of single races. However, in order to understand the structure of a species, one has to consider all the parameters indicated in Fig. 1. This has implications not only for basic research but also a practical relevance for breeding purposes. Breeding of industrially important fungi has been done mainly by mutation and selection. The application of genetic understanding of breeding systems would permit a systematic use of hybridization as with animals and higher plants.

SPECIAL ASPECTS OF BREEDING SYSTEMS

The detailed genetic analysis of breeding systems and their numerous mechanisms has brought many problems to our attention which, on first glance, seem not to be correlated with their main purpose of controlling recombination. A report on breeding systems, however, would be incomplete without mentioning these special aspects. In spite of their correlation with other phenomena they serve the same goal of controlling recombination, as we hope to make clear with the following examples.

Heterogenic incompatibility and the problem of physiological coexistence

In the brief descriptions of homogenic and heterogenic incompatibility given (pp. 90, 91) we have not discussed the physiological action of these breeding systems. It is not surprising that the fundamental difference between them with respect to genetic control is also reflected in a divergent physiological realization. Although in both cases there are no definite biochemical mechanisms known at the molecular level, there is circumstantial evidence of their mode of action. In the homogenic system there is no inhibitory interaction between incompatible partners, but a complementary stimulation occurs when there is incompatibility. In the heterogenic system the inverse is the case. There is no interaction in the compatible combination, but with incompatible combinations there is mutual inhibition due to genetic differences, as will be shown with the mating behaviour of the coprophilous ascomycete *Podospora anserina*.

In *P. anserina*, within each race, the mating relations are controlled by the homogenic system, i.e. there are two hermaphrodite mating types + and −, which react as in Fig. 1. With different races, however, one observes in most cases a heterogenic incompatibility, manifesting itself as presented in Plate 1, which illustrates the action of two different incompatibility mechanisms:

1. The allelic mechanisms, caused independently by a heterogeneity of the two unlinked loci *t* and *u*, in which there is only a vegetative

incompatibility manifested by a demarcation line. In this zone, there is, after hyphal fusion, a mutual influence between the heterogenic nuclei, which finally leads to the death of the hyphal compartments concerned. This becomes evident as the unpigmented barrage zone. The sexual compatibility between + and – mating types is not affected by this mechanism of heterogenic incompatibility.

2. The non-allelic incompatibility mechanism which involves the interaction of specific allelic configurations of two separate loci. There are two such mechanisms known, with identical effects. Both provoke a vegetative as well as a sexual incompatibility. The latter takes the form of a non-reciprocal incompatibility, i.e. one of the reciprocal crosses does not lead to fruiting-body production. In the inverse combination both non-allelic mechanisms overlap and prevent any fructification.

The mutual inhibition of the nuclei carrying different incompatibility genes can be demonstrated easily by studying the corresponding heterokaryons or recombinant types as exemplified in Plate 2, fig. 1, from which it may be deduced that alleles which are incompatible cannot coexist. This is so whether they are located in a common nucleus or in different nuclei in a common cytoplasm. The visible result of this interaction is either poor growth of homokaryons or internal segregation into heterokaryons. In both types of reaction the mutual inhibition finally leads to death of the nuclei concerned, i.e. by the cessation of growth of the a_1b mycelia (the same occurs in the other mechanism in the combination c_1v) and by the formation of barrage zones with dead cells between t and t_1 homokaryons.

A mutual inhibition of this kind or of a similar nature has been found in all cases of heterogenic incompatibility which have been checked in this respect (Esser & Blaich, 1973). This interaction, however, is not restricted to nuclei – it can also occur between extrachromosomal genetic elements, as in the midge *Culex pipiens* (Laven, 1957, 1959). In this insect there exist similar crossing barriers between male and female individuals of different races, as in *Podospora*, but controlled by cytoplasmic elements.

Thus *heterogenic incompatibility* can be given a very general and also very simple definition by saying it *is the consequence of interaction between genetic elements which cannot coexist in close proximity to each other*. From this definition it becomes obvious that heterogenic incompatibility is more than just a breeding system controlling recombination. It is a very general biological phenomenon which does not concern only mating, karyogamy and recombination, but also a broad spectrum of biological structures living in physiological association, ranging from isolated genetic material to highly specialized and differentiated tissues, as may be shown by the following examples.

Restriction of genetic material in prokaryotes

Some twenty years ago it was observed that in *E. coli* some bacteriophages are adapted to a specific host strain. When these viruses are transferred to another strain their DNA is not able to propagate in the 'unknown' host (= restriction). Immediately after infection the phage DNA becomes abortive, with exception of a very few survivals (about 10^{-4}). These few phages can lyse the bacterium and have become adapted to the 'new' strain (= modification), but when brought back to the strain they have previously grown on, they are no longer able to lyse these bacteria. The analysis of this phenomenon has revealed that restriction and also modification are under genetic control. Furthermore, it has been shown that restriction is a rather common mechanism in micro-organisms, because it concerns also the coexistence of other genetic elements such as *F*-episomes, plasmids and fragments of the bacterial chromosome (Dussoix & Arber, 1962; Arber & Linn, 1969; Boyer, 1971; Meselson, Yuan & Heywood, 1972; Haberman, Heywood & Meselson, 1972; Hedgpeth, Goodman & Boyer, 1972). Since there is a rather close similarity between the gene mechanisms for one of the best analysed examples of restriction and for heterogenic incompatibility of *Podospora*, we will discuss further the details of bacterial restriction.

It has been found that the phage DNA, after infecting the unknown bacterial cell, is 'cut in pieces' by bacterial endonucleases which attack specific nucleotides in the genome of the phage. In the cases of the few survivals these sites have been modified by methylation of the purine or pyrimidine bases, which is brought about by a bacterial methylase. The modified phages are then adapted to the new host. Restriction (= dissection of DNA) and modification (protection of DNA by methylation) are both controlled by genetical sites (r^+ and m^+, respectively) which, with a very high degree of probability, belong to adjacent loci in the bacterial genome. Thus the two enzymes (the endonuclease and the methylase) coded by this region are probably aggregated in one complex. In connection with heterogenic incompatibility, the properties of some mutants are of interest: the r^-m^+ mutant is not able to restrict genetic material (no endonuclease present), but can modify a restrictable genome (methylase present); the double mutant r^-m^- has lost both enzyme activities. The third type of mutant to be expected (r^+m^-) has never been found. One may assume that this mutation is lethal, because such strains would destroy their own DNA due to the absence of the protecting enzyme methylase. The analogy between bacterial restriction in prokaryotes and heterogenic incompatibility in eukaryotes therefore concerns not only the phenomenon in general, but also the details of gene action, since, in addition to the inability of heterogeneous genetic material to coexist, the mutual interaction between the a_1 and b genes in *Podospora* is paralleled by the incompatibility between the r^+ and the m^- factors. We conclude that *bacterial restriction can also be understood as a mechanism of heterogenic incompatibility*, because it ultimately diminishes the chances for recombination of the genetic material. This conclusion seems to be justified by the following definition of restriction (Rieger, Michaelis & Green, 1968), which holds true also for heterogenic incompatibility: 'The restriction system may be understood as a defense mechanism against introduced nucleic acid, which prevents the expression or integration of foreign DNA by rapidly degrading it without hindering genetic exchange among cells of the same strain.'

Histoincompatibility in animal tissue

For a long time biologists have been aware that in numerous cases animal tissue from different individuals (except identical twins) is incompatible when brought into physiological contact by transplantation. The best known examples are organ transplantations in human beings (Rapaport, 1966). Despite the fact that histoincompatibility in mammals is realized by a complicated immune response mechanism and heterogenic incompatibility requires at least cytoplasmic contact of the genetically different elements, there is no difference in principle between these two phenomena, for the following reasons: (*a*) As mentioned above there is no unique mechanism for heterogenic incompatibility, but a great variety of processes ranging from the action of single nuclear genes through cytoplasmic elements to the restriction system of prokaryotes. (*b*) There is evidence that histoincompatibility may sometimes be caused by mechanisms which are intermediate between the immune response and the cytoplasmic contact of genetic material necessary for heterogenic incompatibility.

In Coelenterata (Gorgonaceae) in combinations of two individuals of the same species, brought about in nature by close contact or artificially by grafts, an interaction may occur leading to mutual destruction. If the tissues brought together differ in size (1:8), only the smaller piece (= target) is destroyed, whereas the larger (= killer) survives.

The death of the target is hindered after it is treated with inhibitors of protein synthesis. The inverse experiment, inhibition of protein synthesis in the killer, does not stop the decease of the target, which would be expected if the mechanism consists of an immunological system comparable to the one of mammals. Theodor (1970) suggests that even a short contact between target and killer is sufficient to initiate an autodestruction of the first through the agency of its own protein synthesis. He considers this system a precursor of the immunological reaction. In a comparable example described by Granger & Kolb (1968), genetically dissimilar mouse lymphocytes show a mutual in-vitro destruction which also cannot be considered as an immunological reaction. Further cases of histoincompatibility in mammals due to non-immunological interaction are reviewed by Amos (1966).

Therefore we adopt the view that, despite having developed separately in evolution, *histoincompatibility and heterogenic incompatibility* can be classified as very similar biological phenomena, because *both result in the inability of genetically different material to coexist in a common physiological system.*

Homogenic incompatibility and the problem of morphogenetic regulation

In numerous higher Basidiomycetes the sexual cycle is controlled by homogenic incompatibility. The controlling elements are two factors called *A* and *B*, each consisting of two closely linked genes. Owing to mutations within these functional units, a great number of factorial specificities exist behaving like allelic configurations, as established primarily by Raper and his associates (for literature see introduction). Since in these fungi the life cycle is initiated by somatogamy according to the general rule of homogenic incompatibility (see Fig. 1), the hyphal fusion and the following morphogenetic steps leading to the formation of a special type of heterokaryon which is called a dikaryon, is only possible when both factors have a different genetic constitution, i.e. $A_x B_x$ and $A_y B_y$ respectively (tetrapolar mechanism of homogenic incompatibility). This dikaryon, in which owing to special morphological structures (clamp connections), a 1:1 distribution of both types of nuclei is guaranteed, is very stable for many cell generations until finally fruiting-body production occurs.

As in all breeding systems the genetic determinants have only a regulatory function, which can be realized only in co-operation with morphogenetic genes. Until recently the generally held view was that the incompatibility factors are only concerned with regulating the morphogenetic steps leading to dikaryon formation. Formation of fruiting bodies, which naturally depends both on environmental conditions and on the action of specific morphogenetic genes, was not considered to be influenced by incompatibility factors. Genetic studies performed with three different species: *Polyporus ciliatus*, *Polyporus brumalis* (Polyporaceae) and *Agrocybe aegerita* (Agaricaceae), however, have revealed that the controlling action of the incompatibility factors is more comprehensive (Esser & Stahl, 1973 and unpublished work).

Dikaryotic wild stocks of these wood-destroying fungi isolated from nature or obtained from fungal stock centres, appeared to be very stable during long periods of vegetative propagation and under suitable conditions regularly formed fruiting bodies of normal shape. However, most monokaryotic mycelia originating from uninucleated basidiospores developed atypical fruiting bodies ranging from amorphous stromata with hymenia to more or less crippled stalked heads (Plate 2, fig. 2). Since these structures also produced basidiospores, genetic analysis was possible, showing that the monokaryotic fruiters mostly had single gene differences when compared with non-fruiting monokaryons. These genes have therefore to be considered as a sequence of morphogenetic genes being in their unmutated form responsible for the formation of normal fruiting bodies. The phenomenon of monokaryotic fruiting is in itself not new, because it has already been observed as a result of environmental or genetic conditions in other Basidiomycetes (see Esser & Stahl, 1973).

Further studies, however, demonstrated a correlation between monokaryotic fruiting and the action of the incompatibility factors, as shown in Plate 3. As may be seen from the figure monokaryotic fruiting is suppressed when different incompatibility factors become integrated in a common cytoplasm. This is so with both a double heterogeneity in the dikaryon $A_x B_x + A_y B_y$ as well as with single gene differences as with heterokaryon having the constitution $A_y B_x + A_y B_y$. In both cases there is no monokaryotic fruiting, but as expected normal fruiting occurs in the dikaryon with the double heterogeneity and no fruiting in the common *A*, unlike *B* heterokaryon. A third combination, unlike *A* and common *B*, cannot be studied since heterokaryon formation cannot occur in the absence of the regulating influence of unlike *B* factors.

These results show that genetic factors determining the breeding system of homogenic incompatibility control the whole sequence of morphogenetic events ranging from hyphal fusion to fruiting-body production. In addition to the above consequences of homogenic incompatibility for recombination and evolution these experiments permit some general conclusions concerning the significance of the fungal dikaryon. The generally accepted view of the biological significance of this special type of heterokaryon is that the dikaryon, although only containing haploid nuclei, may phenocopy the diploid phase on a physiological level, due to 1:1 distribution of its participating nuclei in all cells. This view needs to be expanded. In contrast to those heterokaryons which are not under the control of incompatibility factors and in which genetically different nuclei are usually not present in equal proportions and in which the formation of monokaryotic sectors is very common, the dikaryon is rather stable. This stability is not solely determined by the complementary effects of the incompatibility factors but as seen above is also due to the ability of these factors to control the expression of the numerous morphogenetic genes which have presumably accumulated as a result of independently occurring mutations during vegetative propagation. The *incompatibility factors* have therefore to be regarded as a *superimposed steering mechanism* for differentiation *channelling all morphogenetic activities in the direction which leads to the formation of only normal fruiting bodies*. All kinds of 'heresy' (abnormal fruiting bodies) are drastically suppressed. In following this line of thought it becomes evident that a fungal *mycelium* is not a multicellular individual, but a *population of numerous nuclei incorporated in a metabolic unit*.

Since nowadays many people, especially students, pose the question of the relevance of basic biological research, it would be amusing to speculate on the analogies between a population of nuclei under the dogmatic and strong government of two factors which suppress all individual impulses, and some current theories of sociology.

CONCLUSIONS

Homogenic incompatibility, by favouring recombination, is one of the major factors responsible for progress in evolution. On the other hand, by creating genetical barriers, heterogenic incompatibility in co-operation with geographical isolation is one of the essentials for speciation, which can lead in turn to the creation of new genera, families and higher taxa. In short, the ramifications of the evolutionary pedigree postulated by Darwin on the basis of the descriptive methods of palaeontology may now be extended and confirmed with experimental methods based on genetics, biochemistry and molecular biology.

The fact that the action of breeding systems are often not seen from a general point of view should now have become obvious, because these parameters are often linked with other phenomena and are thus regarded in a more limited way. In particular, the wide gap between eukaryotes and prokaryotes has not encouraged generalizations of this kind. However, we think that although we have only quoted the phenomenon of restriction as in prokaryotes as simulating the isolating effect of a breeding system in eukaryotic micro-organisms, a further study of the literature would probably reveal further analogies between prokaryote and eukaryote breeding systems. It may be an oversimplification to trace back the versatility of living beings to a simple scheme like the one that we have presented in this paper, but these ideas may at least provide a basis for further research work to prove or disprove their general value.

Since there are available more facts concerning breeding systems and evolution, than many people have been aware of, we are able to confirm at least partially the postulate of Crick quoted in the introduction. In all living beings, nature has 'evolved rather special and ingenious mechanisms' which control recombination and hence evolution; these are the breeding systems, which act in many different ways but with the same goal in both eukaryotes and prokaryotes.

I wish to thank Dr Michael Carlile for a helpful discussion of both scientific and literary points. The work of the author and his collaborators quoted in this chapter was supported by grants from the Deutsche Forschungsgemeinschaft (Bad Godesberg) and the Landesamt für Forschung (Düsseldorf).

REFERENCES

AMOS, B. (1966). Biological defense and recognition mechanisms. *Annals of the New York Academy of Sciences*, **129**, 730–57.

ARASU, N. T. (1968). Self-incompatibility in angiosperms: a review. *Genetica*, **39**, 1–24.

ARBER, W. & LINN, S. (1969). DNA modification and restriction. *Annual Review of Biochemistry*, **38**, 467–500.

BOYER, H. W. (1971). DNA restriction and modification mechanisms in bacteria. *Annual Review of Microbiology*, **25**, 153–76.

BRIEGER, F. (1930). *Selbststerilität und Kreuzungssterilität im Pflanzenreich und Tierreich*. Berlin: Springer-Verlag.

BURNETT, J. H. (1956). The mating systems of fungi. I. *New Phytologist*, **55**, 50–90.

CARLILE, M. J. (1973). Cell fusion and somatic incompatibility in Myxomycetes. *Berichte der Deutschen Botanischen Gesellschaft*, **86**, 123–39.

CATEN, C. E. (1972). Vegetative incompatibility and cytoplasmic infection in fungi. *Journal of General Microbiology*, **72**, 221–9.

CRICK, F. (1970). Molecular biology in the year 2000. *Nature, London*, **228**, 613–15.

DAVIS, R. H. (1966). Mechanisms of inheritance. Heterokaryosis. In *The Fungi*, ed. G. C. Ainsworth & A. S. Sussman, vol. 2, pp. 567–88. London: Academic Press.

DUSSOIX, D. & ARBER, W. (1962). Host specificity of DNA produced by *Escherichia coli*. II. Control over acceptance of DNA from infecting phage. *Journal of Molecular Biology*, **5**, 37–49.

EAST, E. M. (1940). The distribution of self-sterility in flowering plants. *Proceedings of the American Philosophical Society*, **82**, 449–518.

EMERSON, S. (1966). Mechanisms of inheritance. Mendelian. In *The Fungi*, ed. G. C. Ainsworth & A. S. Sussman, vol. 2, pp. 513–66. London: Academic Press.

ESSER, K. (1971). Breeding systems in fungi and their significance for genetic recombination. *Molecular and General Genetics*, **110**, 86–100.

ESSER, K. & BLAICH, R. (1973). Heterogenic incompatibility in plants and animals. *Advances in Genetics*, **17**, 107–52.

ESSER, K. & KUENEN, R. (1967). *Genetics of fungi*. Berlin: Springer-Verlag.

ESSER, K. & RAPER, J. R., eds. (1965). *Incompatibility in Fungi*. Berlin: Springer-Verlag.

ESSER, K. & STAHL, U. (1973). Monokaryotic fruiting in the Basidiomycete *Polyporus ciliatus* and its suppression by incompatibility factors. *Nature, London*, **244**, 304–5.

ESSER, K. & STRAUB, J. (1958). Genetische Untersuchungen an *Sordaria macrospora* Auersw., Kompensation und Induktion bei genbedingten Entwicklungsdefekten. *Zeitschrift für Vererbungslehre*, **89**, 729–46.

FINCHAM, J. R. S. & DAY, P. R. (1971). *Fungal Genetics*, 3rd edition. Oxford: Blackwell.

GRANGER, G. A. & KOLB, W. P. (1968). Lymphocyte in vitro cytotoxicity: Mechanisms of immune and non-immune small lymphocyte mediated target L cell destruction. *Journal of Immunology*, **101**, 111–20.

GRELL, K. G. (1968). *Protozoologie*, 2nd edition. Berlin: Springer-Verlag.

HABERMAN, A., HEYWOOD, J. & MESELSON, M. (1972). DNA modification methylase activity of *Escherichia coli*: restriction endonucleases K and P. *Proceedings of the National Academy of Sciences, USA*, **69**, 3138–41.

HEDGPETH, J., GOODMAN, H. M. & BOYER, H. W. (1972). DNA nucleotide sequence restricted by the RI endonuclease. *Proceedings of the National Academy of Sciences, USA*, **69**, 3448–52.

HORENSTEIN, E. A. & CANTINO, E. C. (1969). Fungi. In *Fertilization*, ed. C. B. Metz & A. Monroy, vol. 2, pp. 95–133. New York: Academic Press.

KNIEP, H. (1928). *Die Sexualität der niederen Pflanzen*. Jena: Fischer.

KOLTIN, Y., STAMBERG, J. & LEMKE, P. A. (1972). Genetic structure and evolution of the incompatibility factors in higher fungi. *Bacteriological Reviews*, **36**, 156–71.

LAVEN, H. (1957). Vererbung durch Kerngene und das Problem der außerkaryotischen Vererbung bei *Culex pipiens*. II. Außerkaryotische Vererbung. *Zeitschrift für induktive Abstammungs-u. Vererbungslehre*, **88**, 478–516.

LAVEN, H. (1959). Speciation by cytoplasmic isolation in the *Culex pipiens*-complex. *Cold Spring Harbor Symposia on Quantitative Biology*, **24**, 166–73.

LEWIS, D. (1954). Comparative incompatibility in angiosperms and fungi. *Advances in Genetics*, **6**, 235–85.

LINSKENS, H. F. & KROH, M. (1967). Inkompatibilität der Phanerogamen. In *Handbuch der Pflanzenphysiologie*, ed. W. Ruhland, vol. 18, pp. 506–30. Berlin: Springer-Verlag.

MESELSON, M., YUAN, R. & HEYWOOD J. (1972). Restriction and modification of DNA. *Annual Review of Biochemistry*, **41**, 447–66.

PAPAZIAN, H. P. (1958). The genetics of Basidiomycetes. *Advances in Genetics*, **9**, 41–69.

RAPAPORT, F. T., ed. (1966). Seventh international transplantation conference. *Annals of the New York Academy of Sciences*, **129**, 1–884.

RAPER, J. R. & ESSER, K. (1964). The Fungi. In *The Cell*, ed. J. Brachet & A. E. Mirsky, vol. 6, pp. 139–244. New York: Academic Press.

RAPER, J. R. & FLEXER, A. S. (1970). The road to diploidy with emphasis on a detour. In *Organization and Control in Prokaryotic and Eukaryotic Cells. Symposia of the Society for General Microbiology*, **20**, 401–32. Ed. H. P. Charles & B. C. J. G. Knight. London: Cambridge University Press.

RIEGER, R., MICHAELIS, A. & GREEN, M. M. (1968). *A Glossary of Genetics and Cytogenetics, Classical and Molecular*, 3rd edition. Berlin: Springer-Verlag.

ROPER, R. A. (1966). Mechanisms of inheritance. The parasexual cycle. In *The Fungi*, ed. G. C. Ainsworth & A. S. Sussman, vol. 2, pp. 589–617. London: Academic Press.

THEODOR, J. L. (1970). Distinction between 'self' and 'not-self' in lower invertebrates. *Nature, London*, **227**, 690–2.

TOWNSEND, C. E. (1971). Advances in the study of incompatibility. In *Pollen: Development and Physiology*, ed. J. Heslop-Harrison, pp. 281–309. London: Butterworths.

WHITEHOUSE, H. L. K. (1949*a*). Heterothallism and sex in the fungi. *Biological Reviews*, **24**, 411–47.

WHITEHOUSE, H. L. K. (1949*b*). Multiple allelomorph heterothallism in the fungi. *New Phytologist*, **48**, 212–44.

EXPLANATION OF PLATES

PLATE 1

Scheme for the action of the genes responsible for the heterogenic incompatibility between races *s* and *M* of *Podospora anserina*. The different alleles are symbolized by small letters, and non-allelic genes which interact are framed with squares. The two parallel lines symbolize vegetative incompatibility, which leads to the macroscopically visible, so-called barrage, which occurs in the zone of contact between two incompatible mycelia. The black dots along the barrage indicate the formation of fruiting bodies. In all combinations + and − strains are confronted, but for reasons of clarity we have omitted the mating type symbols. For further explanation see text. From Esser & Blaich (1973).

PLATE 2

Fig. 1. Mycelia of *Podospora anserina* demonstrating the interaction of the genes responsible for heterogenic incompatibility. Genes of the allelic mechanism: (*a*) heterokaryon containing t and t_1 nuclei shows dissociation and internal barrage formation; (*b*) normal heterokaryon as control. Genes of the non-allelic mechanism: (*c*) homokaryon containing the incompatible combination a_1b shows poor growth; (*d*) *ab* homokaryon as control; age of (*c*) and (*d*) two days.

Fig. 2. Various types of fruiting bodies of *Polyporus ciliatus*. Top: normal fruiting bodies from the dikaryotic stock. Below: abnormal fruiting bodies from various monokaryotic strains. From Esser & Stahl (1973).

PLATE 3

Crosses of monokaryotic strains. (*a*) Cross between strains 70 and 72 which are compatible for both mating type factors. After inoculation of the dish the lateral parts of both partners have been separated with sterile aluminium foil to prevent nuclear migration. Whereas the middle section has become dikaryotic and forms unquestionably normal fruiting bodies, both of the lateral sections show monokaryotic fruiting. (*b*) Cross between strains 63 and 64 having common *A* and different *B* factors. Arrangements as in (*a*). Since nuclear migration has taken place in the middle section, there is no fruiting-body production, whereas in the lateral sections monokaryotic fruiting occurs. From Esser & Stahl (1973).

PLATE 1

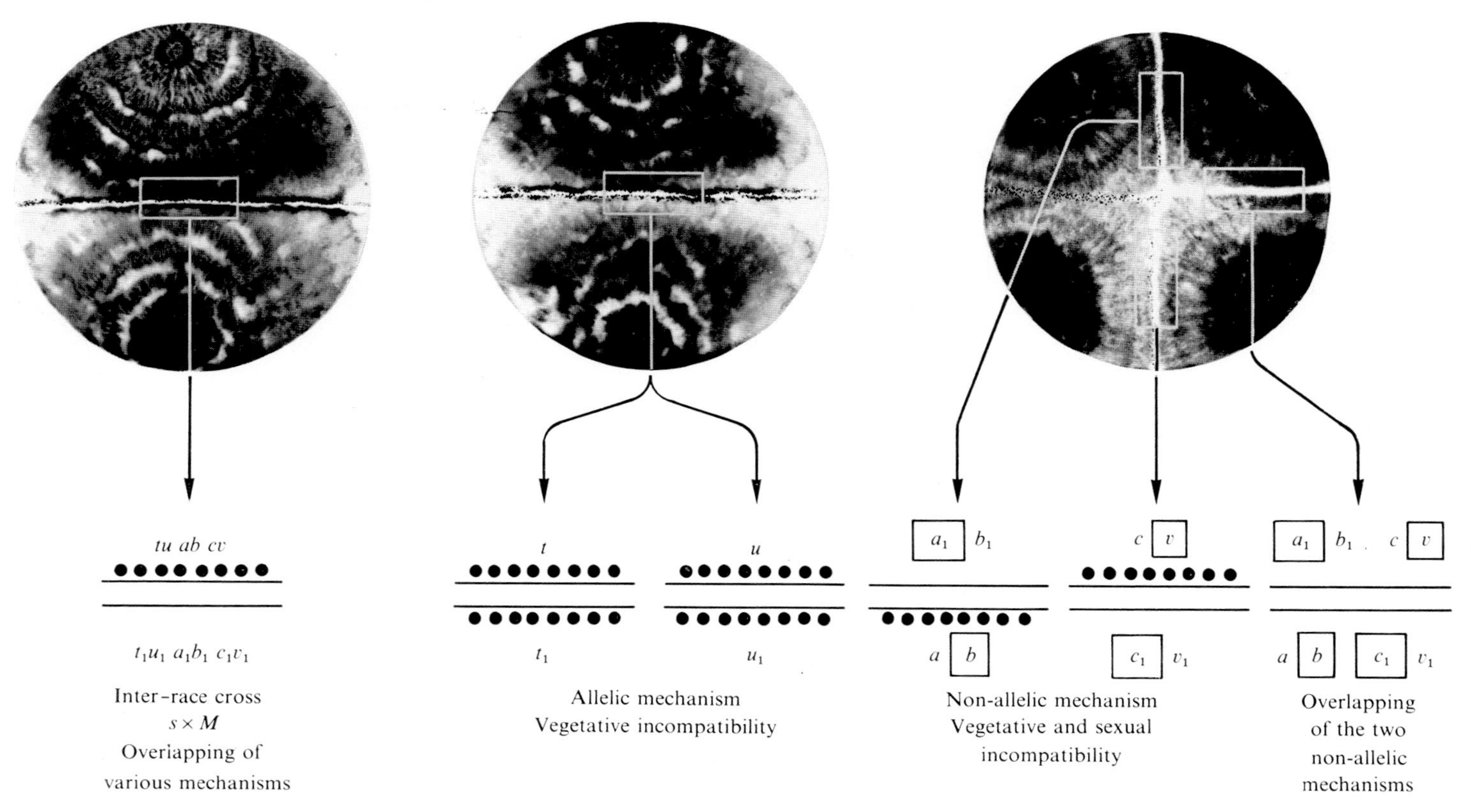

Inter-race cross $s \times M$ Overiapping of various mechanisms

Allelic mechanism Vegetative incompatibility

Non-allelic mechanism Vegetative and sexual incompatibility

Overlapping of the two non-allelic mechanisms

(*Facing p.* 104)

PLATE 2

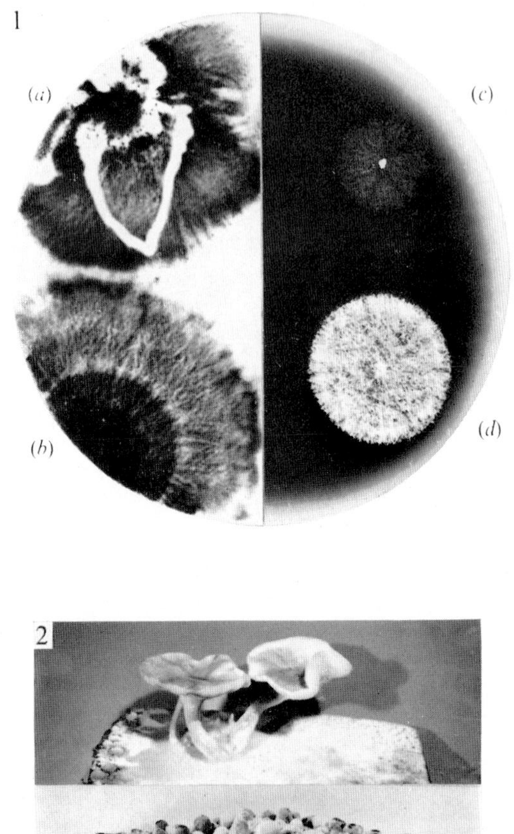

PLATE 3

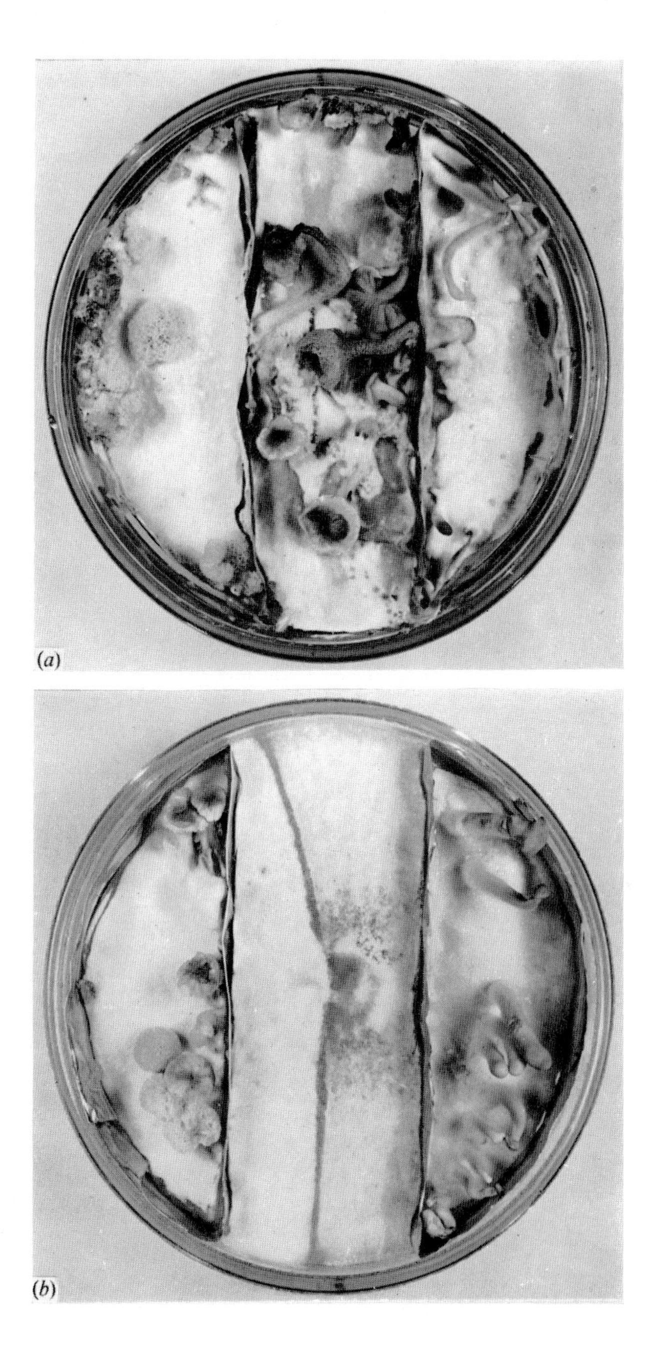
(a)

(b)

OPERATION OF SELECTION PRESSURE ON MICROBIAL POPULATIONS

H. E. KUBITSCHEK

Division of Biological and Medical Research, Argonne National Laboratory, Argonne, Illinois 60439, *USA*

Selection is anything tending to produce a systematic, heritable change in populations between one generation and the next.

George Gaylord Simpson, 1953

Selection, whether in mortality, mating or fecundity, applies to the organism as a whole and thus to the effects of the entire gene system rather than to single genes. A gene which is more favorable than its allelomorph in one combination may be less favorable in another. Even in the case of cumulative effects, there is generally an optimum grade of development of the character and a given plus gene will be favorably selected in combinations below the optimum but selected against in combinations above the optimum. Again the greater the number of unfixed genes in a population, the smaller must be the average effectiveness of selection for each one of them. The more intense the selection in one respect, the less effective it can be in others. The selection coefficient for a gene is thus in general a function of the entire system of gene frequencies. As a first approximation, relating to a given population at a given moment, one may, however, assume a constant net selection coefficient for each gene.

Sewall Wright, 1931

INTRODUCTION

Natural selection was a key factor of Darwin's theory of evolution, which asserted that populations are composed of individuals with different heritable characteristics, and that some individuals are more likely to reproduce than others because of selective pressures. An organism must remain viable to reproduce, and there is no doubt that grossly unadapted organisms are eliminated by selection. But as late as 1953 it was noted that 'theories as to the role and importance of selection range from belief that it has only this broadly limiting effect to belief that it is the only essential factor in evolution' (Simpson, 1953). Difficulties in deciding between those two extremes were increased by the absence of experimental evidence bearing upon weak selection. Furthermore, Fisher (1930) concluded that environmental fluctuations bring about significant changes in selection values, making it difficult to design experiments to discriminate between the alternatives.

Despite tremendous increases in knowledge of the nature of the gene and its action in protein synthesis and the much more sophisticated reasoning that this knowledge has led to, present theories still seem to cover the same broad spectrum with regard to the importance of selection. Controversy is now centred on the significance of 'neutral mutations', that is, genetic changes leading to so little difference between organisms that selection for or against them is absent or negligible. At one extreme, 'non-Darwinian' evolutionists believe that neutral mutations account for the great bulk of evolutionary change, while at the other, 'neo-Darwinians' assert that much of evolutionary change is attributable to natural selection. Crow (1972) has given an excellent summary of the history of 'neutral evolution' and a lucid account of the evidence and arguments for neutral mutation. Since both selected and neutral mutations have been observed in chemostat experiments on bacterial evolution, I shall assume that both mechanisms operate and will describe a relationship between them.

Every heritable character or trait is presumably encoded in the DNA of the genetic system of each organism or its gene-carrying symbionts, and variants arise because of alterations in the code. From this viewpoint, every structure or function of any organism is mutable, and selection acts to preserve some DNA sequences at the expense of others. Presumably, then, it is possible to select for any desired character or trait by choosing the proper environmental system – one which provides an advantage to the organism carrying the trait. Indeed, to some microbiologists selection is simply the method of obtaining new and useful mutant strains, while to others, especially those concerned with mutagenesis, selection is a nuisance which must be corrected for or avoided.

Originally, Darwin's concept of natural selection was one of differential mortality, although it soon became obvious that other factors such as mating probability and fecundity would enter. Theoretically, selection is of several types, all of which are important for population genetics theory (Wright, 1955; Crow & Kimura, 1970; Dobzhansky, 1970). Some selection pressures, such as constant migration or any other constant selection pressure, cause unidirectional evolutionary changes. A second type of selection results from random fluctuations, or changes in gene frequencies occurring from random assortment of gametes, in populations composed of small numbers of individuals. A third type of selection is that which occurs because of some unique incident, such as an environmental catastrophe, or the duplication or deficiency of a nucleotide sequence, gene chromosome, or nucleus. These latter changes, if they increase the amount of genetic material significantly, may well lead to rapid evolutionary progress, especially when they make large numbers of redundant genes available for mutation, and will be called *epochal* events. One example is the origin of the common bread wheat by hybridization between distantly related strains and subsequent doubling of the entire chromosome complement from both parents (amphidiploidy) in daughter cells (Mangelsdorf, 1953).

There has been little quantitative treatment of such epochal events because they lead to evolutionary leaps that may occur only once in the history of a species. Nevertheless, the occurrence of such events lead to predictable consequences (Koch, 1972). Epochal events that increase genetic content can be expected not only to change the relative proportions of existing genes, but also to provide the raw material for the progressive evolution of new genes. Usually, selection acts upon phenotypes, by changing differential reproductive success. However, the survival of a major genetic accretion to constitute an epochal event would depend upon internal gene balance within the cell and would therefore be subject to selection processes operating directly upon the genetic material, which might be even more rigorous than those selection pressures operating upon the phenotype of the organism.

SPECIFIC AND NON-SPECIFIC SELECTION

Specific selection for single genes

The simplest examples of selection are those for single mutant characters in a large population of haploid micro-organisms growing at constant density and rate in a continuous culture. Large populations tend to free us from the distracting issues of chance and of the random and infrequent occurrence of mutations that would occur in small numbers of cells. This is not to say that random fluctuations are not important to evolutionary processes, but that we can describe the effects of selection more simply when fluctuations are absent or negligible. This convenience of large populations is reflected in mathematical procedures. The relatively simple and often used deterministic equations below suffice to describe large populations, while small populations require the use of stochastic methods which, although they describe the actual processes more accurately, are technically more difficult to apply.

For a given locus the number of mutants m depends not only upon the total number of cells N and the rate of mutation ν per cell per unit time, but also upon the rates of cell division of both the mutant and the parent strain, α and β, respectively. We shall assume that there is no cell death, that immigration of cells into the continuous culture does not occur, and that mutation rates per cell are independent of population density. Then, if the mutant population is so small that we may neglect reverse mutations, the rate of accumulation of mutants depends upon their rate of creation νN, upon their rate of increase αm, and upon their rate of loss from the population by dilution βm,

$$\mathrm{d}m/\mathrm{d}t = \nu N + (\alpha - \beta)\, m. \tag{1}$$

Although we shall consider the case of continuous cultures (Novick & Szilard, 1950; Kubitschek, 1970), the same equations describe mutation in cultures increasing exponentially according to $N = N_0\, \mathrm{e}^{\beta t}$ provided that all mutant concentrations are normalized to their corresponding values at the initial population number N_0. That is, in terms of mutant frequencies m/N, Eqn. 1 applies to steady-state cultures of all kinds.

Although Eqn. 1 describes the rate of accumulation of mutants when mutation depends only upon time t, mutation in continuous cultures usually seems to depend upon cell replication and therefore upon the number of elapsed generations g, so that it is more appropriate to rewrite Eqn. 1 in terms of this variable. After steady-state growth conditions have been reached in a continuous culture the generation time T is constant, and $g = t/T$. Thus the rate of accumulation of mutants per generation, from Eqn. 1, is

$$\mathrm{d}m/\mathrm{d}g = \nu NT + (\alpha - \beta)\, mT. \tag{2}$$

Moreover, the mutation rate μ per cell per generation is

$$\mu = \nu T, \tag{3}$$

and as shown earlier (Moser, 1958; Kubitschek, 1970),

$$T = (\ln 2)/\beta, \tag{4}$$

so

$$\mathrm{d}m/\mathrm{d}g = \mu N + \frac{\alpha - \beta}{\beta}\, m \ln 2. \tag{5}$$

The pattern of accumulation of mutants depends upon the value of the selection coefficient s, defined here as

$$s \equiv \frac{\alpha - \beta}{\beta}. \tag{6}$$

If initially at $g = 0$ there are m_0 mutants in the population then the complete solution of Eqn. 5 for the accumulation of mutants under selection s at any number g of elapsed generations is given by

$$m = m_0\, 2^{sg} - \frac{\mu N}{s \ln 2}(1 - 2^{sg}), \quad s > -1. \tag{7}$$

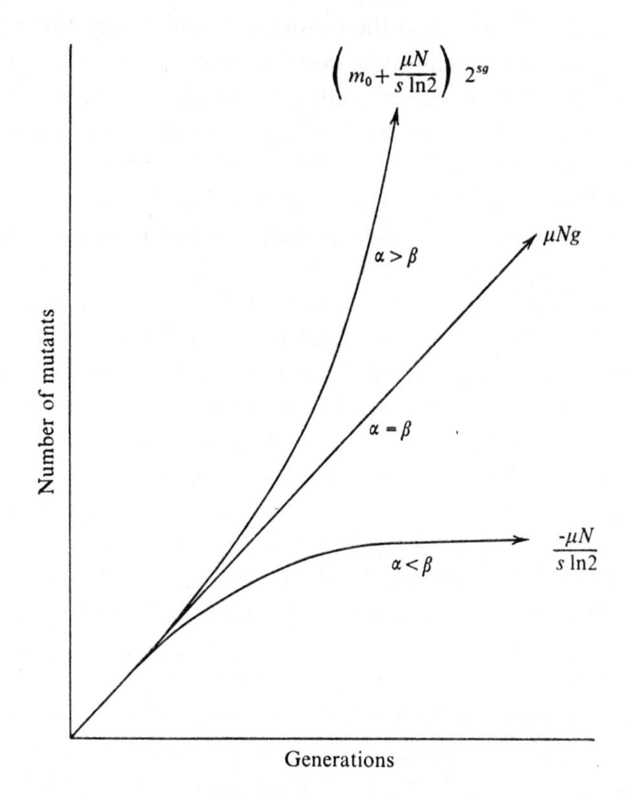

Fig. 1. Patterns of mutant accumulation. α, growth rate of mutant; β, growth rate of parent; μ, mutation rate per cell generation; N, total number of cells; s, selection coefficient; g, number of generations elapsed. The asymptotic dependences of mutant accumulation after many generations are indicated for the three cases where the selection coefficient s is positive ($\alpha > \beta$), zero ($\alpha = \beta$), or negative ($\alpha < \beta$).

When $\alpha > \beta$, s is positive and the rate of increase of mutants is always positive and rapidly approaches an exponential increase proportional to 2^{sg}, as shown in Fig. 1. This increase culminates in the fixation of the mutant gene and the extinction of the original population except for a small fraction of cells arising by reverse mutations. Thus it is an overt evolutionary step, and will be discussed again later.

When $\alpha < \beta$, s is negative, the mutation is deleterious, and mutant numbers depend upon the degree of selection. Initially, when m is small, the rate of increase is essentially constant with the value μN (Fig. 1). As mutants continue to accumulate, the rate of increase approaches zero as m approaches the equilibrium value m_∞, where

$$m_\infty = \frac{-\mu N}{s \ln 2}, \quad -1 \leqslant s < 0. \tag{8}$$

When $\alpha = \beta$, s is zero, and the mutation is selectively neutral. Then, from Eqn. 5, mutants accumulate continuously at the rate μN, linearly with time or numbers of generations. In principle, it is possible for a selectively neutral mutation to displace the original allele from the population, but this occurs by chance and with low probability, through random genetic drift (Crow & Kimura, 1970).

Examples of positive selection abound in the microbial literature, since they arise from the innumerable selections used to obtain new mutant strains of all kinds. One of the most famous selection techniques is that of penicillin enrichment, used to select for mutants unable to grow in the absence of some required factor (Davis, 1948; Lederberg, 1950). There are also many examples of positive selection in continuous or nearly continuous cultures, of which but a few will be cited in order to present the reader with the variety of selection processes already available. In nutrient broth cultures, *E. coli* B was displaced by a mutant resistant to bacteriophage T3 (Cocito & Bryson, 1958) that produced an inhibitory colicin (Bryson, 1959). In cultures of *Brucella abortus*, Braun and his colleagues obtained evidence for an inhibitory substance, that was traced to the excretion of D-alanine from one of the variants (Braun *et al.* 1951; Goodlow, Braun & Mika, 1951). In turbidostat cultures, Cocito & Vogel (1958) selected for cells with reduced levels of the enzyme acetylornithinase in the presence of the repressor L-ornithine, while Weiner (see Horiuchi, Tomizawa & Novick, 1962) selected for constitutive mutants for β-galactosidase production in chemostat cultures limited with lactose.

Negative selection has also been observed in continuous cultures and in serial transfer cultures. Novick & Szilard (1950) found that T4-resistant mutants of a tryptophan-requiring strain of *E. coli* B/1, *trp* were under adverse selection. Atwood, Schneider & Ryan (1951*a*) observed, in studies of mutation of a histidine-requiring strain to histidine independence, that equilibrium numbers of *his*$^-$ mutants were reached within 100 to 130 generations.

Two apparently neutral mutations were found in chemostat cultures, resistant to T5 and to T6 bacteriophages. Novick & Szilard demonstrated linear increases in both mutant concentrations when *E. coli* B/1, *trp* was limited with tryptophan. That selection of T5-resistant mutants was absent or negligible was verified in reconstruction experiments (Kubitschek & Bendigkeit, unpublished). Evidence was also obtained that selection is absent even during the period between mutagenesis and phenotypic expression of T5-resistance (Kubitschek, 1970, pp. 132–4). In addition, no selection was observed when T5-resistant

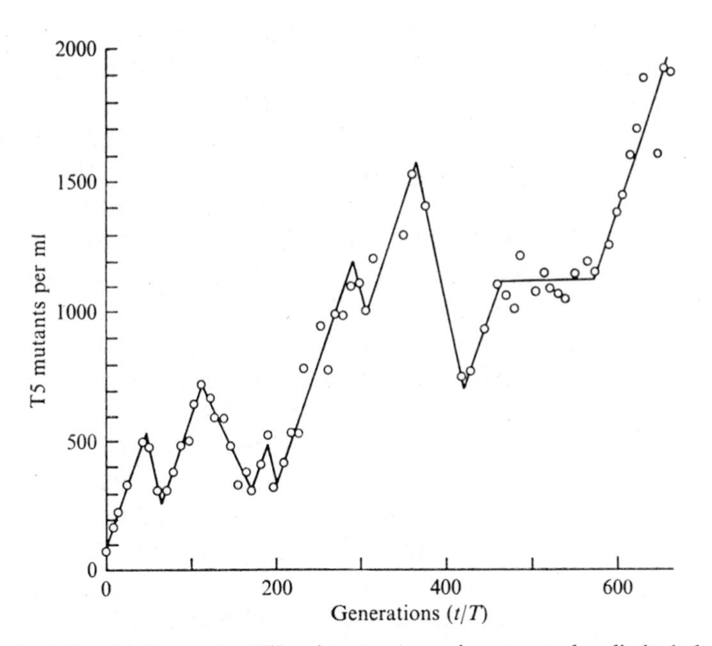

Fig. 2. Periodic selection against T5-resistant mutants in a tryptophan-limited chemostat culture of *E. coli* B/r/1, *trp*. Cell generation time (T), 2.8 h, corresponding to the observed culture generation time (τ), 4.0 h. Cell concentration, 2.5×10^8 cells ml^{-1}. From Novick (1958). Originally published by the University of California Press; reprinted by permission of The Regents of the University of California.

mutants were produced in cultures limited with glucose, succinate, or phosphate (Kubitschek & Bendigkeit, 1964), or with methionine (Kubitschek & Gustafson, 1964). All of these, taken together with Novick's results (1958, Fig. 6) indicate that if selection for T5 resistance occurs, it must be less than 10^{-3} per generation.

The selection discussed above is called *specific selection* because it operates directly upon the particular mutant under study. That is, positive or negative selection occurs because the mutation is pleiotropic and leads to a concomitant alteration of growth rate.

Periodic selection

In their studies of mutation in bacterial cultures, Stocker (1949), Novick & Szilard (1950) and Atwood, Schneider & Ryan (1951*a*, *b*) found that mutant accumulation followed patterns like those shown in Fig. 1 for many generations. Over longer periods, however, the accumulation of mutants was sporadically punctuated by abrupt decreases in mutant frequency, followed again by a similar mutant accumulation, and giving rise to a saw-tooth pattern of the kind shown in Fig. 2 (Novick, 1958). These abrupt decreases, called 'change-overs'

(Novick & Szilard, 1950) or 'adaptive leaps' (Atwood *et al.* 1951*b*), occur with the appearance of a new population that has a selective advantage over the originally predominant type. The new population displaces all of the more slowly growing cells in the original population, including their slowly growing mutants. This is due to the fact that mutants usually are present in very low frequencies and the adaptive mutation is much more likely to occur in the original genotype. The displacement of the original population is followed by the accumulation of mutants again in the new populations.

When selection results in such a change-over, or adaptive leap, it is called *non-specific* or *indirect selection* because mutant frequencies are altered by selection for an unrelated mutant. As shown in Fig. 2, non-specific selection is deleterious for T5-resistant mutants. In principle, however, neutral or advantageous non-specific selection may also occur, although rarely. For example, suppose one of the T5-resistant cells were to mutate to an increased rate of growth and division. Then the progeny of this cell would give rise to a T5 resistant population that would rapidly displace the original population of T5-sensitive cells. Such a mutation, as Morton (1972) and Koch (to be published) pointed out, would, when occurring in the relatively small population of bacteriophage-resistant cells, correspond to a 'jackpot' mutation, and lead to substitution of the allele for bacteriophage resistance into the population. As discussed later, it may be that this process is the major route for fixation of neutral or nearly neutral alleles.

SELECTION AND GENETIC DRIFT

Evolution by selection

Koch (1972) examined evolutionary kinetics for haploid organisms assuming that evolution operated through adaptive changes in DNA and that these occurred in the absence of recombinational events. He employed deterministic equations similar to Eqn. 1 to calculate proportions of mutant alleles after gene duplication and subsequent environmental changes made a redundant gene available for mutation. In order to determine which mutational processes might lead to the most rapid evolutionary rates, he used values for observed mutation rates in *E. coli* and bacteriophage T4 when these were available for specific mutational processes, and estimated the selective advantages expected for reduction in DNA or protein synthesis after deletions or chain-termination events.

If the production of some particular enzyme was limiting for the growth rate of an organism, he argued, then selection for advantageous point mutations and single minor genetic deletions or additions would quickly (on an evolutionary scale) lead to the point where further improvements by these processes would be negligible. He suggested that gene duplication would then permit evolution to proceed, since when selection for the enzyme later became relaxed, one of the duplicate genes would provide a new source of genetic material available for mutation while the other was retained for its original function.

Assuming that gene duplication occurs one-tenth as frequently as typical major deletions, Koch concluded that the time to delete one of the duplicates when its function no longer limits growth is much longer than the time originally required to establish the duplicate in the population. Depending upon the size of the population and the initial fraction of mutant cells, duplicate genes would be fixed in the entire population in approximately ten to twenty days if the growth rate were two generations per day, while loss of cells with complete gene duplications would require periods of five to fifty years, depending upon the degree of selective advantage obtained by eliminating unnecessary replication of DNA or production of enzyme. Losses of a gene by deletions, or of its function by promoter, initiator, or chain-terminating mutations would lead to mixed, polymorphic populations. Koch concluded that the calculated rates were reasonable for enzyme evolution in the Precambrian era, assuming that recombinational mechanisms had not then evolved, but that in present organisms recombination would reduce those rates to unacceptable levels.

Koch (to be published) has since pointed out that this analysis did not take periodic selection into account. Periodic selection would greatly reduce calculated rates because of the tremendous loss of genetic variance at each change-over. The progeny of only a single organism remain after each adaptive leap, and the accumulated mutations in all of the remaining organisms are diluted out with those organisms during change-overs. Thus, change-overs drastically lower the number of deleterious and neutral mutant alleles, and the residual number of mutant alleles fluctuates around some very low level set by the frequency of periodic selection.

Evolution by random genetic drift

The possibility of evolution by genetic drift had been considered as early as 1931 by Wright (1931, 1966) but was generally neglected until experimental evidence was obtained for molecular mutation rates. The very high overall rates of mutation implied by evolutionary rates of amino acid substitutions in proteins led Kimura (1968) to propose that these must be due mainly to neutral or nearly neutral mutations and that evolution proceeds largely by the incorporation of neutral alleles by a mechanism of random genetic drift. Although individual rates of substitution of amino acids in proteins are minute, of the order of one amino acid per codon per 10^9 years, when those rates were extended to the entire chromosomal complement Kimura calculated that a nucleotide substitution occurs in small mammals approximately once every two years. This value is far too large to be accounted for by Haldane's earlier estimate for 'standard rate' (horotelic) evolution of one allele every 300 generations or so, which was in accord with evolutionary rates at the gross phenotypic level. Later, Ohta & Kimura (1971*b*) increased this estimate to an even larger value for man, eight substitutions per genome per year.

Natural selection also fails to account for the remarkable genetic polymorphism in *Drosophila* or man (Kimura, 1968; King & Jukes, 1969; Kimura & Ohta, 1971*a*; Ohta & Kimura, 1971*a*); individual *Drosophila*, for example, are heterozygous at some 12 % of their loci (Lewontin & Hubby, 1966). According to Dobzhansky's calculations (1970, p. 224), natural populations of *Drosophila* are polymorphic for about 3000 genes, and human populations are polymorphic for about 6000 genes. If these had arisen by natural selection they would have led to unbearable genetic loads (Kimura, 1968; Kimura & Ohta, 1971*a*; Dobzhansky, 1970). This was the primary reason for Kimura's proposal that most mutations must be neutral or nearly neutral, since only for these can the resulting genetic loads be small or negligible.

Natural selection also fails to account for the great uniformity of the rate of molecular evolution in individual proteins. This uniformity is consistent with the incorporation of neutral mutations, because the rate of substitution of neutral mutations is equal to their rate of occurrence and is therefore independent of population size (King & Jukes, 1969; Kimura, 1969; Crow, 1969; Ohta & Kimura, 1971*a*). On the other hand, the rate k of substitution of mutations by selection is proportional to the effective population number N_e (essentially the reproductive population, Dobzhansky, 1970), the selective advantage s, and the rate of mutagenesis u_a, as shown in Eqn. 9 below. For this product to remain constant, any change in one of these three variables would have to be compensated for by reciprocal variations in one or both of the others, and there is no reason to expect such reciprocal responses or evidence for their occurrence.

Further evidence for neutral mutation is provided by the experiment of Cox & Yanofsky (1967), who studied the effect of the Treffers' mutator gene. This gene causes transitions from an AT base pair to a CG base pair at the rate of 3.5×10^{-6} per generation, or about seven molecular mutations per bacterium per generation. After 1200–1600 generations, they observed an increase in CG content of about 0.5 %, in agreement with the value estimated from the mutation rate. If even a small fraction of these are lethal or semilethal mutations, then the strain should be strongly selected against. Yet, in experiments with the same mutator strain, Gibson, Scheppe & Cox (1970) found that although this strain had a slight growth rate disadvantage initially as compared to a co-isogenic normal strain, the mutator gene conferred a selective advantage during later generations. Their interpretation that mutations to increased fitness occurred more frequently in the mutator population was supported by similar results with another mutator strain (Nestman & Hill, 1973).

Substitution of neutral alleles by periodic selection

Although evidence was presented in the previous section for abundant substitution of neutral mutations, the evidence for evolution by natural selection is also irrefutable. Which, if either, is the primary mechanism of evolution? Are the two processes related, and if so, how? In asexual populations, the two processes appear to be irrevocably linked: neutral and nearly neutral mutations are substituted into asexual populations by natural periodic selection (Kubitschek, to be published). Since neutral mutations occur more frequently than advantageous mutations, an advantageous mutation usually occurs in a background of new neutral alleles. During selection for the favorable character, these associated neutral alleles will also be carried to fixation.

The average number of alleles thus substituted into the population at each adaptive leap depends upon the relative values of the individual substitution rates for neutral and advantageous mutations. For diploid organisms, each with a total mutation rate u_a per genome per generation, the rate K of substitution of a gene with selective advantage s in a population of actual number N and effective number N_e is given by Kimura & Ohta (1971*b*) as

$$K = 4N_e s u_a \tag{9}$$

provided that s is positive, $s \ll 1$, and that $4N_e s \gg 1$. For asexual haploid micro-organisms, which contain only a single copy of each gene,

$$K = N s u_a \tag{10}$$

The corresponding rate k of substitution of neutral alleles into a population depends only upon the total mutation rate u_n per generation per organism, and is given by

$$k = u_n \tag{11}$$

both for random drift in diploids (Wright, 1931; Kimura & Ohta, 1972; Crow, 1969) and for genetic drift by periodic selection in haploid organisms (Kubitschek, to be published).

The ratio of the two rates for asexual haploid organisms is

$$k/K = u_n/Nsu_a. \quad (12)$$

In terms of the average period L between adaptive leaps,

$$k/K = u_n L/s. \quad (13)$$

This ratio provides estimates of the average number of neutral genes swept to fixation during selection for a favorable mutant in cells containing single chromosomes and for which meiotic recombination is absent or occurs at most very rarely. From Eqn. 13, the phenomenon of fixation of neutral genes by periodic selection is of greatest importance in those fitter populations where adaptive mutations occur at very low frequencies, allowing long periods of accumulation of neutral mutations.

The phenomenon of genetic drift by periodic selection is much less effective in sexual diploids because the extensive meiotic recombination that occurs in these organisms causes favorable mutations to reach random or almost random recombination with almost all neutral mutations. That is, most of the group of neutral alleles originally linked to any favorable mutation is replaced by other neutral alleles, with the result that much of the original spectrum of loosely linked neutral alleles in the population is preserved. Only extremely closely linked mutant alleles survive recombination and are substituted into the entire population during change-overs. These differences in linkage along the adaptive chromosome lead to the retention of a high degree of genetic polymorphism for loci distant from the favorable mutation, and monomorphism is retained only for loci very near to the favorable mutation.

To estimate the number of mutant alleles swept to substitution at an adaptive leap in a culture of haploid micro-organisms, let us assume that the value of 2×10^{-10} mutants per base pair replication in *E. coli* (Drake, 1969) applies to neutral mutations. Since there are 4×10^6 base pairs per genome (Cooper & Helmstetter, 1968; Kubitschek & Freedman, 1971), the neutral mutation rate u_n per cell per generation is approximately 10^{-3}. If change-overs occur every 1000 generations, then the average number of neutral genes fixed at each change-over is, approximately, $1/s$. Assuming values of s from 10^{-1} to 10^{-3}, the corresponding number of neutral mutant alleles substituted per change-over would vary from 10 to 1000.

The possibility that smaller values of s may occur, and therefore that numbers of substitutions per change-over are even larger, suggests that neutral mutation rates may be only a small fraction of the total mutation rate (Crow, 1972; Bodmer, 1970), and/or that much of the bacterial DNA is inert, i.e. not informational for protein or RNA synthesis.

To estimate the amount of non-informational DNA, consider that there are about 460 known genes for *E. coli* (Taylor & Trotter, 1972). If we assume that the average cistron is composed of approximately 10^3 nucleotide pairs then these genes account for about 10 % of the genome content. A more liberal estimate (Watson, 1970) places the maximum number of small metabolites in known pathways at approximately 800. There are, in addition, an unknown number of genes that specify structural proteins of the cell, as well as a minor fraction for the production of ribosomal and transfer RNA. If the addition of these genes doubles the known functional fraction, then the amount of functional DNA still fails to account for half the total DNA. Thus, unless these bacteria ultimately are found to contain many more genes, much of their DNA must be informationally inert.

Crow (1972) suggested a similar presence of non-informational DNA in *Drosophila melanogaster*. Complementation tests seem to indicate perfect correspondence between salivary chromosome band number and the number of complementation units. There are about 6000 chromosome bands, and even if we were to assume that each band contained some ten genes, on average, there would be no more than about 60000 genes. This number, although higher than most estimates, is again only a small fraction of the DNA content, which is equivalent to about 300000 genes (Dobzhansky, 1970).

Ohta & Kimura (1971*b*) concluded that mammalian DNA is also largely non-informational, and that informational DNA accounted for an even smaller fraction of the total. They suggested that duplication of genetic material occurred during evolution, and that degenerative differentiation of one of the duplicate genes, or sets of genes, is the usual fate of the added material. Further mutations in such non-informational DNA might tend to be selectively neutral, since gene function of the unaltered duplicate would be maintained.

Fixation of deleterious mutations by periodic selection would rarely occur because the strong selection against such mutants greatly reduces their numbers. But it is interesting that mutants under very weak positive or negative selection are swept to substitution almost as effectively as neutral mutants. For any particular mutation, the degree of fixation depends not only upon the selection coefficient s, but also upon the average number of generations L between adaptive leaps. From Eqn. 8, it can be seen that if $|s| \gg 1/L$ then frequencies of such deleterious mutants and their probabilities of substitution during periodic selection will remain low. However, all mutants with selection factors such that $|s| < 1/L$ will accumulate essentially linearly between adaptive leaps and therefore will be substituted like neutral mutants. If, for example, L is 1000 generations, then all deleterious mutants with values of $|s| \leqslant 10^{-3}$ will accumulate in essentially the same manner as neutral mutations.

This 'piggyback substitution' of neutral and nearly neutral alleles during periodic selection provides a picture of evolution different from either described in the previous two sections. In the absence of chromosomal recombination, neutral mutations are not accumulated gradually by random walk, but abruptly and in groups, during selection for a single associated advantageous allele. In haploid populations the accumulated variability of all but the mutant progenitor will be lost during the change-over. Over long evolutionary periods the frequency of a particular neutral allele would shift from extremely low values in the population to nearly complete fixation, and about as rarely, back again. The average lengths of these intervals are approximately equal to the reciprocals of the mutation rates for substitutions and loss of this allele, and therefore are very much longer than the average interval between adaptive leaps.

But neutral mutations, by definition, would appear to have no effect on evolution since they are under no selective pressure. What role might they play during evolution? One possibility is that neutral alleles can become adaptive when the environment is changed. Another is that neutral mutations may serve as a substrate for further, selected mutations. A third possibility is that the collection of neutral mutants in the local genetic environment of an adaptive gene alters its function, providing an increased or decreased selective advantage. In addition, as Wright (1959) pointed out, a non-selective process is *required* for a genetic system to move from one adaptive peak to another. He also pointed out that 'the joint operation of directed and random processes tends to result in greater progress than the directed process by itself'.

EVOLUTION IN MICROBIAL POPULATIONS

Measurements of rates of adaptive evolution in microbial populations

Selection factors and the rate of adaptive evolution can both be estimated in populations under periodic selection. Let us begin by considering the effect of a change-over on a selectively neutral mutation.

As before, we shall assume that there is no cell death and that cell numbers are maintained constant by continuous dilution of cells from the culture at the same rate at which they multiply. The equations will also apply to any population in the steady state of exponential increase in numbers, by normalizing the number of cells to a constant value. We shall also assume, in agreement with Novick & Szilard's (1950) observations, that rates of cell division in both the old and the new population depend upon the residual concentration c of a limiting growth factor, and that during steady-state growth the rates of division in each population are proportional to c.

Suppose that the total number N of cells in the culture at time t is composed of n_1 cells with the slower rate α_1 of cell division and n_2 cells dividing at the more rapid rate α_2. Since these rates are proportional to the concentration c of the limiting nutrient, we have $\alpha_1 = a_1 c$ and $\alpha_2 = a_2 c$, where a_1 and a_2 are the specific growth rates per unit concentration of the limiting nutrient. Also, as before, both cell populations are subject to the constant rate of dilution, β. Then, the rate of change of n_1 is

$$\mathrm{d}n/\mathrm{d}t = (a_1 c - \beta)\, n_1, \qquad (14)$$

and of n_2 is

$$\mathrm{d}n/\mathrm{d}t = (a_2 c - \beta)\, n_2. \qquad (15)$$

Solving each equation for c and equating,

$$\frac{1}{a_1 n_1}\frac{\mathrm{d}n_1}{\mathrm{d}t} + \frac{\beta}{a_1} = \frac{1}{a_2 n_2}\frac{\mathrm{d}n_2}{\mathrm{d}t} + \frac{\beta}{a_2}. \qquad (16)$$

Imposing the condition $N = n_1 + n_2$, and assuming that the first cell with specific growth rate a_2 appears at time $t = 0$, then

$$a_2 \ln \frac{n_1}{N-1} - a_1 \ln (N - n_1) = (a_1 - a_2)\, \beta t. \qquad (17)$$

Alternatively, to good approximation,

$$n_2 \left(1 - \frac{n_2}{N}\right)^{-a_1/a_2} = \mathrm{e}^{s_i \beta t}, \qquad (18)$$

where s_i is the indirect selection coefficient

$$s_i \equiv \frac{a_2 - a_1}{a_1}. \qquad (19)$$

From Eqn. 18, it can be seen that early values of n_2, when $n_2 \ll N$, increase exponentially according to

$$n_2 = \mathrm{e}^{s_i \beta t} \qquad (20)$$

and final values of n_1 decrease exponentially according to

$$n_1 = N \frac{a_2 + a_1}{a_2} e^{-s_i \beta t}. \qquad (21)$$

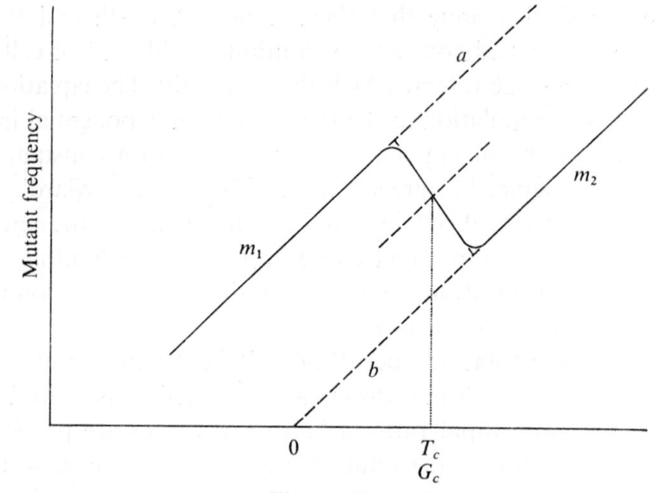

Fig. 3. Analysis of a change-over. The accumulation of mutants m_1 follows line a, before the change-over. The time T_c or number of generations G_c for half displacement of the original mutants (m_1) is given by the mid-point of the line segment connecting lines a and b, as shown. The intersection of line b with the abscissa provides the fiducial point from which T_c or G_c is measured.

The value of the indirect selection coefficient s_i can be determined if we can measure the duration of the change-over period T_c, that is, the period between the appearance of the first cell of the more rapidly growing population and the time at which its progeny account for half of the cells in the culture. At $t = T_c$, then, $n_1 = n_2 = N/2$, and from Eqn. 18,

$$s_i = \frac{\ln N}{\beta T_c - \ln 2}. \tag{22}$$

The corresponding number of generations G_c during this period is $G_c = T_c/T$, where T is the generation time, as before. From Eqn. 4,

$$s_i = \frac{(\ln N)/(\ln 2)}{G_c - 1}. \tag{23}$$

When mutant phenotypes are expressed with little or no delay, the value of T_c (or G_c) can be estimated directly from data for periodic selection, from graphical determinations of the effective time $t = 0$ for appearance of the first rapidly growing variant (birth of the new population) and the time $t = T_c$ for its progeny to displace half of the cells with the original growth rate. As shown in Fig. 3, which displays hypothetical data for concentrations of a neutral mutant during the course of a change-over, the effective time of appearance of the first variant cell is determined by the intersection of the abscissa with the straight-line portion of mutant increase (m_2) after completion of the change-over (line *b*). The time at which the progeny displace 50 % of the more slowly growing cells is determined from the mid-point of the straight-line segment connecting the extended linear mutant increase before (line *a*) and after (line *b*) the changeover. The difference between these two times is T_c in terms of elapsed time, or G_c in terms of elapsed generations. The method can also be used for mutants under negative selection, provided that their initially linear increases can be determined accurately.

When the delay period D between mutagenesis and expression of the mutant phenotype is of significant duration, then the value of G_c must be increased by D to obtain the proper value for the number of generations to obtain half replacement of the culture. This is due to the fact that the apparent time of birth of the new population is delayed by the period D. That is, the true position of line *b* in Fig. 3 is found by shifting it to the left by D generations. The time of half displacement is essentially unchanged, however, because the displacement rate is so rapid near $n_1 = N/2$ that mutants are washed out of the population with essentially the same kinetics as cells of the original unmutated population.

These methods for computing effective times of appearance of the first variant and half displacement of the population may appear invalid at first because the changes in n_2 are extremely non-linear, and because they employ extrapolations based on results obtained for mutant concentrations m_2 long after change-over, when $n_2 = N$, in order to estimate values during change-over at a time when n_2 was the minority population (or even zero). The validity of this extrapolation of m_2 back to zero depends upon the absence of selection for these mutants. This can be shown by considering the number of new mutants m_2 present at any time in the new population. If the rate of mutation is ν per cell per unit time and the rate of division of these mutants is α_2, then from Eqn. 1,

$$\mathrm{d}m_2/\mathrm{d}t = \nu n_2 + (\alpha_2 - \beta)\, n_2. \tag{24}$$

But the wild-type cells have the same division and dilution rates, so

$$\mathrm{d}n_2/\mathrm{d}t = (\alpha_2 - \beta)\, n_2, \tag{25}$$

and Eqn. 24 may be rewritten as

$$\frac{\mathrm{d}}{\mathrm{d}t} \ln \frac{m_2}{n_2} = \nu \frac{n_2}{m_2}. \tag{26}$$

The solution of this equation for mutants accumulated from the time (zero) of appearance of the original progenitor cell to arbitrary time t is

$$m_2/n_2 = \nu t. \tag{27}$$

Thus, no matter how n_2 changes, the mutant frequency m_2/n_2 increases at the constant rate ν; that is, mutant frequencies depend only upon the elapsed time regardless of the rate of increase of n_2. Therefore, the extrapolated intercept gives the time at which the original mutation appeared, even though the method is based on the operationally convenient assumption that $n_2 = N$ immediately after appearance of the mutant.

On the other hand, if mutant frequencies are proportional to the number of elapsed generations rather than to chronological time, then the intercept cannot indicate the time of appearance of the original mutation because the progeny of this cell grow more rapidly than the original population, and a great many more actual generations may elapse for this variant population than are tallied up by the number of chemostat generations during the change-over. Nevertheless, the method again gives the actual number of generations required for half-displacement of the population in terms of the number of elapsed chemostat generations.

With this method of measuring G_c we can estimate the value of the indirect selection coefficient s_i at each adaptive leap. We can also estimate *rates* of adaptive evolution from the frequency of periodic selection and the increment in adaptation s_i at each change-over. If the adaptive leap occurred L generations after the previous one, then the average rate A of increase in adaptation during this interval is $A = s_i/L$.

It is emphasized that this method applies to actual rates of evolution, as observed. It does not measure rates of mutation to favorable genes because it does not take account of the probability of extinction of new mutant genes. The great majority of new mutations, favorable as well as neutral, will be lost within a few generations, even if it is assumed that mutagenesis confers no division delay. For mutation of a daughter cell at division, for example, the probability of loss is 50 % by the very next division. Thus, change-over frequency is a poor estimate of the rate of occurrence of favorable mutations. It is, however, a valid measure of the actual frequency of evolutionary advance.

Progressive evolution

The proposal of Kimura (1961), that progressive evolution results from an increase in the amount of genetic information, seems a natural one. This thesis was proposed by Ohno (1970) in the more limited sense that gene duplication is itself the major force of evolution, and it was used as the basis of Koch's (1972) calculations for enzyme evolution.

The importance of small chromosomal duplications in providing new material for evolution was recognized long ago by Bridges (1935) and by Metz (1947). As stated by Wright (1970) there is no doubt that gene duplication by crossing over followed by differentiation 'has been a general phenomenon of great importance in evolution. Because of this process and of inversions, each chromosome tends to become subdivided into regions of more or less similar material, within which crossing-over is greatly reduced, bounded by unconformities. All of this, however, has to do with the raw material for evolution, not with the dynamics of adaptive transformation.'

The fact that organisms become better adapted to their environments by natural selection also supports Kimura's view of progressive evolution. Suppose one had a population growing in a stable environment that could be maintained to any desired degree of constancy. Then, mutations produced by base alterations, deletions, or recombinations (including translocations, inversions, or transpositions) must persist less and less frequently as the organism becomes better adapted to its environment. These mutations should also occur less frequently as the amount of genetic material is reduced by adaptive deletion mutations. Only by the addition of new genetic material could the organism acquire the renewed variability needed to extend its physiological controls so that it might again increase its adaptedness. Of course, it can be correctly argued that actual environments are not constant and their very variability has played a major role in evolution. But this misses the point that increases in the ability of any organism to respond to environmental challenges are limited by its finite genetic capacity, which can only be extended by increasing the amount of DNA per cell.

Progressive evolution of this kind might occur in varying degrees by a variety of processes, from simple nucleotide pair addition through a frameshift mutation to epochal mutations involving much larger increases in DNA content. The relative increase in raw material for evolution would depend upon the genetic content just before addition. During very early evolutionary stages when amounts of genetic material were relatively small, duplication of a stretch of DNA might have been such an epochal mutation, whereas in present organisms epochal mutations might occur as a result of gene duplication, partial diploidization, addition of one or several chromosomes, duplication of the entire chromosomal set, or symbiotic associations that lead to the fusion of two cells or the incorporation of one by another. Since all of these processes are unstable to a greater or lesser degree, upsetting gene balance, the altered or combined genome becomes stable again only if it survives further periods of mutation and selection. If so, the added

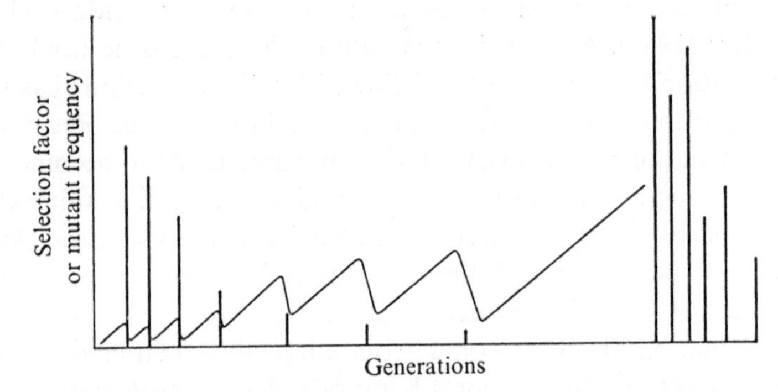

Fig. 4. Mutant frequencies and selection factors for a hypothetical continuous culture. The saw-tooth curve describes the accumulation of a neutral mutant over a period of thousands of generations. The relative amplitude of the indirect selection factor is shown for each change-over by the magnitude of the vertical bar located at the mid-point of the change-over. The increased rate of evolution that might occur again after an epochal mutation is indicated by the vertical bars at the right.

genetic material or its original counterpart may be available for mutation. According to Fisher's Fundamental Theorem of Natural Selection, the rate of increase in fitness of any organism at any time is equal to its genetic variance in fitness. Thus epochal mutations could greatly increase the potential variability and the rate of evolution of the organism.

One would predict, therefore, that the effect of placing an adaptable population in a new environment would be to lead to a response of the kind shown in Fig. 4. First, if some previously advantageous genes now became superfluous they would provide new raw material for mutation and an increased rate of periodic selection over that in the original environment. Second, the greater range of variability available to the organism in the new environment would provide the opportunity for relatively larger increases in selective advantage than that obtainable in the original environment, to which it was well adapted. Third, both the average rate of periodic selection and the average values of the selective advantages at change-overs should decrease with time, as the organism became better adapted. With the occurrence of an epochal event, however, both variability and rate of evolution would increase again.

The early papers on bacterial evolution in chemostat cultures and in serial transfer experiments provide partial support for these predictions. In one of the serial transfer experiments reported by Atwood *et al.* (1951*b*, their Fig. 1) a change-over occurred after about 1500 generations. The geometric mean number of cells was 2.5×10^9 and G_c was approximately 200 generations, yielding a value for s_i of approximately 0.16. This small adaptive increase might be expected for cells already

Table 1. *Rates of evolution in a tryptophan-limited culture of* E. coli *B/r/1*,trp*

	Change-over					
	1	2	3	4	5	6
Generations (G_a) from start of culture to origin of new mutant	30	(111)‡	163	206	343	458
Generation (G_b) from start of culture to half-displacement of previous mutant	60	144	194	294	390	515
Age of previous mutant culture in generations ($G_c = G_b - G_a$) of growth at time of half-displacement†	35	38	36	93	52	62
Indirect selection coefficient (S_i) from Eqn. 23	0.95	0.87	0.92	0.35	0.63	0.53
Generations since previous leap (L)	60	84	50	100	96	125
Average rate of increase in adaption ($A = S_i/L$)	0.0158	0.0104	0.0184	0.0035	0.0066	0.0042

* Based on Fig. 2. $T = 2.8$ h; cell density $= 2.5 \times 10^8$ cells ml^{-1}; $N = 5 \times 10^9$ cells, assuming a culture volume of 20 ml.

† Values include a mean delay period of 5 cell generations for expression of T5 resistance (Kubitschek, 1970).

‡ Because of data variability, the initial value for decrease in mutant frequency is used here.

partially adapted to serial transfer experiments by their long evolutionary history of a feast or famine existence (Koch, 1971),

An analysis of the six change-overs in Fig. 2 gives larger values (Table 1). The number of cell generations between change-overs is 50–84 for the first three change-overs and 96–125 for the last three. The corresponding values for s_i are 0.87–0.95 for the first group and 0.35–0.63 for the second. The average rate of adaptation per generation during the first three periods is $\bar{A} = 0.0141$, and during the last three is 0.0047. These results, although limited, are in agreement with decreasing rates of evolution. Increases in selective advantage were markedly greater in glucose-limited chemostat cultures of *E. coli* B/r/1, *trp*. Values of s_i ranged from about two for spontaneous mutation to about six for photodynamic mutations (data kindly supplied by Dr R. B. Webb). Apparently, *E. coli* is less well adapted to continuous culture in glucose-limited media than to a feast or famine existence in the same media.

SUMMARY AND CONCLUSIONS

The course of evolution in microbial populations depends upon two very different types of selection for individual genes. The first, specific selection, operates continually, while the second, non-specific or indirect selection, operates only periodically as adaptive mutations occur and are fixed in the population. In asexual haploid organisms, periodic selection leads to abrupt decreases in the frequencies of the great majority of neutral or weakly deleterious mutants in the population, and to the replacement of the original population by one having greatly reduced genetic variance and carrying a different spectrum of neutral mutations.

Because periodic selection represents adaptive evolution, studies of evolution and of evolutionary processes can readily be carried out in controlled microbial populations. With the new analytical methods presented here, values for the intensity of selection during an adaptive leap and the average rate of adaptive mutation can now be determined in such cultures. These simple analytical methods permit tests of some of the most searching predictions of theoretical population genetics as applied to asexual haploid organisms. For example, such experiments should provide information on rates of 'spontaneous' adaptive evolution which, according to the previous section, should depend upon the amount of superfluous DNA available for mutation and should decrease with the degree of adaptation of a culture to its new environment. Since the particular environment determines which genes become superfluous, it is clear that spontaneous rates of adaptive mutation must also depend upon the environment.

Because rates of evolution are proportional to genetic variability and inversely proportional to the degree of adaptation, these quantities can be determined and compared for cultures placed in new environments. It should be possible to prepare media that obviate the need for one or more genes and thereby examine rates of evolution as a function of the kind or number of genes. Similarly, it should be possible to stimulate progressive evolution by first employing an environment that selects for gene duplication, as in Weiner's experiments mentioned earlier, and then relaxing that selection. In principle, rates of evolution can be compared before and after gene duplication in the same organism.

Also deserving examination is the prediction of increased evolutionary rates when both random and directed processes are present (Wright, 1959). The increased action of random processes might well be mimicked by the use of an oscillating environment in the laboratory. Of course, such an environment might also cause selection coefficients to oscillate for otherwise neutral alleles, but some indication of these and other effects might be obtained by varying the amplitude of oscillation. Controlled repetitive environmental changes should be studied, because environmental fluctuations have long been called upon to explain difficult points in evolutionary theory, yet there is little experimental information on the effect of particular environmental fluctuations upon selection coefficients.

The use of cultures of *E. coli* to study neutral mutations and their effects upon evolution was suggested earlier by Milkman (1972), since measurements of allozyme variation would provide much-needed information on genetic polymorphism in this organism. But Milkman's assertion that if non-Darwinian evolution is important it implies that there must be scores of allozymes at each of many loci in *E. coli* was made before the consequences of periodic selection were generally known. Because of gross reduction of genetic variance that should occur in asexual haploids during periodic selection, one would now conclude that if even modest genetic polymorphism occurs in *E. coli*, then non-Darwinian evolution must be a very important phenomenon in higher organisms.

Finally, it would be instructive to compare allozymes and rates of adaptive evolution in mutator gene systems with the corresponding enzymes and rates in their co-isogenic normal strains. According to the results of Cox and his collaborators, and of Nestman and Hill, isozyme frequencies almost certainly must increase more rapidly in mutator strains. But the major question is, would the extensive accumulation of neutral mutations bring about associated increases in rates of adaptive evolution? That is, does adaptive evolution occur because of accumulated neutral mutations?

This work was supported by the US Atomic Energy Commission.

REFERENCES

ATWOOD, K. C., SCHNEIDER, L. K. & RYAN, F. J. (1951*a*). Periodic selection in *Escherichia coli*. *Proceedings of the National Academy of Sciences, USA*, **37**, 146–55.

ATWOOD, K. C., SCHNEIDER, L. K. & RYAN, F. J. (1951*b*). Selective mechanisms in bacteria. *Cold Spring Harbor Symposia on Quantitative Biology*, **16**, 345–355.

BODMER, W. F. (1970). The evolutionary significance of recombination in prokaryotes. In *Organization and Control in Procaryotic and Eukaryotic Cells. Symposia of the Society for General Microbiology*, **20**, 279–94. Ed. H. P. Charles & B. C. J. G. Knight. London: Cambridge University Press.

BRAUN, W., GOODLOW, R. J., KRAFT, M., ALTENBERN, R. & MEAD, D. (1951). The effects of metabolites upon interactions between variants in mixed *Brucella abortus* populations. *Journal of Bacteriology*, **62**, 45–52.

BRIDGES, C. B. (1935). Salivary chromosome maps with a key to the banding of chromosomes of *Drosophila melanogaster*. *Journal of Heredity*, **26**, 60–4.

BRYSON, V. (1959). Application of continuous culture to microbial selection. In *Recent Progress in Microbiology*, ed. G. Tunevall, pp. 371–80. Springfield: C. C. Thomas.

COCITO, C. & BRYSON, V. (1958). Properties of colicine from *E. coli* strain B. *Bacteriological Proceedings*, p. 38.

COCITO, C. & VOGEL, H. J., JR. (1958). Heritable lowering of an enzyme level and enzyme repressibility observed upon continuous culture of *Escherichia coli* in the presence of a represser. *Proceedings of the Tenth International Congress of Genetics*, vol. 2, Abstracts, p. 55.

COOPER, S. & HELMSTETTER, C. E. (1968). Chromosome replication and the division cycle of *Escherichia coli* B/r. *Journal of Molecular Biology*, **31**, 519–40.

COX, E. L. & YANOFSKY, C. (1967). Altered base ratios in the DNA of an *Escherichia coli* mutator strain. *Proceedings of the National Academy of Sciences, USA*, **58**, 1895–902.

CROW, J. F. (1969). Molecular genetics and population genetics. *Proceedings of the XII International Congress of Genetics*, **3**, 105–13.

CROW, J. F. (1972). Darwinian and non-Darwinian evolution. In *Proceedings of the Sixth Berkeley Symposium on Mathematical Statistics and Probability*, ed. L. M. Le Cam, J. Neyman & E. L. Scott, vol. 5, pp. 1–22. Berkeley: University of California Press.

CROW, J. F. & KIMURA, M. (1970). *An Introduction to Population Genetics Theory*. New York: Harper & Row.

DAVIS, B. D. (1948). Isolation of biochemically deficient mutants of bacteria by penicillin. *Journal of the American Chemical Society*, **70**, 4267.

DOBZHANSKY, T. (1970). *Genetics of the Evolutionary Process*. New York: Columbia University Press.

DRAKE, J. W. (1969). Comparative rates of spontaneous mutation. *Nature, London*, **221**, 1132.

FISHER, R. A. (1930). *The Genetical Theory of Natural Selection*. Oxford: Clarendon Press.

GIBSON, T. C., SCHEPPE, M. L. & COX, E. C. (1970). Fitness of an *Escherichia coli* mutator gene. *Science*, **169**, 686–8.

GOODLOW, R. J., BRAUN, W. & MIKA, L. A. (1951). Metabolite-resistance and virulence of smooth *Brucella* variants isolated after prolonged cultivation. *Proceedings of the Society of Experimental Biology and Medicine*, **76**, 786–8.

HORIUCHI, T., TOMIZAWA, J. & NOVICK, A. (1962). Isolation and properties of bacteria capable of high rates of beta-galactosidase synthesis. *Biochimica et Biophysica Acta*, **55**, 152–63.

KIMURA, M. (1961). Natural selection as the process of accumulating genetic information in adaptive evolution. *Genetical Research*, **2**, 127–40.

KIMURA, M. (1968). Evolutionary rate at the molecular level. *Nature, London*, **217**, 624–6.

KIMURA, M. (1969). The rate of evolution considered from the standpoint of population genetics. *Proceedings of the National Academy of Science, USA*, **63**, 1181–8.

KIMURA, M. & OHTA, T. (1971*a*). On the rate of molecular evolution. *Journal of Molecular Evolution*, **1**, 1–17.

KIMURA, M. & OHTA, T. (1971*b*). *Theoretical Aspects of Population Genetics*. Princeton: Princeton University Press.

KIMURA, M. & OHTA, T. (1972). Population genetics, molecular biometry, and evolution. In *Proceedings of the Sixth Berkeley Symposium on Mathematical Statistics and Probability*, ed. L. M. Le Cam, J. Neyman & E. L. Scott, vol. 5, pp. 43–68. Berkeley: University of California Press.

KING, J. L. & JUKES, T. H. (1969). Non-Darwinian evolution. *Science*, **164**, 788–98.

KOCH, A. L. (1971). The adaptive responses of *Escherichia coli* to a feast and famine existence. *Advances in Microbial Physiology*, **6**, 147–217.

KOCH, A. L. (1972). Enzyme evolution. I. The importance of untranslatable intermediates. *Genetics*, **72**, 297–316.

KUBITSCHEK, H. E. (1970). *Introduction to Research with Continuous Cultures.* Englewood Cliffs, New Jersey: Prentice-Hall, Inc.

KUBITSCHEK, H. E. & BENDIGKEIT, H. E. (1964). Mutation in continuous cultures. I. Dependence of mutational response upon growth-limiting factors. *Mutation Research*, **1**, 113–20.

KUBITSCHEK, H. E. & FREEDMAN, M. L. (1971). Chromosome replication and the division cycle of *Escherichia coli* B/r. *Journal of Bacteriology*, **107**, 95–9.

KUBITSCHEK, H. E. & GUSTAFSON, L. A. (1964). Mutation in continuous cultures. III. Mutational responses in *Escherichia coli. Journal of Bacteriology*, **88**, 1595–7.

LEDERBERG, J. (1950). Isolation and characterization of biochemical mutants of bacteria. In *Methods in Medical Research*, ed. J. H. Comrie, Jr, vol. 3, pp. 5–22. Chicago: Year Book Publishers.

LEWONTIN, R. C. & HUBBY, J. L. (1966). A molecular approach to the study of genic heterozygosity in natural populations. II. Amount of variation and degree of heterozygosity in natural populations of *Drosophila pseudoobscura. Genetics*, **54**, 595–609.

MANGELSDORF, P. C. (1953). Wheat. *Scientific American*, **189**, 50–9.

METZ, C. W. (1947). Duplication of chromosome parts as a factor in evolution. *American Naturalist*, **81**, 81–103.

MILKMAN, R. (1972). How much room is left for non-Darwinian evolution? In *Evolution of Genetic Systems*, ed. H. H. Smith, pp. 217–29. New York: Gordon and Breach.

MORTON, N. E. (1972). The future of human population genetics. *Medical Genetics*, **8**, 103–24.

MOSER, H. (1958). *The Dynamics of Bacterial Populations Maintained in the Chemostat*, Publication 614. Washington: Carnegie Institution.

NESTMAN, E. R. & HILL, R. F. (1973). Population changes in continuously growing mutator cultures of *Escherichia coli. Genetics*, **73** (Supplement), 41–4.

NOVICK, A. (1958). Some chemical bases for evolution of micro-organisms. In *Perspectives in Marine Biology*, ed. A. A. Buzzati-Traverso, pp. 533–46. Berkeley: University of California Press.

NOVICK, A. & SZILARD, L. (1950). Experiments with the chemostat on spontaneous mutations of bacteria. *Proceedings of the National Academy of Sciences, USA*, **36**, 708–19.

OHNO, S. (1970). *Evolution by Gene Duplication.* Berlin: Springer-Verlag.

OHTA, T. & KIMURA, M. (1971*a*). On the constancy of the evolutionary rate of cistrons. *Journal of Molecular Evolution*, **1**, 18–25.

OHTA, T. & KIMURA, M. (1971*b*). Functional organization of genetic material as a product of molecular evolution. *Nature, London*, **233**, 118–19.

SIMPSON, G. G. (1953). *The Major Features of Evolution.* New York: Columbia University Press.

STOCKER, B. A. D. (1949). Measurements of rate of mutation of flagellar antigenic phase in *Salmonella typhimurium. Journal of Hygiene* **47**, 398–413.

TAYLOR, A. L. & TROTTER, C. D. (1972). Linkage map of *Escherichia coli* strain K12. *Bacteriological Reviews*, **36**, 504–24.

WATSON, J. D. (1970). *Molecular Biology of the Gene*, p. 99. New York: Benjamin.

WRIGHT, S. (1931). Evolution in Mendelian populations. *Genetics*, **16**, 97–159.

WRIGHT, S. (1955). Classification of the factors of evolution. *Cold Spring Harbor Symposia on Quantitative Biology*, **20**, 16–24.

Wright, S. (1959). Physiological genetics, ecology of populations, and natural selection. *Perspectives in Biology and Medicine*, **3**, 107–51.

Wright, S. (1966). Polyallelic random drift in relation to evolution. *Proceedings of the National Academy of Sciences, USA*, **55**, 1074–81.

Wright, S. (1970). Random drift and the shifting balance theory of evolution. In *Mathematical Topics in Population Genetics*, ed. K. Kogima, pp. 1–31. New York: Springer-Verlag.

EVOLUTIONARY IMPLICATIONS OF DOUBLET ANALYSIS

J. H. SUBAK-SHARPE, R. A. ELTON AND G. J. RUSSELL

Medical Research Council Virology Unit, Institute of Virology, University of Glasgow

INTRODUCTION

The work to be described in this article arose out of consideration of problems of polypeptide synthesis which were likely to affect host/virus relationships. A concept was developed which was basic to our subsequent studies and approach. It appeared to make genetic sense that the translation apparatus of the cell, and in particular the population of transfer RNAs (regarded as consisting of distinct molecular species by virtue of their codon–anticodon recognition), would have developed a fairly tight, optimally adapted relationship with the cell's nuclear DNA as a consequence of natural selection. The translation apparatus would, generation after generation, have been selected to provide optimal conditions for the translation of messenger RNAs originating from the nuclear DNA. Given that the translation apparatus imposed characteristic limitations on the efficient and effective translation of codon sequences, it followed that it might not be well adapted for the translation of messenger RNAs copied from foreign DNA. As investigations into this concept progressed, it was inferred (Subak-Sharpe, 1967) that once the ancestral cells had fixed a particular population of transfer RNAs, then the further evolution of polypeptide-specifying DNA would be restricted by this component of selection pressure. As a consequence it would be expected that even relatively small subfractions of the polypeptide-specifying DNA would have to resemble the average nuclear DNA in terms of sequence characteristics related to this restriction. In order to investigate these hypotheses, a method of characterising DNA (and RNA) had to be applied.

In 1961, Josse, Kaiser & Kornberg published an elegant technique which allowed estimation of the frequency distribution of nearest neighbour base sequences in any DNA (that is, the characterisation of a DNA in terms of the sixteen possible two-base (doublet) sequences). For convenience, this technique is referred to as 'doublet analysis'. This, and a series of subsequent studies by Kornberg's colleagues and others, established: (1) that the method was reliable and gave highly characteristic results; (2) that the polymerase reaction yielded a faithful, if not necessarily perfect, copy of the supplied template DNA; (3) that the product was formed from copying both strands of the template DNA; and (4) that the doublet frequencies obtained were non-random and characteristic of the DNA (Josse *et al.* 1961; Swartz, Trautner & Kornberg, 1962; Josse & Swartz, 1963; Skalka, Fowler & Hurwitz, 1966).

Using this technique, Josse *et al.* (1961) and Swartz *et al.* (1962) were able to show that DNA from the closely related vertebrates had very similar, if not identical, doublet frequencies, whereas the DNA of distantly related organisms including vertebrates, echinoderms, arthropods, *Tetrahymena* and *E. coli* possessed different doublet frequencies. This suggested that the method was likely to prove an excellent tool for examining more distant evolutionary relationships than can normally be studied by presently available techniques like peptide sequence comparisons or molecular hybridisation of nucleic acids. Since correlations exist between three-base sequences (that is codons) in messenger RNA and two-base sequences in the DNA coding for that messenger RNA (Subak-Sharpe, 1969*a*), we were therefore encouraged to start to investigate DNAs by doublet analysis.

The initial work involved animal cells and various animal viruses (Subak-Sharpe *et al.* 1966; Subak-Sharpe, 1967, 1969*a*, *b*; Morrison, Keir, Subak-Sharpe & Crawford, 1967; Hay & Subak-Sharpe, 1968). The studies showed that the different groups of *small* animal viruses have characteristic and distinct doublet frequencies but that they all resembled, to a greater or lesser extent, the highly characteristic pattern of vertebrate nuclear DNA. *Large* double-stranded DNA viruses have totally different patterns which showed no similarity to that of the animal host. One group, the adenoviruses with DNA of intermediate molecular weight, exhibit doublet frequencies which, though different from the host, have sufficient similarity to place them in an intermediate category.

From the onset it was clear that, while doublet frequencies gave highly characteristic patterns, these frequencies were related to the overall G+C content of the DNAs. This made comparisons of DNAs of differing G+C content extremely difficult. A method of 'normalisation' was therefore developed (Subak-Sharpe *et al.* 1966) from which it was possible to make pictorial representations of DNAs as a function of their deviation from the random expectation for their particular G+C contents. The term 'general design' was used for these patterns

(Subak-Sharpe, 1967). With this technique it was then possible to compare directly DNAs of differing G+C contents. For example, Maitra, Cohen & Hurwitz (1966) published doublet frequency data on the two halves of phage lambda (broken by hydrodynamic shear and separated by buoyant density centrifugation). When the general designs were derived from these data it was shown that the two halves, though differing by 12 % in G+C content, were virtually indistinguishable (Subak-Sharpe, 1969*b*).

Having the tool, not only to characterise DNAs in terms of doublet frequencies, but also to compare directly DNAs of differing G+C content, we were able to investigate an extensive range of organisms. These included vertebrates, invertebrates, protozoa, angiosperms, ferns, fungi, algae, bacteria and also (in addition to the animal viruses) plant and bacterial viruses. Fractionated DNAs, including mitochondrial DNA, chloroplast DNA, satellite DNAs, renaturation rate fractions and density fractions have also been analysed. The present discussion, however, is confined to recently published results with prokaryotes, and will also review the findings with regard to viruses.

DOUBLET ANALYSIS

Methodology

The initial papers on nearest neighbour frequency determinations gave very full and detailed descriptions of the method (Josse *et al.* 1961; Swartz *et al.* 1962; Josse & Swartz, 1963). As only very minor modifications have been adopted (Subak-Sharpe *et al.* 1966; Russell, Follett, Subak-Sharpe & Harrison, 1971), the technique will not be discussed here. The information gained from doublet analysis is the proportions of the sixteen two-base sequences in the product DNA copied from the template DNA (as shown in Fig. 1*a* for the slow renaturing fraction of guinea-pig nuclear DNA). This can be represented in histogram form (Fig. 1*b*). In this diagram the random expectation with respect to the sixteen doublets is indicated by three straight lines, whose position will of course be dependent on the G+C content of the particular DNA used in the analysis. The values for the random expectation are derived from the approximate base compositions gained by taking the average of A and T and the average of G and C. To remove the effect of G+C content, so as to compare DNAs of different G+C content, the deviation of each doublet from random expectation is determined by dividing each observed experimental value by its theoretical random expectation. The theoretical random expectation is gained from the

(a)

	pA	pT	pG	pC
Ap	95	80	69	49
Tp	71	109	74	61
Gp	58	54	47	39
Cp	69	71	8	46

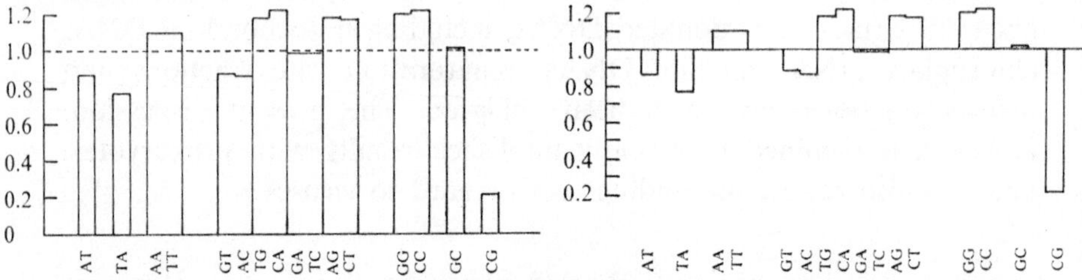

Fig. 1. Derivation of the general design pattern of the slow renaturing fraction of guinea-pig nuclear DNA.

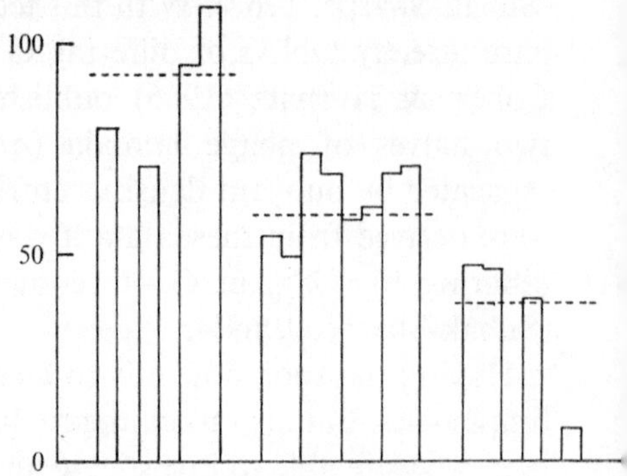

(*a*) Experimentally determined doublet frequencies expressed in parts per thousand.

(*b*) Histogram of doublet frequencies. The dashed lines represent the random expected frequencies for a DNA of 39 % G+C content, where A = T and G = C.

(*c*) Histogram of the ratios of the observed frequencies to the expected random frequencies. The expected random frequencies are gained from the products of the experimentally determined base compositions. The dashed line at unity represents the expected ratios of a random DNA.

(*d*) The ratios of the observed frequencies to the expected random frequencies expressed as deviations from unity, i.e. deviation from random expectation – the 'general design'.

product of the experimentally derived base composition values for the two bases involved in each doublet. These ratios can again be plotted in histogram form as shown in Fig. 1*c*. Values at random expectation would, of course, give a ratio of one indicated by the horizontal line. Deviations from this line therefore show the proportional excess or shortage relative to the random expectation. This is shown more dramatically in Fig. 1*d*, where the ratios are plotted as deviations from *unity* or random expectation giving the general design as defined by Russell, McGeoch, Elton & Subak-Sharpe (1973). This is the same pictorial representation as given by the normalisation technique originally used (Subak-Sharpe *et al.* 1966), differing only in the values in the ordinate. It must be stressed that these ratios are determined from the incorporation factors of the component bases of each doublet internally derived from the doublet frequency data (Josse *et al.* 1961). This has the effect of reducing any systematic experimental error, thus giving better agreement between the theoretically equal complementary doublets than is obtained in the original frequencies (e.g. the difference between AA and TT is completely removed in the example in Fig. 1). This effect has been observed in several scores of analyses in this laboratory and with the data originating from other laboratories.

Stability of the general design

An empirical observation from the large number of duplicated analyses now performed is that the general design is a reliable and highly characteristic measure of any DNA. Moreover, we find that the general design is not easily affected by experimental error variation. From theoretical considerations it can be shown (Subak-Sharpe, 1967, 1969*a*) that the doublet pattern of a DNA is a very stable attribute and that this resistance to change through random mutational events is directly proportional to the size of the genome. It has been calculated that 65000 separate directionally consistent, selected, mutational events would be necessary to change the frequency of a doublet like CG by 1 % in a bacterial genome 1.3×10^7 nucleotides long (Subak-Sharpe, 1967).

Clearly such a burden of random changes in polypeptide-specifying DNA must, in the coded polypeptides, result in a very large (easily calculated) number of consequential amino acid changes of which a large proportion would be deleterious. One illustration of the stability of the pattern can be seen from the analyses of the DNAs of the vertebrate mitochondria (Russell, Schutgens, Borst & Subak-Sharpe, in preparation). These analyses show that the mitochondrial DNAs of guinea pig, rat, mouse and chicken have doublet patterns giving general designs which are indistinguishable from one another but which differ from the general design of vertebrate nuclear DNA. The results prove that even a piece of DNA as small as mitochondrial DNA (9×10^6 daltons or 15000 base pairs) has maintained the characteristic pattern and undergone no appreciable change in general design during its evolutionary history since birds and rodents diverged, even though the G+C contents of these DNAs differ measurably between chicken (48 %) and rodent (41 %). Other considerations relevant to the stability of the doublet pattern and also of the selective forces which could overcome the innate stability of the doublet pattern have been discussed and reviewed (Subak-Sharpe, 1967, 1969*a*, *b*) and will not be considered here.

PROKARYOTES

The general designs of bacterial DNAs

Among the DNAs analysed by Josse *et al.* (1961) were those of six bacterial species, and it was pointed out (Subak-Sharpe *et al.* 1966) that their general designs could be grouped into two major classes. The first is shared by *Micrococcus luteus* and *Mycobacterium phlei*, while the second contains *Haemophilus influenzae*, *Bacillus subtilis*, *Escherichia coli* and *Aerobacter aerogenes*. A third group was revealed by the data of Skalka *et al.* (1966), who measured the doublet frequencies of *Clostridium pasteurianum*. This suggested that the prokaryotes might perhaps have only a few distinct classes of general designs, and to test this hypothesis we have looked at a further twenty-four prokaryote DNAs (seventeen as detailed by Russell *et al.* (1973) and the remaining seven as yet unpublished). These included the DNAs of four blue-green algae. The spectrum of species used in our analyses was selected only because their cells or DNAs were made available to us by other colleagues, and not because of any preconceived ideas concerning possible bacterial evolutionary relationships.

The thirty-one species of prokaryotes for which doublet analyses have now been carried out are listed in Table 1. Fig. 2 shows a cluster analysis of the general designs of these DNAs, in which DNAs with closely similar general designs are closely linked on the distance scale. As can be clearly discerned from the dendrogram in Fig. 2, there are indeed a small number of fairly discrete classes found among the different species. Fig. 3 shows an example of each of the five major classes of general designs that can be identified in the data. Briefly, these are a high-G+C group typified by *M. luteus*, a low-G+C group typified by *Mycoplasma* (sp. kid), the photosynthetic bacteria (*Rhodospirillium rubrum*), *Thermus aquaticus* constituting a group on its own, and finally the large group of species typified by *E. coli*. These five groups fall into two categories, which are considered separately.

Code limit species

As Table 1 indicates, two of the five general design groups have DNAs with G+C contents which are at the extremes so far observed in the DNAs of whole organisms (the low- and high-G+C groups, at about 25 % and 70 % G+C respectively). These two groups afford us an especially favourable opportunity to demonstrate the polypeptide-specifying nature of DNA, since it is possible to explain their doublet frequencies by a relatively simple hypothesis involving the concept of

Table 1. *List of thirty-one prokaryote organisms divided into five groups on the basis of similarity in general design patterns from doublet frequency analysis of their DNAs* (see Fig. 2)

A. Group of extreme low G+C content.
B. The *E. coli* group – large group whose DNAs span the range of 33–64 % G+C content and possess similar general designs to *E. coli*. The four blue-green algae also share this general design.
C. *Thermus aquaticus* – sole member of the group.
D. Photosynthetic bacteria group.
E. Group of extreme high G+C content.

(A) *Mycoplasma pulmonis* (25)
M. orale (25)*
Mycoplasma (sp. kid) (26)
Clostridium perfringens (28)
C. pasteurianum (33)‡

(B) *Staphylococcus aureus* (33)
Bacillus megaterium (34)
B. thuringiensis (34)*
Proteus vulgaris (37)
Vibrio '01' (37)
Haemophilus influenzae (38)†
Photobacterium phosphoreum (39)
Streptococcus faecalis (40)*
S. pneumoniae (40)
Bacillus subtilis (44)†
Escherichia coli (50)†
Peptostreptococcus elsdenii (50)*
Aerobacter aerogenes (56)†
Serratia marcescens (57)
Pseudomonas aeruginosa (64)
Blue-green algae:
Anabaena cylindria (39)*
A. variabilis (42)
Chlorogloea fritschii (43)
Anacystis nidulans (54)*

(C) *Thermus aquaticus* (64)*

(D) *Rhodospirillium rubrum* (62)
Rhodopseudomonas capsulata (67)
R. sphaeroides (68)

(E) *Mycobacterium phlei* (67)†
Sarcina lutea (68)
Micrococcus luteus (71)†

The numbers in parentheses are the G+C contents derived from doublet analyses.
Unless indicated below, all doublet frequency data and general designs are in Russell *et al.* (1973).
* Unpublished results.
† Doublet frequency data from Josse *et al.* (1961) and general designs from these data are in Russell *et al.* (1973).
‡ Doublet frequency data from Skalka *et al.* (1966) and the general design from these data is in Russell *et al.* (1973).

strong selective pressure towards extremity of G+C content acting on protein-coding DNA (Woese, 1967; Elton, 1973*b*).

The amino acid composition of bulk and individual proteins in micro-organisms can be shown to correlate with the G+C content of their DNA (Sueoka, 1961; Elton, 1973*a*) in a way that is consistent with known genetic code assignments (for example, species of low G+C content have low percentages of alanine, which has G+C-rich codons).

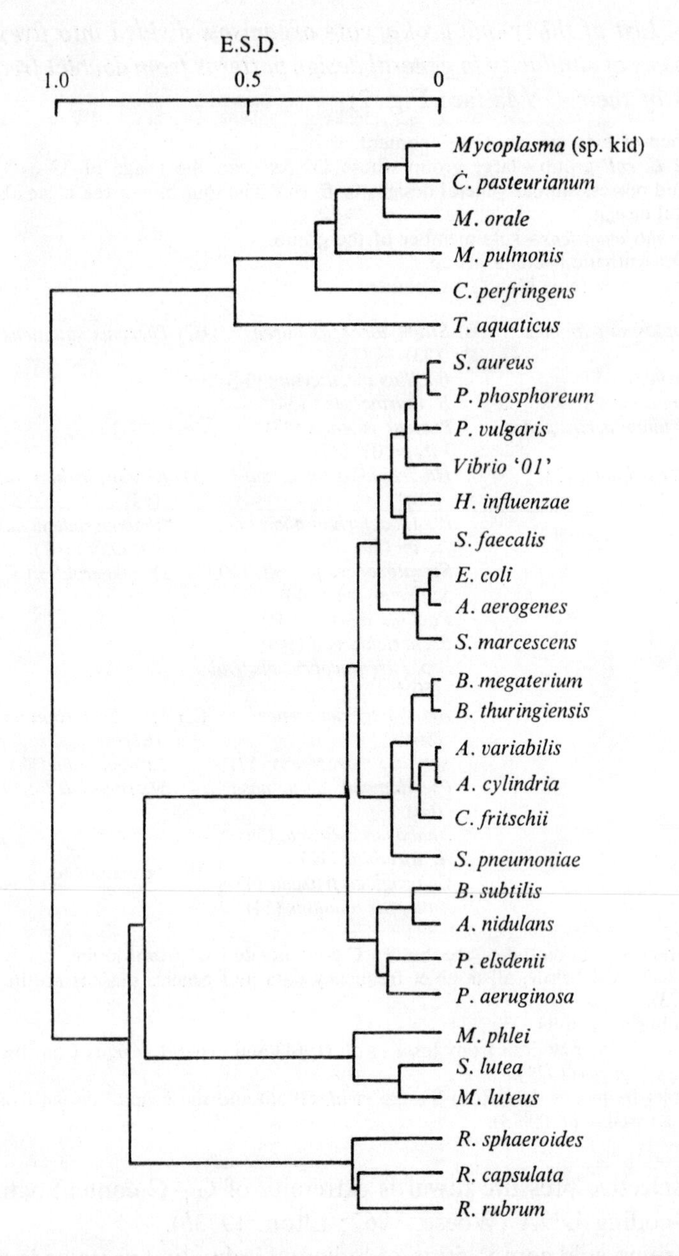

Fig. 2. Cluster analysis of the general designs of thirty-one prokaryote DNAs. The unweighted average linkage method is used (Sokal & Sneath, 1963), with euclidean squared distance as the metric.

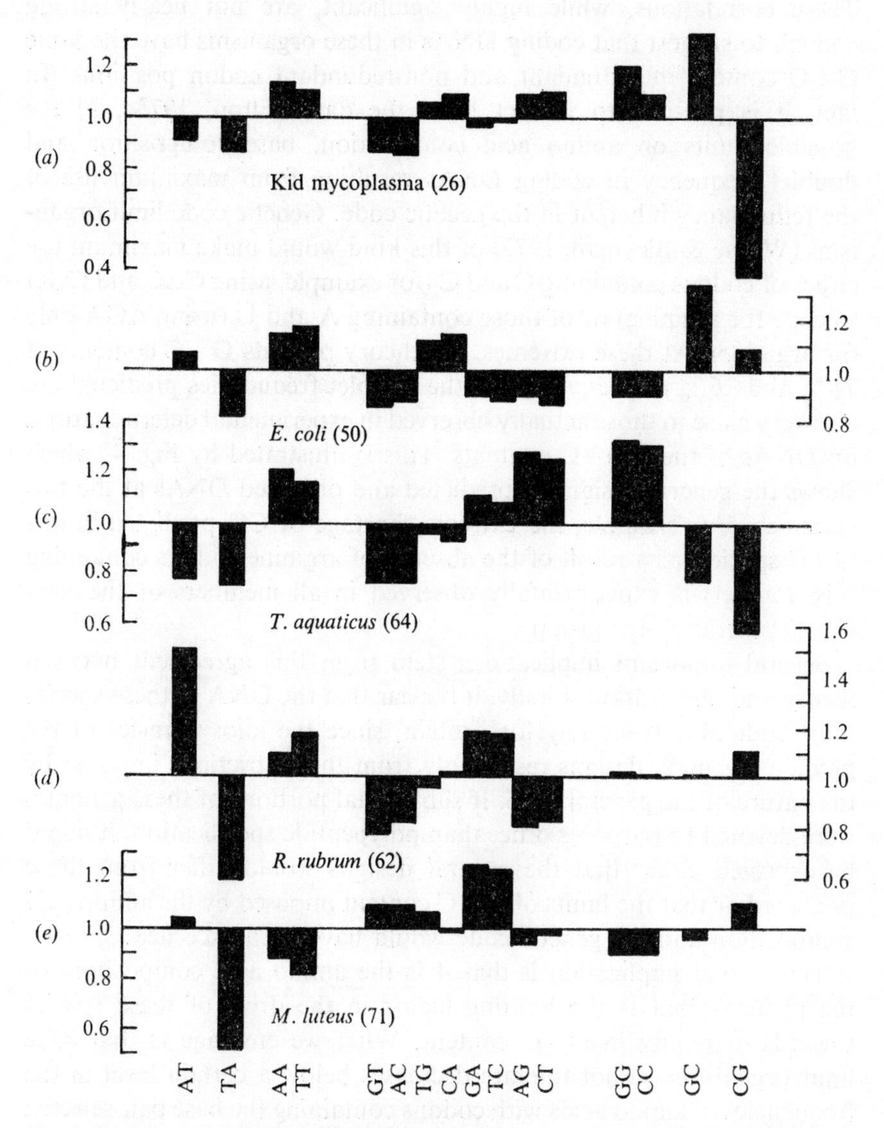

Fig. 3. Representative patterns of the five general design groups of bacterial DNAs.
(*a*) *Mycoplasma* (sp. kid) from the extreme low G+C content group.
(*b*) *Escherichia coli* – primary representative of large group of bacterial DNAs spanning the range 33–64 % G+C content.
(*c*) *Thermus aquaticus* – sole member of group.
(*d*) *Rhodospirillium rubrum* from the photosynthetic bacteria group.
(*e*) *Micrococcus luteus* from the extreme high G+C content group.
Numbers in parentheses are the G+C contents determined from doublet frequency data.

These correlations, while highly significant, are not nearly strong enough to suggest that coding DNAs in these organisms have the same G+C content in redundant and non-redundant codon positions. In fact, it is possible to predict from the data (Elton, 1973*a*, *b*) the possible limits on amino acid composition, base composition and doublet frequency in coding DNAs resulting from maximum use of the redundancy inherent in the genetic code. Genetic code limit organisms (Woese & Bleyman, 1972) of this kind would make maximum use either of codons containing C and G (for example, using CGC and CGG to code for arginine) or of those containing A and U (using AGA only for arginine). At these extremes, the theory predicts G+C contents of 71 % and 26 % respectively, and the doublet frequencies predicted are also very close to those actually observed in experimental determinations on DNAs of these G+C contents. This is illustrated by Fig. 4, which shows the general designs of predicted and observed DNAs at the two extremes. For example, the extreme shortage of CG predicted in low G+C species (as a result of the absence of arginine codons containing this doublet) is experimentally observed in all members of the low-G+C general design group.

Several important implications stem from this agreement between theory and observation. Firstly, it is clear that the DNA of these species must code almost entirely for protein, since the idiosyncrasies of the predicted general designs result only from the restrictions imposed by the nature of the general code. If substantial portions of these genomes were devoted to purposes other than polypeptide specification, it might be expected either that the general designs would differ from those predicted or that the limits of G+C content imposed by the amino acid composition and the genetic code would have been exceeded.

The second implication is that it is the amino acid composition of the proteins that is the limiting factor in the drive of these species towards extremity in G+C content. What we envisage is that code limit organisms cannot tolerate decreases below a certain level in the frequencies of amino acids with codons containing the base pair selected against, and that this limitation on the extremity of amino acid composition accounts for the observed limits on the preponderance of one type of base pair over the other in natural DNAs.

Thus it is clear that selection towards extremes of G+C content exists, and that it operates apparently exclusively through the translation apparatus in moulding the doublet patterns of the DNAs. It is also clear that convergent evolution of the doublet patterns of code limit species has occurred. At the low-G+C end of the scale, not only do

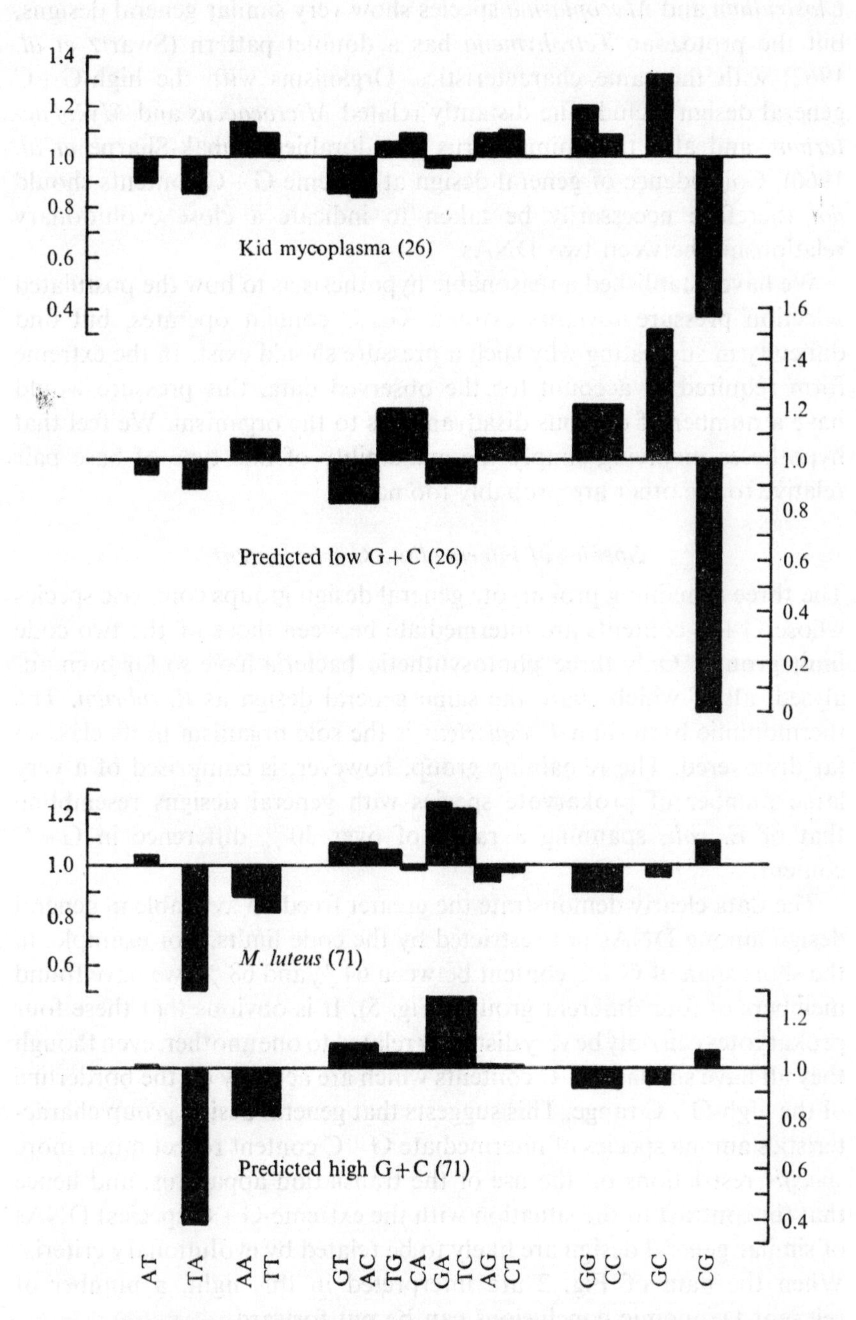

Fig. 4. General designs of predicted and observed DNAs at the two extremes of G+C content. The predicted doublet frequencies are derived by Elton (1973*b*). The numbers in parentheses are the G+C contents.

Clostridium and *Mycoplasma* species show very similar general designs, but the protozoan *Tetrahymena* has a doublet pattern (Swartz *et al.* 1962) with the same characteristics. Organisms with the high-G+C general design include the distantly related *Micrococcus* and *Mycobacterium*, and also the animal virus pseudorabies (Subak-Sharpe *et al.* 1966). Coincidence of general design at extreme G+C contents should *not* therefore necessarily be taken to indicate a close evolutionary relationship between two DNAs.

We have established a reasonable hypothesis as to how the postulated selection pressure towards extreme G+C content operates, but find difficulty in suggesting why such a pressure should exist. In the extreme form required to account for the observed data, this pressure would have a number of obvious disadvantages to the organism. We feel that hypotheses involving simply the availability of one type of base pair relative to the other are probably too naive.

Species of intermediate G+C content

The three remaining prokaryote general design groups comprise species whose G+C contents are intermediate between those of the two code limit groups. Only three photosynthetic bacteria have so far been analysed, all of which share the same general design as *R. rubrum.* The thermophilic bacterium *T. aquaticus* is the sole organism in its class so far discovered. The remaining group, however, is comprised of a very large number of prokaryote species with general designs resembling that of *E. coli*, spanning a range of over 30 % difference in G+C content.

The data clearly demonstrate the greater freedom available in general design among DNAs not restricted by the code limits. For example, in the short span of G+C content between 64 % and 68 %, we have found members of four different groups (Fig. 5). It is obvious that these four prokaryotes can only be very distantly related to one another, even though they all have similar G+C contents which are actually on the borderline of the high-G+C range. This suggests that general design group characteristics among species of intermediate G+C content reflect much more *specific* restrictions on the use of the translation apparatus, and hence that (in contrast to the situation with the extreme-G+C species) DNAs of similar general design are likely to be related by evolutionary criteria. When the data of Fig. 2 are interpreted in this light, a number of relevant taxonomic conclusions can be put forward.

It is encouraging to note first that in several cases congeners (*Mycoplasma, Clostridium, Bacillus, Anabaena*) show close general design simi-

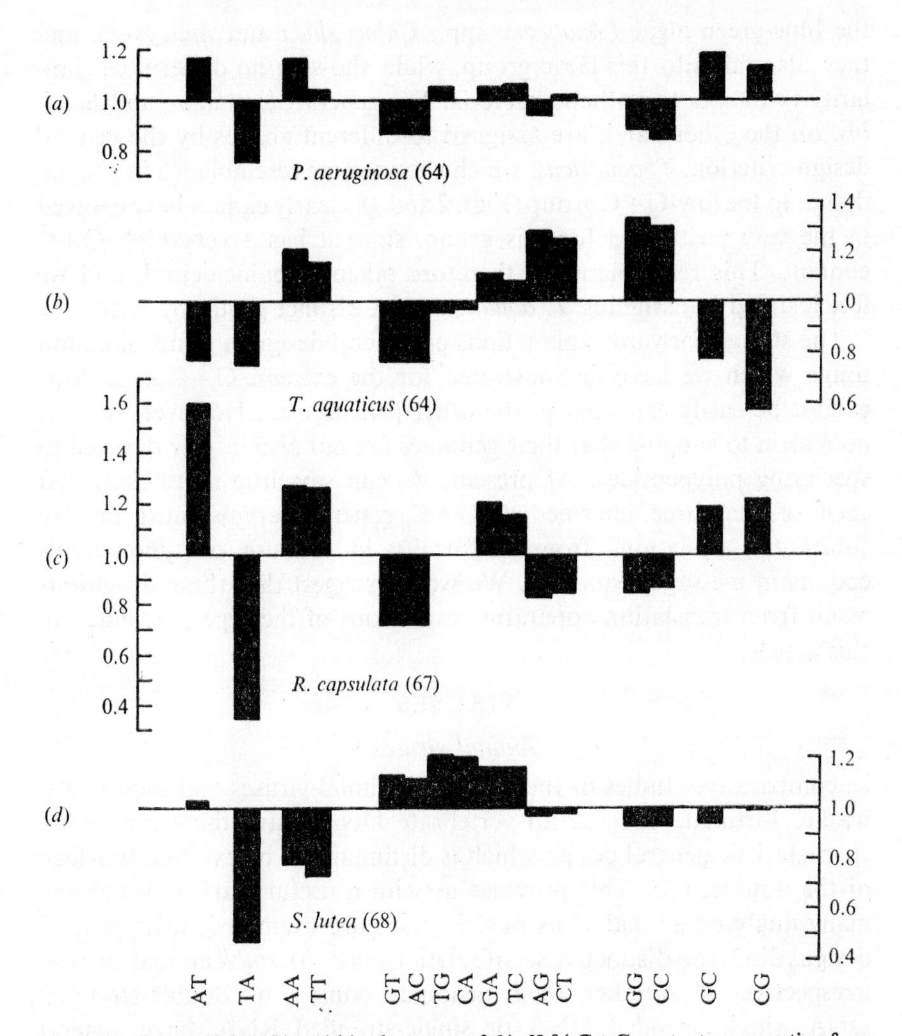

Fig. 5. Four bacterial DNAs within the range of 64–68 % G+C content representing four of the general design groups.

(*a*) *Pseudomonas aeruginosa* from the *E. coli* group.
(*b*) *Thermus aquaticus* – sole member of group.
(*c*) *Rhodopseudomonas capsulata* from the photosynthetic bacteria group.
(*d*) *Sarcina lutea* from the extreme high G+C content group.

Numbers in parentheses are the G+C contents determined from doublet frequency data.

larity; this is in agreement with conventional taxonomy. The Enterobacteriaceae are all placed in the same group, but this also contains a number of other species from different families and even orders. For example, the pseudomonads represented in our sample all belong to the *E. coli* general design group. Four of the DNAs analysed are from the blue-green algae (*Anabaena* spp., *Chlorogloea* and *Anacystis*), and they also fall into this large group, while showing no discernible similarity to the photosynthetic bacteria. The genera *Clostridium* and *Bacillus*, on the other hand, are assigned to different groups by the general design criterion. *T. aquaticus*, which shows some resemblance in general design to the low-G+C group (Figs. 2 and 3), clearly cannot have evolved in the way postulated for this group, since it has a very high G+C content. This resemblance is therefore taken as coincidental, and we feel justified in assigning *T. aquaticus* to a distinct group by itself.

The straightforward explanations of general design in terms of codon usage which we have demonstrated for the extreme-G+C organisms cannot be easily extended to the other prokaryotes. However, there is no reason to suppose that their genomes are not also largely devoted to specifying polypeptides. At present, we can say little more than that each of the three intermediate-G+C general designs must involve substantial deviations from uniformity in the use of synonymous codons in message sequences. We would suggest that these deviations result from translation apparatus restrictions of the type postulated in this article.

VIRUSES

Animal viruses

In comparative studies of the DNAs of animal viruses and their hosts, we are fortunate because all vertebrate DNAs have the same highly characteristic general design which is distinguished by extreme shortage of the doublet CG. This provides us with a useful marker. When the many analysed animal virus nucleic acids are examined, their general designs disclose a distinctive set of relationships. All *small* animal viruses, irrespective of whether their genomes consist of double-stranded DNA, single-stranded DNA or single-stranded RNA, have general designs which resemble, in considerable detail and degree, the highly characteristic general design of the vertebrate host cell DNA. This conclusion covers viruses of the papovavirus group (Subak-Sharpe *et al.* 1966; Morrison *et al.* 1967), the parvovirus group (McGeoch, Crawford & Follett, 1970), the picornavirus group (Hay & Subak-Sharpe, 1968), the myxovirus group (Scholtissek, 1969), the paramyxovirus group (Scholtissek & Rott, 1969) and the leucovirus group (S. Spiegelman, personal communication). In contrast, large animal DNA viruses ($100–160 \times 10^6$ daltons), like members of the herpesvirus group and poxvirus group, possess general designs which show no resemblance whatsoever to that of the host (Subak-Sharpe *et al.* 1966).

The adenoviruses, which are of intermediate size ($19–23 \times 10^6$ daltons), have a general design which shows only some limited resemblance to that of the vertebrate host (Morrison *et al.* 1967).

Plant virus

The DNA plant virus, cauliflower mosaic virus, has also been shown to have a similar general design to its host (Russell *et al.* 1971). However, the similarity is not as close as the animal virus/host relationship and there is no extreme characteristic like the CG shortage. There is also evidence (Shapiro & August, 1965) that tobacco mosaic virus RNA has a different general design from that of its host.

Bacteriophage

The general designs of a number of coliphages have been investigated by Josse *et al.* (1961) and Swartz *et al.* (1962), and also by ourselves (unpublished results). In some cases (ϕX174, T1, lambda), the general designs are similar to that of the normal host bacterium, *E. coli*, while in others (T-even, T5, T7) the resemblance is less marked. As with the animal viruses, there is some correlation between small genome size and identity with host general design, although in this case there is no clear-cut definition into small- and large-size classes. Also, as in the plant system, Shapiro & August (1965) have shown that the small RNA phage f2 has doublet frequencies which give a general design differing from that of the host. However, it is again difficult to make meaningful comparisons when the general designs lack extreme characteristics.

Origin of viruses

As already discussed, the small animal viruses closely resemble their hosts in general design, unlike the larger animal viruses. This prompted us to propose (Subak-Sharpe, 1967, 1969*a*, *b*) that all these small viruses have had their origins from within the polypeptide-specifying stretches of the DNA of ancestral host nuclei or very closely related nucleic acid. It is envisaged that different small viruses could have taken their origin from within the host nucleic acid at any time during the hosts' evolutionary history. This does not imply that these different types of genome all originated from a particular stretch of ancestral host DNA, nor does it mean that the genomes of single- and double-stranded RNA and DNA viruses arose by essentially the same processes or by a common sequence of events. The hypothesis can be taken further to suggest that the DNAs of large mammalian viruses, whose general designs show no discernible resemblance to that of vertebrate DNA, must have originated from different lines of descent. These large viruses probably retain information which allows them to modify their host's translation apparatus in each generation, thus creating a changed internal host environment. Modification of the host translation apparatus could be achieved either by virus-coded transfer RNAs, by modification of the pre-existing cell-specified transfer RNAs through virus-coded enzymes, or by differential stimulation of the host to produce those transfer RNAs needed to redress the balance.

While there is no positive evidence for such changes in the large animal viruses (Bell, Wilkie & Subak-Sharpe, 1971), the large bacterial viruses T4 and T5 do specify or modify transfer RNA (Scherberg & Weiss, 1970; Yudelevich, 1971). However, the argument becomes more difficult to apply in the case of the bacterial viruses, because of the difficulty in contrasting the general designs of the virus and host. Hypotheses about the possible evolution of plant viruses are premature because of the lack of sufficient characteristic data.

Implicit in our inferences about animal viruses has been the assumption that the vertebrate DNA general design reflects the translation apparatus restrictions occurring in the cell. Several lines of circumstantial evidence favour the hypothesis that the part of vertebrate DNA which specifies polypeptides has a general design similar to that observed for the whole DNA. These include so far unpublished analyses of fractionated vertebrate DNAs and also the finding of Bullock & Elton (1972) that dipeptide frequencies in vertebrate proteins indicate a shortage of the doublet CG in messenger RNA sequences. In addition, the animal viruses provide strong evidence themselves. Few people would contest that virtually all the genome in small animal viruses like SV40 or EMC must code for polypeptides (see, for example, discussions by Walter, Roblin & Dulbecco, 1972, or Laskey, Gurdon & Crawford, 1972). Thus the general designs of the small viruses are of polypeptide-specifying nucleic acid, implying that those of all vertebrate DNAs are also essentially of polypeptide-specifying vertebrate DNA.

DISCUSSION

In the light of these results, how does doublet analysis emerge as a tool for discerning evolutionary relationships? The first point that has been established is the stability of the general design. This persuades us that comparisons of general designs provide a good criterion for discerning ancient relationships. In the prokaryotes, we already recognise at least five major general design classes which can only be distantly related to one another. For example, we feel that the differences in the general designs of the blue-green algae and the photosynthetic bacteria indicate that they cannot have a recent common evolutionary origin.

There are two main exceptions to the converse conclusion that organisms which have similar general designs are related. The first of these is in the extreme-G+C case, where we have shown that organisms falling within one class are not necessarily closely related but may have acquired the same general design by convergent evolution. Secondly, coincidental similarity in general design may occur, as illustrated by the example of *T. aquaticus* and the low-G+C group. This type of coincidence occurs frequently when attempts are made to infer taxonomic relationships from G+C content alone. When classifications are made according to a parameter with just one degree of freedom, it is to be expected that unrelated species will often happen to have similar parameter values. The general design of a double-stranded DNA has essentially six degrees of freedom, which very much reduces the chances of purely fortuitous coincidence. Nevertheless, this possibility should be borne in mind when drawing conclusions about evolutionary relationships.

The first objection cannot hold true for the *E. coli* group, since it has intermediate G+C content. We also feel that the large number of species showing this general design argues strongly against more than a small proportion of the members of the group having evolved this pattern coincidentally. This therefore suggests that members of this group are more closely related to one another in evolutionary terms than to members of other classes. Another type of general design similarity has been shown in the small animal viruses and their hosts. Here it is argued that the similarity is due to direct relationship arising from the evolution of these viruses from within the polypeptide-specifying regions of the ancestral host genome as opposed to the large DNA viruses where origin from different lines of descent is inferred.

Finally, one general point should perhaps be made about the basic concept underlying this work. Selection, of course, ultimately takes place at the level of the whole individual organism. Selective advantage or disadvantage is measured in terms of sum total survival potential encoded in the organism's genetic information, most of which is finally expressed in terms of active or functional polypeptides. We consider natural selection to be working by refining and improving the biological survival value of these polypeptides to the organism as a whole. What we are envisaging is that at a different level the translation apparatus must, of necessity, exert a restraint on the possible responses of the organism to selection on its protein constitution. The translation apparatus, because its flexibility is limited, must have generated a special and consistent selection pressure of its own. However, because of the degeneracy of the genetic code, this is unlikely to have caused real problems at the level of response to natural selection of the organism in terms of its polypeptides. Instead, it will have resulted in a stable and entrenched characteristic use made by the organism of the possibilities offered by the degenerate genetic code. As a consequence, response to this selection pressure will have primarily affected the general design of the organism's DNA, and only in a secondary way is it likely to have carried over its effect by changing the amino acid sequence of polypeptides. This secondary effect has actually been detected in vertebrates as a non-randomness in the distribution of certain dipeptide sequences to avoid using the doublet CG for coding for proteins (Bullock & Elton, 1972). It should also be pointed out that the transfer RNAs (change in any of which would immediately affect the translation of virtually all the proteins coded by the organism) must, of necessity, be very limited in their ability to accept mutational changes without serious consequences for the organism. It may well be for this reason that many of the differences that have been observed are in terms of modified bases rather than normal base substitutions.

With regard to future work, there are a number of obvious lines along which these investigations should be extended, some of which are listed here:

1. Analysis of many more bacteria should confirm the existing groupings and investigate the possibility of any further general design classes. Attempts will be made to isolate and analyse single strands of prokaryote DNAs. It is hoped that this will provide further clues as to codon utilisation, especially in the *E. coli* group.

2. The present coverage of bacterial and plant viruses is inadequate, and should also be extended. In addition, analysis of the viruses of eukaryotes other than the vertebrates should provide much useful information as to the host/virus relationship. A further question that should be tested is whether sub-fractions of the DNAs of tumour viruses like the adenoviruses all have the same general design, or whether a part of the adenovirus genome has a design which is much more closely related to that of the host. This might be expected where this part of the DNA is involved in integration.

3. Similar investigations on fractionated DNAs from the much more complex vertebrate genome are in progress (unpublished results). We hope to extend these to gain more insight into the relation of sequence characteristics to function in DNA.

4. A number of different ways of elucidating DNA sequence properties are currently being used in many laboratories. It is hoped that the doublet analysis technique itself can be further developed (specifically, to yield information on triplet frequencies) to complement these other methods in extending our knowledge of the way evolutionary forces act on the genetic material.

REFERENCES

BELL, D., WILKIE, N. M. & SUBAK-SHARPE, J. H. (1971). Studies on arginyl transfer ribonucleic acid in herpes virus infected baby hamster kidney cells. *Journal of General Virology*, **13**, 463–75.

BULLOCK, E. & ELTON, R. A. (1972). Dipeptide frequencies in proteins and the CpG deficiency in vertebrate DNA. *Journal of Molecular Evolution*, **1**, 315–25.

ELTON, R. A. (1973*a*). The relationship of DNA base composition and individual protein composition in micro-organisms. *Journal of Molecular Evolution*, **2**, 263–76.

ELTON, R. A. (1973*b*). Doublet frequencies in the DNA of genetic code limit organisms. *Journal of Molecular Evolution*, **2**, 293–302.

HAY, J. & SUBAK-SHARPE, H. (1968). Analysis of nearest neighbour base frequencies in the RNA of a mammalian virus: encephalomyocarditis virus. *Journal of General Virology*, **2**, 469–72.

JOSSE, J., KAISER, A. D. & KORNBERG, A. (1961). Enzymatic synthesis of deoxyribonucleic acid. VIII. Frequencies of nearest-neighbour base sequences in deoxyribonucleic acid. *Journal of Biological Chemistry*, **236**, 864–75.

JOSSE, J. & SWARTZ, M. (1963). Determination of frequencies of nearest-neighbour base sequences in DNA. *Methods in Enzymology*, **6**, 739–51.

LASKEY, R. A., GURDON, J. B. & CRAWFORD, L. V. (1972). Translation of encephalomyocarditis viral RNA in oocytes of *Xenopus laevis*. *Proceedings of the National Academy of Sciences, USA*, **69**, 3665–9.

MCGEOCH, D. J., CRAWFORD, L. V. & FOLLETT, E. A. C. (1970). The DNAs of three parvoviruses. *Journal of General Virology*, **6**, 33–40.

MAITRA, U., COHEN, S. N. & HURWITZ, J. (1966). Specificity of initiation and synthesis of RNA from DNA templates. *Cold Spring Harbor Symposia on Quantitative Biology*, **31**, 113–22.

MORRISON, J. M., KEIR, H. M., SUBAK-SHARPE, H. & CRAWFORD, L. V. (1967). Nearest neighbour base sequence analysis of the deoxyribonucleic acids of a further three mammalian viruses: simian virus 40, human papilloma virus and adenovirus type 2. *Journal of General Virology*, **1**, 101–8.

RUSSELL, G. J., FOLLETT, E. A. C., SUBAK-SHARPE, J. H. & HARRISON, B. D. (1971). The double-stranded DNA of cauliflower mosaic virus. *Journal of General Virology*, **11**, 129–38.

RUSSELL, G. J., MCGEOCH, D. J., ELTON, R. A. & SUBAK-SHARPE, J. H. (1973). Doublet frequency analysis of bacterial DNAs. *Journal of Molecular Evolution*, **2**, 277–92.

SCHERBERG, N. H. & WEISS, S. B. (1970). Detection of bacteriophage T4- and T5-coded transfer RNAs. *Proceedings of the National Academy of Sciences, USA*, **67**, 1164–71.

SCHOLTISSEK, C. (1969). Synthesis *in vitro* of RNA complementary to parental viral RNA by RNA polymerase induced by influenza virus. *Biochimica et Biophysica Acta*, **179**, 389–97.

Scholtissek, C. & Rott, R. (1969). Ribonucleic acid nucleotidyl transferase induced in chick fibroblasts after infection with Newcastle disease virus. *Journal of General Virology*, **4**, 565–70.

Shapiro, L. & August, J. T. (1965). Replication of RNA viruses. II. The RNA product of a reaction catalyzed by a viral RNA-dependent RNA polymerase. *Journal of Molecular Biology*, **11**, 272–84.

Skalka, A., Fowler, A. V. & Hurwitz, J. (1966). The effect of histones on the enzymatic synthesis of ribonucleic acid. *Journal of Biological Chemistry*, **241**, 588–96.

Sokal, R. R. & Sneath, P. H. A. (1963). *Principles of Numerical Taxonomy*, p. 182. San Francisco: Freeman.

Subak-Sharpe, J. H. (1967). Doublet patterns and evolution of viruses. *British Medical Bulletin*, **23**, 161–8.

Subak-Sharpe, J. H. (1969*a*). The doublet pattern of the nucleic acid in relation to the origin of viruses. In *Handbook of Molecular Cytology*, ed. A. Lima-de-Faria, pp. 67–87. Amsterdam: North-Holland.

Subak-Sharpe, J. H. (1969*b*). The doublet pattern of the nucleic acid of oncogenic and non-oncogenic viruses and its relationship to that of mammalian DNA. In *Proceedings of the 8th Canadian Cancer Research Conference*, ed. J. F. Morgan, pp. 242–60. Oxford: Pergamon Press.

Subak-Sharpe, H., Burk, R. R., Crawford, L. V., Morrison, J. M., Hay, J. & Keir, H. M. (1966). An approach to evolutionary relationship of mammalian DNA viruses through analysis of the pattern of nearest neighbour base sequence. *Cold Spring Harbor Symposia on Quantitative Biology*, **31**, 737–48.

Sueoka, N. (1961). Correlation between base composition of deoxyribonucleic acid and amino acid composition of protein. *Proceedings of the National Academy of Sciences, USA*, **47**, 1141–9.

Swartz, M. N., Trautner, T. A. & Kornberg, A. (1962). Enzymatic synthesis of deoxyribonucleic acid. XI. Further studies on nearest neighbour base sequences in deoxyribonucleic acid. *Journal of Biological Chemistry*, **237**, 1961–7.

Walter, G., Roblin, R. & Dulbecco, R. (1972). Protein synthesis in simian virus 40-infected monkey cells. *Proceedings of the National Academy of Sciences, USA*, **69**, 921–4.

Woese, C. R. (1967). *The Genetic Code*, p. 70. New York: Harper and Row.

Woese, C. R. & Bleyman, M. A. (1972). Genetic code limit organisms – do they exist? *Journal of Molecular Evolution*, **1**, 223–9.

Yudelevich, A. (1971). Specific cleavage of an *Escherichia coli* leucine transfer RNA following bacteriophage T4 infection. *Journal of Molecular Biology*, **60**, 21–9.

ENZYME FAMILIES

B. S. HARTLEY
Medical Research Council, Laboratory of Molecular Biology, Hills Road, Cambridge CB2 2QH

Lessons from enzyme structure

Over the last decade, determination of the tertiary structures of several enzymes has initiated a revolution in biochemistry. These structures have been carefully studied for what they can tell us about the specificity and catalytic activity of enzymes, but we have only just begun to explore some of the other evidence that they offer. They have a lot to teach us about the folding rules by which a particular amino acid sequence adopts a particular tertiary structure, but they also offer lessons about more traditional biological problems: for example, the molecular basis of evolution. Myoglobin and the haemoglobins have always been considered close relatives because of their similar physiological activity. Nevertheless, I well remember the excitement that I felt in a seminar in the Cavendish Laboratory, Cambridge, in 1961 when it first became apparent that the tertiary structures of the α- and β-subunits of haemoglobin and that of myoglobin were essentially similar (Perutz, 1963). As the primary structures of these proteins were gradually determined it became apparent that a surprising degree of difference in amino acid sequence was consistent with this conservation of conformation. It looked as though natural selection had conserved tertiary structure rather than primary structure.

How far can we push this lesson gained from the oxygen binding proteins: that relationships in mechanism are reflected in anatomical similarity? The 'serine proteases' were so named because a serine residue in their catalytic sites reacts uniquely with organo-phosphorus compounds like di-isopropyl phosphorofluoridate or sulphonyl fluorides such as phenylmethane sulphonyl fluoride. This mechanistic identity is not paralleled by closely similar specificities. Table 1 shows a selection of mammalian serine proteases. Five pancreatic genes specify zymogens of serine proteases that are secreted into the intestine via pancreatic juice and there triggered by the activation of trypsinogen. Three of the enzymes are chymotrypsins with broad specificity for aromatic side chains, whereas trypsin prefers basic residues and elastase splits bonds adjacent to small hydrophobic residues. The liver secretes into plasma zymogens of serine proteases concerned with the complex cascade

Table 1. *Mammalian serine proteases*

Enzyme	Specificity	% Sequence homologies
Pancreas		
Chymotrypsin-A	Phe-, Tyr-, Trp-, Leu-	53
Chymotrypsin-B		49
Chymotrypsin-C		?
Trypsin	Lys-, Arg-	100
Elastase	Ala-, Val-	48
Plasma		
Thrombin	Fibrinogen	38
Factor Xa	Prothrombin, Lys-, Arg-	50
Plasmin	Fibrin, Lys-	50

Homologies represent % chemical similarities with respect to bovine trypsin for sequences aligned according to Hartley *et al.* (1972). All enzymes are bovine except for elastase (porcine) and plasmin (human). Factor Xa comparisons (Titani *et al.* 1972) are for the N-terminal 37 residues and a 25 residue active centre serine sequence. For plasmin (Robbins, Bernabe, Arzadon & Summaria, 1972) the comparison is for the N-terminal 20 residues.

conversion of fibrinogen to fibrin in blood clotting. These enzymes are of high specificity: thrombin appears to act only on fibrinogen and Factor Xa only on prothrombin, although both can split small synthetic lysyl or arginyl amides. Platelets produce plasminogen, the activation of which to plasmin dissolves blood clots. Its specificity is predominantly towards lysyl bonds (Iwanago *et al.* 1967).

We would expect structural similarities in these enzymes because of their similar catalytic mechanisms, but how much difference is needed to accommodate the different specificities? The primary structures offer only small clues. The isoenzymes chymotrypsin-A and -B are identical in 78 % of their sequences but Table 1 shows that the homology drops to around 50 % when we compare enzymes of different specificity. Nor is there any significant difference between the pancreatic enzymes and the plasma enzymes. All are clearly members of a single family, but the sequence changes could allow appreciable differences in tertiary structure. It is clear that primary structures can tell us whether proteins belong to a closely related family, but the divergence that has been selected by the evolution of new enzyme specificities will be apparent only at the three-dimensional level.

Determination of the tertiary structure of bovine α-chymotrypsin (Matthews, Sigler, Henderson & Blow, 1967; Birktoft & Blow, 1972) put these sequence homologies in a new light. Of the side chains that are internal in chymotrypsin, 60 % are identical or chemically similar in trypsin, elastase and thrombin, whereas only 10 % of the surface residues are so conserved (Hartley & Shotton, 1971). Model building seemed justified on the assumption that the conformations of the chains were identical and this revealed no structural problems (Hartley, 1970*b*). The insertions and deletions that were necessary to maximise the sequence homologies lay at external loops of the chain, and no distortion of the chain was needed to accommodate the new internal residues.

Subsequently the tertiary structures of bovine trypsin (Stroud, Kay & Dickerson, 1971) and porcine elastase (Shotton & Watson, 1970) have been determined to high resolution. These structures confirm the essential conclusions of the model building. In Table 2 the sequences have been aligned according to a computer analysis by D. M. Blow (personal communication) of the α-C co-ordinates in these three enzymes. Residues vertically beneath each other have α-C atoms within 3 Å when the structures are superimposed and rotated to closest fit. It is clear that the conformations of the chains are amazingly similar; deletions and insertions cluster in areas where the chain travels along the external surface. Trypsin is the smallest enzyme and Fig. 1 demonstrates the way in which the deletions relative to chymotrypsin are distributed.

We are now able to examine the changes that evolution has wrought on these molecules and to try to relate these to the biological activity. As expected, we see an identical arrangement of the catalytic groups involved in the charge-relay mechanism, Asp-102, His-57 and Ser-195, but the changes which explain the wide variations in specificity are surprisingly small! Of those residues in contact with the substrate only Ser-189 of chymotrypsin changes to Asp in trypsin, allowing lysyl or arginyl side chains to form a salt bridge within the substrate binding pocket. In elastase Gly-216 becomes Val- and Gly-226 becomes Thr-. These bulky side chains block the entrance to the binding pocket, allowing entry only to substrates with small hydrophobic side chains. It looks as though a single mutation would be sufficient to convert chymotrypsin to trypsin and two to change chymotrypsin to elastase.

Why then is the surface away from the binding site more than 70 % different in these enzymes? Some of this surface is involved in binding the activation peptides of trypsinogen and pro-elastase and the A-chain of chymotrypsin – the external loop between residues 36–38 of chymotrypsin which is deleted in trypsin is one example – but it is difficult to find structural or catalytic rationalisations for most of the rest of the surface.

The concept of neutral amino acid substitutions that have been fixed by 'genetic drift' has been put forward to explain sequence differences

Table 2. *Structural alignment of pancreatic proteases*

	16	17	18	19	20	21	22	23	24	25		26	27	28	29	30	31		32	33
CA	ILE	VAL	Asn	GLY	Glu	GLU	ALA	Val	Pro	Gly	----	SER	TRP	PRO	TRP	GLN	VAL	----	SER	LEU
	7	8	9	10	11	12	13	14	15	16		17	18	19	20	21	22		23	24
T	ILE	VAL	GLY	GLY	Tyr	Thr	Cys	Gly	Ala	Asn	----	THR	Val	PRO	TYR	GLN	VAL	----	SER	LEU
	1	2	3	4	5	6	7	8		9	10	11	12	13	14		15	16	17	18
E	VAL	VAL	GLY	GLY	Thr	GLU	ALA	Gln	----	Arg	Asn	SER	TRP	PRO	Ser	----	GLN	ILE	SER	LEU
	34	35	36	37				38	39	40	41	42	43	44	45	46	47	48	49	50
CA	GLN	Asp	LYS	THR	----	----	----	GLY	PHE	HIS	PHE	CYS	GLY	GLY	SER	LEU	ILE	ASN	GLU	ASN
	25	26						27	28	29	30	31	32	33	34	35	36	37	38	39
T	ASN	Ser	----	----	----	----	----	GLY	TYR	HIS	PHE	CYS	GLY	GLY	SER	LEU	ILE	ASN	Ser	GLN
	19	20	21	22	23	24	25	26	27	28	29	30	31	32	33	34	35	36	37	38
E	GLN	Tyr	ARG	SER	Gly	Ser	Ser	Trp	Ala	HIS	Thr	CYS	GLY	GLY	THR	LEU	ILE	Arg	GLN	ASN
	51	52	53	54	55	56	57	58	59	60			61	62	63	64	65		66	67
CA	TRP	VAL	VAL	THR	ALA	ALA	HIS	CYS	Gly	VAL	----	----	THR	Thr	SER	Asp	VAL	----	VAL	VAL
	40	41	42	43	44	45	46	47	48	49			50	51	52	53	54		55	56
T	TRP	VAL	VAL	SER	ALA	ALA	HIS	CYS	Tyr	Lys	----	----	SER	Gly	Ile	Gln	VAL	----	Arg	Leu
	39	40	41	42	43	44	45	46		47	48	49	50	51	52	53	54	55	56	57
E	TRP	VAL	Met	THR	ALA	ALA	HIS	CYS	----	VAL	Asp	Arg	Glu	Leu	THR	Phe	Arg	Val	VAL	VAL
	68	69	70	71	72	73	74	75	76	77	78	79	80	81		82	83	84	85	86
CA	Ala	GLY	GLU	Phe	ASP	Gln	Gly	Ser	Ser	Ser	GLU	Lys	Ile	GLN	----	Lys	LEU	Lys	ILE	Ala
		57	58	59	60	61	62	63		64	65	66	67	68	69	70	71	72	73	74
T	----	GLY	GLN	Asp	ASN	ILE	ASN	Val	----	Val	GLU	GLY	Asn	GLN	GLN	PHE	ILE	Ser	Ala	Ser
		58	59	60	61	62	63	64		65	66	67	68	69	70	71	72	73	74	
E	----	GLY	GLU	His	ASN	LEU	ASN	Gln	----	Asn	Asn	GLY	Thr	GLU	GLN	TYR	VAL	Gly	VAL	----
	87		88	89	90	91	92	93	94	95	96		97	98		99	100	101	102	103
CA	LYS	----	VAL	Phe	Lys	Asn	Ser	Lys	TYR	ASN	SER	----	Leu	THR	----	ILE	ASN	ASN	ASP	ILE
	75		76	77	78	79	80	81	82	83	84		85	86		87	88	89	90	91
T	LYS	----	Ser	ILE	VAL	HIS	PRO	Ser	TYR	ASN	SER	----	ASN	THR	----	LEU	ASN	ASN	ASP	ILE
	75	76	77	78	79	80	81	82	83	84	85	86	87	88	89	90	91	92	93	94
E	Gln	Lys	ILE	VAL	VAL	HIS	PRO	Tyr	TRP	ASN	THR	Asp	ASP	Val	Ala	Ala	Gly	Tyr	ASP	ILE
	104	105	106	107	108	109	110	111	112	113	114	115	116	117	118	119	120	121	122	123
CA	Thr	LEU	LEU	LYS	LEU	Ser	THR	ALA	ALA	SER	Phe	SER	GLN	Thr	VAL	Ser	Ala	VAL	Cys	LEU
	92	93	94	95	96	97	98	99	100	101	102	103	104	105	106	107	108	109	110	111
T	Met	LEU	ILE	LYS	LEU	Lys	SER	ALA	ALA	SER	LEU	ASN	SER	Arg	VAL	Ala	Ser	ILE	Ser	LEU
	95	96	97	98	99	100	101	102	103	104	105	106	107	108	109	110	111	112	113	114
E	Ala	LEU	LEU	ARG	LEU	Ala	Gln	Ser	Val	THR	LEU	ASN	SER	Tyr	VAL	Gln	Leu	Gly	Val	LEU
	124	125	126	127	128	129	130	131	132	133	134	135	136	137	138	139	140	141	142	143
CA	PRO	SER	ALA	SER	Asp	Asp	Phe	ALA	Ala	Gly	Thr	Thr	CYS	Val	Thr	THR	GLY	TRP	GLY	LEU
	112	113		114	115	116	117	118	119	120	121		122	123	124	125	126	127	128	129
T	PRO	THR	----	SER	Cys	Ala	Ser	ALA	Gly	Thr	Gln	----	CYS	Leu	ILE	SER	GLY	TRP	GLY	Asn
	115	116	117	118	119	120	121	122	123	124	125	126	127	128	129	130	131	132	133	134
E	PRO	Arg	ALA	Gly	Thr	Ile	Leu	ALA	Asn	Asn	Ser	Pro	CYS	Tyr	ILE	THR	GLY	TRP	GLY	LEU

Table 2 (*cont.*)

```
   144 145 146 147 148 149 150      151 152 153 154 155 156 157 158 159 160 161 162
CA THR-ARG-Tyr-THR-ASN-Ala-ASN-----Thr-PRO-ASP-Arg-LEU-GLN-GLN-ALA-Ser-LEU-PRO-LEU-

   130 131 132      133 134 135 136 137 138 139 140 141 142 143 144 145 146 147 148
T  THR-LYS-SER-----Ser-GLY-Thr-Ser-Tyr-PRO-ASP-Val-LEU-Lys-Cys-Leu-Lys-Ala-PRO-ILE-

   135 136 137 138 139 140  141   142 143 144     145 146 147 148 149 150 151 152
E  THR-ARG-THR-ASN-GLY-GLN---Leu---Ala-Gln-Thr-----LEU-GLN-GLN-ALA-Tyr-LEU-PRO-Thr-

   163 164 165 166 167 168 169 170              171 172 173 174 175 176 177 178 179 180
CA LEU-SER-ASN-THR-Asn-CYS-LYS-Lys---------TYR-TRP-GLY-THR-Lys-ILE-LYS-ASP-Ala-MET-

   149 150 151 152 153 154 155 156              157 158 159 160 161 162 163 164 165 166
T  LEU-SER-ASN-SER-Ser-CYS-LYS-SER---------Ala-TYR-Pro-Gly-Gln-ILE-Thr-Ser-Asn-MET-

   153 154 155 156 157 158 159 160 161 162 163 164 165 166 167 168 169 170 171 172
E  Val-Asp-Tyr-Ala-Ile-CYS-Ser-SER-Ser-Ser-TYR-TRP-GLY-SER-Thr-VAL-LYS-ASN-Ser-MET-

   181 182 183 184  185    186      187    188 189 190 191     192 193 194 195 196 197
CA ILE-CYS-ALA-GLY--Ala---Ser-----GLY---Val-Ser-SER-CYS---Met-GLY-ASP-SER-GLY-GLY-

   167 168 169 170  171 172 173 174 175 176 177 178 179     180 181 182 183 184 185
T  Phe-CYS-ALA-GLY--Tyr-Leu-Glu-GLY-Gly-LYS-Asp-SER-CYS---GLN-GLY-ASP-SER-GLY-GLY-

    173 174 175 176 177      178 179 180  181  182 183 184 185 186 187 188 189 190
E  -VAL-CYS-ALA-GLY-Gly-----Asn-GLY-Val--ARG--SER-GLY-CYS-GLN-GLY-ASP-SER-GLY-GLY-

   198 199 200    201 202 203 204 205 206 207 208 209 210  211  212 213 214 215 216
CA PRO-LEU-VAL---CYS-Lys-Lys-ASN-GLY-Ala-TRP-Thr-LEU-Val--GLY--ILE-VAL-SER-TRP-GLY-

   186 187 188    189 190                    191 192 193 194  195  196 197 198 199 200
T  PRO-Val-VAL---CYS-Ser-----------------Gly-Lys-LEU-Gln--GLY--ILE-VAL-SER-TRP-GLY-

   191 192 193 194 195 196 197  198  199 200 201 202    203 204 205 206 207 208 209
E  PRO-LEU-His-CYS-Leu-Val-ASN--GLY--Gln-TYR-Ala-Val---His-GLY-VAL-Thr-SER-PHE-Val-

   217 218     219 220      221      222 223 224      225 226 227 228 229 230 231 232
CA SER-Ser-----Thr-CYS-----Ser-----THR-Ser-Thr-----PRO-GLY-VAL-TYR-Ala-ARG-VAL-THR-

   201 202             203 204 205      206 207 208      209 210 211 212 213 214 215 216
T  SER-Gly---------CYS-Ala-GLN-----Lys-Asn-LYS-----PRO-GLY-VAL-TYR-THR-LYS-VAL-Cys-

   210 211 212 213 214      215 216 217      218 219 220 221 222 223 224 225 226 227
E  SER-Arg-Leu-Gly-CYS-----ASN-Val-THR-----ARG-Lys-PRO-Thr-VAL-PHE-THR-ARG-VAL-SER-

   233 234 235 236 237 238 239 240 241 242 243 244 245
CA ALA-Leu-VAL-Asn-TRP-VAL-GLN-GLN-THR-LEU-ALA-Ala-ASN

   217 218 219 220 221 222 223 224 225 226 227 228 229
T  Asn-TYR-VAL-SER-TRP-ILE-Lys-GLN-THR-ILE-ALA-SER-ASN

   228 229 230 231 232 233 234 235 236 237 238 239 240
E  ALA-TYR-ILE-SER-TRP-ILE-ASN-ASN-VAL-ILE-ALA-SER-ASN
```

The sequences described in Hartley & Shotton (1971) have been aligned according to a computer calculation of D. M. Blow (private communication) in which the α-carbon co-ordinates of bovine trypsin (T) (Stroud, Kay & Dickerson, 1971) and porcine elastase (E) (Shotton & Watson, 1970) are rotated to give optimum fit. Residues which lie vertically beneath each other have α-carbon co-ordinates with 3 Å. Chemically similar residues are in capitals and identities are underlined. Numbers underlined show side chains in chymotrypsin that are partially buried and double underlining shows completely buried side chains.

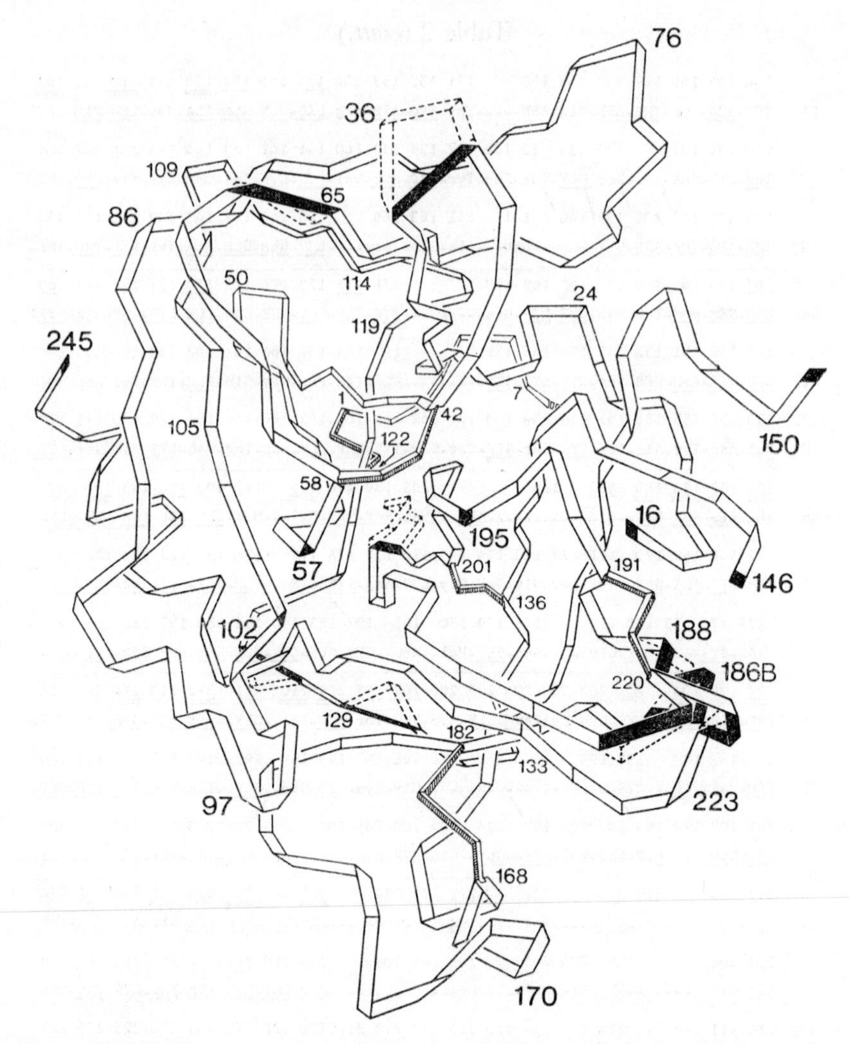

Fig. 1. Conformation of trypsin and chymotrypsin. Creases in the tape show α-carbon positions. Dotted lines show peptide bonds in chymotrypsin that have been changed in trypsin (shaded).

in a given protein which apparently conserves exact function in several different species (Kimura, 1969; King & Jukes, 1969). Cytochrome *c* is a favourite example because of the large number of sequences that are available from widely separated species (Smith, 1970). Opponents of 'neutral mutations' point out that selective advantage must reside in factors other than catalytic activity. In the case of cytochrome *c*, for example, interactions with the appropriate electron donors and acceptors in an ordered membrane must select a great deal of the surface in a fashion that would escape normal assay methods. The surface of intracellular enzymes must be under selective constraints to bind or to avoid binding other cellular components, but the surface of the pancreatic enzymes cannot be moulded by such specific factors, since they operate in a relatively uncontrolled environment. Barnard, Cohen, Gold & Kim (1972) have pointed out that surface residues, as well as internal residues, are subject to protein-folding constraints, since changes may influence the relative stability of alternative folding arrangements or introduce 'kinetic traps' into the physiological folding process. There is much force in this argument, but it cannot be a potent factor for every amino acid residue since it is possible to denature and renature chemically modified proteins such as polyalanyl ribonuclease (Anfinsen, 1966). Selective factors may also operate at the DNA and mRNA level, in order to maintain their necessary structures. Despite these caveats, however, it is difficult to believe that all of the residues in these enzymes are uniquely appropriate for their biological role.

Some of the changes in the interiors of these enzymes are, nevertheless, difficult to square with the hypothesis of stepwise neutral mutation. Most of these changes in the pancreatic proteases are extremely conservative, as I have already pointed out. Fig. 2, however, shows a cluster of side chains that form an internal hydrophobic globule which appears to dictate the folding of the chains which wrap around it. As we have seen, the conformations of the chains are identical in trypsin, chymotrypsin and elastase, and as expected the side chains are almost identical in trypsin, chymotrypsin and thrombin. In elastase, however, some radical changes have occurred in adjacent residues in part of this cluster. An alternative answer to the packing problem appears to have evolved by multiple compensating mutations. Any single-step pathway to this alternative arrangement would leave holes or excrescences that would surely distort the chain and upset the folding interactions. It is therefore difficult to see how the protein could evolve except through inactive intermediates.

Gene-doubling in evolution

To explain these phenomena, I return to an argument that I advanced some years ago (Hartley, 1966) (Fig. 3). Most mutations in a single gene will be disadvantageous and screened out by natural selection. Rare 'neutral' or advantageous mutations may eventually appear as species differences. If a family of enzymes has a common ancestor, gene multiplication of the ancestral gene must be the first evolutionary step, induced perhaps by a need for higher enzyme activity. If this selective pressure relaxes, subsequent 'lethal' mutations in one copy can be

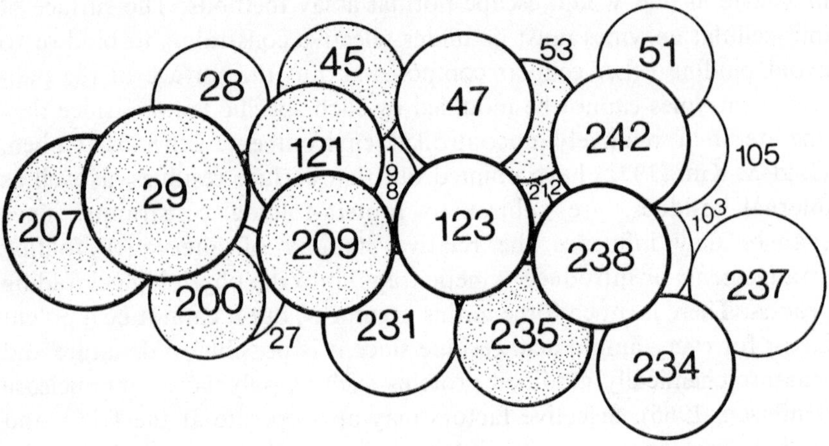

	C	T	Th	E
237	Trp	Trp	Trp	Trp
234	**LEU**	Tyr	**LYS**	Tyr
238	**VAL**	Ile	Ile	Ile
242	LEU	Ile	Ile	Ile
51	Trp	Trp	Trp	Trp
105	Leu	Leu	Leu	Leu
103	Ile	Ile	Ile	Ile

	C	T	Th	E
235	Val	Val	LEU	ILE
123	Leu	Leu	Leu	Leu
47	Ile	Ile	Ile	Ile
53	Val	Val	LEU	**MET**
212	Ile	Ile	Ile	VAL
231	Val	Val	Val	Val

	C	T	Th	E
209	Leu	Leu	**GLN**	VAL
121	Val	Val	Val	**GLY**
45	Ser	Ser	Ser	THR
198	Pro	Pro	Pro	Pro
200	Val	Val	Val	**HIS**

	C	T	Th	E
29	Trp	Trp	Trp	**SER**
28	Pro	Pro	Pro	Pro
27	Trp	**VAL**	**SER**	Trp
207	Trp	—	Trp	**TYR**

Fig. 2. Hydrophobic core of serine proteases. The spheres roughly represent contacts between side chains that are completely or partially buried in bovine α-chymotrypsin (C). The shaded spheres show residues that have changed appreciably in the structure of elastase. The table shows the sequence homologies in bovine trypsin (T) and thrombin (Th) and porcine elastase (E). Differences are shown in capitals and major differences in bold lettering. The conformation of the chain surrounding this 'core' is identical in chymotrypsin, trypsin and elastase.

carried by the viable partner. This viable copy is now subject to the full conservative discipline of natural selection, but what happens to the defective copy? Every mutational or recombinational event is as likely to be fixed as any other. The population therefore rapidly becomes heterogeneous in the progeny of this 'silent gene'. At any stage, however, a reversion of the original lesion or a repair mutation can restore the activity of the enzyme, but this will carry the mutational history of the gene in its 'silent' phase. If the protein is now just viable, compensating mutations will be selected to remove minor defects. We might then see a pair of isoenzymes such as chymotrypsin-A and -B with

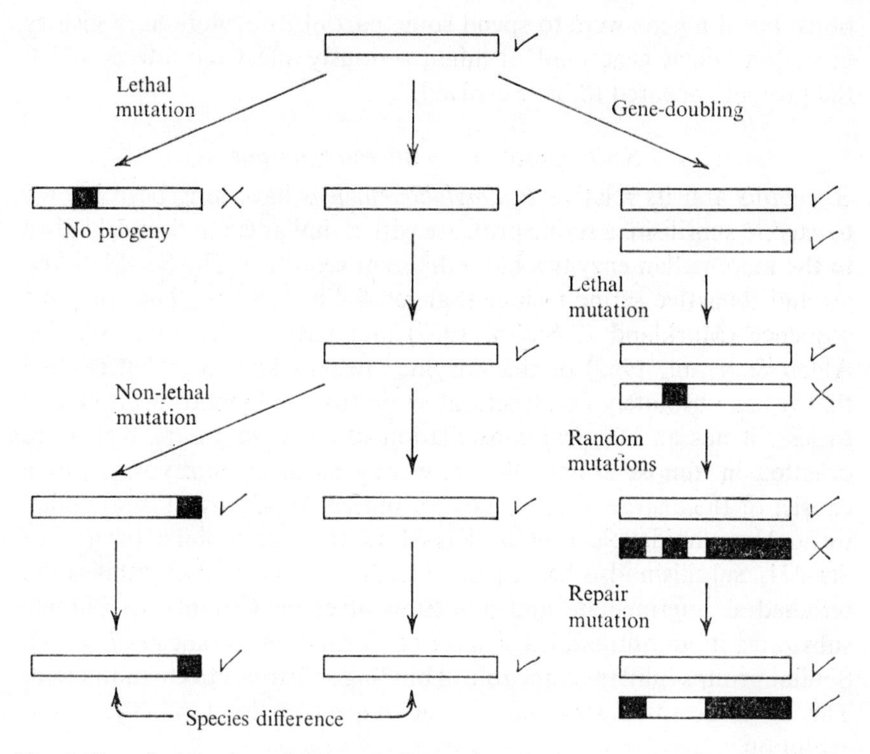

Fig. 3. Gene-doubling in evolution. ✓ indicates an active enzyme, × an inactive enzyme. Shaded areas show mutations in the genes.

identical activities but considerable surface variation. If, on the other hand, selection demands a new enzyme specificity, nature can go fishing by means of such repair mutations in this hidden heterogeneous gene pool to find a foldable protein with appropriate side-chain substituents.

If this particular evolutionary mechanism is accepted, then the evolutionary trees produced by protein chemists masquerading as 'molecular palaeontologists' become somewhat suspect. Such trees invariably postulate a 'minimum mutational distance' between hypothetical ancestors. If a gene were to spend an appreciable part of its history in the 'silent gene pool' that I postulate, it would be freed from the selective constraints that conserve the sequence of normal genes. On recall to useful life it must be repolished by natural selection and these 'compensatory mutations' are a consequence of the particular mutations which occurred in the silent phase. The whole event would appear as a single evolutionary step. I don't suppose that the shape of the evolutionary trees would be radically changed by such considerations, but if a gene were to spend some part of its evolutionary history in such a 'silent gene pool' it might seriously affect the rate at which the protein appeared to have evolved.

Serine proteases in micro-organisms

B. subtilis and its relative *B. amyloliquefaciens* have long been known to excrete subtilisin, a serine protease with a similar catalytic mechanism to the mammalian enzymes but a different sequence, Thr-*Ser*-Met-Ala, around its active serine residue (Sanger & Shaw, 1960). The complete sequence (Markland & Smith, 1967) and tertiary structure (Wright, Alden & Kraut, 1969) of this enzyme are now known and it is clear that it has absolutely no structural similarity to chymotrypsin. Nevertheless, it has an identical constellation of catalytic groups which are oriented in almost exactly the same way as in chymotrypsin. These consist of the charge-relay groups – a buried Asp-32 hydrogen-bonded through an imidazole ring of His-64 to the nucleophilic oxygen of Ser-221. Subtilisin also has a pair of hydrogen bonds that stabilise the tetrahedral intermediate and a section of chain that orients peptide substrates in an antiparallel β-sheet conformation (Kraut *et al.* 1971). Similar groups fulfil the same role in binding substrates to chymotrypsin. This is a beautiful example at the molecular level of convergent evolution.

Another bacterium, *Proteus vulgaris*, and a mould, *Aspergillus oryzae*, have long been known to contain enzymes having an active centre serine sequence, Thr-*Ser*-Met-Ala, identical to that in subtilisin (Shaw, 1962). It was therefore tempting to speculate that the subtilisin class would be typical of serine enzymes in micro-organisms and that the chymotrypsin type is a later evolutionary development. More recently, however, sequence studies on serine proteases from *Myxobacter sorangium* (Olson *et al.* 1970) and *Streptomyces griseus* (Johnson & Smillie, 1972) have established unequivocally that they are closely related to the mammalian enzymes.

One of the *M. sorangium* enzymes, the α-lytic protease, has a specificity rather like elastase (Kaplan, Symonds, Dugas & Whitaker, 1970) so I have described it in Table 3 as an 'elastase'. Two of the several enzymes in 'Pronase' preparations from *S. griseus* have also been fully sequenced by Smillie and his colleagues. One of these, the A-protease, has elastase-like specificity whereas another is almost identical to trypsin in activity. The sequence homologies between these enzymes have been discussed at length elsewhere (Olson *et al.* 1970; Hartley *et al.* 1972) and are summarised in Table 3, in which the sequences are com-

Table 3. *Species differences in serine proteinases*

Enzyme	Species	Sequence homologies (%)	Disulphide bridges
Trypsin	Cow	100	6
	Dogfish	69	6
	S. griseus	43	3
Elastase	Pig	48	4
	M. sorangium	26	3
	S. griseus	24	2
Subtilisin	*B. subtilis*	0	0
	B. amyloliquefaciens	0	0

Homologies are chemical similarities as illustrated in Table 2. Sequence references: bovine trypsin (Mikeš, Tomášek, Holeyšovsky & Sorm, 1966; Hartley, 1970*a*); dogfish trypsin (partial sequence) (Neurath, Bradshaw & Arnon, 1960); *Streptomyces griseus* trypsin (L. B. Smillie, private communication); porcine elastase (Shotton & Hartley, 1970); *Myxobacter* elastase (Olson *et al.* 1970); *Streptomyces griseus* elastase (Johnson & Smillie, 1972); subtilisins (Markland & Smith, 1967).

pared with bovine trypsin. The two 'elastases' are considerably different from trypsin and in order to maximise homologies a considerable number of large 'insertions' and 'deletions' must be arbitrarily introduced. Nevertheless model-building (McLachlan & Shotton, 1971) makes it clear that these enzymes probably have a central core of a disulphide-bridged 'histidine loop' and 'serine loop' identical to that in chymotrypsin. The conformations of the chains wrapped around this central core must differ from chymotrypsin, but the positions of 'insertions' and 'deletions' make it likely that the two bacterial 'elastases' resemble each other in conformation even though their sequence identity is only 30 %. These two bacterial elastases, therefore, have a structure with a typical 'mammalian' core surrounded by unique 'bacterial' loops. Does this place them as evolutionary ancestors of the pancreatic enzymes? The picture is complicated by the presence in *Streptomyces griseus* of a gene which codes for a trypsin-like enzyme. This shows high homology with bovine trypsin (43 %), which is in fact higher than that between bovine trypsin and bovine thrombin (38 %). Careful examination of the sequence shows deletions at external loops, such as are shown in Fig. 1, that are typical of trypsin and not found in elastase or chymotrypsin. Here then is the fully evolved 'mammalian' trypsin conformation present in the same cell as a putative 'bacterial' elastase ancestor! Confusion is worse confounded by the discovery in *Streptomyces griseus* of yet another serine protease with a Thr-*Ser*-Met-Ala sequence at its active site, like that in subtilisin (Awad *et al.* 1972)! A single *Streptomyces griseus* cell therefore contains examples of the three classes of serine protease structure that I have described. The obvious conclusion is that each of these structures is a consequence of great evolutionary conservatism, arising from some common ancestor of *Streptomyces* and *B. subtilis*, but I think that we ought to be prepared to consider other possibilities such as my previous suggestion that the bacterium has been infected by a cow (Hartley, 1970*a*)!

In the following article Pat Clarke suggests that transfer of metabolic genes from one species of *Pseudomonas* to another may be an important feature of their biochemical versatility. We are also familiar with episomal transfer of drug resistance genes from one species to another. In the laboratory, genes have been transferred from *Salmonella* and *Proteus* to *E. coli* (Baron, Gemski, Johnson & Wohlhieter, 1968) and from *Klebsiella* to *E. coli* (Dixon & Postgate, 1972). Should we not be prepared to envisage mechanisms of genetic transfer from a 'higher' species to a 'lower', perhaps via episomes or viruses, as a potential evolutionary tool? The fact that such episomes may flit from branch to branch of the evolutionary tree, conducting genes from one species to another, may be upsetting to biological systematists but I'm afraid it is something we'll have to watch out for. Proteins which appear to have conserved their structure as closely as *Streptomyces* trypsin and bovine trypsin are under such suspicion.

The evolution of disulphide bridges

I have another argument in favour of the cow infecting the bacterium, namely that disulphide bridges are a relatively late evolutionary development. I cannot adequately document this argument here, but it goes essentially as follows:

(*a*) Disulphide bridges are thermodynamically unstable within the glycolytic compartment of the cell (which is also the protein-synthesising compartment) because the redox potential maintained by NADH and mediated by glutathione pushes the equilibrium in favour of reduction. Within this compartment, therefore, disulphide bridges cannot *stabilise* tertiary structure. With the exception of lipoic dehydrogenase where a disulphide bridge is part of the catalytic mechanism (Massey, 1963), I have never come across a well documented example of an intracellular disulphide-bridged protein.

(*b*) Outside the reducing compartment, or in the membrane, in the presence of oxygen, disulphide bridges provide a huge pool of energy to stabilise protein conformation.

(*c*) Disulphide-bridge proteins such as ribonuclease are predominantly random coil when reduced (Tamburro, Boccu & Celotti, 1970): the oxidation of the bridges *determines* the conformation, though it is equally true that side-chain interactions determine the particular pattern of bridges. Therefore if ribonuclease were to be made within the glycolytic compartment it would be inactive. This is of obvious biological advantage.

(*d*) If the extracellular enzyme evolved from an intracellular ancestor, the latter must adopt its conformation without the benefit of bridges. Any appropriate pair of cysteine residues would, however, stabilise the structure *after export*.

(*e*) Mutations in such a protein subsequent to the evolution of the bridge might prevent the intracellular folding, but could be compensated by the folding energy provided by the extracellular bridges.

(*f*) If our hypothetical enzyme were a nuclease or protease, the cell must evolve mechanisms to protect its intracellular RNA or nascent protein from attack. *B. amyloliquefaciens* ribonuclease has no disulphide bridges but an intracellular inhibitor exists to protect the cell against enzyme which might be accidentally made intracellularly (Smeaton & Elliott, 1967). The evolution of bridges followed by destabilising mutations would provide an alternative mechanism.

If this hypothesis is correct it follows that selection for disulphide bridges could occur only in enzymes that the cell has learned to export. Few bacteria secrete large numbers of extracellular enzymes for most of their growth cycle, so the evolutionary pool is limited. Multicellular organisms secrete a larger number of proteins and so would have a greater evolutionary pool.

Another consequence of this hypothesis is that proteins with more bridges have the potential of becoming less stable in the reduced form. We might expect to see defects in the hydrophobic or hydrogen bonding relative to proteins with less bridges or no bridges at all. In this respect it is interesting to note that X-ray crystallography reveals a considerable number of buried water molecules in chymotrypsin (Birktoft & Blow, 1972), whereas there are none within the myoglobin molecule (Watson, 1969) or the chains of haemoglobin (Perutz, 1969). These latter proteins fold without the benefit of disulphide bridges, and it may be that the internal water molecules in chymotrypsin represent the 'holes' postulated above.

Other enzyme families

Will other enzyme families with a common catalytic mechanism prove to be as structurally similar as the serine proteases? Preliminary sequence studies of the acid proteases, pepsin and chymosin (rennin) suggested high homology (Hartley, 1970*b*), and now that the complete sequence of porcine pepsin (Tang, Sepulveda, Marciniszyn & Liu, 1973) and most of chymosin (Foltmann, Kauffman, Parl & Andersen, 1973) have been determined it is clear that they are about as closely related as the chymotrypsin family. A fascinating feature of the pepsin structure has been revealed by the complete sequences; when the sequences containing the two aspartyl residues that form the catalytic site (Chen & Tang, 1972) are aligned, an extensive internal homology appears, suggesting that a large part of the pepsin sequence has arisen by gene duplication and fusion. One might imagine that the ancestral enzyme was a symmetrical dimer using an aspartyl residue from each subunit. Gene fusion would clearly stabilise such a dimer in an external environment by preventing dissociation.

Perhaps the most encouraging result for those of us who would like to see these structural homologies spreading throughout textbooks of enzymology is the observation by Brew, Vanaman & Hill (1967) of sequence homologies between chicken lysozyme and bovine α-lactalbumin. Subsequently this protein was shown to be the modifier B protein of lactose synthetase (Brodbeck, Denton, Tanahashi & Ebner, 1967). Model-building, assuming identical conformations of the chain, was as convincing as in the case of elastase and trypsin mentioned above (Browne *et al.* 1969).

At the recent International Congress of Biochemistry we were treated to a feast of new enzyme structures, among them horse liver alcohol dehydrogenase (Bränden *et al.* 1973), lobster glyceraldehyde phosphate dehydrogenase (Buehner, Ford, Olsen & Rossman, 1973) and horse phosphoglycerate kinase (Blake & Evans, 1973). The corridors were buzzing with the speculation that a similar constellation of helices and β-structure was involved in the coenzyme binding in all these structures, in the same fashion as in dogfish lactate dehydrogenase (Adams *et al.* 1972), pig malate dehydrogenase (Hill, Tsernoglou, Webb & Banaszak, 1973) and flavodoxin (Watenpaugh *et al.* 1972; Rao & Rossmann, 1973). There has been insufficient time to evaluate the depth of this similarity, but it is one which would have been by no means obvious from sequence comparisons. Moreover, appreciable differences exist between the structures of the dehydrogenases in the substrate binding regions, so we have to be prepared to accept them as a less close family than the others that I have discussed.

Aminoacyl-tRNA synthetases, however, are a family of enzymes where one might legitimately expect a high degree of structural homology. They each react with a specific L-amino acid and ATP to form a 5′-aminoacyl adenylate, which then transfers the aminoacyl group to the 3′-hydroxyl of a particular tRNA. Transfer RNAs are believed to have similar secondary and tertiary structures (Cramer, 1971) as might be expected from their binding to the same site in the ribosome. The specificity for a particular amino acid and a particular tRNA must have been conserved since the dawn of life, since these enzymes are responsible for translating the genetic code and the code seems to be similar in all species.

Conversely, however, one might argue that such enzymes would be prime candidates for convergent evolution. The primaeval proteins that first began to supply some degree of specificity to the translation process may have been a very heterogeneous lot that would supply varied models for evolutionary improvement. Co-ordinate adaptation of the transfer RNAs to these enzymes, and vice versa, might leave us with a mixture of divergent and convergent classes within the enzymes of a single cell. For any one enzyme in different species, however, we might reasonably expect great conservation of structure within the binding sites in both the enzyme and the tRNA.

At first sight, the evidence for structural homology looks unpromising. Table 4 summarises the molecular weights and subunit structures of a number of purified synthetases. They fall into four broad classes:

(1) Monomers with subunits mol. wt. 100000–120000 with a single binding site for amino acid and tRNA, e.g. *E. coli* isoleucyl tRNA synthetase.

(2) Dimers with 70000–90000 mol. wt. subunits with one binding site for amino acid and tRNA per chain, e.g. *E. coli* methionyl-tRNA synthetase.

(3) Dimers of 40000–50000 mol. wt. subunits, e.g. *E. coli* tryptophanyl enzyme.

(4) Tetramers of $\alpha_2\beta_2$ type, e.g. Yeast phenylalanyl enzyme or *E. coli* glycyl enzyme.

This diversity seems to point to the mixture of convergent and divergent evolution that I have discussed above.

The extent to which evolution has conserved the sites responsible for recognition of a particular tRNA by its cognate enzyme also seems to be relatively low. There are large changes between primary structures of the same transfer RNA in different species. For example, wheat $tRNA^{Phe}$ differs 20 % in base sequence from that in baker's yeast and 42 % from that in *E. coli*; and $tRNA^{Val}$ from baker's yeast differs 5 % from that in *Torula* yeast and 51 % from that in *E. coli* (Dayhoff, 1972).

Table 4. *Subunits of aminoacyl-tRNA synthetases*

Enzyme	Species	Subunits (mol. wt. $\times 10^{-3}$)	Reference
Met	*E. coli*	2×90	1
Met	*B. stearothermophilus*	2×66	2
Ala	*E. coli*	2×70	3
Lys	Yeast	2×69	4
Pro	*E. coli*	2×47	3
Ser	*E. coli*	2×47	3
Ser	Yeast	2×47	3
Tyr	*E. coli*	2×47	5
Tyr	*B. subtilis*	2×44	3
Tyr	*B. stearothermophilus*	2×44	2
Trp	*E. coli*	2×37	5
Trp	Bovine	2×54	3
Trp	*B. stearothermophilus*	2×35	2
Ile	*E. coli*	1×112	3
Val	*E. coli*	1×110	3
Val	Yeast	1×112	3
Val	*B stearothermophilus*	1×110	2
Leu	*E. coli*	1×105	3
Leu	Yeast	2×60	6
Leu	*B. stearothermophilus*	1×110	2
Gln	*E. coli*	1×70	3
Arg	*E. coli*	1×70	3
Tyr	Yeast	1×45	9
Phe	*E. coli*	4×45	3
Phe	Yeast	$(2\times 75)+(2\times 63)$	7
Gly	*E. coli*	$(2\times 80)+(2\times 33)$	3
Glu	*E. coli*	$(1\times 56)+(1\times 46)$	8

References

1. G. Koch (unpublished results).
2. Koch, Boulanger & Hartley (1973).
3. Loftfield (1972).
4. Lagerkvist, Rymo, Lindqvist & Andersson (1972).
5. Muench & Joseph (1971).
6. Chirikjian, Wright & Fresco (1972).
7. Schmidt, Wang, Stanfield & Reid (1971).
8. Lapointe & Söll (1972).
9. Beikirch, Vonderhaar & Cramer (1972).

The overall secondary structures are, of course, base paired in the typical clover-leaf, but the differences distribute themselves throughout the structure. Indeed there are fewer differences between *E. coli* $tRNA^{Phe}$ and *E. coli* $tRNA^{Met}$ (43 %) than between the baker's yeast and *E. coli* $tRNA^{Phe}$s (46 %) and it would be hard to recognise this latter pair as isoaccepting species from their sequence alone, neglecting the anticodon. If, despite this sequence variability, nature had conserved the sites in a tRNA that interacts with its appropriate enzyme, we should expect that enzymes from one species would aminoacylate the cognate tRNA from another species. Very many such experiments have been performed, seldom alas with pure components under controlled conditions, but the general picture that emerges is of a very patchy spectrum of recognition in which there is about an even chance that any synthetase would aminoacylate a cognate tRNA from another species (Jacobson, 1971). A few cases of mischarging have been observed, for example, phenylalanyl-tRNA synthetase from *Neurospora crassa* forms Phe-$tRNA^{Val}$ and Phe-$tRNA^{Ala}$ with transfer RNA from *E. coli* (Barnett & Jacobson, 1964). It appears that there has been a lot of co-ordinate variation in the enzyme-tRNA binding sites during speciation. The extent of structural homology within this family of enzymes may therefore be less than one would at first expect.

Gene-doubling and fusion in aminoacyl-tRNA synthetases

To begin to determine the structural homology we have begun to sequence the methionyl-tRNA synthetase of *E. coli*, since X-ray studies of crystals of an active fragment of this enzyme are in progress in Paris (Waller, Risler, Monteilhet & Zelwer, 1971). We have also developed a purification procedure that yields 100–200 mg quantities of pure leucyl-, methionyl-, tryptophanyl- and valyl-tRNA synthetases from 5 kg of the thermophile *B. stearothermophilus*. These enzymes are particularly stable, which helps during the lengthy purification procedure and also allows flexibility in attempts to crystallise. Small crystals of the tryptophanyl enzyme have been obtained (R. Jakes, unpublished evidence) but the tyrosyl enzyme is particularly promising. Large crystals of suitable space-group with the monomer as the asymmetric unit have been obtained, which diffract out to 3 Å (Reid *et al.* 1973); and a suitable heavy atom derivative has been found. Sequence studies have also begun on this enzyme. Yeast leucyl-tRNA synthetase is also under crystallographic study elsewhere (Chirikjian, Wright & Fresco, 1972) so there is a good prospect of a definite solution to the problems of homology in this family.

In the meantime, some preliminary sequence studies of the *B. stearothermophilus* enzymes have revealed features that increase the prospect of structural homology (Koch, Boulanger & Hartley, 1973). Careful studies of the molecular weights of the native and denatured enzymes gave the subunit structures shown in Table 4. Extensive peptide mapping of tryptic and other digests of the four enzymes showed that the number of peptides from the leucyl, methionyl and valyl enzymes was about half that expected from the amino acid analysis (Table 5). It appeared that large portions of each polypeptide chain were extensively internally duplicated.

Confirmation was obtained by attempting to isolate quantitatively several tryptic peptides from the leucyl, methionyl and valyl enzymes.

Table 5. *Tryptic peptides from* B. stearothermophilus *aminoacyl-tRNA synthetases*

Aminoacyl-tRNA synthetase	Number of peptides containing: Arg	His	Cys	Total
Tyrosyl	22 (28)	10 (9)	2 (2)	45–50 (56)
Methionyl	19 (34)	6 (12)	3 (5)	40–45 (80)
Valyl	32 (69)	12 (31)	3 (8)	50–55 (138)
Leucyl	24 (47)	10 (26)	2–3 (6)	40–45 (113)

The figures in parentheses show the total number of peptides expected from the amino acid analysis, based on the chain lengths shown in Table 4. The number of peptides was determined from tryptic digests of oxidized or ^{14}C-carboxymethylated proteins separated by ion-exchange followed by paper electrophoresis and chromatography, and stained with appropriate reagents.

Peptides were chosen which purified readily in single-step ion-exchange chromatography. From the leucyl enzyme four peptides including a 27-residue peptide and a 9-residue peptide and a 5-residue peptide were isolated in yields of 1.67–1.91 mol mol^{-1} protomer. Five peptides from the methionyl enzyme, including an octapeptide, gave yields between 1.14 and 1.39 mol mol^{-1}. The valyl enzyme gave a pentapeptide (1.79 mol mol^{-1}), a hexapeptide (1.91 mol mol^{-1}) and an octapeptide (1.79 mol mol^{-1}). Each of these peptides, chosen more or less at random, must occur at least twice in each polypeptide chain. The distribution of repeating sequences in each chain was determined by exploiting the susceptibility of these enzymes to 'nicking' by extremely mild digestion with proteases. For example, trypsin splits the single chain of the valyl enzyme (mol. wt. 110000) to give 67000 and 40000 fragments that separate from each other only after the protein is denatured. There are eight cysteine molecules in the native enzyme, four in the large fragment and three in the small fragment. Peptide maps of tryptic digests of the ^{14}C-carboxymethylated proteins show three radioactive peptides in the native enzyme that are present in *both* the 67000 and 40000 fragment. It looks as though the repeating sequences are distributed symmetrically in each half of the chain. The methionyl enzyme (mol. wt. 2×66000) is 'nicked' by trypsin to give a fully active single chain of mol. wt. 55000 which has the same N-terminal sequence as the native enzyme. About seventeen tryptic peptides are removed from the C-terminus in this process, but peptide maps of the native enzyme and fragment show almost complete identity apart from differences in intensity of a few spots. The 'missing' C-terminal peptides must be repeated in the N-terminal fragment.

Similar results have been obtained with the methionyl-tRNA synthetase from *E. coli*, a dimer of identical 90000 mol. wt. chains* (Bruton, Jakes & Hartley, to be published). Peptides liberated from the C-terminus during formation of the active 64000 mol. wt. tryptic fragment (Cassio & Waller, 1971) are also present in the fragment, and many tryptic peptides have been isolated in yields greater than 1 mol mol^{-1} chain. Moreover the preliminary sequence studies are entirely consistent with a primary structure that contains two almost identical repeats per chain.

The large chains of the leucyl, methionyl and valyl enzymes therefore seem to consist largely, if not entirely, of two almost identical repeats. The situation is reminiscent of the haptoglobins (Black & Dixon, 1968) which appear to have arisen by fusion of adjacent doubled genes. If this proves to be the case with these large synthetases, it will provide a new set of problems as to why the sequence has been so carefully conserved in each half. Post-translational fusion remains, of course, a possibility. Whatever the answer the results provide encouragement that these large enzymes may yet show structural homology with the shorter chains of the tryptophanyl and tyrosyl enzymes.

EXPERIMENTAL ENZYME EVOLUTION

The feast of speculation in which I have so far indulged is, I am afraid, typical of much written by protein chemists about evolution. We realise, of course, that hundreds of structures may have to be examined before anything approaching a biological principle can be established, and that even then all the pitfalls of comparative anatomy may apply to our conclusions. We hope, however, that history will forgive our naive enthusiasm after discovering these first protein structures, as it has forgiven the wilder speculations of our nineteenth-century ancestors who discovered the first dinosaur bones.

Some discipline would be imposed on our imagination if only there were such things as genuine fossil proteins, but almost all of these are, alas, long since decomposed into their constituent biological elements. However, by studying the present-day evolution of new enzyme activities in micro-organisms we can hope to reach more rigorous conclusions about some of the ways in which evolution moulds protein structures. In the following article Pat Clarke reviews many of the current approaches to this problem. In her own elegant work, for example, one can see the possibility of describing step by step the evolution of a new enzyme structure.

* Studies by G. Koch (unpublished evidence) disagree with the tetrameric structure, 4×45000 suggested by Lemoine, Waller & Van Rapenbusch (1968).

Evolution of ribitol dehydrogenase in continuous cultures of Klebsiella aerogenes

Our own contribution to this problem has been to try to evolve the enzyme ribitol dehydrogenase of *Klebsiella aerogenes* towards a new specificity for xylitol (Hartley *et al.* 1972; Rigby, Burleigh & Hartley, 1973). As Dr Clarke describes (following article – Fig. 1 and pp. 192–3), a mutant of this organism constitutive for ribitol dehydrogenase grows on xylitol by utilising a side specificity of this enzyme to produce xylulose, but whereas the effective K_m (app) for ribitol is about 1 mM the K_m for xylitol is around 1 M. We have shown that at low concentrations of xylitol as limiting carbon nutrilite the activity of this enzyme is rate-limiting for the growth of the organism. In these circumstances this enzyme can be looked on as a 'bad' xylitol dehydrogenase, and evolutionary pressure can be applied to improve its activity.

We chose to use continuous culture in chemostats as our selective tool, because the theory of the chemostat predicts that any mutant that arises with a relatively small increase in growth rate has a high probability of taking over the entire culture (Powell, 1958). By continuously monitoring the steady-state biomass and the effluent substrate concentration a series of small evolutionary steps could thus be detected. Philosophical considerations also prejudiced us towards the chemostat. It represents a closed ecological niche in which the steady-state biomass and the residual limiting nutrilite in the effluent directly reflect the biological fitness of the organism for growth in this niche. Perfect adaptation would be signalled by complete utilisation of the limiting nutrilite and a conversion of food to biomass that approached a thermodynamic maximum. Moreover, the selective pressure due to the external xylitol concentration can be precisely controlled by adjusting the dilution rate of the vessel. Parallel chemostats operating under identical selective constraints might allow us to discover the degree of determinism in our particular evolutionary process: do events always follow the same pattern or are there many alternative ways of responding to the selective pressure?

As our parental organism we chose a strain of *Klebsiella aerogenes*, hereafter called strain A, which is constitutive for ribitol dehydrogenase (Strain X1 of Wu, Lin & Tanaka, 1968). To test the selective power of our chemostats, we grew an arginine-requiring auxotroph of strain A on xylitol (0.2 % w/v) as limiting carbon source in minimal salts medium supplemented with traces of arginine and guanine. A steady state was established containing approximately 10^{12} cells. At the point indicated

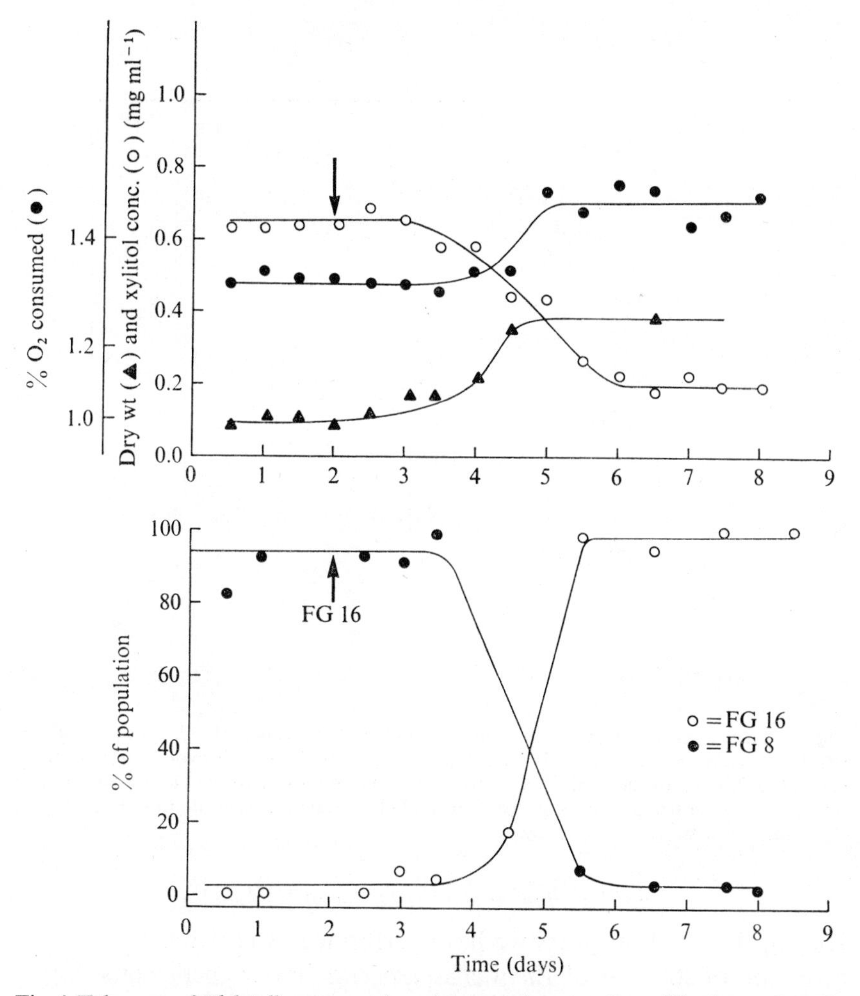

Fig. 4. Take-over of *Klebsiella aerogenes* in a chemostat by an evolvant (Hartley *et al.* 1972). At the point indicated by the arrow approximately twenty cells of strain 1B (FG 16, a guanine auxotrophe of a mutant with improved xylitol dehydrogenase) were added to a steady-state culture containing approximately 10^{12} cells of strain 2A (FG 8, an arginine auxotroph of the ancestral strain) grown on xylitol. The composition of the population was determined by plating on appropriate media.

by the arrow in Fig. 4, twenty cells of strain B were added. This was a guanine-requiring auxotroph of the mutant strain X2 of Wu *et al.* (1968) which has a roughly two-fold improvement in V_{max} and a three-fold improvement in K_m (app) at 1 mM NAD and 5 mM xylitol (Hartley *et al.* 1972). Fig. 4 shows that strain B effectively displaced strain A from the chemostat over a period of seven days. It was obvious, therefore, that our chemostat was well able to select improvements of this magnitude.

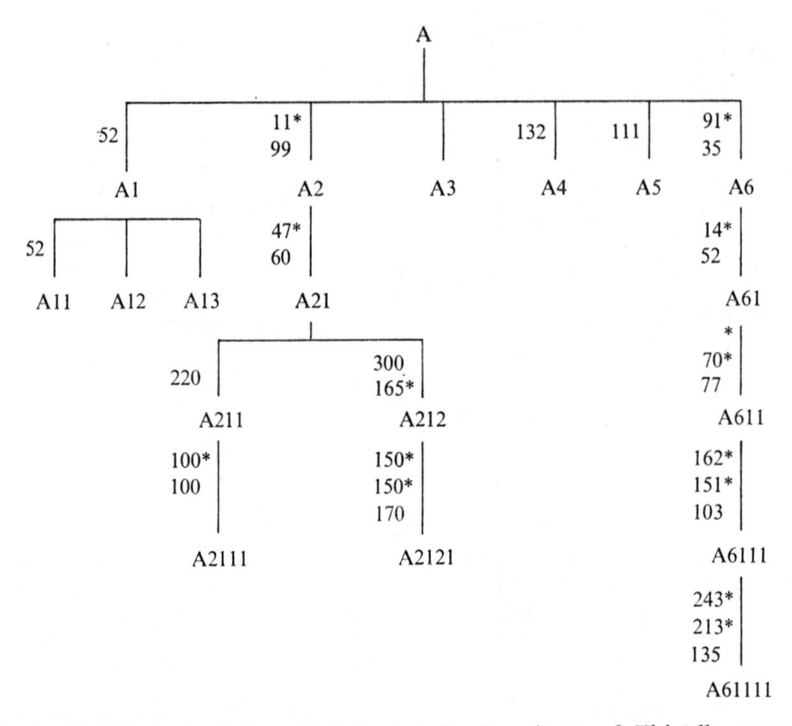

Fig. 5. Genealogy of spontaneous and u.v.-induced evolvants of *Klebsiella aerogenes* (Rigby, Burleigh & Hartley 1973). The strain numbers indicate their evolutionary history, e.g. A1, A2, etc., are derived directly from A; A11, A12, etc. from A1 and so on. The numbers between strains are the generations that elapsed before isolation of the new strain. * Indicates a step of u.v.-mutagenesis.

Gene-doubling in the evolution of Klebsiella

During the last four years we have performed extensive studies of the evolution of strain A of *Klebsiella aerogenes* in our chemostats. Fig. 5 summarises the nomenclature and genealogy of the evolvants that were detected by significant changes in the steady-state concentration of the effluent xylitol. Cell-free extracts were made from each of the strains listed in Fig. 5 and kinetic constants were determined for the ribitol dehydrogenase and xylitol dehydrogenase activity.

At first we investigated the spontaneous events. In no case was there any significant change in the ratio of these activities or the K_m (app) for ribitol or xylitol, but in each step a large increase in the total ribitol dehydrogenase activity of the extracts was observed. The enzyme was purified to homogeneity from some of these strains and appeared to be identical to the ancestral enzyme in kinetic behaviour and electrophoretic mobility. It was clear that each evolutionary step represented a change in the amount of enzyme within the cell rather than a change in the enzyme itself.

Similar experiments were carried out under conditions in which the rate of mutation was increased by irradiating the cultures for a short period *in situ* with u.v. light. These conditions were unlikely to cause very extensive mutagenesis because a significant proportion of such operations failed to yield a detectable evolvant. Once again, all of the events observed were increases in the total amount of protein in the cell. The extent of this increase is impressive: the constitutive ancestor A produces ribitol dehydrogenase at about 1 % of its total soluble protein in late log phase, whereas this enzyme is more than 20 % of the total protein in late evolvants such as A6111!

What is the molecular basis for this superproduction? Constitutive mutations are ruled out, because the ancestor itself was constitutive. Promoter mutations, catabolite de-repression or gene amplification remain among the options discussed by Dr Clarke. We decided to test some of these strains for the possible presence of multiple copies of the RDH gene. Shortage of tools for adequate genetic analysis of *Klebsiella aerogenes* persuaded us to investigate the frequency (P) at which this gene mutates to RDH^- in different evolvants. If any strain possessed two copies it should mutate at frequency P^2 under identical mutagenic conditions. In this way strain A1 was shown *not* to be gene-doubled, but the step from A1 to A11 probably involved gene multiplication because A11 produced RDH^- progeny (lacking ribitol dehydrogenase) at negligible frequency relative to A and A1.

Further evidence for gene doubling was obtained by studying the rate at which strains producing large amounts of enzyme segregated clones of low producers. Point mutations in a promoter, for example, should revert spontaneously at frequencies of around 10^{-8}, but authentic cases of organisms with doubled genes have been shown to segregate progeny with single copies at much higher frequency: 3–20 % for an *E. coli* glycyl-tRNA synthetase gene in a doubled section of chromosome involving about 4.5 minutes of the genetic map (Folk & Berg, 1971) or 0.0001 % for adjacent tyrosyl-tRNA genes of *E. coli* (Russell *et al.* 1970), for example. For strain A1 we found no segregants, but strain A11 segregated low producers at 0.14 % frequency, confirming our conclusions that gene doubling was involved in the step from A1 to A11 but not in the step A to A1. The extent of segregation suggests that the A11 duplication covers several genes, and we have observed on polyacrylamide gels of extracts of this strain that a second protein in addition to the ribitol dehydrogenase increases appreciably in intensity relative to the ancestral strain A. We suspect that the superproducer strain A211 carries *three* copies of the gene, because it segregates small colonies at overall frequency of 0.68 % which are a mixture of both medium producers like A11 and low producers like A.

The conclusions that we draw from the work described so far (Rigby *et al.* 1973) are as follows. Under conditions of spontaneous or mild u.v. mutagenesis, thirty evolutionary steps have occurred in the order and at the frequency pictured in Fig. 5. All of these steps have been events that increase the intracellular concentration of ribitol dehydrogenase to surprisingly high levels. An appreciable proportion of these events seem to be due to duplication of the gene for this enzyme. A total number of 10^{14} organisms have been screened through our selection system, but no evolvants have arisen with improved enzyme kinetics for xylitol. We estimate that every amino acid replacement in the RDH sequence possible by a single base change would occur with a frequency of 10^{-9}. Any such event that improved significantly the activity towards xylitol would be detected with 10 % efficiency by our chemostat screening system. It follows that there is no way in which the efficiency of the enzyme can be improved by a single-step amino acid substitution, since we would have screened 10^4 such events.

This conclusion appears to conflict remarkably with the evidence of Dr Clarke and her colleagues on the evolution of new specificities in her *Pseudomonas* amidase (following article). In their case, a whole galaxy of new enzyme specificities appear by point mutations at the drop of a hat! I suspect that the answer to this apparent contradiction is that they have chosen to ask a much simpler question of a less complex enzyme in a more biochemically versatile organism. We are trying to change the stereospecificity for a hydroxyl group adjacent to the carbon atom involved in hydride transfer to NAD. Both substrate, coenzyme and catalytic groups in the enzyme must be correctly oriented. In the amidase, only binding and correct orientation of the carbonyl group may be involved. Even so, I feel that the remarkable changes in specificity that she has elicited require further explanation: it almost looks as though the *Pseudomonas* enzyme was designed to be adaptable! We have speculated (Rigby *et al.* 1973) that a multimeric enzyme, in which the catalytic site lay at the junction between subunits with some side chains contributing to specificity on the adjacent subunit, would have a high capacity to evolve new specificities by point mutation. Fig. 6 shows a hypothetical example. Mutation of a single surface residue within the interface between subunits could allow a new subunit interaction that would change the quaternary structure without appreciable

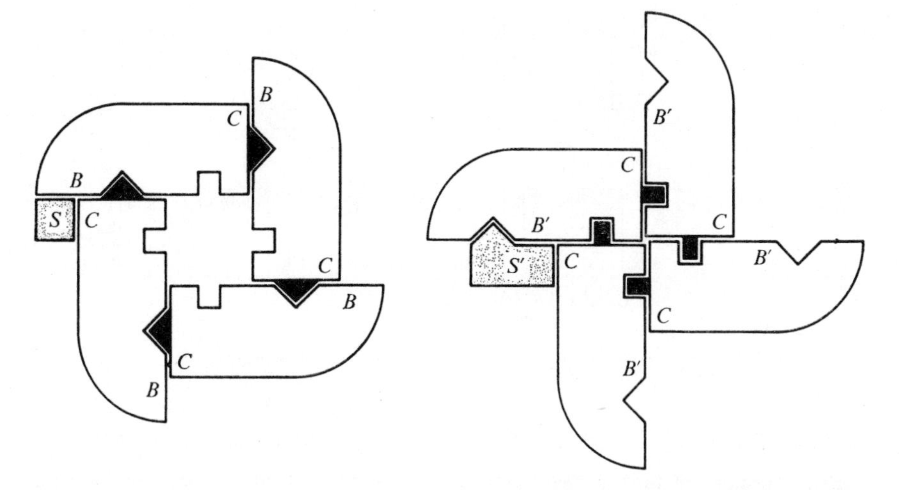

Fig. 6. Schematic illustration of how a single mutation affecting quaternary structure might lead to drastic changes in specificity. B,B' are substrate binding sites for substrates S and S'. C indicates catalytic groups. The mutation in the interface between subunits is heavily shaded.

effect on tertiary structure. This would expose the catalytic site to a new constellation of groups on the adjacent subunit, allowing radically different substrates to bind. Models such as this may be necessary to explain the catalytic versatility of point mutants of the *Pseudomonas* amidase which is, as Dr Clarke points out, a hexamer.

Evolvants with changed substrate specificity

If point mutants of the ribitol dehydrogenase of *Klebsiella aerogenes* refuse to take over chemostats grown on xylitol, how did the specificity mutant isolated by Wu *et al.* (1968) arise? Is it because such mutants are invariably less temperature-stable and lose their biological advantage in cultures grown at 37 °C? Fig. 7 (Dothie, Rigby & Hartley, unpublished evidence) shows a few experiments conducted with strain A and strain B at 27 °C. In each case the evolvant was a superproducer.

Having introduced some 'loosening' of the enzyme structure in strain B, will further specificity improvements now become more probable? Fig. 7 shows that at both 27 °C and 37 °C superproducers were the next evolutionary steps.

Strain B had been obtained by Wu *et al.* (1968) following nitroso-guanidine mutagenesis. Is nitrosoguanidine a 'magic' mutagen for enzyme improvement? Fig. 7 shows that low levels of nitrosoguanidine mutagenesis (sufficient to allow a few non-events when screened in the chemostat) also led to higher enzyme production. These necessary

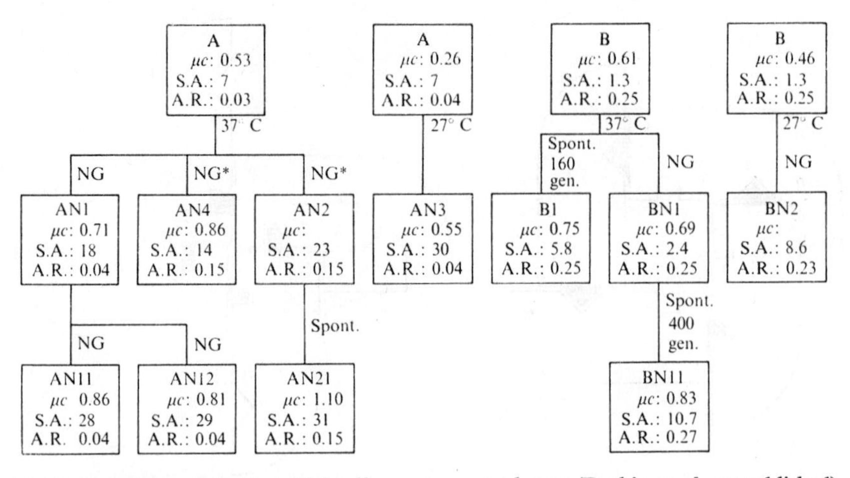

Fig. 7. Genealogy of further *Klebsiella aerogenes* evolvants (Dothie *et al.*, unpublished). Chemostat cultures of strains A or B, grown on limiting xylitol at 27 or 37 °C, were subjected to mild (NG) or severe (NG*) mutagenesis. μc indicates the maximum growth rate of the strain; S.A., the specific activity of ribitol dehydrogenase in cell free extracts and A.R. the ratio of xylitol dehydrogenase to ribitol dehydrogenase activity.

experiments all fortified our conclusion that there was no way in which the enzyme could change its specificity in a single step.

When, however, we increased the mutagenesis with nitrosoguanidine to a level where multiple mutations might be expected, we obtained evolvants with new enzyme specificity in a majority of cases, e.g. AN2, AN4 and AN21 in Fig. 7. Each of these proved to have a lowered K_m (app) for xylitol and an increased xylitol:ribitol activity ratio.

We have recently determined the complete sequence of ribitol dehydrogenase from strain A (C. H. Moore, S. S. Taylor, M. J. Smith & B. S. Hartley, in preparation). Preliminary peptide maps of elastase digests of the oxidised proteins from strains A211, B and AN21 show multiple differences that suggest that the mutations responsible for the change in specificity are different in B and AN21 and probably number between two and six in each. We intend to identify the sequence changes responsible.

The evolutionary lessons to be learned so far are that single-step improvement of this enzyme for growth on xylitol is impossible, but that gene multiplication is a frequent alternative response of the organism to the selective pressure. Referring back to Fig. 2, we now have evidence for the first step in our postulated evolutionary pathway. We hope to continue this study pursuing the stepwise evolution of the enzyme under severe mutagenic conditions, but also to examine the response of a gene-doubled strain of the organism to fluctuating selective pressure, by alternating xylitol and glucose as limiting carbon nutrilites in the chemostat. Fig. 2 would predict that evolution would be more rapid in the 'silent gene pool' that may then arise.

An example of recent enzyme evolution?

Our attempts to evolve new enzymes in the laboratory can always be criticised as of little biological significance because they do not reflect the multiple and various selective pressures that must occur in a real ecological niche. We must aim to keep our conditions as 'biological' as possible and be prepared to compare our results with examples from real life. In this latter respect, it would be an advantage to know the changes that have occurred in an enzyme that has evolved only recently.

The enzymes coded by the episomal R factors of enteric bacteria, that are responsible for the phenomenon of multiple transmissible drug resistance, are a possible source of 'recent' enzymes. Many of the drugs have been discovered or synthesised this century. Where did the enzymes that detoxify them come from?

Shaw (1971*a*) has discussed evolutionary aspects of bacterial resistance to chloramphenicol. The R factors contain genes that, after infection, cause production of chloramphenicol acetyl transferase (CAT) which catalyses acetyl transfer from acetyl-CoA to the 3-hydroxyl group of chloramphenicol to give an inactive derivative. The enzyme is specific for the D-threo isomer of chloramphenicol and has been purified to homogeneity. It is a tetramer with four apparently identical subunits, each of mol. wt. 19000.

Wild-type *Proteus mirabilis* is chloramphenicol sensitive, but Shaw (1971*b*) obtained conventional chromosomal mutants that were resistant to chloramphenicol but not to other antibiotics normally detoxified by cells harbouring R factors. A chloramphenicol acetyl transferase was purified from these mutants that differs in K_m (app) for chloramphenicol (18 μM) from the *E. coli* R-factor enzyme (6.8 μM) and is electrophoretically distinct. It is, however, a tetramer of 4×20000 mol. wt. and cross-reacts strongly with rabbit antibodies to the R-factor enzyme. Shaw therefore returned to the wild-type *Proteus mirabilis* and isolated from it an enzyme with an extremely weak chloramphenicol acetyl transferase activity, less than 1 % of that of the R-factor enzyme, that also cross-reacted with antibodies to the latter.

The precursor protein is probably a normal metabolic transacetylase of *Proteus mirabilis* with another specificity (X-transacetylase). It certainly has no biological value in detoxifying chloramphenicol. Since it can mutate to a useful chloramphenicol transacetylase, it may be identical to or analogous to the evolutionary ancestor of the R-factor enzyme. To test this hypothesis, W. V. Shaw, B. D. Burleigh & B. S. Hartley (unpublished results) have determined the amino acid sequence of the *E. coli* R-factor enzyme, and experiments are in progress to compare with this the sequence of the putative ancestral enzyme and the chromosomal mutant of *Proteus mirabilis*. The extent of difference may tell us something about an evolutionary event that has occurred within our own lifetime.

CONCLUSION

The tertiary structures of proteins that are available so far have only begun to whet our appetite. Every new structure brings a new surprise, so we are a long way from the systematic molecular anatomy that must underlie any firm opinions about how these molecules evolved. The structures do, however, allow us to ask new questions about the factors that are involved in their evolution. Some of these questions can be answered by evolving new enzymes in the laboratory, and the tools of molecular biology now allow us to follow this process step by step at the atomic level. I'm sure that Darwin would have wished that he were alive today.

REFERENCES

ADAMS, M. J., BUEHNER, M., CHANDRASEKHAR, K., FORD, G. C., HACKERT, M. L., LILJAS, A., LENTZ, P., RAO, S. T., ROSSMAN, M. G., SMILEY, I. E. & WHITE, J. L. (1972). Subunit interactions in lactic dehydrogenases. In *Protein–Protein Interactions*, **23**, Colloqium der Gesellschaft für Biologische Chemie, pp. 139–56. Heidelberg: Springer-Verlag.

ANFINSEN, C. B. (1966). The formation of the tertiary structure of proteins. In *The Harvey Lectures*, vol. 61, pp. 95–116. New York: Academic Press.

AWAD, W. M., SOTO, A. R., SIEGEL, S., SKIBA, W. E., BERNSTROM, G. G. & OCHOA, M. S. (1972). The proteolytic enzymes of the κ-1 strain of *Streptomyces griseus* obtained from a commercial preparation (Pronase). *Journal of Biological Chemistry*, **247**, 4144–5.

BARNARD, E. A., COHEN, M. S., GOLD, M. H. & KIM, J. K. (1972). Evolution of ribonuclease in relation to polypeptide folding mechanisms. *Nature, London*, **240**, 395–8.

BARNETT, W. E. & JACOBSON, K. B. (1964). Evidence for degeneracy and ambiguity in interspecies aminoacyl-tRNA formation. *Proceedings of the National Academy of Sciences, USA*, **51**, 642–7.

BARON, L. S., GEMSKI, P., JOHNSON, E. M. & WOHLHIETER, J. A. (1968). Intergeneric bacterial matings. *Bacteriological Reviews*, **32**, 362–9.

BEIKIRCH, H., VONDERHAAR, F. & CRAMER, F. (1972). Tyrosyl-tRNA synthetase from baker's yeast. *European Journal of Biochemistry*, **26**, 182–90.

BIRKTOFT, J. J. & BLOW, D. M. (1972). The atomic structure of tosyl-α-chymotrypsin at 2 Å resolution. *Journal of Molecular Biology*, **68**, 187–240.

BLACK, J. A. & DIXON, G. H. (1968). Amino acid sequence of α-chains of human haptoglobins. *Nature, London*, **218**, 736–41.

BLAKE, C. C. F. & EVANS, P. R. (1973). X-ray studies on horse-muscle phosphoglycerate kinase. *Abstracts, 9th International Congress of Biochemistry*, 2Sb4.

BRÄNDEN, C. I., EKLUND, H., ZEPPEZAUER, E., NORDSTRÖM, B., BOIWE, T., SÖDERLUND, G. & OHLSSON, I. (1973). Structure and function of horse liver alcohol dehydrogenase. *Abstracts, 9th International Congress of Biochemistry*, 2Sa4.

BREW, K., VANAMAN, T. C. & HILL, R. L. (1967). Comparison of the amino acid sequence of bovine α-lactalbumin and hens egg white lysozyme. *Journal of Biological Chemistry*, **242**, 3747–8.

BRODBECK, U., DENTON, W. L., TANAHASHI, N. & EBNER, K. E. (1967). The isolation and identification of the B protein of lactose synthetase as α-lactalbumin. *Journal of Biological Chemistry*, **242**, 1391–7.

BROWNE, W. J., NORTH, A. C. T., PHILLIPS, D. C., BREW, K., VANAMAN, T. C. & HILL, R. L. (1969). A possible three-dimensional structure of bovine α-lactalbumin based on that of hens egg white lysozyme. *Journal of Molecular Biology*, **42**, 65–86.

BUEHNER, M., FORD, G. C., OLSEN, K. W. & ROSSMAN, M. G. (1973). The structure of D-glyceraldehyde 3-phosphate dehydrogenase at 3 Å resolution. *Abstracts, 9th International Congress of Biochemistry*, 2g3.

CASSIO, D. & WALLER, J. P. (1971). Modification of methionyl-tRNA synthetase by proteolytic cleavage and properties of the trypsin modified enzyme. *European Journal of Biochemistry*, **20**, 283–300.

CHEN, K. C. S. & TANG, J. (1972). Amino acid sequence around the epoxide-reactive residues in pepsin. *Journal of Biological Chemistry*, **247**, 2566–74.

CHIRIKJIAN, J. G., WRIGHT, H. T. & FRESCO, J. R. (1972). Crystallisation of tRNA Leu-synthetase from baker's yeast. *Proceedings of the National Academy of Sciences, USA*, **69**, 1638–41.

CRAMER, F. (1971). Three-dimensional structure of tRNA. *Progress in Nucleic Acid Research and Molecular Biology*, **11**, 391–421.

DAYHOFF, M. O. (1972). In *Atlas of Protein Sequence and Structure*, p. D 382. Washington, D.C.: National Biomedical Research Foundation.

DIXON, R. A. & POSTGATE, J. R. (1972). Genetic transfer of nitrogen fixation from *Klebsiella pneumoniae* to *E. coli*. *Nature, London*, **237**, 102–3.

FOLK, W. R. & BERG, P. (1971). Duplication of the structural gene for glycyl-tRNA synthetase in *E. coli*. *Journal of Molecular Biology*, **58**, 595–610.

FOLTMANN, B., KAUFFMAN, D., PARL, M. & ANDERSEN, P. M. (1973). Comparison between the primary structures of chymosin, pepsin, and of their zymogens. *Netherlands Milk and Dairy Journal*, in press.

HARTLEY, B. S. (1966). Enzymes are proteins. *The Advancement of Science*, May 1966, pp. 47–54.

HARTLEY, B. S. (1970*a*). Homologies in serine proteinases. *Philosophical Transactions of the Royal Society of London* B, **257**, 77–87.

HARTLEY, B. S. (1970*b*). Strategy and tactics in protein chemistry. *Biochemical Journal*, **119**, 805–22.

HARTLEY, B. S., BURLEIGH, B. D., MIDWINTER, G. G., MOORE, C. H., MORRIS, H. R., RIGBY, P. W. J., SMITH, M. J. & TAYLOR, S. S. (1972). Where do new enzymes come from? In *Enzymes: Structure and Function, 8th FEBS Meeting* 1972, vol. 29, ed. J. Drenth, R. A. Oosterbaan & C. Veeger, pp. 151–76. Amsterdam: North-Holland.

HARTLEY, B. S. & SHOTTON, D. M. (1971). Pancreatic elastase. In *The Enzymes*, vol. 3, 3rd edition, ed. P. D. Boyer, pp. 323–73. New York: Academic Press.

HILL, E., TSERNOGLOU, D., WEBB, L. & BANASZAK, L. J. (1973). Polypeptide conformation of cytoplasmic malate dehydrogenase. *Journal of Molecular Biology*, **72**, 577–91.

IWANAGO, S., WALLÉN, P., GRÖNDAHL, N. J., HENSCHEN, A. A. BLOMBÄCK, B. (1967). Isolation & characterisation of N-terminal fragments obtained by plasmin digestion of human fibrinogen. *Biochimica et Biophysica Acta*, **147**, 606–9.

JACOBSON, K. B. (1971). Reaction of aminoacyl-tRNA synthetases with heterologous tRNAs. *Progress in Nucleic Acid Research*, **11**, 461–88.

JOHNSON, P. & SMILLIE, L. B. (1972). A proposed structure for *Streptomyces griseus* protease A. *Biochemical Journal*, **130**, 36P.

KAPLAN, H., SYMONDS, V. B., DUGAS, H. & WHITAKER, D. R. (1970). Specificity of α-lytic protease from *Myxobacter sorangium*. *Canadian Journal of Biochemistry*, **48**, 649.

KIMURA, M. (1969). The rate of molecular evolution considered from the standpoint of population genetics. *Proceedings of the National Academy of Sciences, USA*, **63**, 1181–3.

KING, J. L. & JUKES, T. H. (1969). Non-Darwinian evolution. *Science*, **164**, 788–98.

KOCH, G., BOULANGER, Y. & HARTLEY, B. S. (1973). Gene-doubling and fusion in aminoacyl-tRNA synthetases. *Nature, New Biology*, in press.

KRAUT, J., ROBERTUS, J. D., BIRKTOFT, J. J., ALDEN, R. A., WILCOX, P. E. & POWERS, J. C. (1971). The aromatic binding site in subtilisin BPN′ and its resemblance to chymotrypsin. *Cold Spring Harbor Symposia on Quantitative Biology*, **36**, 117–24.

LAGERKVIST, U., RYMO, L., LINDQVIST, O. & ANDERSSON, E. (1972). Some properties of crystals of lysine transfer ribonucleic acid ligase from yeast. *Journal of Biological Chemistry*, **247**, 3897–9.

LAPOINTE, J. & SÖLL, D. (1972). Glutamyl tRNA synthetase of *E. coli*. *Journal of Biological Chemistry*, **247**, 4966–74.

LEMOINE, F., WALLER, J. P. & VAN RAPENBUSCH, R. (1968). Studies on methionyl-tRNA synthetase. *European Journal of Biochemistry*, pp. 213–21.

LOFTFIELD, R. B. (1972). The mechanism of aminoacylation of transfer RNA. *Progress in Nucleic Acid Research*, **12**, 87–128.

MCLACHLAN, A. D. & SHOTTON, D. M. (1971). Structural similarities between α-lytic protease of *Myxobacter* 495 and elastase. *Nature, New Biology*, **229**, 202–5.

MARKLAND, F. S. & SMITH, E. L. (1967). Subtilisin BPN′. Isolation of cyanogen bromide peptides and the complete amino acid sequence. *Journal of Biological Chemistry*, **242**, 5198–211.

MASSEY, V. (1963). Lipoyl dehydrogenase. In *The Enzymes*, vol. 7, ed. P. D. Boyer, H. Lardy & K. Myrbäck, pp. 275–306. New York: Academic Press.

MATTHEWS, B. W., SIGLER, P. B., HENDERSON, R. & BLOW, D. M. (1967). Three-dimensional structure of tosyl-α-chymotrypsin. *Nature, London*, **214**, 652–6.

MIKEŠ, O., TOMÁŠEK, V., HOLEYŠOVSKÝ, V. & SORM, F. (1966). Covalent structure of bovine trypsinogen. The position of the remaining amides. *Biochimica et Biophysica Acta*, **117**, 281–5.

MUENCH, K. H. & JOSEPH, D. R. (1971). Quaternary structure and substrate binding of the aminoacyl-tRNA synthetases. In *Nucleic Acid–Protein Interactions*, ed. D. W. Ribbons, J. F. Woessner & J. Schultz, pp. 172–83. Amsterdam: North-Holland.

NEURATH, H., BRADSHAW, R. A. & ARNON, R. (1970). Homology and phylogeny of proteolytic enzymes. In *Structure–Function Relationships of Proteolytic Enzymes*, ed. P. Desnuelle, H. Neurath & M. Ottesen, pp. 113–34. Copenhagen: Munksgaard.

OLSON, M. O. J., NAGABHUSHAN, N., DWINZIEL, M., SMILLIE, L. B. & WHITAKER, D. R. (1970). Primary structure of α-lytic protease. *Nature, London*, **228**, 438–42.

PERUTZ, M. F. (1963). X-ray analysis of haemoglobin. *Science*, **140**, 863–9.

PERUTZ, M. F. (1969). The haemoglobin molecule. *Proceedings of the Royal Society*, B, **173**, 113–40.

POWELL, E. O. (1958). Criteria for the growth of contaminants and mutants in continuous culture. *Journal of General Microbiology*, **18**, 259–68.

RAO, S. T. & ROSSMANN, M. G. (1973). Super-secondary structures in proteins. *Journal of Molecular Biology*, **76**, 241–56.

REID, B. R., KOCH, G. L. E., BOULANGER, Y., HARTLEY, B. S. & BLOW, D. M. (1973). Crystallisation and preliminary X-ray diffraction studies on tyrosyl-tRNA synthetase from *B. stearothermophilus*. *Journal of Molecular Biology*, in press.

RIGBY, P. W. J., BURLEIGH, B. D. & HARTLEY, B. S. (1973). An experimental approach to enzyme evolution. *Science*, in press.

ROBBINS, K. C., ARZADON, L., BERNABE, P. & SUMMARIA, L. (1972). Amino-terminal sequences of human plasminogen. *Federation Proceedings, Federation of American Societies for Experimental Biology*, **31**, 446.

ROBBINS, K. C., BERNABE, P., ARZADON, L. & SUMMARIA, L. (1972). The amino terminal sequences of human plasminogen and the S-carboxy-methyl heavy (A) and light (B) chain derivatives of plasmin. *Journal of Biological Chemistry*, **247**, 6757–62.

RUSSELL, R. L., ABELSON, J. N., LANDY, A., GEFTER, M. L., BRENNER, S. & SMITH, J. D. (1970). Duplicate genes for tyrosine tRNA in *E. coli*. *Journal of Molecular Biology*, **47**, 1–13.

SANGER, F. & SHAW, D. C. (1970). Amino acid sequence about the reactive serine of a proteolytic enzyme from *B. subtilis*. *Nature, London*. **187**, 872–3.

SCHMIDT, J., WANG, R., STANFIELD, S. & REID, B. R. (1971). Yeast phenylalanyl tRNA synthetase. Purification, molecular weight and subunit structure. *Biochemistry*, **10**, 3264–8.

SHAW, D. C. (1962). Radiochemical techniques for the determination of amino acid sequences in proteins. Ph.D. thesis, University of Cambridge.

SHAW, W. V. (1971*a*). Evolutionary aspects of bacterial resistance to chloramphenicol. *Transactions of the Association of American Physicians*, **84**, 190–9.

SHAW, W. V. (1971*b*). Comparative enzymology of chloramphenicol resistance. *Annals of the New York Academy of Sciences*, **182**, 234–42.

SHOTTON, D. M. & HARTLEY, B. S. (1970). Amino acid sequence of porcine pancreatic elastase and its homologies with other serine proteinases. *Nature, London*, **225**, 802–6.

SHOTTON, D. M. & WATSON, H. C. (1970). Three-dimensional structure of tosyl-elastase. *Nature, London*, **225**, 811–16.

SMEATON, J. R. & ELLIOTT, W. H. (1967). Isolation and properties of a specific bacterial ribonuclease inhibitor. *Biochimica et Biophysica Acta*, **145**, 547–560.

SMITH, E. L. (1970). Evolution of enzymes. In *The Enzymes*, 3rd edition, vol. 1, ed. P. D. Boyer, pp. 267–339. New York: Academic Press.

STROUD, R. M., KAY, L. M. & DICKERSON, R. E. (1971). The crystal and molecular structure of DIP-inhibited bovine trypsin at 2.7 Å resolution. *Cold Spring Harbor Symposia on Quantitative Biology*, **36**, 125–40.

TAMBURRO, A. M., BOCCU, E. & CELOTTI, L. (1970). The role of disulfide bonds in the protein structure. Conformational studies on reduced ribonuclease and lysozyme. *International Journal of Protein Research*, **2**, 157–64.

TANG, J., SEPULVEDA, J., MARCINISZYN, J. & LIU, D. (1973). Amino acid sequence of porcine pepsin. *Abstracts, 9th International Congress of Biochemistry*, 2a11.

TITANI, K., HERMODSON, M. A., FUJIKAWA, K., ERICSSON, L. H., WALSH, K. A., NEURATH, H. & DAVIE, E. W. (1972). Bovine Factor X_{1a} (Activated Stuart Factor). Evidence of homology with mammalian serine proteases. *Biochemistry*, **11**, 4899–902.

WALLER, J. P., RISLER, J. L., MONTEILHET, C. & ZELWER, C. (1971). Crystallisation of a trypsin modified methionyl-tRNA synthetase from *E. coli*. *FEBS Letters*, **16**, 186–8.

WATENPAUGH, K. D., SIEKER, L. C., JENSEN, L. H., LEGALL, J. & DUBOURDIEU, M. (1972). Structure of the oxidised form of a flavodoxin at 2.5 Å resolution. *Proceedings of the National Academy of Sciences, USA*, **69**, 3185–8.

WATSON, H. C. (1969). The stereochemistry of the protein myoglobin. In *Progress in Stereochemistry*, ed. B. J. Aylett & M. M. Harris, pp. 299–333. London: Butterworth.

WRIGHT, C. S., ALDEN, R. A. & KRAUT, J. (1969). Structure of subtilisin BPN′ at 2.5 Å resolution. *Nature, London*, **221**, 235–42.

WU, T. T., LIN, E. C. C. & TANAKA, S. (1968). Mutants of *Aerobacter aerogenes* capable of growth on xylitol. *Journal of Bacteriology*, **96**, 447–56.

THE EVOLUTION OF ENZYMES FOR THE UTILISATION OF NOVEL SUBSTRATES

PATRICIA H. CLARKE

Department of Biochemistry, University College London, London WC1E 6BT

But with bacteria constant evolutionary changes occur under our eyes and can be controlled and imitated in the laboratory.

M. Stephenson (1949)

INTRODUCTION

Adaptation to new environments

Micro-organisms, particularly bacteria, have long been known to be able to acquire new phenotypic characters, including the capacity to grow in media which did not support the growth of the original isolates. Early description referred to the 'training of bacteria' and this encompassed such changes as growth in the absence of amino acids required for the parent strain, fermentation of sugars and resistance to drugs. Until it was possible to differentiate clearly between effects due to phenotypic response to alterations in the chemical composition of the environment, and those due to mutations conferring permanent changes in the genetic characters of the population, the discussion of these phenomena was confused.

It is unlikely that anyone would now disagree with the statement made by Spiegelman in 1953 during the discussion at the Third Symposium of the Society on 'Adaptation in Micro-organisms'.

Changes in enzymatic constitution of microbial populations may therefore in some cases be due to any one of the following: (1) the selection of a mutant type; (2) the induced synthesis of enzyme in all, or a majority of the cells, of a genetically homogeneous population; (3) the simultaneous functioning of both mechanisms.

We are now able to look more closely at the relationships between the genetic complement of an organism and its actual and potential enzymatic activities. Mutations can be assigned to definite gene loci and although *Escherichia coli* remains pre-eminent in this field, there are now methods available for genetic exchange and chromosomal mapping in other genera. The molecular basis of enzyme induction and repression has advanced from speculation to fact, for at least some enzyme systems. A new method of enzymatic adaptation is suggested by recent findings that the genes for some catabolic enzymes, as well as those for drug resistance, may be carried on infectious plasmids (Chakrabarty, 1972; Dunn & Gunsalus, 1973; Rheinwald, Chakrabarty & Gunsalus, 1973). This provides a method by which genetic information carried by only a few cells of population may spread very rapidly when conditions are favourable.

The enzymes necessary for an organism to grow on any compound can be divided into three functional groups. The biosynthetic enzymes determine the synthesis of all the cell constituents from the intermediates of the metabolic pool; the central metabolic pathway enzymes are required to a greater or lesser extent whatever the primary growth substrate; the peripheral enzymes convert the primary growth substrate to compounds which can enter the central metabolic pathways. In attempting to examine the evolution of enzymes for the utilisation of novel substrates, it is the enzymes of the latter group which are the most interesting. These peripheral catabolic enzymes, whose function is to initiate the attack on organic compounds in the environment, are normally inducible and are almost always subject to catabolite repression by other carbon compounds.

It is possible to isolate mutants which are constitutive for the peripheral catabolic enzymes, but most if not all strains isolated from nature are inducible. There will be a definite growth advantage for an inducible strain compared with a constitutive strain, since unnecessary protein synthesis will be avoided, with a consequent saving in energy and in amino acids. Natural selection will operate against constitutive mutants which may arise in a population. A similar growth advantage would be possessed by strains which had evolved a catabolic repression control whereby synthesis of certain peripheral enzymes was repressed in the presence of compounds metabolised more directly. Adaptation to new growth substrates may require changes in regulator controls as well as, or instead of, changes in the enzymes themselves.

Comparative and experimental evolution

Comparative studies based on cell morphology are of relatively little importance to the study of microbial evolution but this can now be reinforced by comparative morphology at the molecular level. The base composition of DNA is now an accepted element of taxonomic grouping and DNA and RNA hybridisation is used for comparisons between species or genera (e.g. Brenner *et al.* 1969; Palleroni, Ballard, Ralston & Doudoroff, 1972). Cocks & Wilson (1972) compared the immunological relationships of alkaline phosphatases produced by thirty-two strains of enterobacteria with a view to understanding the evolution of this enzyme. Stanier, Wachter, Gasser & Wilson (1970) observed immunological relationships between two of the enzymes of the β-ketoadipate pathway which confirmed other findings on these *Pseudomonas* species and indicated possible evolutionary divergence.

Although metabolic pathways may be similar in widely divergent micro-organisms the regulatory patterns may be very different. The β-ketoadipate pathway occurs in *Acinetobacter* as well as in *Pseudomonas*, but while the regulation is the same in all the *Pseudomonas* species it is quite different in *Acinetobacter* (Ornston, 1971). For the biosynthetic pathways the enterobacteria characteristically possess isoenzymes at branch points which are independently regulated by feedback inhibition. In other genera the activity of branching biosynthetic pathways is controlled by other mechanisms such as concerted feedback inhibition of single enzymes (Cohen, Stanier & Le Bras, 1969; Jensen, Nasser & Nester, 1967). The regulatory patterns appear to be characteristic of well defined taxa and, since they may have arisen later than the enzymes they control, may be of evolutionary significance.

Detailed comparison of amino acid sequences has been applied to relatively few bacterial proteins compared with the very large number now known for eukaryotic organisms. Some recent reports suggest that the differences in sequences of homologous proteins in closely related prokaryotes may be surprisingly high. Ambler & Wynn (1973) compared the sequences of the cytochromes *c*-551 from *Pseudomonas aeruginosa*, *P. stutzeri*, *P. mendocina* and *P. fluorescens* biotype C. The differences between pairs of sequences ranged from 22 % to 39 % although it is known that these species are very similar (Stanier *et al.* 1966; Palleroni *et al.* 1970). The polypeptide chains were of the same length and detailed comparisons left no doubt that they were homologous proteins, but the divergence was so great that Ambler & Wynn (1973) considered that homologous proteins of more widely separated bacterial species might have diverged to such an extent that a common evolutionary origin would not be obvious. This is particularly startling since the amino acid sequences of the eukaryotic cytochromes *c* are so similar. Li, Denney & Yanofsky (1973) compared the amino acid sequences of the tryptophan synthetase α-chains of *Escherichia coli*, *Salmonella typhimurium* and *Klebsiella aerogenes* and also the extent of specific DNA–RNA hybridisation and concluded that during the course of evolution synonymous codon changes have accumulated in the α-chain structural genes. More proteins will become available for such analysis and it is clear that comparative studies of protein evolution in prokaryotes have much to offer.

A totally different approach is to take an organism and attempt to change it in the laboratory. The bacteria are ideal subjects for such manipulation. The short generation time, the possibility of working with very large numbers, the availability of methods for increasing the mutation rate, and the wealth of knowledge now available on biochemical pathways and regulatory systems offer very hopeful prospects for 'directed laboratory evolution'. There will, of course, be no guarantee that this will follow the same course as evolution by selection in nature, but if an adaptation can be shown to happen under controlled conditions then there is at least a chance that something similar did happen, or will happen, in the natural environment. Accounts of earlier experiments in this area were reviewed by Hegeman & Rosenberg (1970) and the following account will deal with more recent work, particularly that which can be related to defined changes in enzymes and regulatory systems.

HOW TO ACQUIRE NEW METABOLIC ACTIVITIES

For a compound to be used as a substrate for growth it must be able to enter the cell at a reasonable rate and be converted by appropriate enzymes to intermediates of the central metabolic pathways. Contrariwise, a compound is not utilised if it is unable to enter a cell, if there are no enzymes which can convert it to suitable metabolic intermediates, if the enzymes are not induced by the compound, if enzymes which can attack the compound, or its products, do so at a rate too low to be effective, or if the compound is an inhibitor of an essential cellular activity. Mutations which affect one or more of these properties may allow growth on a compound which cannot be metabolised by the parent strain. However, since growth depends on the controlled interlocking of rates of very many enzyme reactions which contribute to metabolic pathways, there are limitations to the types of evolutionary adaptation which might be expected. *Pseudomonas* species which are capable of metabolising phenols and cresols could be expected to extend the range of their metabolic activities so that they would grow much more rapidly on these compounds, or grow on related compounds, but adaptive changes of this sort would not be expected in *Lactobacilli*. Adaptation to growth on novel substrates has been shown to occur as a result of mutations affecting the following characteristics.

Regulator genes – induction and repression

The specificity of an enzyme is seldom absolute, so that an organism may already possess an enzyme which could attack compound *X* if some arrangement could be made to synthesise it in sufficient quantity. However, the specificities of an enzyme and its regulation system are usually similar but not identical and a compound may be a substrate, even though a very poor one, yet be totally lacking in inducer activity. In a wild-type inducible strain the basal level of an inducible catabolic enzyme is very low and this increases about 1000-fold in a fully induced culture. The same high enzyme levels are found in constitutive mutants. This means that even a compound for which the enzyme has low affinity, and which it attacks slowly, could be a potential growth substrate if a mutation occurred in the regulatory system. A negative control of the *lac* type depends on a regulator gene producing a repressor protein which binds specifically to the operator and prevents transcription (Reznikoff *et al.* 1969). Mutants which lack effective repressor proteins are constitutive and synthesise high levels of β-galactosidase in the absence of inducer.

Mutations leading to constitutivity need not be at specific sites, since all that is required is a change which impairs repressor function. Some of the *Klebsiella aerogenes* mutants which utilise 'unnatural' pentoses and pentitols were found to have mutations resulting in the constitutive synthesis of enzymes whose original substrates were chemically related compounds normally metabolised by the parent strains (Oliver & Mortlock, 1971*a*, *b*).

Mutants are also known in which a mutation in a regulator gene has changed the inducer specificity so that a compound for which an enzyme has some affinity is now able to induce enzyme synthesis. This has occurred in the selection of mutants of *Escherichia coli* which are able to grow on D-arabinose (LeBlanc & Mortlock, 1971*b*). Some of the mutants of *Pseudomonas aeruginosa* which are able to grow in succinate + formamide medium have regulator mutations which allow amidase to be induced by formamide at a much higher rate than in the wild-type strain (Brammar, Clarke & Skinner, 1967).

Promotor mutations

Mutations in the promotor region of the *lac* operon were first identified in *lac*⁻ strains which had intact *i*, *z* and *o* genes but were able to synthesise β-galactosidase only at very low rates. It was concluded that these mutations were in the DNA region which corresponded to the site of attachment of RNA polymerase and the initiation of transcription of the *lac* mRNA (Scaife & Beckwith, 1966). This led to the idea that, while by definition there should be a promotor region for each structural gene, or group of structural genes transcribed together, some promotors might be much more efficient than others. Most genes for catabolic enzymes would be expected to have efficient promotors which would permit a high rate of transcription in the presence of the inducer. At the other extreme would be genes which were transcribed very slowly and there would be a very low constitutive level of the proteins which they determined. An example of a structural gene with a very inefficient promotor would be the *lac i* gene which has so slow a rate of initiation of transcription that the bacteria synthesise only about 5–10 molecules per generation. Müller-Hill, Crapo & Gilbert (1968) isolated mutants which make more *lac* repressor and these i^q mutants are thought to have more efficient promotors which allow a higher rate of initiation of transcription with an increase of at least fivefold in the amount of *lac* repressor synthesised.

Some enzymes are normally required only at low concentrations and are synthesised constitutively at low rates since that is a simple and economical method of control. Pardee *et al.* (1971) found that nicotinamide deamidase was a microconstitutive enzyme in *E. coli*, active in the cyclical salvage pathway, but capable of hydrolysing only 3 nmol nicotinamide min^{-1} mg^{-1} protein. They were able to isolate mutants which were hyperconstitutive for the deamidase, producing sufficient enzyme to be able to utilise nicotinamide as the nitrogen source for growth. These were thought to be mutants with more efficient promotors. Microconstitutive enzymes might be overlooked in a wild-type strain and new enzyme activities which are difficult to relate to known enzymes in the cell may in some cases be due to promotor mutations of this type. Mutations which can be described as 'up-promotor' have been described in *Salmonella typhimurium* for the histidine catabolic pathway (Brill & Magasanik, 1969) and for the proline catabolic pathway (Newell & Brill, 1972) or as 'super-promotor' mutants for the amidase of *P. aeruginosa* (Smyth & Clarke, 1972).

Catabolite repression

The promotor region is also considered to be the site of interaction between catabolite repression and gene transcription (Silverstone *et al.* 1969), Pastan & Perlman (1968) showed that glucose repression of β-galactosidase synthesis could be relieved by cyclic AMP and a cAMP-binding (CAP or CRP) protein was shown to be essential for in-vivo and in-vitro transcription of the *lac* and *ara* operons (Emmer, De Crombrugghe, Pastan & Perlman, 1970; Zubay, Schwartz & Beckwith, 1970; De Crombrugghe *et al.* 1971). A mutation in the promotor may confer resistance to catabolite repression and this could enable a mutant to use a compound as a nitrogen source in the presence of a carbon source which normally represses the synthesis of the enzyme required to liberate the available nitrogen. Some of the histidine pathway mutants of *S. typhimurium*, and amidase mutants of *P. aeruginosa*, are thought to be promotor mutants of this type. A promotor-like mutant of *E. coli* produces high levels of glycerol kinase and is resistant to catabolite repression (Berman-Kurtz, Lin & Richey, 1971).

Not all catabolite repression-resistant mutants can be ascribed to the promotor region. There are some which have pleiotropic effects on the synthesis of several enzymes and these may be due to mutations in the gene for the CRP protein. Mutations in genes for enzymes determining the rates of functioning of metabolic pathways could indirectly determine the extent of catabolite repression and therefore affect the usefulness of enzymes with marginal activity towards the potential growth substrate.

Enzymes with new activities

Mutations in enzyme structural genes, which alter the affinities of enzymes for the substrates, or change the rates of reactions, or modify catalytic activities, must by their nature be specific and will therefore be less common than regulator mutations. In devising selection methods for detecting and isolating altered enzyme mutants, it may be necessary to use an indirect approach. Lerner, Wu & Lin (1964) isolated a mutant with an altered ribitol dehydrogenase by selecting for growth on xylitol. Mutant X1 was able to grow on xylitol and was constitutive for ribitol dehydrogenase. From this they obtained mutant X2 which grew faster on xylitol and produced the altered enzyme. All the *Pseudomonas aeruginosa* mutants producing altered amidases, discussed in a later section, were isolated from regulator mutants. Continuous culture offers a very good system for screening large numbers of bacteria to select for mutants which have acquired some growth advantage over the original population. In a medium in which growth is limited by a substrate-inducer, such as lactose, as the carbon source, the mutants isolated are constitutive since they will grow faster than the inducible parent on lactose and there is no selection pressure against them under these conditions. This is used as a general method for the isolation of constitutive mutants, but it will not operate if the organism has a very high affinity for the substrate-inducer and an efficient permease system.

Francis & Hansche (1972) used continuous culture to select an altered acid phosphatase from *Saccharomyces cereviseae*. Growth was limited by providing β-glycerophosphate as the sole source of phosphate, and the selection pressure was increased by buffering the growth medium at pH 6.0. This reduced the effective activity of the acid phosphatase by 70 %. After 400 generations they observed a change in the growth of the culture and this was the result of a mutation in the acid phosphatase structural gene altering the pH optimum of the enzyme from 4.2 to 4.8 and restoring 40 % of the activity lost by buffering the medium at pH 6.0. This is a very elegant example of combining the advantages of continuous culture with direct selection pressure on the enzyme.

Transport systems

None of the above mutations will be of practical value unless the compound can enter the cell. However, the specificity of many transport systems is low, and there are often several transport systems for related compounds with overlapping specificities. A transport system for a compound may differ in specificity from the enzyme which attacks it, and from the induction system. This situation gives a range of possibilities for the potential substrate. If it can enter the cell by an existing constitutive permease, then it might be attacked by an existing constitutive enzyme of low specificity. Alternately, the appropriate permease may be induced by the compound although the enzyme is not. The more rapid utilisation of xylitol by mutant X3 isolated by Wu, Lin & Tanaka (1968) depended on a third mutation which resulted in a constitutive arabitol permease. Xylitol can be transported into the cells by arabitol permease although it does not induce it. Thus, mutant X3 owes its growth advantage on xylitol to (*a*) the constitutive ribitol dehydrogenase mutation of X1, (*b*) the mutation increasing the affinity of the constitutive ribitol dehydrogenase for xylitol occurring in X2, and (*c*) the mutation producing a constitutive arabitol permease occurring in X3.

Pseudomonas putida cannot utilise β-carboxy-*cis,cis*-muconate, one of the intermediates of the protocatechuate branch of the β-ketoadipate pathway, and is therefore 'cryptic' for this compound. Mutants can be isolated which have an inducible uptake system for this compound (Meagher, McCorkle, Ornston & Ornston, 1972). An apparently new metabolic activity may therefore be due to an alteration in cellular permeability.

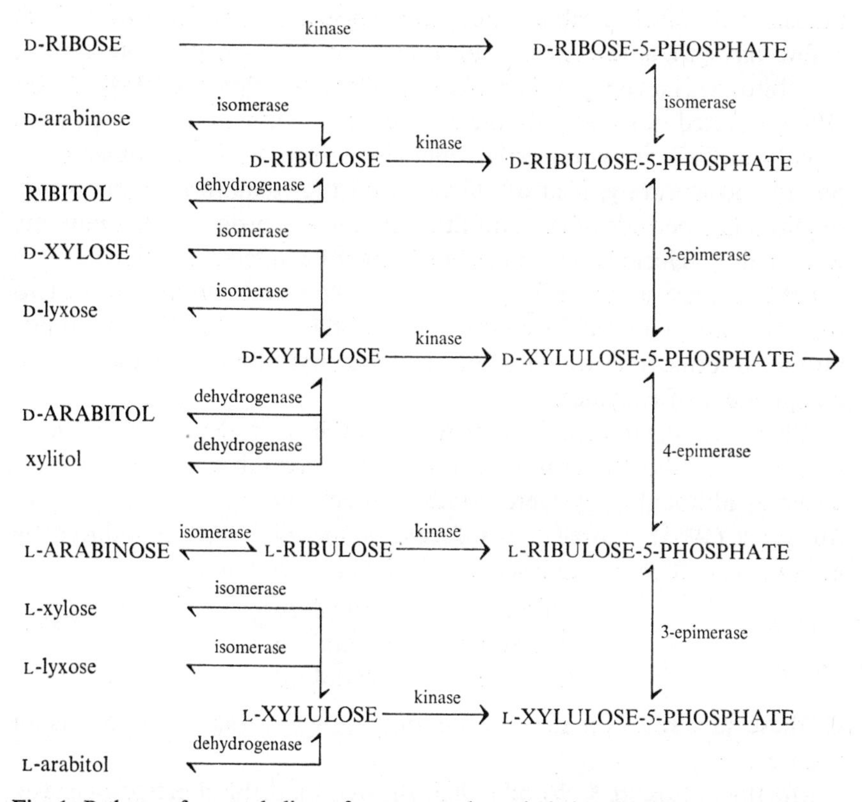

Fig. 1. Pathways for metabolism of pentose and pentitols in *Klebsiella aerogenes* (from Wood, 1966). The compounds listed in lower case are rarely found in the natural environment. Reproduced, with permission, from 'Carbohydrate Metabolism', *Annual Review of Biochemistry*, vol. 35, p. 539.

THE UTILISATION OF PENTOSES AND PENTITOLS BY ENTEROBACTERIA

The sugars and polyalcohols are used as the basis of fermentation media to discriminate between various enteric bacteria. The Emden–Meyerhof glycolytic pathway, or the pentose pathway, will be employed for any of these compounds which can be metabolised, so the differences between species are likely to be due to differences in the peripheral catabolic enzymes. The chemical similarities between many of the compounds would suggest that modifications in the properties of the enzymes, or their regulation systems, might increase the range which could be utilised for growth by a particular bacterial strain.

Wood and Mortlock and their colleagues have carried out many such investigations over a number of years. They showed that *K. aerogenes* strain PRL-R8 grew readily on D-xylose, D-ribose, L-arabinose, D-arabitol or ribitol, and after lag periods, which varied from a few days to a few weeks, they observed growth on D-arabinose, L-arabitol, D-lyxose, xylitol or L-xylose (Mortlock & Wood 1964). Wood (1966) pointed out that only the five pentoses and pentitols which were rapidly utilised are commonly found in nature, while the other compounds occur rarely, if at all. Growth on the 'unnatural' compounds required lag periods of such duration that it was probable that mutants were being selected. The possible metabolic interconversions which could occur are shown in Fig. 1 and it can be seen that only one or two reactions are required to convert one of the 'unnatural' compounds into an intermediate which would be recognised by one of the normal complement of enzymes.

When the enzyme specificities were examined in detail it was found that some of the 'unnatural' compounds were substrates for existing enzymes although they were attacked much more slowly. Camyre & Mortlock (1965) showed that L-fucose isomerase, which is induced by growth on L-fucose, catalysed the following reactions:

$$\text{L-fucose} \rightleftharpoons \text{L-fuculose}$$
$$\text{D-arabinose} \rightleftharpoons \text{D-ribulose}$$
$$\text{L-xylose} \rightleftharpoons \text{L-xylulose}$$

Of these substrates for the enzyme, only L-fucose was able to act as an inducer.

Mortlock, Fossitt & Wood (1965) showed that ribitol dehydrogenase, induced by growth on ribitol, catalysed these reactions:

$$\text{Ribitol} + \text{NAD} \rightleftharpoons \text{D-ribulose} + \text{NADH}$$
$$\text{Xylitol} + \text{NAD} \rightleftharpoons \text{D-xylulose} + \text{NADH}$$
$$\text{L-arabitol} + \text{NAD} \rightleftharpoons \text{L-xylulose} + \text{NADH}$$

These overlapping specificities were very important in the acquisition of new metabolic potential. Mutants of *K. aerogenes* selected for ability to grow on D-arabinose (or L-xylose) were found to be constitutive for L-fucose isomerase (Camyre & Mortlock, 1965). Oliver & Mortlock (1971*a*, *b*) showed that the constitutive strain grew on D-arabinose after a very short lag, and that if the wild-type inducible strain had been previously grown on L-fucose so that isomerase was already induced, then it too was able to grow on D-arabinose with only a short lag. Further metabolism of D-arabinose requires the phosphorylation of D-ribulose by ribulokinase. This enzyme is induced by growth on ribitol, but the actual inducer for both ribitol dehydrogenase and D-ribulokinase is D-ribulose, so that as soon as D-arabinose has been isomerised to D-ribulose the way is clear for it to be metabolised by the ribitol pathway. This is a very interesting example of the acquisition of a new metabolic pathway by a patchwork process of putting together two sequential catalytic steps which originated from two separate metabolic pathways. The new metabolic pathway of the D-arabinose utilising mutant of *K. aerogenes* is therefore:

D-arabinose
↑ ↓ L-*fucose isomerase* (constitutive)
D-ribulose
↓ D-*ribulokinase* (induced by D-ribulose)
D ribulose-5-phosphate
↑ ↓
D-xylulose-5-phosphate

Oliver & Mortlock (1971*b*) isolated a 'fitter' D-arabinose mutant from the strain which was constitutive for L-fucose isomerase. This mutant produced larger colonies on plates with D-arabinose as carbon source, and the L-fucose isomerase produced by one of these mutants had a higher relative activity for D-arabinose and a slightly lower *Km* for both pentoses. Thus, from a regulator mutant they were able to isolate a mutant, with an altered enzyme. We have already seen that the X3 xylitol-utilising mutant of *K. aerogenes* isolated by Wu *et al.* (1968) was constitutive for a ribitol dehydrogenase which was altered in activity, and that xylitol was transported into the cell by a constitutive arabitol permease.

A different solution to the problem of growth on D-arabinose was found in mutants of *Escherichia coli.* It was known that growth on L-fucose induced three enzymes which were capable of acting on D-arabinose and its metabolites. A mutant was isolated by LeBlanc & Mortlock (1971*a*, *b*) in which these enzymes were induced by D-arabinose as well as by L-fucose. The pathway for D-arabinose metabolism is therefore thought to be as follows:

D-arabinose
⇅ L-*fucose isomerase*
D-ribulose
↓ L-*fuculokinase*
D-ribulose-1-phosphate
⇅ L-*fuculose*-1-*phosphate aldolase*
Dihydroxyacetone phosphate + glycolaldehyde

In this case there has been a change in inducer specificity and enzymes with some activity for D-arabinose and its products are now inducible by D-arabinose as well as by L-fucose. Since all the enzymes are induced by both pentoses it is probable that they form a single regulation unit.

EVOLUTIONARY POTENTIAL OF *PSEUDOMONAS* SPECIES

Pseudomonas species are widely distributed in nature and are well known to be able to attack many different organic compounds. In their survey of the characteristics of the aerobic pseudomonads, Stanier *et al.* (1966) list the origins of some of the strains examined as 'polluted sea water', 'soil by naphthalene enrichment', 'soil by camphor enrichment' and 'clay suspended in kerosene for 3 weeks', which indicates that these species must have evolved a battery of peripheral catabolic enzymes for converting some rather exotic growth substrates into intermediates for the central metabolic pathways. Bacterial taxonomists have from time to time complained about the variability of pseudomonads (Rhodes, 1971), but as Stanier pointed out in 1953, 'The taxonomist's misfortune is the evolutionist's opportunity.' The array of biochemical information now available on the catabolic pathways of pseudomonads (Ornston, 1971), and the existence of genetic systems in *P. aeruginosa* and *P. putida* (Holloway, Krishnapillai & Stanisich, 1971), make them a very suitable group for comparative and experimental evolutionary studies.

The genes for biosynthetic enzymes, unlike those of *Escherichia coli*, are scattered around the chromosomes in small gene clusters. However, the genes for some catabolic enzymes have a remarkable degree of supra-operonic clustering so that a transduction group may include a number of separate gene clusters of a catabolic pathway. Further, genes for unrelated catabolic pathways may also be found in the same transduction linkage group (Wheelis & Stanier, 1970; Leidigh & Wheelis, 1973). In some strains the genes for a particular catabolic pathway appear to be on the chromosome while in others they are carried on a plasmid (Dunn & Gunsalus, 1973). If the plasmid can be introduced into a strain unable to metabolise one set of compounds, it may displace a resident plasmid (Dunn & Gunsalus, 1973; Chakrabarty, Chou & Gunsalus, 1973). This method of storing genetic information in small packets which can be readily passed on from cell to cell may be important for the biochemical versatility of the *Pseudomonas* group.

THE DIRECTED EVOLUTION OF THE ALIPHATIC AMIDASE OF *PSEUDOMONAS AERUGINOSA*

$$CH_3CONH_2 + H_2O \longrightarrow CH_3COOH + NH_3$$

The aliphatic amides provide a homologous series of compounds which, if hydrolysed by a suitable enzyme, could be used as carbon and nitrogen sources for growth. The amidase produced by wild-type strains of

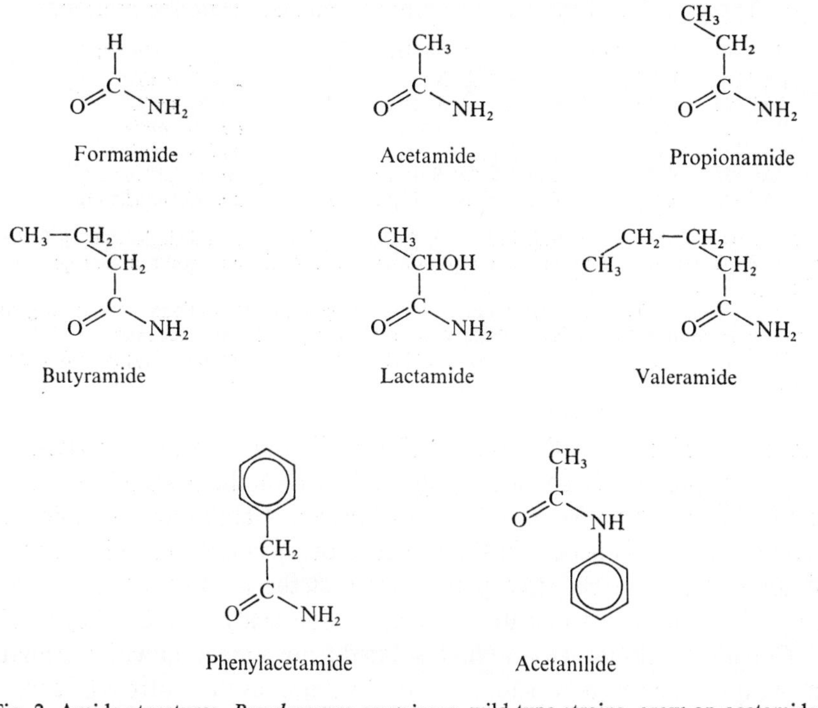

Fig. 2. Amide structures. *Pseudomonas aeruginosa*, wild-type strains, grow on acetamide, propionamide and lactamide as sole carbon and nitrogen sources. Mutants can utilise other amides.

P. aeruginosa permits growth on acetamide and propionamide. These two amides are good inducers and good substrates of the enzyme; formamide is a poor substrate and a poor inducer; butyramide is a very poor substrate and moreover does not induce but acts as an amide analogue repressor and prevents induction by other amides. Some substituted amides can interact with the system in various ways. Glycollamide is a good substrate and a good inducer while lactamide is a good inducer but a poor substrate; some *N*-substituted amides, like *N*-methylacetamide and *N*-acetylacetamide, can be used as non-substrate inducers, while *N*-phenylacetamide and cyanoacetamide compete with inducers and act as amide analogue repressors (Fig. 2 compares some amide structures). By exploiting the differences in the specificities of the enzyme and its regulatory system, it has been possible to devise methods for the isolation of regulatory mutants and also mutants producing enzymes with altered substrate specificities. The genetic system is relatively simple, with the amidase structural gene closely linked to a regulatory gene, and although it has not yet been possible to

Table 1. Pseudomonas aeruginosa *amidase regulator mutants*

Strain no.	Series no.	Phenotype	Genotype
PAC1	Wild type	Ind S/F$^-$A$^+$B$^-$	*amiR$^+$ amiE$^+$*
PAC101	C1	Con S/F$^+$A$^+$B$^+$But-r	*amiR1 amiE$^+$*
PAC111	C11	Con S/F$^+$A$^+$B$^-$But-s	*amiR11 amiE$^+$*
PAC153	F6	F-ind S/F$^+$A$^+$B$^-$	*amiR43 amiE$^+$*
PAC128	CB4	Con S/F$^+$A$^+$B$^+$But-r	*amiR11,37 amiE$^+$*
PAC142	L10	Con S/F$^+$A$^+$B$^+$But-r Crp-r	*amiR53 amiE$^+$ crp-7*

Phenotype abbreviations: Ind, inducible; Con, constitutive; F-ind, inducible by formamide; But-r, resistant to butyramide repression; But-s, sensitive to butyramide repression; Crp-r, resistant to catabolite repression.

Genotype abbreviations: amiR, amidase regulator gene; *amiE*, amidase structural gene; *crp-7*, gene, unlinked to amidase genes, concerned in repression by succinate.

References: Brammar, Clarke & Skinner (1967); Brown & Clarke (1970); Betz & Clarke (1972).

test for dominance by the construction of diploids, the characteristics of the system appear to be most easily explained on the basis of a negative control of the type found for the *lac* operon. Catabolite repression of amidase synthesis occurs in the presence of succinate and related compounds. There is preliminary evidence that the level of cyclic AMP is related to the extent of catabolite repression (Smyth & Clarke, 1972).

Constitutive mutants were first isolated from a medium with succinate providing the carbon source and formamide the nitrogen source (S/F medium). This medium also selects F mutants which are inducible by formamide (Brammar *et al.* 1967). Selection on succinate + lactamide plates (S/L medium) gives L mutants which are resistant to catabolite repression, and occasionally double mutants are selected which are both constitutive and resistant to catabolite repression. Table 1 lists the properties of the wild-type strain and some of the regulator mutants. These strains have been assigned series numbers as well as the usual laboratory strain numbers. The series number indicates the class of mutant and its origin and will be used where appropriate to indicate more clearly the sequence of evolutionary events which may have occurred (e.g. C11 is a constitutive mutant isolated on S/F medium).

Butyramide-utilising B and CB mutants

The wild-type strain can hydrolyse butyramide at about 2% of the acetamide rate under the standard assay conditions and it could be predicted that if enough of the enzyme were present then growth might occur on butyramide. However, the obstacles to growth on butyramide include low enzyme activity, lack of induction and also the amide analogue repressor activity of butyramide. Not only is amidase induction of the wild-type strain repressed by butyramide, but some constitutive mutants are severely repressed by butyramide, although they can produce high amidase levels in its absence. We pointed out (Clarke & Lilly, 1969) that there was a critical value (*P*) relating the amount of amidase synthesised in unit time (*E*) and the rate of activity of the enzyme on butyramide (*B*), above which growth could occur and below which the rate of hydrolysis of butyramide would be too slow. For any marginal growth substrate such as butyramide, it should be possible to achieve the threshold value of *P* either by increasing the amount of enzyme or by increasing its activity. We have obtained butyramide-utilising CB mutants altered in their regulation which are able to synthesise amidase at high levels in the presence of butyramide, and B mutants which synthesise an altered enzyme.

The first CB mutants isolated result from a second mutation in the regulator gene of the constitutive mutant C11, which itself was unable to grow on butyramide because it was severely repressed. It was found later that CB mutants could be isolated directly from the wild type in a single mutational step, and that some of the C mutants isolated from S/F medium were partially resistant to butyramide repression and could therefore grow on butyramide (Brown & Clarke, 1970). This is clearly a common type of mutation and it can be envisaged that mutations at many different sites could result in a regulator protein which had lost the capacity to bind butyramide, or could bind butyramide but could no longer take up the correct conformation for combining with the operator.

The other class of butyramide-utilising mutants, the B group, produce altered enzymes and have been isolated only from the constitutive mutant C11. Brown, Brown & Clarke (1969) concluded that six independent isolates all produced the same altered enzyme, B amidase, which differed from wild-type A amidase in electrophoretic mobility, K_m and V_{max}, but was similar in thermal stability and gave a complete cross-reaction with the antiserum prepared against purified A amidase. The two proteins were therefore very alike but the difference was sufficient to allow butyramide to be hydrolysed at 30 % of the acetamide rate under the standard assay conditions. The B amidase could be described as a more efficient enzyme than A amidase since, although the rates of acetamide hydrolysis were similar for both enzymes, the rate of propionamide hydrolysis by B amidase was about 1.5 times that of A amidase, while the rate of butyramide hydrolysis was ten-fold greater. It is doubtful, however, whether this would have any advantage in nature. Such an alteration in an enzyme protein requires a specific change in the amino acid sequence and it may well be that it involves a particular change at a particular site and that no other mutation could produce an enzyme with B amidase properties in a single mutational step from the wild-type gene. The derivation of the two classes of butyramide-utilising mutants is shown in Fig. 3. The growth response in butyramide medium is shown in Fig. 4. The constitutive mutant CB4 producing A amidase grows on butyramide, but the rate of growth is limited by the rate of hydrolysis of butyramide which disappears slowly from the medium. The altered enzyme mutant B6 soon produces enough of the B amidase to hydrolyse the butyramide which soon disappears from the medium so that the repression of enzyme synthesis is relieved.

Valeramide-utilising V mutants

The wild-type enzyme has no activity on valeramide and this compound acts as an amide analogue repressor. It was found that mutant B6 had trace activity for valeramide hydrolysis and this strain was used to isolate a group of valeramide-utilising V mutants (Fig. 3). No V mutants were obtained directly from the wild type, or from the butyramide-sensitive strain C11. The V mutants were a heterogeneous group and it was clear that more than one type of alteration could produce an enzyme which could hydrolyse valeramide at a rate sufficient for growth to occur. Several of the V mutants had higher specific activities for butyramide hydrolysis than their parent strain B6, and we have suggested that, in selecting for an enzyme to attack a particular compound belonging to homologous series, it might be worth carrying out the selection with the next compound up in the series. The V enzymes were less stable under all conditions than the A and B amidases, but we have not attempted to repeat this isolation to see if we could find any stable V amidases. Some of them, in gaining the ability to grow on valeramide, had lost the ability to grow on acetamide, thus losing a characteristic property of the wild-type strain in the process of gaining the new mutant character. The continued identity of the enzyme protein was demonstrated by immunodiffusion cross-reactions with antisera to A and B amidases, but the reactions were incomplete and the bands diffuse, which may have indicated dissociation into subunits. Activity was rapidly lost from cell-free extracts and some of the enzymes are cold-labile (unpublished observations).

Phenylacetamide-utilising Ph mutants

Phenylacetamide was not a very promising growth substrate. None of the mutant enzymes previously isolated appeared to attack it and it was known to be an amide analogue repressor. However, some of the

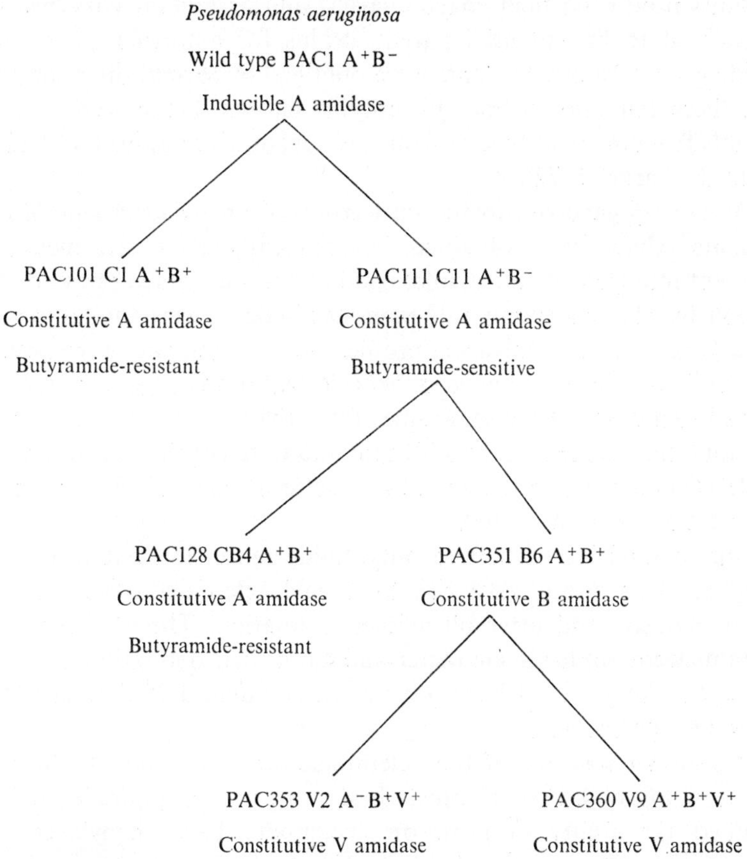

Pseudomonas aeruginosa

Wild type PAC1 A^+B^-

Inducible A amidase

PAC101 C1 A^+B^+

Constitutive A amidase

Butyramide-resistant

PAC111 C11 A^+B^-

Constitutive A amidase

Butyramide-sensitive

PAC128 CB4 A^+B^+

Constitutive A amidase

Butyramide-resistant

PAC351 B6 A^+B^+

Constitutive B amidase

PAC353 V2 $A^-B^+V^+$

Constitutive V amidase

PAC360 V9 $A^+B^+V^+$

Constitutive V amidase

Fig. 3. Derivation of butyramide and valeramide-utilising mutants. Regulator mutants, resistant to butyramide repression (C and CB), produce A amidase. Mutant B6 produces B amidase and mutants V2 and V9 produce V-type amidases.

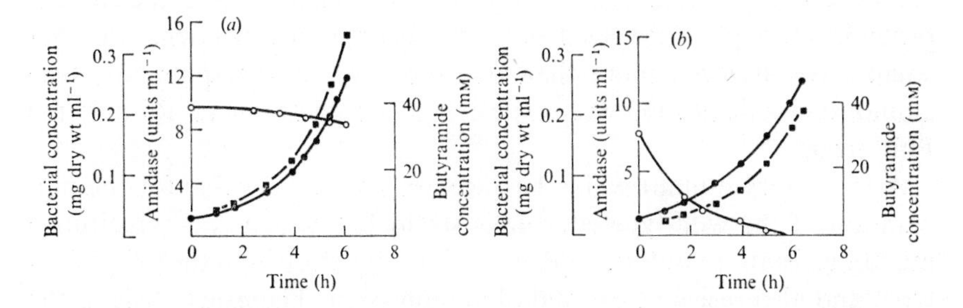

Fig. 4. Growth of butyramide-utilising mutants in minimal medium containing 40 mM-butyramide as carbon source. (*a*) Mutant CB4; (*b*) mutant B6. ●—●, Bacterial growth (mg dry wt ml^{-1}); ■—■, amidase (units ml^{-1}); ○—○, butyramide concentration. (From Brown & Clarke, 1970.)

mutants producing high enzyme levels, and/or mutant enzymes, were considered to be potential parent strains for a further extension of amide growth range. Mutants were obtained by several different routes and these fell into distinct phenotypic classes. None were obtained directly from the wild type or from the constitutive strains C11 and L10 (Betz & Clarke, 1972).

Mutant B6 gave rise to the PhB group of phenylacetamide-utilising mutants which were obtained independently after treatment with different mutagens or ultraviolet irradiation, and some appeared spontaneously. The sixteen mutants examined in detail were similar in amide growth range and in substrate profile. No recombinants were obtained in transduction crosses between them, so that it was concluded that they all had mutations at the same sites. Since the parent strain already had one mutation (at least), the PhB mutants were considered to have two (at least) mutations in the amidase structural gene and two amino acid substitutions were expected.

Mutant CB4 is resistant to butyramide repression but produces A amidase. This strain gave rise to phenylacetamide-utilising mutants spontaneously, and after ultraviolet irradiation. Three of these PhF mutants were similar in character and from their derivation it was predicted that they would have one (at least) amino acid difference from the wild-type enzyme.

Mutant V9 was one of the valeramide-utilising mutants which had a reasonably high activity on its best substrate, butyramide, and had retained the ability to grow on acetamide. Two phenylacetamide-utilising mutants were isolated from V9 and both these occurred spontaneously. They differed in phenotype and while PhV1 was relatively thermostable and resembled PhB3, the other mutant PhV2 was both thermolabile and cold-labile. In each case the mutational changes which had occurred could have led to three amino acid changes in the enzyme protein, although these need not have been at independent sites but could have involved more than one mutation at the same site. One amino acid substitution was expected to be common to B6 and the PhB group.

A regulatory mutant which has been of particular value for the preparation of the wild-type enzyme is strain L10 which is a constitutive mutant resistant to butyramide repression (therefore with the CB phenotype) and also resistant to catabolite repression. The exact basis of the latter mutation is not known but it is in a gene which is unlinked to the amidase structural and regulator genes. From strain L10 were isolated a number of butyramide-negative mutants, some of which grow slowly

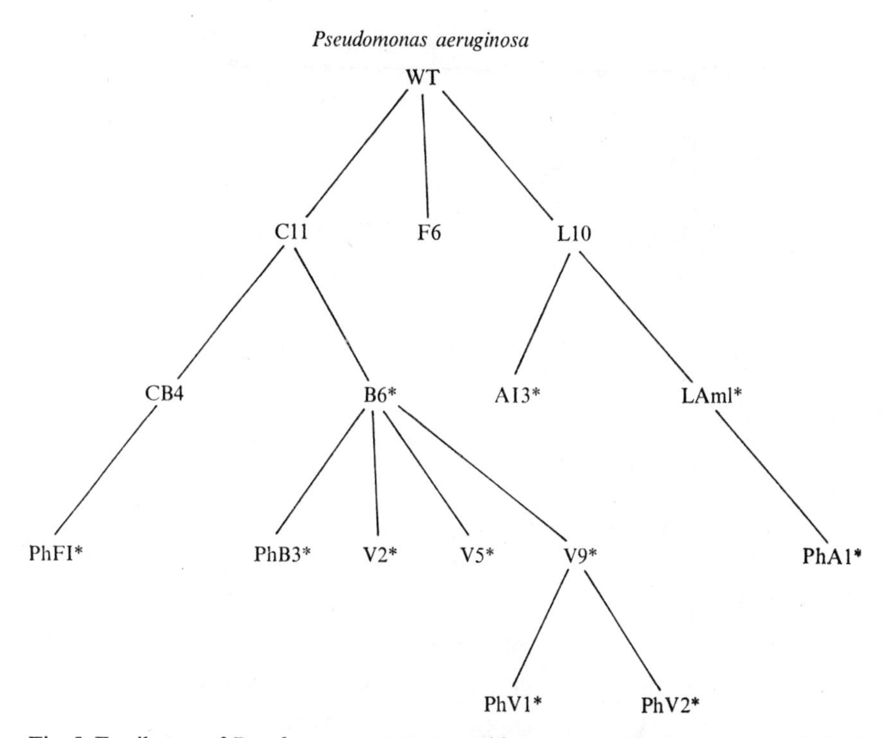

Fig. 5. Family tree of *Pseudomonas aeruginosa* amidase mutants. Each mutant was derived by a single mutational step from the strain which precedes it. Mutants producing altered amidase proteins are marked with an asterisk, the unmarked mutants are regulator mutants. Further details are given in Table 1 and Fig. 3 and in the text.

on acetamide. The rationale for the isolation of this class of mutant was that it would give a series of mutant proteins with defective catalytic properties but with enough residual activity to allow them to be isolated and studied. The regulatory mutations were expected to ensure that large amounts of the mutant proteins would be produced. Strain L10 itself did not give rise to any phenylacetamide-utilising mutants but one of the defectives gave rise to the phenylacetamide-utilising mutant PhA1. The enzyme produced by PhA1 was very thermolabile. From the mutational history it might be expected to have two amino acid changes and neither of these need to be related to those of the other Ph mutants. The derivation of the phenylacetamide-utilising mutants is shown in Fig. 5. The two phenotypic classes are: Group I, PhB mutants and PhV1 which are unable to grow on acetamide and produce thermostable amidases; Group II, PhV2, PhF mutants and PhA1 which grow very slightly on acetamide and produce thermolabile amidases (Fig. 6). The amidases produced by both groups of Ph mutants hydrolyse

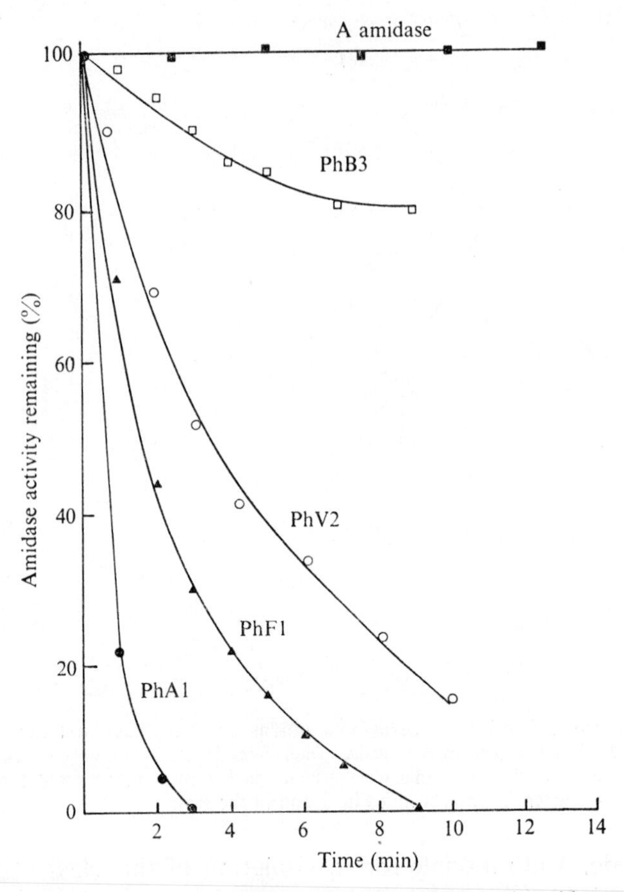

Fig. 6. Effect of heating at 60 °C on activity of pure wild-type A amidase and Ph mutant amidases in crude extracts. ■—■, A amidase; □—□, strain PAC377 (PhB3); ○—○, strain PAC389 (PhV2); ●—●, strain PAC391 (PhA1); ▲—▲, strain PAC392 (PhF1). (From Betz & Clarke, 1972.)

valeramide rather better than phenylacetamide under the standard conditions, and some have been shown to be most active on hexanoamide. The specific activities are low compared with the A and B amidases but nevertheless enough enzyme is synthesised to allow a reasonable rate of growth on plates and in liquid medium. Phenylacetamide is a rather insoluble compound and its solubility would limit the available concentration even in the most favourable environment. It was interesting that the apparent K_m for phenylacetamide for mutants PhB3, PhV1 and PhF1 was 1–3 mM whereas the apparent K_m for butyramide was about 200 mM for PhB3 and PhV1 and 38 mM for PhF1. All the Ph amidases gave cross-reactions with antisera to wild-type A amidase.

Table 2. *Amide growth range of* Pseudomonas *species and mutants*

Strain no.		A	B	V	H	Pyr/Ph	Ph
P. aeruginosa	PAC1	+	−	−	−	−	−
	PAC377 (PhB3)	−	+	+	+	+	−
P. putida	A87	+	+	−	−	+	+
	A87C/3	−	−	−	−	+	+
	A90	−	−	−	−	+	+
P. cepacia	716	+	−	+	+	+	+
	716/5	−	−	+	+	+	+
P. acidovorans	NC1B 9681	+	+	−	−	+	+

Growth media: Minimal salt agar + amides % (w/v): A, acetamide 0.3; B, butyramide 0.2; valeramide, 0.2; hexanoamide, 0.1; Pyr/Ph, pyruvate, 0.5 + phenylacetamide, 0.1; Ph, phenylacetamide 0.1. (*P. aeruginosa* is unable to grow on phenylacetate but the other species are able to do so.) *P. putida* A87C/3 and *P. cepacia* 716/5 were isolated as acetamide-negative mutants (Betz & Clarke, 1973). *P. aeruginosa* PAC377 (PhB3) was isolated as a phenylacetamide-utilising mutant (Betz & Clarke, 1972).

Occurrence of phenylacetamidases in nature

In a comparative study of amidases in other *Pseudomonas* species it became apparent that some strains of *P. putida*, *P. acidovorans* and *P. cepacia* were able to grow on phenylacetamide and that enzymes capable of hydrolysing phenylacetamide were induced by phenylacetamide but not by acetamide. It thus appeared that some strains produced two distinct amidases, since growth on acetamide resulted in the synthesis of acetamidases but not phenylacetamidases. A strain of *P. putida*, A90 (Stanier *et al.* 1966), which has been widely used for the study of the aromatic pathway enzymes, is able to grow on phenylacetamide but does not grow on acetamide. It thus resembled the *P. aeruginosa* mutant PhB3 in phenotype, with the exception that its phenylacetamidase is inducible. *P. putida* A87 grows both on acetamide and phenylacetamidase and produces two amidases with different specificities, but mutants were obtained which had lost acetamidase activity but retained phenylacetamidase activity. These now resembled the wild-type strain A90 (Betz & Clarke, 1973) in amide growth phenotype. Table 2 compares the amide growth range of *Pseudomonas* species and some of the mutants. *P. cepacia* appears to be the most versatile of the wild-type strains. While the acetamidase attacks only acetamide (formamide and propionamide) the second amidase attacks valeramide, hexanoamide and phenylacetamide, and this activity is retained unchanged in mutants selected for inability to grow on acetamide. The *P. acidovorans* acetamidase allows growth on butyramide and the phenylacetamidase activity is more restricted, so that growth does not occur on valeramide or hexanoamide. The growth of *P. acidovorans* and *P. putida* A87 on butyramide is dependent on the synthesis of an aliphatic amidase with very similar properties to the aliphatic amidase produced by strains of *P. aeruginosa*. However, in the case of *P. acidovorans* and *P. putida*, butyramide is able to induce the synthesis of the aliphatic amidase, while in *P. aeruginosa* strains it represses synthesis (Clarke, 1972).

During natural evolution there has been a divergence of regulatory properties of the aliphatic amidases produced by *Pseudomonas* species. Further, three species of *Pseudomonas* have evolved phenylacetamidases which resemble the phenylacetamidases produced by the phenylacetamide-utilising strains of *P. aeruginosa* obtained by directed evolution in the laboratory. All our phenylacetamide-utilising mutants are constitutive and this is a direct consequence of the methods by which they were derived. From the point of view of overall cell economy in protein synthesis, the next step would be to acquire a mutation which would result in an inducible phenotype. We have selected for changes in a single gene but if we had been able to retain the original amidase structural and regulator genes we might have arrived at a mutant of *P. aeruginosa* producing two distinct amidases like *P. putida* A87.

Acetanilide-utilising AI mutants

$$C_6H_5\text{-}NHCOCH_3 + H_2O \longrightarrow C_6H_5\text{-}NH_2 + CH_3COOH$$

Mutants capable of growth on acetanilide (*N*-phenylacetamide) were isolated by Brown & Clarke (1972) from the constitutive, catabolite repression-resistant strain L10. These mutants were capable of good growth with acetanilide as a carbon source and could also utilise the aniline produced by hydrolysis of acetanilide as a nitrogen source, but growth was then much slower and depended on the induction of enzymes for the dissimilation of aniline. The amidases produced by the AI mutants could be distinguished from both A and B amidases by their electrophoretic mobility on starch gels. They were more thermolabile than A, B, PhB3 and PhV1 amidases, but not as sensitive to heating at 60 °C as the Group II phenylacetamidases. The enzyme was stable in cell-free extracts and retained about 20 % of the specific hydrolase activity for acetamide possessed by the parental A amidase. The cross-reaction with antiserum to A amidase was complete. The relative ease of selection of this strain suggested that only one mutation with a single amino acid substitution had taken place.

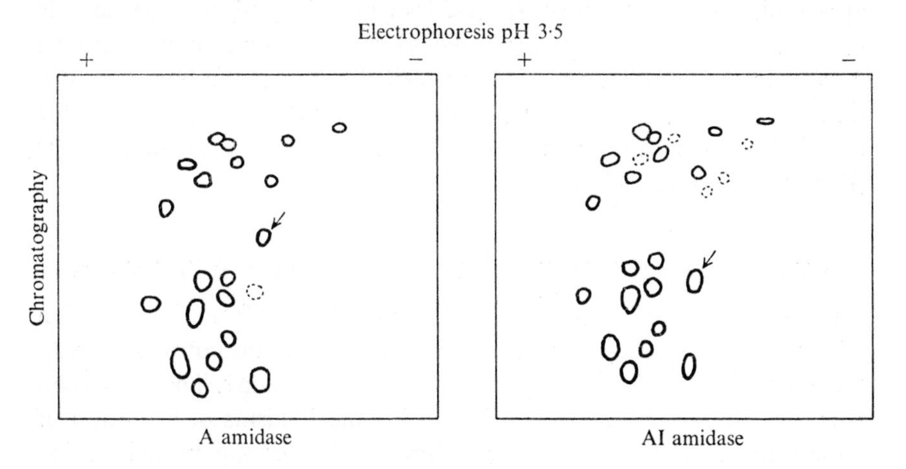

Fig. 7. Fingerprint of neutral peptides obtained after trypsin + chymotrypsin digestion of A amidase purified from strain PAC142 (L10) and AI amidase purified from strain PAC366 (AI3). (From Brown & Clarke, 1972.)

Table 3. *Amino acid substitution in mutant amidase*

Strain no.		Sequence of hexapeptide
PAC366	AI3 amidase	Ser-Leu-*Ile*-Gly-Glu-Arg
PAC142	Wild-type amidase	er-Leu-*Thr*-Gly-Glu-Arg

Strain PAC366 (AI3) was isolated on acetanilide (*N*-phenylacetamide) from a catabolite-derepressed magnoconstitutive strain PAC142 (L10).

Amino acid substitutions in the mutant amidases

Mutations leading to alterations in substrate specificity would be expected to leave the catalytic site of the enzyme unaltered but to result in amino acid changes in regions which determine the accommodation of the side chains of the amide substrates. These could be adjacent to the catalytic site of the enzyme but need not necessarily be so, since the folding of the enzyme is likely to bring together amino acid sequences from various parts of the polypeptide chain. The limited number of aliphatic amides which can be accommodated by the wild-type enzyme had indicated that the size of the side chain was of the greatest importance in determining which amides could be hydrolysed. When the purified amidase from strain AI3 was subjected to trypsin + chymotrypsin digestion and compared with a preparation of the wild-type amidase it was found that a single peptide was displaced in the map of the neutral peptides (Fig. 7). The arrow shows the peptide in the map from A amidase which is missing from the AI3 enzyme and the new peptide which appeared. Both these were hexapeptides which differed in sequence only in the substitution of isoleucine in the AI peptide for a threonine residue present in the A amidase. It is suggested that this is the only substitution in this mutant enzyme and that this is sufficient to account for the difference in substrate specificities of the two enzyme proteins (Brown & Clarke, 1972).

EVOLUTION OF NEW ENZYMES AND NEW PATHWAYS

The theoretical approach to enzyme evolution starts from the evolution of the primitive catalysts of the primordial soup and continues up to the enzymes of the higher organisms of the present day. The discussion here will be concerned only with possible mechanisms of evolution of biological catalysts in micro-organisms well after the period during which they became contained within defined cellular structures and would have been recognisable as polypeptides with binding sites for substrates and co-factors. The emphasis will be on those models for enzyme evolution which can be related to the changes in enzyme structure, function and regulation which have been observed in the laboratory.

Evolution of more efficient enzymes

It is generally assumed that the first catalysts which could be recognised as belonging to living organisms must have been very inefficient compared with the enzymes we now recognise. Koch (1972) has attempted to extrapolate back from what we know about mutation, recombination and selection of enzymes in *Escherichia coli* to the evolution of more efficient enzymes in the Precambrian era. During the long period of time before biological activity had made significant changes to the environment of the planet, he suggests that mutation rate of these primitive organisms might have been very high with more intense radiation in the atmosphere and in the absence of DNA repair mechanisms. He suggests that gene duplication was essential for evolution to occur. Gene duplication alone can result in an immediate increase in enzyme activity if the transcription and translation system is active enough for the amount of enzyme synthesised to be directly related to the number of gene copies. If the enzyme was rate-determining for growth then the organisms with multiple gene copies would have a growth advantage and be selected. At a later time when the selection pressure was less intense, one gene copy might accumulate mutations and become more efficient and effectively supersede the original form of the gene. Koch (1972) suggests that this might involve a series of intermediate steps in which one gene copy was made non-functional and non-translatable by chain termination, deletions or other mutations such as promotor mutations. These non-functional forms of the gene might persist in a quasi-stable state for a long period of time during which they would accumulate mutations. If not lost altogether by a large-scale deletion event these silent genes could eventually be rendered translatable again by a reversion mutation to allow translation of the modified gene. Koch (1972) is suggesting that this sequence of mutational events would be very rapid compared with a route involving simultaneous multiple changes in a single gene. His premise is that 'The evolution of polypeptide structure is favoured by long periods of time during which a gene whose function is temporarily not needed can be maintained in the population.' When the environmental conditions again require high activity of the enzyme, then although the same duplication of the original gene could occur again, it is the enzyme coded for by the reversion of the untranslatable gene which will take over if during the latent period it has acquired mutations which code for a 'better enzyme'.

Gene duplication as a solution to the problem of increasing the amount of a particular enzyme has been observed in the selection of hyper strains for β-galactosidase production in *Escherichia coli* grown in a chemostat with limiting lactose (Horiuchi, Tomizawa & Novick, 1962). Hyper strains carrying multiple copies of the *lac* genes tend to lose the extra copies when the selection pressure is removed and there is no record of any of the partial diploids giving rise to more efficient β-galactosidases. However, this enzyme has a long evolutionary history and may have already attained the maximum possible efficiency. A very interesting gene duplication (amplification) was reported by Folk & Berg (1971) for the structural gene of glycyl-tRNA synthetase in *Escherichia coli*. Some glycine auxotrophs had been shown to produce an altered synthetase with a lower affinity for glycine and it was concluded that the rate of esterification of the mutant $tRNA^{Gly}$ with endogeneously synthesised glycine was too low for growth. Revertants to glycine independence were isolated at a high frequency ($> 10^{-5}$) which still retained the mutant synthetase gene, but had more than one copy. If an additional copy of the mutant allele of the glycyl-tRNA synthetase gene could be maintained during many generations of growth in the presence of glycine, when the additional gene copy would be superfluous, then this might be a suitable experimental system for testing Koch's theory of the evolution of more efficient enzyme proteins via untranslatable intermediates.

Evolution of regulatory proteins

Gene duplication has been suggested as the origin of regulatory proteins controlling either enzyme activity or the rate of enzyme synthesis. There is as yet no experimental evidence for this, but comparisons could be made between amino acid sequences of regulatory subunits and catalytic subunits of polymeric proteins, and between enzymes and repressor proteins. Engel (1973) found that two extended sequences of bovine glutamate dehydrogenase showed considerable homology. This enzyme is a hexamer of identical subunits but has separate binding sites for its substrates and for nucleotides which modify its activity. Engel (1973) suggests that the polypeptide sequence with regulatory binding sites could have been derived by partial duplication of the structural gene in the region around the catalytic site. If such a partial gene duplication were to be followed by a chain-terminating mutation it could result in a protein with two different types of subunits, of which the regulatory subunit would be expected to be smaller than the catalytic subunit.

Since the *lac* repressor was isolated in 1966 by Gilbert & Müller-Hill, several other repressor proteins have been isolated and partially purified including the *ara* repressor (Wilcox, Clemetson, Santi & Englesberg, 1971) and the *trp* repressor (Zubay *et al.* 1972). The repressor proteins are able to bind compounds similar to the substrates of the enzymes whose synthesis they control. Since it is probable that the repressor proteins will soon become available in adequate quantities it will be of interest to compare sequences of these functionally related proteins. The earliest enzymes were probably inefficient catalysts whose synthesis was uncontrolled, and at a later date the regulatory proteins emerged whose function was to repress the synthesis of catabolic enzymes in the absence of the substrate, and to repress biosynthetic enzymes in the presence of the product. One alternative to the evolution of regulator proteins by partial duplication and mutation of the structural genes for the enzymes they now control, would be that they evolved from proteins which were able to bind to DNA and in that case there might be similarities in a group of regulator proteins controlling different enzyme systems.

Evolution of enzymes with new activities

The evolution of an enzyme with activities similar to, but distinct from, another enzyme in the cell would seem to provide a case in which gene duplication followed by mutational divergence would be the most likely mechanism. It is tempting to speculate that the two amidases which occur in *P. acidovorans*, *P. cepacia* and some strains of *P. putida*, were derived from a common ancestral gene determining an amidase with a rather broad specificity. The wild-type *P. aeruginosa* might not have undergone the gene-duplication step or have lost the second copy at an early stage. *P. putida* A90 could have evolved both amidases, and at a later stage lost the amidase attacking the low molecular weight amides. There is no experimental evidence for any homology between the naturally occurring aliphatic amidase and the phenylacetamidases, but the studies on the directed evolution of *P. aeruginosa* amidase have shown that relatively few amino acid substitutions in the polypeptide sequence are required to change the wild-type aliphatic amidase into a phenylacetamidase resembling that of *P. putida* A90 in catalytic properties and substrate profile.

A very convincing example of gene duplication with divergence to produce two enzymes with dissimilar but overlapping properties was described by Khan & Haynes (1972). The inducible fermentation of α-methyl glucoside and isomaltose is controlled by a single gene determining α-methylglucoside permease, and two genes MGL1 and MGL3 which are structural genes for enzymes attacking both α-methylglucoside and isomaltose. Enzymes were purified from strains containing only one of the two possible genes and it was found that the enzymes had the same molecular weight, and that there was no detectable difference in the immunodiffusion cross-reactions. However, the MHL3 enzyme was more thermostable than the MGL1 enzyme, and while the substrate profiles were the same the specific activities differed, as did the apparent K_m values for *p*-nitro-α-D-glucopyranoside. This suggested a genetic redundancy which might lead to the evolution of an enzyme with a new metabolic function. The existence of a genetic redundancy of this type is considered to have led to the existence of isoenzymes in higher organisms, and duplications and divergence as an evolutionary mechanism is discussed by Ohno (1970).

Evolution of metabolic pathways

Horowitz (1945, 1965) suggested that the biosynthetic pathways could have evolved by a retrograde stepwise process of tandem gene duplication and divergence. The primordial soup is thought to have been rich in organic compounds such as amino acids, so that only as it became depleted would there be any pressure to acquire the biosynthetic machinery. It is more difficult to see how catabolic pathways could have evolved in this way. An alternative mechanism, suggested by the results obtained by Mortlock & Wood and their colleagues and Wu *et al.*

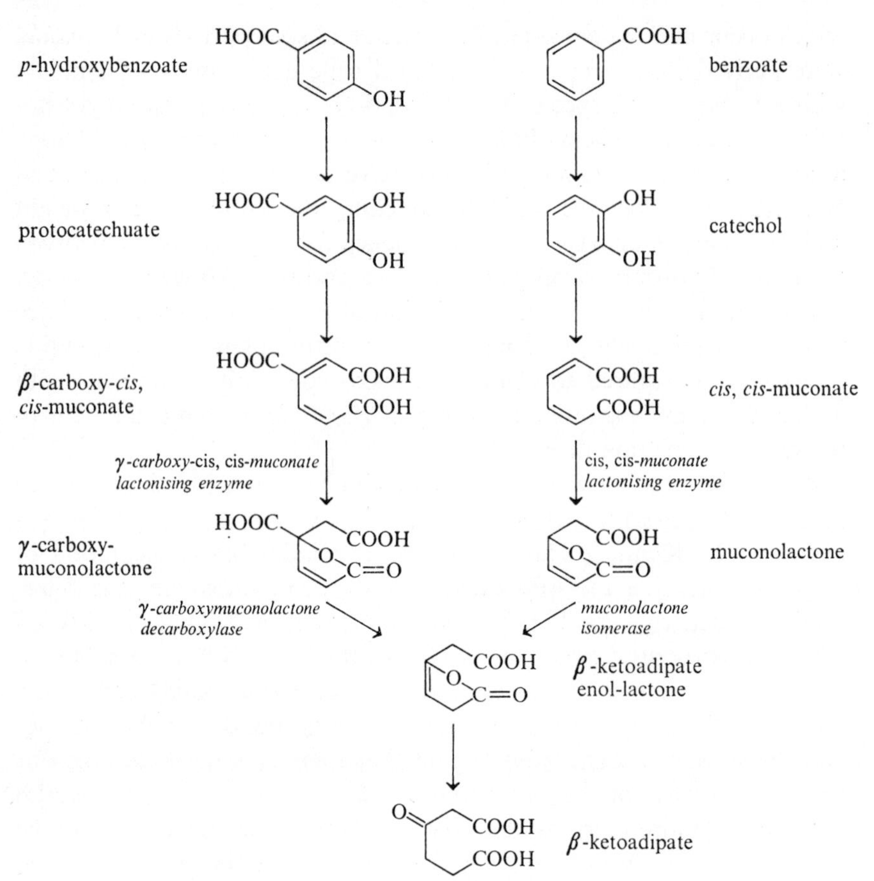

Fig. 8. The β-ketoadipate pathway in *Pseudomonas putida*.

(1968), is that catabolic pathways evolved by patchwork assembly of catalysts with low activity for the new reactions followed by gene duplication and enzyme 'improvement'. Ornston (1971) pointed out the consequences of these opposing theories for the properties of present-day enzymes. If retrograde evolution had occurred, then there should be some homology between an enzyme and the next enzyme in the sequence. If evolution proceeded by duplication and acquisition of new substrate specificities, then there would be expected to be some homology between enzymes carrying out similar catalytic functions. Stanier *et al.* (1970) found that there was no immunological cross-reaction between two adjacent enzymes of the β-ketoadipate pathway of *Pseudomonas putida*, the muconate lactonising enzyme and muconolactone isomerase. In a more detailed study, Meagher & Ornston (1973) and Patel, Meagher & Ornston (1973) showed that the enzymes

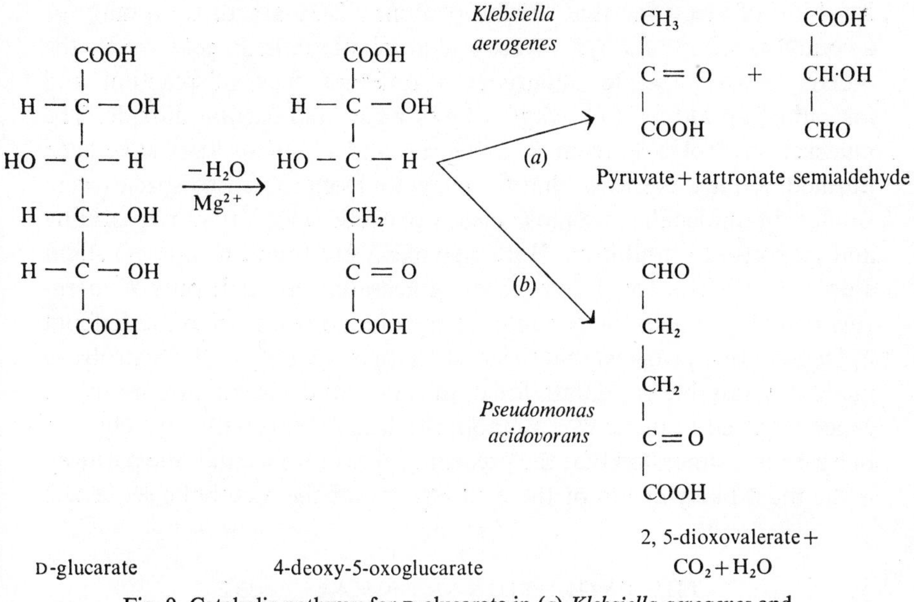

Fig. 9. Catabolic pathway for D-glucarate in (*a*) *Klebsiella aerogenes* and (*b*) *Pseudomonas acidovorans*.

which carry out the corresponding functions in the two convergent parts of the β-ketoadipate pathway were similar. The muconate lactonising enzyme and the carboxymuconate lactonising enzyme (Fig. 7) were similar in molecular weight (about 200000 daltons), had the same size subunits (about 40000 daltons), the same crystal form and the same first two N-terminal amino acids. The next two enzymes of each sequence, muconolactone isomerase and carboxymuconolactone decarboxylase resembled each other and not the preceding enzyme of their pathways. This would suggest resemblance between enzymes carrying out the same catalytic function and not between successive enzymes of the pathway.

One of the objections to the theory of retrograde evolution is that successive steps in a metabolic pathway frequently involve quite different chemical transformations. Jeffcoat & Dagley (1973) pointed out that this objection could be overcome if the enzyme mechanisms were similar although the chemical changes were different. In the degradation of D-glucarate by *Escherichia coli* and *Klebsiella aerogenes* the action of a hydrolyase is followed by an aldolase (Fig. 9*a*). They suggest that mechanism of action of both these enzymes require similar electron shifts. The formation and cleavage of 4-deoxy-5-oxoglucarate by *K. aerogenes* has been shown to require Mg^{2+} ions and to be initiated by shifts of electrons that release protons. They are thus arguing for a common origin of a hydrolyase and an aldolase. In *P. acidovorans* the 4-deoxy-5-oxoglucarate undergoes a different type of reaction and instead of an aldolase cleavage it loses water and carbon dioxide. The glucarate hydrolyases from *K. aerogenes* and *P. acidovorans* have been purified and it was shown that the enzymes from these two species were similar in molecular weights, electrophoretic mobilities, pH optima and response to inhibitors. Both enzymes were found to possess slight aldolase activities for 4-deoxy-5-oxoglucarate, although only *K. aerogenes* employs the aldolase route for the breakdown of glucose. Jeffcoat & Dagley (1973) suggest that the weak aldolase activity of the hydrolyase made it a suitable candidate for duplication and the mutational divergence required to produce a suitable aldolase for this pathway. Thus, in this case it is suggested that the preceding enzyme of a catabolic pathway is the most likely origin of the next enzyme of the metabolic sequence.

THE EVOLUTIONARY CHALLENGE

It would hardly be appropriate to draw firm conclusions from an account of experimental evolution since 'to make an end is to make a beginning. The end is where we start from' (Eliot, 1942). Instead, we might gaze into the future and ask if we can make any use of what we have already learnt about the ways in which bacteria can acquire new enzymes and new metabolic activities and what more we should do. Whether or not our experiments in the laboratory truly imitate events in nature, it is certainly the case that directed evolution in the laboratory can provide strains with new properties. These new strains in themselves may be useful irrespective of any light they may throw on the evolutionary process.

The enzymologists ask about enzyme mechanisms and employ devices to speed up, slow down and detect intermediates of catalysis. One of the main tools of their experiments is the use of substrate analogues for assay systems and as inhibitors. If they could be provided with a range of mutant enzymes they could try the other approach and use the same substrate but use variants of the original enzyme. A mutant enzyme which has activity but is changed in only one or two amino acids can answer directly questions about the importance of particular parts of the sequence for substrate binding and enzyme specificity. In return, the enzymologists could suggest suitable enzymes for which they have already worked out the mechanisms of catalysis, for experimental evolutionists to test the hypothesis of evolution of new activities suggested by Jeffcoat & Dagley (1973).

The enzyme technologist has long demanded organisms producing more of an enzyme, or more of the product of a particular pathway, and will no doubt continue to do so. But however efficient and suitable the present day enzymes may be to the organisms which contain them, they may not be the most useful for enzyme reactors or other industrial processes. It may be that we could tailor enzymes which could be more readily attached to immobilising supports, and perhaps retain more activity, or catalyse reactions which were not characteristic of the original enzyme.

But the greatest challenge to the micro-organisms comes from the activities of man as an organic chemist. It is so obvious that quite complex compounds which are synthesised biologically are also degraded biologically, that the possibility that man-made chemicals added to the environment might persist almost indefinitely was very slow to be appreciated. At a symposium on 'Degradation of synthetic organic molecules in the biosphere' it was pointed out by Dagley (1972) that 'we blithely scatter novel compounds over the face of the earth while leaving in comparative neglect a study of the capabilities and limitations of the microbes necessary for their removal'. What we do know about the metabolic capabilities of micro-organisms should help to distinguish between those compounds which are likely to be degraded and those which are not. Aromatic compounds can be broken down completely by several groups of micro-organisms and some of these can also tackle various substituted aromatic compounds. For example *p*-toluene sulphonate is metabolised by a *Pseudomonas* species via 2,3-dihydroxy-4-sulphotoluene, from which sulphate is eliminated leaving 3-methylcatechol (Focht & Williams, 1970), whereas benzene sulphonate is metabolised by another *Pseudomonas* which eliminates the sulphonate as sulphite. Johnston & Cain (1973) found a strain in which the genes for benzene sulphonate degradation appeared to be carried on a transmissible plasmid. Halogenated hydrocarbons can be metabolised with the release of the halogen substituent. 2-fluorobenzoate is metabolised by a *Pseudomonas* species as if it were benzoate, and fluorine is eliminated during the reaction in which catechol is produced, apparently by a mechanism similar to that in which hydrogen is released from the same position (Goldman, Milne & Pignataro, 1967).

Other reactions are known in which chlorine atoms are removed, but some chlorinated hydrocarbons are more susceptible to attack than others. Dagley (1972) suggests that of all the herbicides introduced in recent years 2,4-dichloro-phenoxyacetic acid (2,4-D) is exactly the type of compound which would have been designed if the metabolic pathway had been understood, since the final products are normal cell metabolites. What we don't know is how many of the metabolic pathways for 'novel' compounds which we can now identify in nature, have evolved in the years since these compounds were added to the environment. The micro-organisms in soil and water are being challenged to increase their biochemical armoury; perhaps we could help them.

REFERENCES

AMBLER, R. P. & WYNN, M. (1973). The amino acid sequences of cytochromes *c*-551 from three species of *Pseudomonas. Biochemical Journal*, **131**, 485–98.

BERMAN-KURTZ, M., LIN, E. C. C. & RICHEY, D. P. (1971). Promotor-like mutant with increased expression of the glycerol kinase operon of *Escherichia coli. Journal of Bacteriology*, **106**, 724–31.

BETZ, J. L. & CLARKE, P. H. (1972). Selective evolution of phenylacetamide-utilizing strains of *Pseudomonas aeruginosa. Journal of General Microbiology*, **73**, 161–74.

BETZ, J. L. & CLARKE, P. H. (1973). Growth of *Pseudomonas* species on phenylacetamide. *Journal of General Microbiology*, **75**, 167–77.

BRAMMAR, W. J., CLARKE, P. H. & SKINNER, A. J. (1967). Biochemical and genetic studies with regulator mutants of the *Pseudomonas aeruginosa* 8602 amidase system. *Journal of General Microbiology*, **47**, 87–102.

BRENNER, D. J., FANNING, G. R., JOHNSON, K. E., CITARELLA, R. V. & FALKOW, S. (1969). Polynucleotide sequence relationships among members of *Enterobacteriaceae. Journal of Bacteriology*, **98**, 637–50.

BRILL, W. & MAGASANIK, B. (1969). Genetic and metabolic control of histidase and urocanase in *Salmonella typhimurium*, strain 15–59. *Journal of Biological Chemistry*, **244**, 5392–402.

BROWN, J. E., BROWN, P. R. & CLARKE, P. H. (1969). Butyramide-utilizing mutants of *Pseudomonas aeruginosa* 8602 which produce an amidase with altered substrate specificity. *Journal of General Microbiology*, **57**, 273–95.

BROWN, J. E. & CLARKE, P. H. (1970). Mutations in a regulator gene allowing *Pseudomonas aeruginosa* 8602 to grow on butyramide. *Journal of General Microbiology*, **64**, 329–42.

BROWN, P. R. & CLARKE, P. H. (1972). Amino acid substitution in an amidase produced by an acetanilide-utilizing mutant of *Pseudomonas aeruginosa. Journal of General Microbiology*, **70**, 287–98.

CAMYRE, K. P. & MORTLOCK, R. P. (1965). Growth of *Aerobacter aerogenes* on D-arabinose and L-xylose. *Journal of Bacteriology*, **90**, 1157–8.

CHAKRABARTY, A. M. (1972). Genetic basis of the biodegradation of salicylate in *Pseudomonas. Journal of Bacteriology*, **112**, 815–23.

CHAKRABARTY, A. M., CHOU, G. & GUNSALUS, I. C. (1973). Genetic regulation of octane dissimilation plasmid in *Pseudomonas* (incompatibility). *Proceedings of the National Academy of Sciences, USA*, **70**, 1137–40.

CLARKE, P. H. (1972). Biochemical and immunological comparison of aliphatic amidases produced by *Pseudomonas* species. *Journal of General Microbiology*, **71**, 241–57.

CLARKE, P. H. & LILLY, M. D. (1969). The regulation of enzyme synthesis during growth. In *Microbial Growth. Symposia of the Society for General Microbiology*, **19**, 113–59. Ed. P. M. Meadow & S. J. Pirt. London: Cambridge University Press.

COCKS, G. T. & WILSON, A. C. (1972). Enzyme evolution in the *Enterobacteriaceae*. *Journal of Bacteriology*, **110**, 793–802.

COHEN, G. N., STANIER, R. Y. & LE BRAS, G. (1969). Regulation of the biosynthesis of amino acids of the aspartate family in coliform bacteria and pseudomonads. *Journal of Bacteriology*, **99**, 791–801.

DAGLEY, S. (1972). In *Degradation of Synthetic Organic Molecules in the Biosphere*, p. 338. Washington, D.C.: National Academy of Sciences.

DE CROMBRUGGHE, B., CHEN, B., ANDERSON, W., NISSLEY, P., GOTTESMAN, M., PASTAN, I. & PERLMAN, R. (1971). *Lac* DNA, RNA polymerase and cyclic AMP receptor protein, cyclic AMP, *lac* repressor and inducer are the essential elements for controlled *lac* transcription. *Nature, New Biology*, **231**, 139–42.

DUNN, N. W. & GUNSALUS, I. C. (1973). A transmissible plasmid coding for naphthalene oxidation in *Pseudomonas putida*. *Journal of Bacteriology*, **113**, 974–9.

ELIOT, T. S. (1942). In *Little Gidding*. London: Faber & Faber.

EMMER, M., DE CROMBRUGGHE, B., PASTAN, I. & PERLMAN, R. (1970). Cyclic AMP receptor protein of *E. coli*: Its role in the synthesis of inducible enzymes. *Proceedings of the National Academy of Sciences, USA*, **66**, 480–7.

ENGEL, P. C. (1973). Evolution of enzyme regulator sites: Evidence for partial gene duplication for amino-acid sequence of bovine glutamate dehydrogenase. *Nature, London*, **241**, 118–20.

FOCHT, D. D. & WILLIAMS, F. D. (1970). The degradation of *p*-toluenesulfonate by a *Pseudomonas*. *Canadian Journal of Microbiology*, **16**, 309–16.

FOLK, W. R. & BERG, P. (1971). Duplication of the structural gene for the glycyl-transfer RNA synthetase in *Escherichia coli*. *Journal of Molecular Biology*, **58**, 595–610.

FRANCIS, J. C. & HANSCHE, P. E. (1972). Directed evolution of metabolic pathways in microbial populations. 1. Modification of the acid phosphatase pH optimum in *S. cerevisiae*. *Genetics*, **70**, 50–73.

GILBERT, W. & MÜLLER-HILL, B. (1966). Isolation of the *lac* repressor. *Proceedings of the National Academy of Sciences, USA*, **56**, 1891–8.

GOLDMAN, P., MILNE, G. W. A. & PIGNATARO, M. T. (1967). Fluorine containing metabolites formed from 2-fluorobenzoic acid by *Pseudomonas* species. *Archives of Biochemistry and Biophysics*, **118**, 178–84.

HEGEMAN, G. D. & ROSENBERG, S. L. (1970). The evolution of bacterial enzyme systems. *Annual Review of Microbiology*, **24**, 429–62.

HOLLOWAY, B. W., KRISHNAPILLAI, V. & STANISICH, V. (1971). *Pseudomonas* genetics. *Annual Review of Genetics*, **5**, 425–46.

HORIUCHI, T., TOMIZAWA, J. & NOVICK, A. (1962). Isolation and properties of bacteria capable of high rates of *β*-galactosidase synthesis. *Biochimica et Biophysica Acta*, **55**, 152–63.

HOROWITZ, N. H. (1945). On the evolution of biochemical syntheses. *Proceedings of the National Academy of Sciences, USA*, **31**, 153–7.

HOROWITZ, N. H. (1965). The evolution of biochemical syntheses – retrospect and prospect. In *Evolving Genes and Proteins*, ed. V. Bryson & H. J. Vogel. New York: Academic Press.

JEFFCOAT, R. & DAGLEY, S. (1973). Bacterial hydrolyases and aldolases in evolution. *Nature, New Biology*, **241**, 186–7.

JENSEN, R. A., NASSER, S. D. & NESTER, E. W. (1967). Comparative control of a branch-point enzyme in microorganisms. *Journal of Bacteriology*, **94**, 1582–93.

JOHNSTON, J. B. & CAIN, R. B. (1973). The control of benzenesulphonate degradation in bacteria. Abstract 137. Dublin: Federation of European Biochemical Societies.

KHAN, N. A. & HAYNES, R. H. (1972). Genetic redundancy in Yeast: Non-identical products in a polymeric gene system. *Molecular and General Genetics*, **118**, 279–85.

KOCH, A. L. (1972). Enzyme evolution. 1. The importance of untranslatable intermediates. *Genetics*, **72**, 297–316.

LEBLANC, D. J. & MORTLOCK, R. P. (1971*a*). Metabolism of D-arabinose: Origin of a D-ribulokinase activity in *Escherichia coli*. *Journal of Bacteriology*, **106**, 82–9.

LEBLANC, D. J. & MORTLOCK, R. P. (1971*b*). Metabolism of D-arabinose: a new pathway in *Escherichia coli*. *Journal of Bacteriology*, **106**, 90–6.

LEIDIGH, B. J. & WHEELIS, M. L. (1973). Genetic control of the histidine dissimilatory pathway in *Pseudomonas putida*. *Molecular and General Genetics*, **120**, 201–10.

LERNER, S. A., WU, T. T. & LIN, E. C. C. (1964). Evolution of a catabolic pathway in bacteria. *Science*, **146**, 1313–15.

LI, S. L., DENNEY, R. M. & YANOFSKY, C. (1973). Nucleotide sequence divergence in the α-chain-structural genes of tryptophan synthetase from *Escherichia coli*, *Salmonella typhimurium*, and *Aerobacter aerogenes*. *Proceedings of the National Academy of Sciences, USA*, **70**, 1112–16.

MEAGHER, R. B., MCCORKLE, G. M., ORNSTON, M. K. & ORNSTON, L. N. (1972). Inducible uptake system for β-carboxy-*cis,cis*-muconate in a permeability mutant of *Pseudomonas putida*. *Journal of Bacteriology*, **111**, 465–73.

MEAGHER, R. B. & ORNSTON, L. N. (1973). Relationships among enzymes of the β-ketoadipate pathway. I. Properties of *cis, cis*-muconate lactonizing enzyme from *Pseudomonas putida*. *Biochemistry*, **12**, 3523–30.

MORTLOCK, R. P., FOSSITT, D. D. & WOOD, W. A. (1965). A basis for the utilization of unnatural pentoses and pentitols by *Aerobacter aerogenes*. *Proceedings of the National Academy of Sciences, USA*, **54**, 572–9.

MORTLOCK, R. P. & WOOD, W. A. (1964). Metabolism of pentoses and pentitols by *Aerobacter aerogenes*. 1. Demonstration of pentose isomerase, pentulokinase and pentitol dehydrogenase enzyme families. *Journal of Bacteriology*, **88**, 838–44.

MÜLLER-HILL, B., CRAPO, L. & GILBERT, W. (1968). Mutants that make more *lac* repressor. *Proceedings of the National Academy of Sciences, USA*, **59**, 1259–64.

NEWELL, S. L. & BRILL, W. J. (1972). Mutants of *Salmonella typhimurium* that are insensitive to catabolite repression of proline degradation. *Journal of Bacteriology*, **111**, 375–82.

OLIVER, E. J. & MORTLOCK, R. P. (1971*a*). Metabolism of D-arabinose by *Aerobacter aerogenes*. Purification of the isomerase. *Journal of Bacteriology*, **108**, 293–9.

OLIVER, E. J. & MORTLOCK, R. P. (1971*b*). Growth of *Aerobacter aerogenes* on D-arabinose: Origin of the enzyme activities. *Journal of Bacteriology*, **108**, 287–93.

ORNSTON, L. N. (1971). Regulation of catabolic pathways in *Pseudomonas*. *Bacteriological Reviews*, **35**, 87–116.

OHNO, S. (1970). *Evolution by gene duplication*. London: Allen & Unwin Ltd.

PALLERONI, N. J., BALLARD, R. W., RALSTON, E. & DOUDOROFF, M. (1972). Deoxyribonucleic acid homologies among some *Pseudomonas* species. *Journal of Bacteriology*, **110**, 1–11.

PALLERONI, N. J., DOUDOROFF, M., STANIER, R. Y., SOLANES, R. E. & MANDEL, M. (1970). Taxonomy of the aerobic pseudomonads. The properties of the *Pseudomonas stutzeri* group. *Journal of General Microbiology*, **60**, 215–31.

PARDEE, A. R., BENZ, E. J., ST PETER, D. A., KRIEGER, J. N., MEUTH, M. & TRIESHMANN, H. W. (1971). Hyperproduction and purification of nicotinamide deamidase, a microconstitutive enzyme of *Escherichia coli* K12 strains. *Journal of Biological Chemistry*, **246**, 6792–6.

Pastan, I. & Perlman, R. L. (1968). The role of the *lac* promotor locus in the regulation of β-galactosidase synthesis by cyclic 3′,5′-adenosine monophosphate. *Proceedings of the National Academy of Sciences, USA*, **61**, 1336–42.

Patel, R. N., Meagher, R. B. & Ornston, L. N. (1973). Relationships among enzymes of the β-ketoadipate pathway. II. Properties of crystalline β-carboxy-*cis,cis*-muconate lactonizing enzyme from *Pseudomonas putida. Biochemistry*, **12**, 3531–7.

Reznikoff, W. S., Miller, J. H., Scaife, J. G. & Beckwith, J. R. (1969). A mechanism for repressor action. *Journal of Molecular Biology*, **43**, 201–13.

Rheinwald, J. G., Chakrabarty, A. M. & Gunsalus, I. C. (1973). A transmissible plasmid controlling camphor oxidation in *Pseudomonas putida. Proceedings of the National Academy of Sciences, USA*, **70**, 885–9.

Rhodes, M. E. (1971). The taxonomy of *Pseudomonas fluorescens* group. *Journal of General Microbiology*, **69**, xi.

Scaife, J. & Beckwith, J. R. (1966). Mutational alteration of the maximal level of *lac* operon expression. *Cold Spring Harbor Symposia on Quantitative Biology*, **31**, 403–8.

Silverstone, A. E., Magasanik, B., Reznikoff, W. S., Miller, J. H. & Beckwith, J. R. (1969). Catabolite sensitive site of the *lac* operon. *Nature, London*, **221**, 1012–14.

Smyth, P. F. & Clarke, P. H. (1972). Catabolite repression of *Pseudomonas aeruginosa* amidase. *Journal of General Microbiology*, **73**, ix.

Spiegelman, S. (1953). In *Adaptation in Micro-organisms. Symposia of Society for General Microbiology*, **3**, 44. Ed. E. F. Gale & R. Davies. London: Cambridge University Press.

Stanier, R. Y. (1953). In *Adaptation in Micro-organisms. Symposia of Society for General Microbiology*, **3**, 8. Ed. E. F. Gale & R. Davies. London: Cambridge University Press.

Stanier, R. Y., Palleroni, N. J. & Doudoroff, M. (1966). The aerobic pseudomonads; a taxonomic study. *Journal of General Microbiology*, **43**, 159–271.

Stanier, R. Y., Wachter, D., Gasser, C. & Wilson, A. C. (1970). Comparative immunological studies of two *Pseudomonas* enzymes. *Journal of Bacteriology*, **105**, 351–62.

Stephenson, M. (1949). In *Bacterial Metabolism*, 3rd edition, p. 312. London: Longmans, Green & Co.

Wheelis, M. L. & Stanier, R. Y. (1970). The genetic control of dissimilatory pathways in *Pseudomonas putida. Genetics*, **66**, 245–66.

Wilcox, C., Clemetson, K. J., Santi, D. V. & Englesberg, E. (1971). Purification of the *ara C* protein. *Proceedings of the National Academy of Sciences, USA*, **68**, 2145–8.

Wood, W. A. (1966). Carbohydrate metabolism. *Annual Review of Biochemistry*, **35**, 521–58.

Wu, T. T., Lin, E. C. C. & Tanaka, S. (1968). Mutants of *Aerobacter aerogenes* capable of utilizing xylitol as a novel carbon source. *Journal of Bacteriology*, **96**, 447–56.

Zubay, G., Morse, D. E., Schrenk, W. J. & Miller, J. H. M. (1972). Detection and isolation of the repressor protein for the tryptophan operon of *Escherichia coli. Proceedings of the National Academy of Sciences, USA*, **69**, 1100–3.

Zubay, G., Schwartz, D. & Beckwith, J. (1970). Mechanism of activation of catabolite-sensitive genes. A positive control system. *Proceedings of the National Academy of Sciences, USA*, **66**, 104–10.

THE ORIGINS OF PHOTOSYNTHESIS IN EUKARYOTES

R. Y. STANIER
Service de Physiologie Microbienne,
Institut Pasteur, Paris 15*e*, *France*

INTRODUCTION

The differences in structure and function between prokaryotic and eukaryotic cells are many and profound, and no contemporary biological group has a cellular organization that can plausibly be interpreted as intermediate. The only possible exception is the dinoflagellates, although only their nuclear properties are exceptional, all other components of the cell being typically eukaryotic. This major evolutionary discontinuity bisects the microbial world, separating the bacteria *sensu lato* from the protozoa, fungi and algae. Nevertheless, one complex property occurs on both sides of the structural gap: a particular machinery and mechanism of photosynthesis, oxygenic ('plant') photosynthesis. It is the only mode of photosynthesis among eukaryotes, and is also characteristic of the largest group of photosynthetic prokaryotes, the cyanobacteria ('blue-green algae'). Mechanistically simpler anoxygenic modes of photosynthesis occur in two other prokaryotic groups, the purple and green bacteria; these organisms are also distinguished from other phototrophs by the nature of their pigment systems, and by certain other peculiarities, structural and chemical, of the photosynthetic apparatus (Table 1).

The restriction of major variations on the theme of photosynthesis to prokaryotes suggests that this metabolic process first arose and evolved in the context of the prokaryotic cell, and that its ultimate version was subsequently transferred to certain eukaryotic cell lines. The possible mechanism of such a transfer is intimately related to a larger evolutionary problem: the origin of the eukaryotic cell. Two different explanatory hypotheses can be envisaged. The stem line from which contemporary eukaryotes arose might have been derived from a group of prokaryotes which possessed the ability to perform oxygenic photosynthesis. In this event, the primary evolution of the eukaryotic cell must have been accompanied by the segregation of the photosynthetic apparatus, together with certain other pieces of cellular machinery, in a special type of membrane-bounded organelle, the chloroplast.

Table 1. *Some functional and chemical attributes of the photosynthetic apparatus in chloroplasts and in prokaryotes*

	Chloro- plasts	Prokaryotes: Cyano- bacteria	Prokaryotes: Purple bacteria	Prokaryotes: Green bacteria
Oxygen production	+	+	−	−
Electron donors	H_2O	H_2O	H_2S, H_2, organic compounds	H_2S, H_2, organic compounds
Major chlorophylls	*a*	*a*	bacterio. *a* or *b*	bacterio. *c* or *d*
Structure of major carotenoids	Alicyclic	Alicyclic	Aliphatic or aryl	Aryl
Specific location of photosynthetic apparatus	Thylakoids	Thylakoids	Cell membrane	Chlorobium vesicles
Presence in photosynthetic apparatus of:				
monogalactosyl diglycerides	+	+	−	+
digalactosyl diglycerides	+	+	−	−
polyunsaturated fatty acids	+	variable	−	−

Recent work on the properties of chloroplasts and mitochondria suggests an alternative hypothesis, based on the supposition that the component parts of the eukaryotic cell had multiple cellular origins. It assumes that the nucleo-cytoplasmic region of the eukaryotic cell evolved in a line characterized by a fermentative mode of energy-yielding metabolism and that the acquisition of both respiratory and photosynthetic function occurred secondarily, by the incorporation into this host cell of two different types of prokaryotic endosymbionts, each specifically endowed with one of these metabolic attributes. Both endosymbioses were subsequently stabilized through a partial loss of the genetic autonomy of the endosymbionts, which gave rise to the objects that we now recognize as mitochondria and chloroplasts. The plausibility of this hypothesis has been greatly increased by the discovery that both chloroplasts and mitochondria contain organelle-specific DNA together with the machinery required for the transcription and translation of genetic messages. Furthermore, these organellar components are prokaryotic in structure, and therefore markedly different from the counterpart components of equivalent function in the nucleo-cytoplasmic region of the eukaryotic cell; see Wilkie (1970) for a discussion of this point. A primary endosymbiotic origin for chloroplasts and mitochondria is also consonant with the fact that the ability to incorporate and maintain cellular endosymbionts is one of the distinctive biological attributes of the eukaryotic cell, not being possessed by prokaryotes (Stanier, 1970). Cellular endosymbionts occur in virtually all major groups of eukaryotes. The diversity and frequency of photosynthetic endosymbionts, mostly unicellular eukaryotic algae, in invertebrates (Buchner, 1965) provides a suggestive contemporary analogy to the postulated evolutionary derivation of chloroplasts.

EXPERIMENTAL APPROACHES TO THE ORIGIN OF PHOTOSYNTHESIS IN EUKARYOTES

The cyanobacteria are the only contemporary prokaryotes which embody the mode of photosynthesis characteristic of eukaryotes. A detailed comparative study of these organisms and of contemporary eukaryotic algae or their chloroplasts therefore seems to offer the most favorable prospect for acquiring further insights into the origin of photosynthesis in eukaryotes. However, this immediately raises an important question: which particular group of algae or their chloroplasts should be selected for the purposes of comparative analysis? The algae are a very large and diverse biological assemblage, and although the mechanism of photosynthesis appears to be uniform throughout the group, its machinery is not. The algae can be subdivided into a series of classes or divisions, distinguished by certain cellular properties, one of which is the specific array of light-harvesting pigments present in the chloroplast. Within any given algal division, the pigment complement is relatively constant, subject to some variation with respect to carotenoids (Stransky & Hager, 1970). The constancy of the chloroplast pigment system is particularly striking in the Chlorophyta, the largest division of algae, and the one from which vascular plants arose. A virtually unmodified complement of pigments has been maintained in the chloroplasts of these algae and of all higher plants. At the prokaryotic level, only one type of pigment system is associated with the performance of oxygenic photosynthesis. The cyanobacterial pigment system is distinctive and relatively uniform, although variations do occur with respect to the complement of carotenoids (Hertzberg, Liaaen-Jensen & Siegelman, 1971).

How is the diversity of the contemporary algal pigment systems to be interpeted in evolutionary terms? If the primary eukaryotic cell line was derived from a particular group of photosynthetic prokaryotes, all chloroplasts must share a common origin. On this interpretation, the various pigment systems characteristic of major algal groups must have arisen through an adaptive radiation which post-dated the emergence of the eukaryotic cell. A different interpretation is possible in terms of the hypothesis that chloroplasts had an endosymbiotic origin.

This type of endosymbiosis could have been initiated on many different occasions, between a variety of host cells and of photosynthetic prokaryotes. The 'successful' symbioses might all have been eventually stabilised by partial loss of the genetic autonomy of the endosymbionts. As Margulis (1968) first suggested, the contemporary diversity of algal pigment systems might thus reflect a much more ancient diversity among prokaryotes that performed oxygenic photosynthesis, the type of chloroplast now characteristic of each major algal group having been derived from a specific type of prokaryotic endosymbiont. If the hypothesis of Margulis is correct, it must be assumed that the prokaryotic progenitors of most types of chloroplasts subsequently became extinct as free-living photosynthetic organisms: only the cyanobacteria appear to have survived. However, one should perhaps not dismiss out of hand the possibility that other types of oxygen-producing photosynthetic prokaryotes may still be extant. A new group of photosynthetic prokaryotes which possess the chlorophylls characteristic of green bacteria but differ from these organisms in many significant respects was very recently discovered (Pierson & Castenholz, 1971). Nature could therefore have other surprises in store for the alert microbiologist.

The characters that distinguish the major groups of algae are by no means exclusively expressed at the level of the chloroplast. Many of the differential features, such as the mechanism of karyokinesis and the structure of the locomotor apparatus, are nucleo-cytoplasmic. These divergences could have occurred in the ancestral host cells, before they acquired photosynthetic endosymbionts. Acceptance of the possibility that the component parts of the eukaryotic cell may have had different cellular origins therefore permits a novel interpretation of the general evolution of the algae, which perhaps provides the most satisfactory explanation for the numerous and clear-cut differences between the cellular features of algal groups.

At all events, it now seems desirable to keep in mind the possibility that chloroplasts may be structures of polyphyletic cellular origin. This implies that the information about organellar properties obtained by the study of a particular type of chloroplast may not necessarily be of general validity. This is a point of some importance, since most of our present information has been derived from intensive research on the chloroplasts of a small number of organisms, mostly belonging to the evolutionary line of chlorophytes and vascular plants.

Table 2. *Pigments of the photosynthetic apparatus in cyanobacteria, rhodophytes and cryptophytes**

	Cyanobacteria	Rhodophytes	Cryptophytes
1. Chlorophylls			
a	+	+	+
c_2	−	−	+
2. Major carotenoids			
α-Carotene	−	+ or −	+
β-Carotene	+	+	−
Zeaxanthin	+	+	−
Cryptoxanthin	−	−	+
Alloxanthin	−	−	+
Echinone	+†	−	−
Carotenoid glycosides	+†	−	−
3. Phycobiliproteins			
Allophycocyanin	+	+	−
Phycocyanin	+	+	+ or −‡
Phycoerythrin	+ or −	+ or −	+ or −‡

* Data compiled from Stransky & Hager (1970), Hertzberg *et al.* (1971), Siegelman, Chapman & Cole (1968), Jeffrey (1969).

† Usually present, but absent from some members of the group.

‡ Either phycocyanin or phycoerythin present.

The distribution of phycobiliproteins

In most photosynthetic organisms, the harvesting of light energy is performed exclusively by chlorophylls and carotenoids, hydrophobic pigments integrated into the membrane system of the photosynthetic apparatus. In cyanobacteria, this function is largely taken over by another class of pigments, the phycobiliproteins, which are water-soluble chromoproteins. Phycobiliproteins are associated with the photosynthetic apparatus in only two other groups, both eukaryotic: the Rhodophyta and the Cryptophyta. However, both in terms of nucleo-cytoplasmic structure and in terms of the overall composition of their pigment systems, rhodophytes and cryptophytes show many differences. When the major photosynthetic pigments characteristic of cyanobacteria, rhodophytes and cryptophytes are compared (Table 2), it is evident that the two former groups, one prokaryotic and one eukaryotic, have broadly similar pigment systems from which the cryptophytan pigment system diverges in numerous respects.

Chemical properties of phycobiliproteins

The phycobiliproteins are strongly fluorescent blue or red pigments, with major absorption bands in the spectral region between 530 and 660 nm. Although intimately associated *in vivo* with the membranous matrix of the photosynthetic apparatus, they are not integrated into its fabric, and pass into solution when the cell is disrupted.

Table 3. *Spectral properties that distinguish the three subclasses of phycobiliproteins*

Subclass	Colour	Colour of fluorescence	Location of absorption maximum in the visible region (nm)
Phycoerythrins	Red	Orange	540–575
Phycocyanins	Blue	Red	615–625
Allophycocyanins	Blue	Red	650

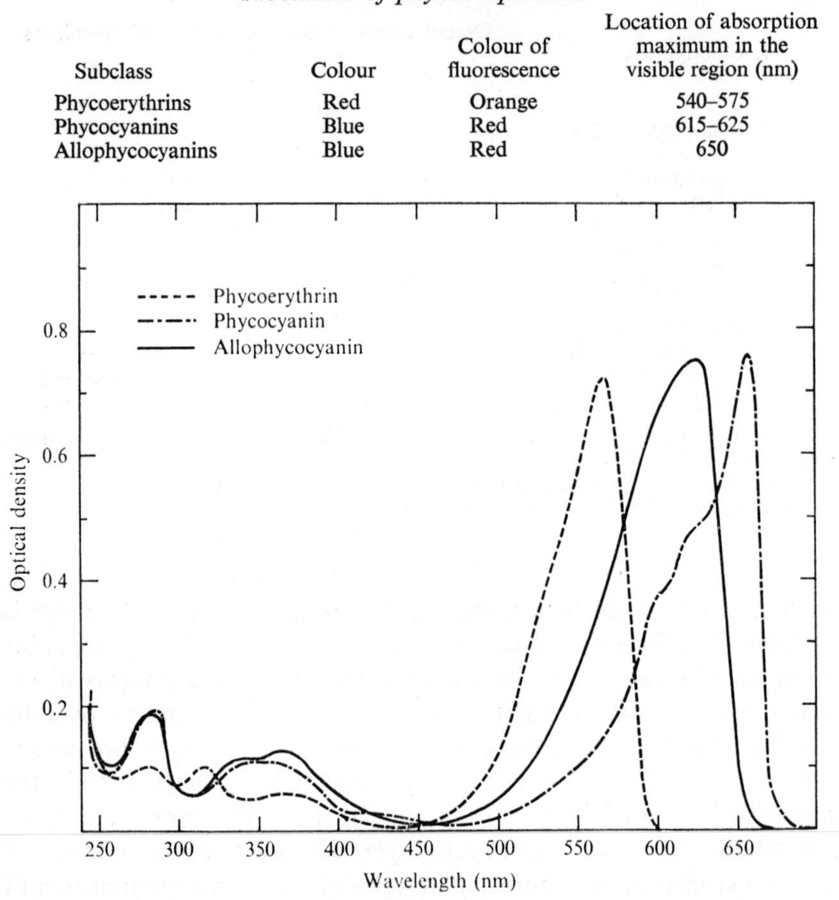

Fig. 1. Absorption spectra of purified phycoerythrin, phycocyanin and allophycocyanin, isolated from the cyanobacterium *Fremyella diplosiphon.* Spectra determined in 0.1 M phosphate buffer (pH 7.0), and adjusted to the same peak heights at the absorption maxima. After Bennett & Bogorad (1971).

Three subclasses of phycobiliproteins – phycoerythrins, phycocyanins and allophycocyanins – can be distinguished on the basis of their characteristic absorption spectra (Table 3 and Fig. 1). The spectral differences among the subclasses are sufficiently great to permit unambiguous assignments, even though the members of a particular subclass isolated from a variety of biological sources do differ to some degree in spectral character (ÓhEocha, 1965). The chromophores of these pigments are covalently linked to the polypeptide chains. They are linear tetrapyrroles, structurally related to the bile pigment, meso-

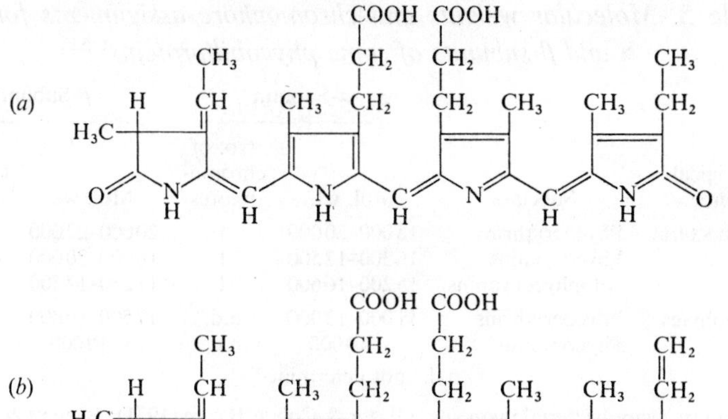

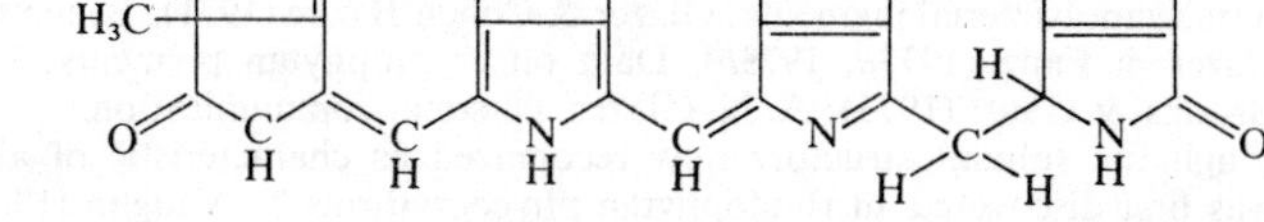

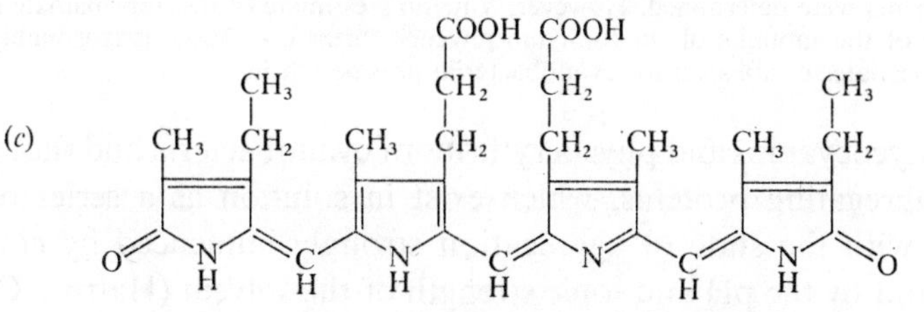

Fig. 2. Structures of phycocyanobilin (*a*) and phycoerythrobilin (*b*), compared with that of the bile pigment, mesobilverdin (*c*).

Table 4. *Reported values for the molecular weight of native phycobiliproteins*

Spectral class	Biological sources	Range of molecular weight	References
Phycoerythrins	Cyanobacteria and rhodophytes	220000–340000	1–3
	Cryptophytes	28000–31000	4–5
Phycocyanins	Cyanobacteria and rhodophytes	180000–360000	6–9
	Cryptophytes	58000	10
Allophycocyanins	Cyanobacteria	96000	11

References: 1, Vaughn (1964); 2, ÓhEocha (1965); 3, Gantt (1969); 4, Nolan & ÓhEocha (1967); 5, Brooks & Gantt (1973); 6, Scott & Berns (1965); 7, Hattori *et al.* (1965); 8, Craig & Carr (1968); 9, Neufeld & Riggs (1969); 10, A. N. Glazer, personal communication; 11, Glazer & Cohen-Bazire (1971).

biliverdin (Lemberg & Bader, 1933). Phycocyanins and allophycocyanin have the same chromophore, phycocyanobilin (Schram & Kroes, 1971). The chromophore of phycoerythrins, phycoerythrobilin (Chapman, Cole & Siegelman, 1968) is an isomer of phycocyanobilin (Fig. 2). Native phycobiliproteins differ greatly in molecular weight (Table 4).

Table 5. *Molecular weights and chromophore assignments for the α and β subunits of some phycobiliproteins*[1,2]

Biological source	Subclass	α-Subunit Mol. wt.	α-Subunit No. of chromophores	β-Subunit Mol. wt.	β-Subunit No. of chromophores
Cyanobacteria	Phycoerythrins	18000–20000	1	20000–22000	2
	Phycocyanins	16300–17500	1	18000–20000	2
	Allophycocyanins	15200–16600	1	17250–17500	1
Cryptophytes	Phycoerythrins	11000–12000	n.d.	17500–19000	n.d.
	Phycocyanin	~ 10000	n.d.	~ 19000	n.d.

n.d.: not determined

1. Data on cyanobacterial pigments: Glazer & Cohen-Bazire (1971); Bennett & Bogorad (1971); Glazer & Fang (1973*a*, 1973*b*). Data on cryptophytan pigments: Glazer *et al.* (1971*a*): Brooks & Gantt (1973); A. N. Glazer, personal communication.
2. Although the subunit structure now recognized as characteristic of all phycobiliproteins was first discovered in rhodophytan phycoerythrins by Vaughn (1964), his data could not be tabulated, since neither precise subunit molecular weights nor chromophore assignments were determined. However, Vaughn's estimate of the approximate molecular weights of the subunits of rhodophytan phycoerythrins (~ 17000) is reasonably close to the values later established for cyanobacterial phycoerythrins.

The phycocyanins and phycoerythrins of cyanobacteria and rhodophytes are aggregating proteins, which exist in solution as a series of oligomers, with the state of aggregation strongly influenced by concentration, and by the pH and ionic strength of the solvent (Hattori, Crespi & Katz, 1965; Scott & Berns, 1965). For these particular proteins, a precise native molecular weight cannot, in consequence, be determined; this explains the wide range of reported values listed in Table 4. Cryptophytan phycoerythrins and phycocyanins, on the other hand, are of low and constant molecular weight, *ca* 30000 and 58000 respectively.

A major advance in the chemistry of phycobiliproteins came with the discovery that they all possess the same kind of subunit structure (Vaughn, 1964; Glazer & Cohen-Bazire, 1971; Bennett & Bogorad, 1971; Glazer, Cohen-Bazire & Stanier, 1971*a*; Brooks & Gantt, 1973). They are composed of two non-covalently associated polypeptide chains (α- and β-chains), of slightly different molecular weight, and each carrying at least one chromophore (Table 5). The native proteins of lowest molecular weight (cryptophytan phycoerythrins) are monomers, with the structure $\alpha_1\ \beta_1$. Cyanobacterial allophycocyanin (mol. wt. 96000) can be construed as a trimer with the structure $(\alpha_1\ \beta_1)_3$ and cryptophytan phycocyanin (mol. wt. 58000) as a dimer, $(\alpha_1\ \beta_1)_2$. The cyanobacterial and rhodophytan phycoerythrins and phycocyanins, of high and variable molecular weight, presumably consist of a series of oligomers, $(\alpha_1\ \beta_1)_n$.

Table 6. *Amino acid composition of the phycocyanin of* Synechococcus 6301, *of its α- and β-subunits, and of the reconstituted native protein: values rounded to nearest whole number (data from Glazer & Fang,* 1973*b)*

	Native protein	α-Subunit	β-Subunit	Sum of two subunits	Reconstituted protein
Lysine	13	7	5	12	13
Histidine	1	1	0	1	1
Arginine	20	7	13	20	20
Aspartic acid	42	18	21	39	39
Threonine	19	10	9	19	18
Serine	24	12	11	22	23
Glutamic acid	22	11	11	22	22
Proline	11	6	4	10	10
Glycine	27	13	13	26	26
Alanine	59	25	32	57	58
Half-cystine	1	n.d.	n.d.	—	n.d.
Valine	21	8	13	21	20
Methionine	3	0	3	3	3
Isoleucine	17	7	10	17	17
Leucine	31	16	14	30	28
Tyrosine	15	10	5	15	14
Phenylalanine	12	6	6	12	11
Tryptophan	n.d.	1	n.d.	—	n.d.

n.d.: not determined

Under denaturing conditions, the α- and β-chains can be separated and purified by electrophoresis or chromatography. Glazer & Fang (1973*b*) have compared the properties of the two subunits prepared from three different cyanobacterial phycocyanins. The biological sources of the three pigments were diverse: a unicellular rod (*Synechococcus* sp.), a unicellular coccus (*Aphanocapsa* sp.), and a filamentous heterocyst-former (*Anabaena* sp.). The two unicellular strains differed considerably in mean DNA base composition: 55 and 36 mol percent guanine + cytosine, for *Synechococcus* and *Aphanocapsa* respectively. This study revealed the following facts:

1. The α- and β-chains differ both quantitatively and qualitatively in amino acid composition (Table 6). Histidine is absent from the β-chain of all three phycocyanins, whereas the α-chain contains either one or two residues of this amino acid. There are, nevertheless, considerable similarities between the α- and β-chains in overall composition; they may therefore be of common evolutionary origin, like the α- and β-chains of hemoglobins.

2. Despite their diverse origins, the three phycocyanins are closely similar with respect to the amino acid composition of subunits. This is shown for the phycocyanins of *Aphanocapsa* and *Synechococcus* in Table 7. Since the mean DNA base compositions of these two organisms

Table 7. *Amino acid composition of the α- and β-subunits of two cyanobacterial phycocyanins: values rounded to nearest whole number (data from Glazer & Fang, 1973b)*

	α-Subunits from:			β-Subunits from:		
	1 *Synechococcus*	2 *Aphanocapsa*	Differences (1−2)	1 *Synechococcus*	2 *Aphanocapsa*	Differences (1−2)
Lysine	7	7	0	5	4	+1
Histidine	1	1	0	0	0	0
Arginine	7	7	0	13	11	+2
Aspartic acid	18	15	+3	21	19	+2
Threonine	10	10	0	9	11	−2
Serine	12	15	−3	11	12	−1
Glutamic acid	11	12	−1	11	11	0
Proline	6	5	+1	4	4	0
Glycine	13	12	+1	13	11	+2
Alanine	25	22	+3	32	30	+2
Half-cystine[1]	n.d.	n.d.	—	n.d.	n.d.	—
Valine	8	6	+2	13	11	+2
Methionine	0	1	−1	3	4	−1
Isoleucine	7	8	−1	10	10	0
Leucine	16	13	+3	14	14	0
Tyrosine	10	9	+1	5	6	−1
Phenylalanine	6	6	0	6	5	+1
Tryptophan	1	n.d.	—	n.d.	n.d.	—
Total per subunit:	158	154	20	170	162	17

[1] Native *Synechococcus* phycocyanin contains 1.3 residues mol^{-1}; native *Aphanocapsa* phycocyanin, 2.5 residues mol^{-1}.
n.d.: not determined.

differ by almost 20 mol percent guanine + cytosine, the compositional similarities of their phycocyanins suggest that the primary structure of these proteins may be very highly conserved.

3. Under appropriate conditions, the isolated α- and β-subunits reassociate in equimolar ratio (Table 6), to produce a renatured phycocyanin which cannot be distinguished physically, chemically or spectrally from the original native protein. The yield of this reaction is high: 40–60 percent. Furthermore, α- and β-subunits derived from two phycocyanins of different origin (e.g. *Synechococcus* and *Aphanocapsa*) will reassociate, with similar yields, to produce 'hybrid' renatured proteins. This constitutes a further striking demonstration of the conservation of these proteins, specifically with respect to the regions of the two polypeptide chains that are involved in the inter-subunit association.

Phycobiliprotein complements of cyanobacteria, rhodophytes and cryptophytes

Among cryptophytes, the phycobiliproteins of any given strain or species belong to only one spectral class: some synthesize phycocyanins, others phycoerythrins. Allophycocyanins have never been detected in members of this algal division (ÓhEocha, 1965; Glazer *et al.* 1971*a*; Brooks & Gantt, 1973).

In cyanobacteria and rhodophytes, the cellular complement of phycobiliproteins is always more complex, and includes pigments of at least two spectral classes. Allophycocyanin seems to be universal (ÓhEocha, 1965; Siegelman *et al.* 1968); however, it is always a minor component, usually less than 5 percent of total phycobiliprotein, and hence cannot reliably be detected except by chromatographic separation. Phycocyanin is probably likewise universal, although its concentration is low in most rhodophytes and in some phycoerythrin-rich cyanobacteria. Only phycoerythrins are sometimes absent: many cyanobacteria, and a few 'blue-green' rhodophytes, do not synthesize pigments of this class.

In some cyanobacteria which contain both phycoerythrin and phycocyanin, the ratio between these two pigments can vary widely in response to changes in environmental conditions such as salinity, light intensity, light quality. The influence of light quality is particularly striking: by growth in light of certain wavelengths, the cells can be induced to synthesize either predominantly phycocyanin (red light) or phycoerythrin (blue-green light). This response, first described by Gaidukov (1902), is known as *chromatic adaptation*. A detailed study on the cyanobacterium *Tolypothrix tenuis* (Fujita & Hattori, 1963) showed that differential stimulation of either phycocyanin or phycoerythrin synthesis can be induced by appropriate chromatic illumination for a few minutes; and recent work suggests that the response is mediated by a special phytochrome-like pigment (Scheibe, 1972).

The fraction of cellular protein represented by phycobiliproteins can be very large. In some cyanobacteria, the two or three proteins of this class account for over half the protein synthesized by the organism.

Location of phycobiliproteins in the photosynthetic apparatus

In the rhodophytan chloroplast, the thylakoids have an unusual arrangement, unlike that in any other type of chloroplast. They are never immediately contiguous, being separated from one another by a rather regular spacing of some 50 nm. As first shown by Gantt &

Conti (1965, 1966*a*) in the unicellular rhodophyte *Porphyridium*, the interthylakoid space contains a regular array of granules, each some 30 nm in diameter, which are attached in rows to the outer surface of each thylakoid (Plate 1, fig. 1). Gantt & Conti (1966*b*) interpreted these novel chloroplast components as aggregates of phycobiliproteins, and christened them *phycobilisomes*. They have since been observed in suitably fixed material of all rhodophytes examined.

The chemical nature of phycobilisomes has been elegantly confirmed by Gantt & Lipschultz (1972), who developed a special extraction procedure which permits their isolation in a structurally intact state from *Porphyridium*. Isolated phycobilisomes (Plate 1, fig. 2) have the approximate shape and dimensions of the granules observed in electron micrographs of thin sections. They contain the full cellular complement of phycobiliproteins (allophycocyanin, phycocyanin and two forms of phycoerythrin). No other major constituents were detected in the isolated material.

The internal structure of the cryptophytan chloroplast is completely different (Plate 2). The thylakoids tend to be associated in pairs or stacks, sometimes separated from one another by a narrow gap (3–5 nm), and do not bear phycobilisomes. An unusual feature of the cryptophytan thylakoids is their expansion, to form a lumen some 20–30 nm wide, which is uniformly filled with electron-dense material. Gantt, Edwards & Provasoli (1971), who have made the most detailed study of chloroplast structure in this group, propose that the phycobiliproteins are intrathylakoidal, being represented by the electron-dense material which fills the lumen of each thylakoid. If this plausible interpretation is correct, the topological relationship between the phycobiliproteins and the membrane-bound pigments of the photosynthetic apparatus in cryptophytes is the inverse of that in rhodophytes.

Like other photosynthetic prokaryotes, cyanobacteria do not contain organelles interpretable as being homologous with chloroplasts. In the cyanobacterial cell (Plate 3), the photosynthetic apparatus is carried on an extensive system of thylakoids which runs through the cytoplasm, and is enclosed only by the cytoplasmic membrane. The thylakoids are separated from one another by a spacing of 30–40 nm; and regular arrays of phycobilisomes attached to the outer thylakoid membranes have been reported in many species (Cohen-Bazire, 1971). Cyanobacterial phycobilisomes are typically somewhat smaller and more densely packed than rhodophytan phycobilisomes. This probably explains the fact that they cannot be resolved clearly in some cyanobacteria (G. Cohen-Bazire, personal communication).

The extreme regularity with which phycobilisomes are arrayed over the thylakoid surface suggests that the thylakoid membrane in cyanobacteria and rhodophytes may contain specific attachment sites for these organelles. Recent observations on freeze-etched material provide some support for this inference (Lefort-Tran, Cohen-Bazire & Pouphile, 1973). As Branton (1966) has shown, the cleavage plane of freeze-etched unit membranes passes between the inner and outer layers, providing an image of the structure of their complementary internal faces. In the thylakoid membranes of cyanobacteria and rhodophytes, one internal face of the unit membrane bears regular arrays of small granules, each some 5 nm in diameter (Plate 4). This structural feature is not evident in the thylakoid membranes of chlorophytes (Goodenough & Staehelin, 1971). The 5 nm granules may therefore be associated with modified regions in the thylakoid membrane which function as attachment sites for phycobilisomes.

The different modes of association between phycobiliproteins and thylakoids which distinguish the cryptophytes from both rhodophytes and cyanobacteria are perhaps reflected in the molecular behavior of the phycobiliproteins isolated from these groups. As already mentioned, the phycocyanins and phycoerythrins of both rhodophytes and cyanobacteria exist in solution as oligomeric aggregates, of variable but relatively high molecular weight. The associative properties of these pigments could well play an important role in maintaining the much more complex intermolecular association which exists in the phycobilisome. Cryptophytan pigments of the same two spectral classes exist in solution entirely as monomers (phycoerythrins) or as dimers (phycocyanin). If, as the electron microscopic evidence suggests, these pigments are packed into the lumen of the thylakoid, a specific intermolecular association is probably not necessary to maintain close molecular packing *in vivo*: the enclosing thylakoid membrane provides an external constraint to prevent disaggregation.

The light-harvesting function

Measurements of the photosynthetic action spectra of cyanobacteria and rhodophytes show that most of the light responsible for oxygen production is captured by phycobiliproteins. Wavelengths absorbed by chlorophyll and carotenoids make little or no contribution (Engelmann, 1883, 1884; Haxo & Blinks, 1950).

Although the detailed action spectra determined by Haxo & Blinks (1950) clearly established the major role played in light capture by the phycocyanins and phycoerythrins of these two groups, the function of

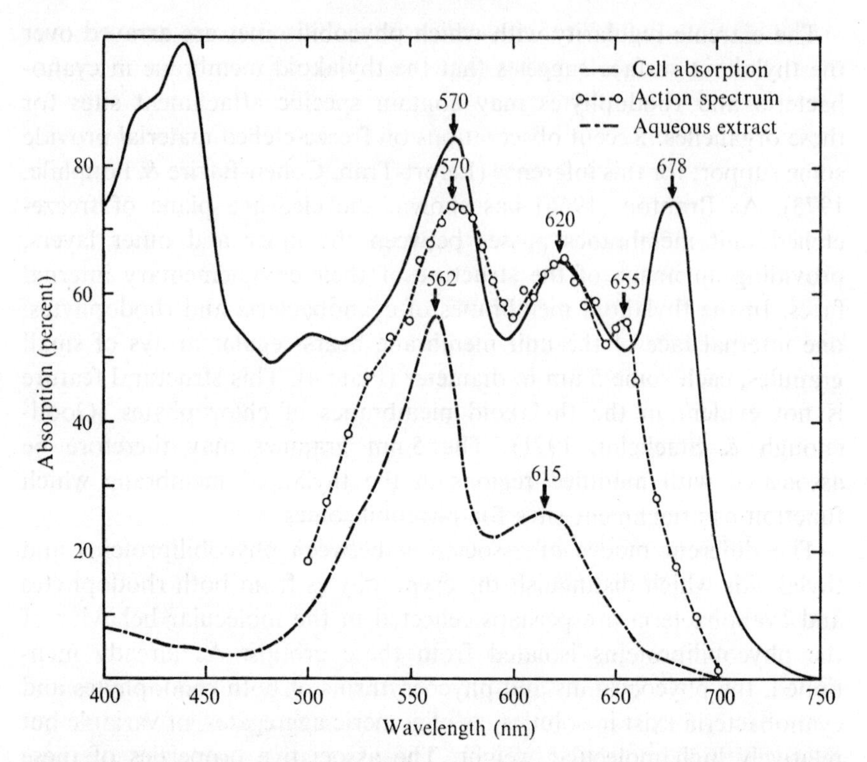

Fig. 3. The photosynthetic action spectrum of a cyanobacterium, *Aphanocapsa* sp., compared with the absorption spectrum of the intact cells and of a crude, aqueous extract of the cellular phycobiliproteins. The positions of the three peaks in the photosynthetic action spectrum correspond closely to the absorption maxima *in vivo* of phycoerythrin (*ca* 570 nm), phycocyanin (*ca* 620 nm) and allophycocyanin (*ca* 655 nm). There is no significant contribution from light absorbed by chlorophyll *a* (absorption max. at 678 nm). In the aqueous extract, the phycobiliprotein peaks are displaced to shorter wavelengths, an effect attributable to the destruction of the phycobilisomes. Data by courtesy of N. Tandeau de Marsac and C. Lemasson.

allophycocyanin in light-harvesting and energy transfer has only recently been demonstrated. In the cellular absorption spectrum, the small allophycocyanin peak at *ca* 650 nm is completely masked by the much greater absorption in this spectral region attributable to phycocyanin and chlorophyll *a*. However, Lemasson, Tandeau de Marsac & Cohen-Bazire (to be published) have found a distinct peak in the action spectra of several cyanobacteria at 650–655 nm (Fig. 3), which provides evidence that allophycocyanin also serves as a light-harvesting pigment. Moreover, the existence of this peak, not detectable in the absorption spectrum, implies that light absorbed by allophycocyanin is transferred to reaction centers with considerably greater efficiency than light absorbed by the two other classes of phycobiliproteins.

Gantt & Lipschultz (1973) have determined the fluorescence emission spectrum of isolated phycobiliproteins. Excitation with wavelengths of light absorbed either by phycoerythrin or by phycocyanin gives rise to fluorescence of a wavelength attributable to allophycocyanin, which indicates that this pigment is involved in energy transfer, as well as light harvesting.

Putting together these two sets of observations, a plausible interpretation of the function of the phycobilisome can now be offered (Fig. 4). All its constituent phycobiliproteins are light-harvesting pigments; and the energy absorbed by phycoerythrin and phycocyanin is channelled (with some loss of efficiency) via allophycocyanin to reaction centers in the thylakoid membrane, these centers being at or near the sites of phycobilisome attachment.

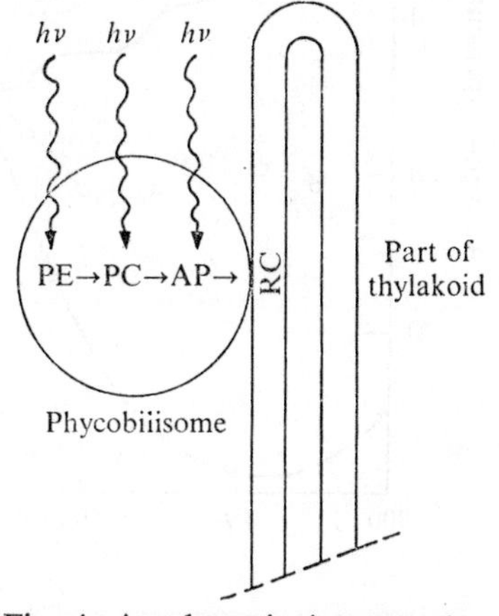

Fig. 4. A schematic interpretation of the path of energy transfer to reaction centers in cyanobacteria and rhodophytes. PE = phycoerythrin; PC = phycocyanin; AP = allophycocyanin; RC = reaction centers.

In cryptophytes, a homologous pathway of energy transfer cannot operate, since allophycocyanin is absent. Nevertheless, published action spectra (Haxo & Fork, 1959) show that cryptophyte phycoerythrin is an effective light-harvesting pigment (Fig. 5). Light absorbed by chlorophyll also contributes with fair efficiency to photosynthetic oxygen production, whereas it generally does not in rhodophytes and cyanobacteria. The two latter groups contain only chlorophyll *a*. However, cryptophytes contain a second chlorophyllous pigment, chlorophyll c_2 (Jeffrey, 1969) which has an absorption maximum near 650 nm; a small peak at this wavelength is also evident in the cryptophytan action spectra. It is therefore conceivable that chlorophyll c_2 plays a role analogous to that of allophycocyanin, serving both as a light-harvesting pigment and as an agent in the transfer of light absorbed by phycoerythrin (or phycocyanin) to reaction centers (Fig. 6).

Comparative immunology of phycobiliproteins

The immunological studies on phycobiliproteins so far published (Vaughn, 1964; Bogorad, 1965; Berns, 1967; Glazer *et al.* 1971*b*) have all been performed by the Ouchterlony double diffusion technique and

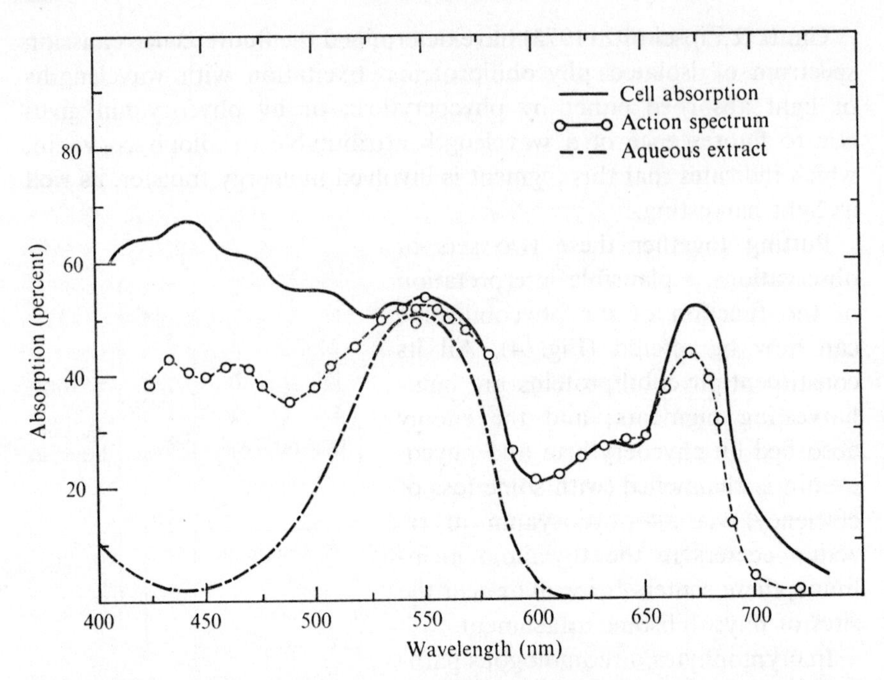

Fig. 5. Photosynthetic action spectrum of a cryptophyte, *Rhodomonas lens*, compared with the absorption spectra of the intact cells and of an aqueous extract. Redrawn from Haxo & Fork (1959).

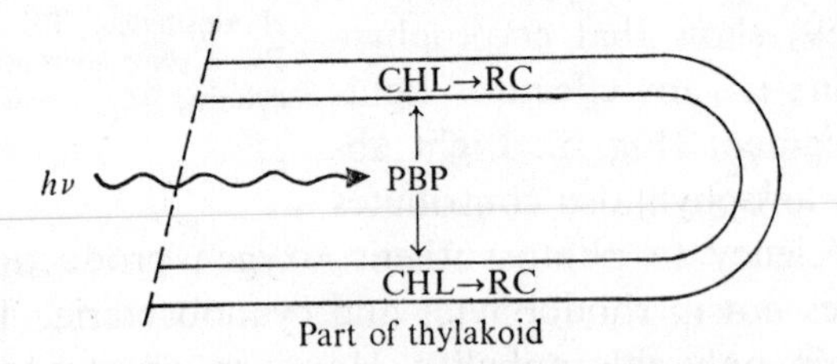

Fig. 6. A schematic interpretation of the possible path of energy transfer to type II reaction centers in cryptophytes. PBP = phycobiliprotein; CHL = chlorophyll c_2; RC = reaction center.

therefore provide only qualitative information. The conclusions derived from this work can be summarized as follows:

1. No cross-reactions can be detected between phycocyanins, allophycocyanins and phycoerythrins (Bogorad, 1965; Glazer *et al.* 1971*b*), even when the immunizing antigens are all derived from the same organism. Since phycocyanins and allophycocyanins have the same chromophore, their failure to cross-react immunologically shows that phycocyanobilin is not a haptenic determinant, presumably because it is not located in an exposed position at the surface of the protein molecule.

Vaughn (1964) has adduced both chemical and immunological arguments in support of the view that the chromophore of phycoerythrins (phycoerythrobilin) is masked in the native molecule and does not contribute directly to the antigenicity of this class of phycobiliproteins.

2. Phycobiliproteins of a given spectral class (namely phycoerythrins, phycocyanins or allophycocyanins) cross-react, usually very strongly, with all other phycobiliproteins of the same class, provided that both the immunizing antigen and the heterologous antigens tested are of either cyanobacterial or rhodophytan origin (Vaughn, 1964; Berns, 1967; Glazer *et al.* 1971*b*). In other words, each of the three spectral classes of phycobiliproteins is homologous throughout both these biological groups, as judged by immunological criteria.

3. Members of each phycobiliprotein class, provided that they are of cyanobacterial or rhodobacterial origin, share a common set of major antigenic determinants (Glazer *et al.* 1971*b*). This follows from the observation that absorption of a specific antiserum (e.g. an anti-phycocyanin) with a heterologous cross-reacting antigen (a phycocyanin from another biological source) eliminates simultaneously all other heterologous cross-reactions, leaving only a weak residual antibody activity, directed specifically against the homologous antigen.

4. The phycocyanins and phycoerythrins of cryptomonads do not cross-react immunologically with any phycobiliproteins of rhodophytes or cyanobacteria (Berns, 1967; Glazer *et al.* 1971*b*). There is accordingly no immunological evidence for homology between the phycobiliproteins of cryptophytes and of the two other biological groups. Two types of cryptophytan phycoerythrins, present in different members of this division, have been well characterized (Brooks & Gantt, 1973). Although of closely similar molecular weight and subunit size, they have markedly different spectral properties, the absorption maxima being located at 566 and 542 nm respectively. An antiserum directed against phycoerythrin 566 nm fails to cross-react with phycoerythrin 544 nm. This observation (G. Cohen-Bazire, personal communication) suggests that the immunological conservatism so characteristic of other phycoerythrins may not be shared by cryptophytan phycoerythrins.

The separation of the α- and β-subunits of phycobiliproteins on a preparative scale (Glazer & Fang, 1973*b*) opens the way to more refined immunological analyses. G. Cohen-Bazire (personal communication) has observed that an antiserum directed against native phycocyanin cross-reacts independently with the isolated α- and β-subunits. The antigenic determinants of the native protein are therefore distributed between the two subunits, none being common to both.

Table 8. *The location, nature and probable role of the phycobiliproteins in cyanobacteria and in rhodophytan and cryptophytan chloroplasts*

		Chloroplasts of:	
	Cyanobacteria	Rhodophyta	Cryptophyta
Location			
In phycobilisomes	+	+	−
Within thylakoids	−	−	+
Nature			
Complement	AP, PC, (PE)	AP, PC, (PE)	PC or PE
Inter-group immunological cross-reactions:			
PE	+	+	−
PC	+	+	−
AP	+	+	Not applicable
Probable role			
Chain of energy transfer:	(PE)→PC→AP→RC	(PE)→PC→AP→RC	PC or PE →Chl. c_2→RC

AP, allophycocyanin; PC, phycocyanin; PE, phycoerythrin; RC, reaction center.

PHYLOGENETIC IMPLICATIONS

In terms both of photosynthetic structure and photosynthetic function, the contemporary cyanobacteria show close affinities to only one group of eukaryotes, the Rhodophyta. The broad resemblances between the pigment systems characteristic of the two groups have been recognized for many years. However, recent work has revealed much more profound resemblances between rhodophytan chloroplasts and cyanobacteria. These include a unique organization of the photosynthetic apparatus, which bears a special light-harvesting organelle, the phycobilisome, a common mechanism of energy capture and transfer in this organelle and the presence in the phycobilisome of three different chromoproteins, each of which is homologous throughout the cyanobacteria and the rhodophytes by immunological criteria. A cyanobacterial origin for the rhodophytan chloroplast, if not for other structures in the rhodophytan cell, therefore now appears virtually certain. However, there is no evidence to suggest a like ancestry for the chloroplast in other algal groups. Although the cryptophytes also use phycobiliproteins as light-harvesting pigments, their chloroplasts differ in many structural and functional respects from those of rhodophytes (Table 8), and cannot plausibly be ascribed a cyanobacterial derivation. Indeed, the data summarized in Table 8 tend to favor the assumption that rhodophytan and cryptophytan chloroplasts and, *a fortiori*, those of the other major algal divisions, are structures or polyphyletic origin.

SUMMARY

Two mechanisms of photosynthesis (i.e. oxygenic and anoxygenic) occur among prokaryotes. Only the more complex variant, oxygenic photosynthesis, exists in eukaryotes. These facts suggest that the metabolic process arose and evolved among the prokaryotes, its terminal version subsequently being transferred to certain eukaryotic groups. The possible mechanisms for such a transfer are discussed in the context of a larger evolutionary problem, the origin of the eukaryotic cell; and the evidence for relationships between the contemporary prokaryotic cyanobacteria ('blue-green algae') and eukaryotic algae (or their chloroplasts) is reviewed.

REFERENCES

Bennett, A. & Bogorad, L. (1971). Properties of subunits and aggregates of blue-green algal biliproteins. *Biochemistry*, **10**, 3625–34.

Berns, D. S. (1967). Immunochemistry of biliproteins. *Plant Physiology*, **42**, 1569–86.

Bogorad, L. (1965). Studies of phycobiliproteins. *Records of Chemical Progress*, **26**, 1–12.

Branton, D. (1966). Fractures faces of frozen membranes. *Proceedings of the National Academy of Sciences, USA*, **55**, 1048–56.

Brooks, C. & Gantt, E. (1973). Comparison of phycoerythrins (542,566 nm) from cryptophycean algae. *Archiv für Mikrobiologie*, **88**, 193–204.

Buchner, P. (1965). *Endosymbiosis of Animals with Plant Microorganisms*. New York: Interscience.

Chapman, D. J., Cole, W. J. & Siegelman, H. W. (1968). A comparative study of the phycoerythrin chromophore. *Phytochemistry*, **7**, 1831–5.

Cohen-Bazire, G. (1971). The photosynthetic apparatus of prokaryotic organisms. In *Biological Ultrastructure: The origin of cell organelles*, ed. P. Harris, pp. 65–90. Corvallis: Oregon State University Press.

Craig, I. W. & Carr, N. G. (1968). C-phycocyanin and allophycocyanin in two species of blue-green algae. *Biochemical Journal*, **106**, 361–6.

Engelmann, T. W. (1883). Farbe und Assimilation. *Botanische Zeitung*, **41**, 1–13.

Engelmann, T. W. (1884). Untersuchungen über die quantitativen Beziehungen zwischen Absorption des Lichtes und Assimilation in Pflanzenzellen. *Botanische Zeitung*, **42**, 81–95.

Fujita, Y. & Hattori, A. (1963). Effects of second chromatic illumination on phycobilin chromoprotein formation in chromatically pre-illuminated cells of *Tolypothrix tenuis*. In *Microalgae and Photosynthetic Bacteria*, *Plant and Cell Physiology* (special issue), pp. 431–40.

Gaidukov, N. (1902). Uber den Einfluss farbigen Lichtes auf die Färbung lebendes Oscillatorien. *Abhandlungen preussischer Akademie Wissenschaften Mathematische Naturwissenschafticher Klasse*, pp. 1–36.

Gantt, E. (1969). Properties and ultrastructure of phycoerythrin from *Porphyridium cruentum*. *Plant Physiology*, **44**, 1629–38.

Gantt, E. & Conti, S. F. (1965). The ultrastructure of *Porphyridium cruentum*. *Journal of Cell Biology*, **26**, 365–81.

GANTT, E. & CONTI, S. F. (1966*a*). Granules associated with the chloroplast lamellae of *Porphyridium cruentum. Journal of Cell Biology*, **29**, 423–34.

GANTT, E. & CONTI, S. F. (1966*b*). Phycobiliprotein localization in algae. *Brookhaven Symposia in Biology* **19**, 393–405.

GANTT, E., EDWARDS, M. R. & PROVASOLI, L. (1971). Chloroplast structure of the Cryptophyceae. Evidence for phycobiliproteins within intrathylakoidal spaces. *Journal of Cell Biology*, **48**, 280–90.

GANTT, E. & LIPSCHULTZ, C. A. (1972). Phycobilisomes of *Porphyridium cruentum*. I. Isolation. *Journal of Cell Biology*, **54**, 313–24.

GANTT, E. & LIPSCHULTZ, C. A. (1973). Energy transfer in phycobilisomes from phycoerythrin to allophycocyanin. *Biochimica et Biophysica Acta* **292**, 858–61.

GLAZER, A. N. & COHEN-BAZIRE, G. (1971). Subunit structure of the phycobiliproteins of blue-gree algae. *Proceedings of the National Academy of Sciences, USA*, **68**, 1398–401.

GLAZER, A. N., COHEN-BAZIRE, G. & STANIER, R. Y. (1971*a*). Characterization of phycoerythrin from a *Cryptomonas* sp. *Archiv für Mikrobiologie*, **80**, 1–18.

GLAZER, A. N., COHEN-BAZIRE, G. & STANIER, R. Y. (1971*b*). Comparative immunology of algal biliproteins. *Proceedings of the National Academy of Sciences, USA*, **68**, 3005–8.

GLAZER, A. N. & FANG, S. (1973*a*). Chromophore content of blue-green algal phycobiliproteins. *Journal of Biological Chemistry*, **248**, 659–62.

GLAZER, A. N. & FANG, S. (1973*b*). Formation of hybrid proteins from the α- and β-subunits of phycocyanins of unicellular and filamentous blue-green algae. *Journal of Biological Chemistry*, **248**, 663–71.

GOODENOUGH, U. W. & STAEHELIN, L. A. (1971). Structural differentiation of stacked and unstacked chloroplast membranes. Freeze-etch electron microscopy of wild-type and mutant strains of *Chlamydomonas. Journal of Cell Biology*, **48**, 594–619.

HATTORI, A., CRESPI, H. L. & KATZ, J. J. (1965). Association and dissociation of phycocyanin and effects of deuterium substitution on the process. *Biochemistry*, **4**, 1225–38.

HAXO, F. T. & BLINKS, L. R. (1950). Photosynthetic action spectra of marine algae. *Journal of General Physiology*, **33**, 389–422.

HAXO, F. T. & FORK, D. C. (1959). Photosynthetically active accessory pigments of cryptomonads. *Nature, London*, **184**, 1051–2.

HERTZBERG, S., LIAAEN-JENSEN, S. & SIEGELMAN, H. W. (1971). The carotenoids of blue-green algae. *Phytochemistry*, **10**, 3121–7.

JEFFREY, S. W. (1969). Properties of two spectrally different components in chlorophyll *c* preparations. *Biochimica et Biophysica Acta*, **177**, 456–67.

LEFORT-TRAN, M., COHEN-BAZIRE, G. & POUPHILE, M. (1973). Les membranes photosynthetiques des algues à biliproteines observées après cryodécapage. *Journal of Ultrastructure Research*, **44**, 199–209.

LEMBERG, R. & BADER, G. (1933). Die Phycobiline der Rot-algen. Ueberführung in Mesobilirubin und Dehydromesobilirubin. *Liebigs Annalen Chemie*, **505**, 151–77.

MARGULIS, L. (1968). Evolutionary criteria in thallophytes: a radical alternative. *Science*, **161**, 1020–2.

NEUFELD, G. & RIGGS, A. F. (1969). Aggregation properties of C-phycocyanin from *Anacystis nidulans. Biochimica et Biophysica Acta*, **181**, 234–43.

NOLAN, D. N. & ÓHEOCHA, C. (1967). Determination of molecular weights of algal biliproteins by gel filtration. *Biochemical Journal*, **103**, 39–40.

ÓHEOCHA, C. (1965). Biliproteins of algae. *Annual Review of Plant Physiology*, **16**, 415–34.

PIERSON, B. K. & CASTENHOLZ, R. W. (1971). Bacteriochlorophylls in gliding filamentous prokaryotes from hot springs. *Nature New Biology*, **233**, 25.

SCHEIBE, J. (1972). Photoreversible pigment; occurrence in a blue-green alga. *Science*, **176**, 1037–9.

SCHRAM, B. L. & KROES, H. H. (1971). Structure of phycocyanobilin. *European Journal of Biochemistry* **19**, 581–94.

SCOTT, E. & BERNS, D. S. (1965). Protein–protein interaction. The phycocyanin system. *Biochemistry*, **4**, 2597–606.

SIEGELMAN, H. W., CHAPMAN, D. J. & COLE, W. J. (1968). The bile pigments of plants. In *Porphyrins and Related Compounds*, ed. T. W. Goodwin, pp. 107–20. New York and London: Academic Press.

STANIER, R. Y. (1970). Some aspects of the biology of cells and their possible evolutionary significance. In *Organization and Control in Prokaryotic and Eukaryotic Cells. Symposia of the Society for General Microbiology*, **20**, 1–38. Ed. H. P. Charles & B. C. J. G. Knight. London: Cambridge University Press.

STRANSKY, H. & HAGER, A. (1970). Das Carotinoidmuster und die Verbreitung des Lichtinduzierten Xanthophyllcyclus in verschiedenen Algenklassen. VI. Chemosystematische Betrachtung. *Archiv für Mikrobiologie*, **73**, 315–23.

VAUGHN, M. H. (1964). Structural and comparative studies of the algal protein phycoerythrin. Dissertation, Massachusetts Institute of Technology.

WILKIE, D. (1970). Reproduction of mitochondria and chloroplasts. In *Organization and Control in Prokaryotic and Eukaryotic Cells. Symposia of the Society for General Microbiology*, **20**, 381–400. Ed. H. P. Charles & B. C. J. G. Knight. London: Cambridge University Press.

EXPLANATION OF PLATES

PLATE 1

Fig. 1. Electron micrograph of a thin section of the unicellular rhodophyte *Porphyridium cruentum*. The large, multilobed chloroplast, with a centrally located pyrenoid (PY), occupies much of the cell volume. Note the regularity of the interthylakoid spacing, and the regular granular arrays of phycobilisomes on the thylakoid surfaces. These features of chloroplast structure are particularly clear in cortical areas (arrows), where the thylakoids occur in relatively dense parallel array. From Gantt & Conti (1965).

Fig. 2. Electron micrograph (negative staining) of a suspension of isolated phycobilisomes from *Porphyridium cruentum*. From Gantt & Lipschultz (1972).

PLATE 2

Electron micrographs of thin sections of a *Cryptomonas* sp., to illustrate the fine structure of the cryptophytan chloroplast.

Fig. 1. Transverse section of a cell showing two lobes of the chloroplast. From Glazer *et al.* (1971*a*).

Fig. 2. Section of a chloroplast viewed at higher magnification, to show the characteristic relatively close packing of the thylakoids and the expanded, electron-dense structure of the thylakoid lumen. Courtesy of Dr Germaine Cohen-Bazire.

PLATE 3

Electron micrograph of a thin section of a unicellular cyanobacterium, *Aphanocapsa* sp. An extensive system of thylakoids fills much of the cytoplasm: note the very regular interthylakoidal spacing, of approx. 40 nm. The arrows point to a region where the array of phycobilisomes which occupy the interthylakoidal space is particularly clear. Courtesy of Dr Germaine Cohen-Bazire.

PLATE 4

A freeze-etched preparation of the cyanobacterium shown in Plate 3. The arrows point to parallel arrays of small granules (diameter *ca* 5 nm), situated on one of the internal fracture faces of the thylakoid membrane. From their position and arrangement, these granules appear to correspond to the attachment sites of phycobilisomes on the outer surface of the thylakoid membrane. Courtesy of Dr Marcelle Lefort-Tran & Dr Germaine Cohen-Bazire.

PLATE 1

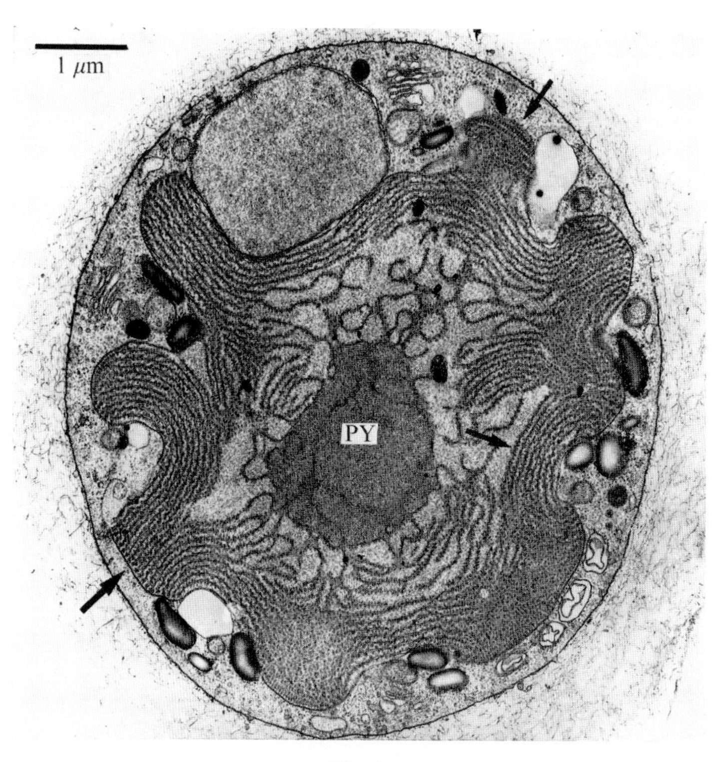

Fig. 1

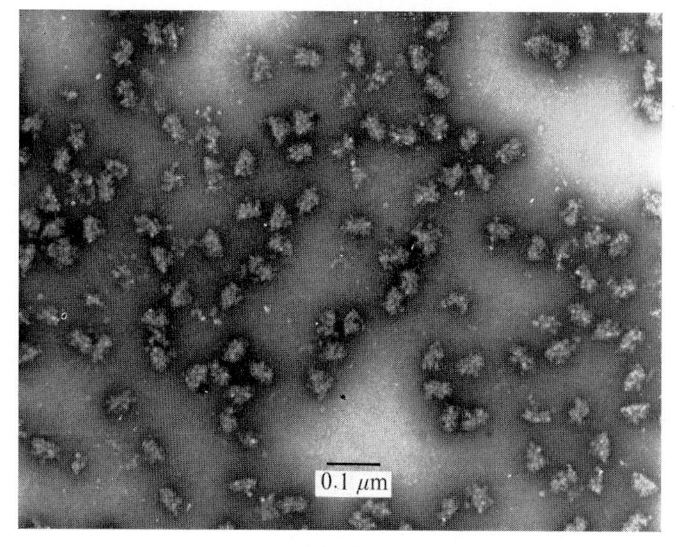

Fig. 2

(*Facing p.* 240)

PLATE 2

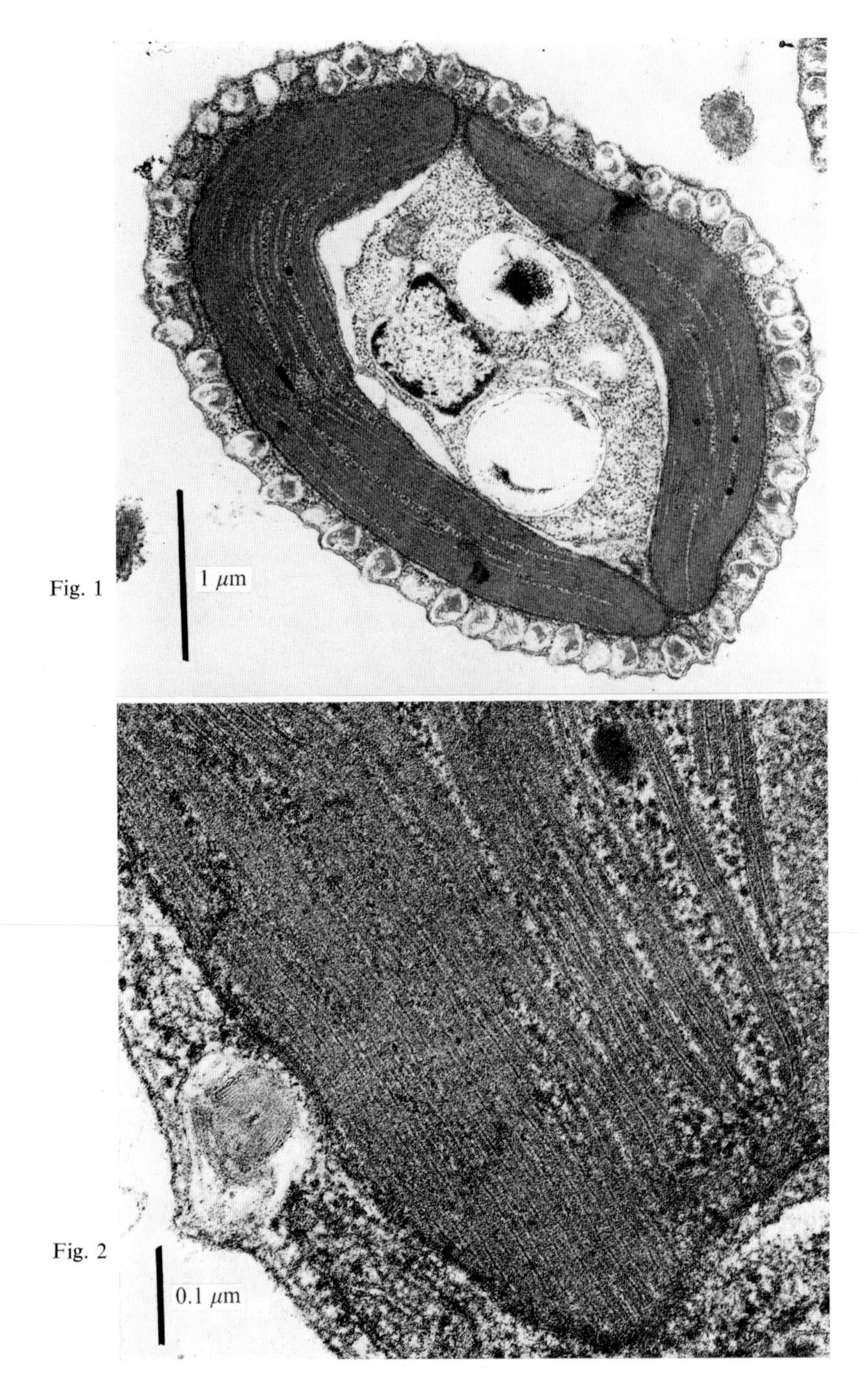

Fig. 1

Fig. 2

PLATE 3

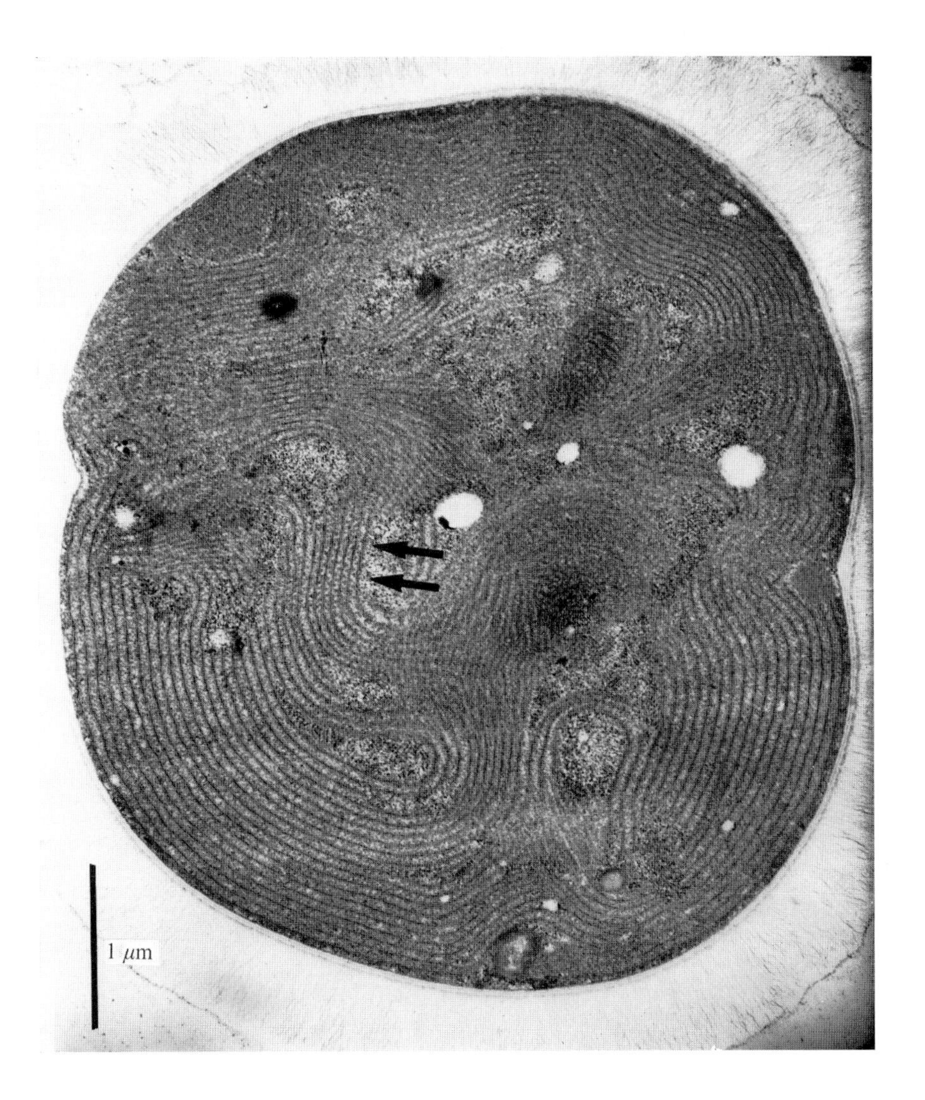

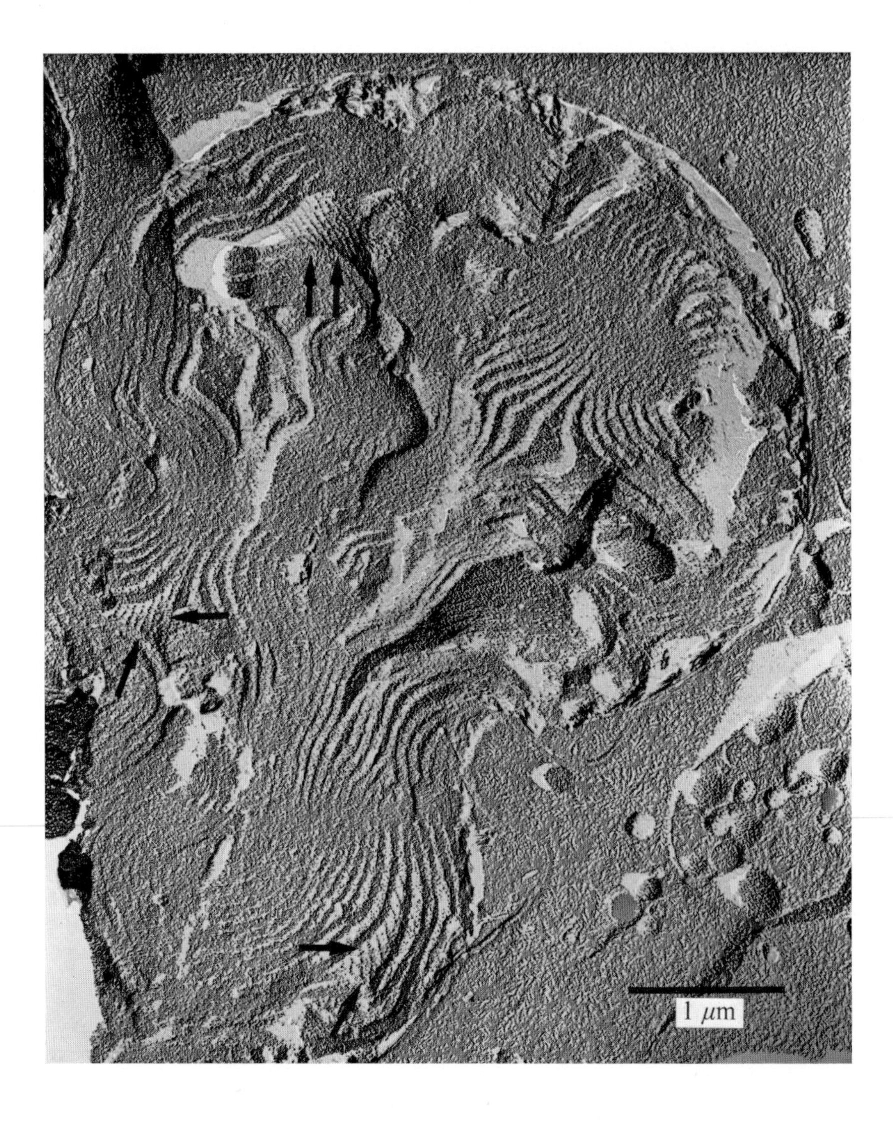
1 μm

THE EVOLUTIONARY SIGNIFICANCE OF INORGANIC SULFUR METABOLISM

H. D. PECK, Jr
Department of Biochemistry, University of Georgia, Athens, Georgia, USA

INTRODUCTION

The ability of micro-organisms to utilize reduced inorganic sulfur compounds as electron donors in respiration or to reduce oxidized inorganic sulfur compounds as their terminal electron acceptors has been regarded as a primitive metabolic characteristic (Klein & Cronquist, 1967; Peck, 1967). Direct evidence for this view has not been convincing. For example the Thiobacilli, which can satisfy their entire energy requirement for growth by the oxidation of reduced sulfur compounds, require molecular oxygen (or an oxidized anion, NO_3^-) and are biosynthetically complete as they are able to utilize carbon dioxide as their sole source of carbon. If one accepts that the primitive bacteria were anaerobes growing in a reduced environment and biosynthetically simple, the Thiobacilli cannot be considered as examples of early forms of life. This idea concerning the primitive nature of sulfur metabolism is probably based on the fact that the most primitive of the photosynthetic bacteria oxidize reduced sulfur compounds during photosynthetic growth under anaerobic conditions. The concept was also introduced into microbiological literature at a time when it was believed that chemoautotrophic bacteria, such as the Thiobacilli, were simple forms of life which could flourish in the absence of organic compounds and that the primitive crust of the earth lacked organic materials. The Haldane–Oparin ideas concerning the abundance of organic compounds in the primitive crust of the earth make it unnecessary to postulate that early organisms were lithotrophic and modern physiological and biochemical investigations have destroyed the idea that chemolithotrophic and autotrophic bacteria are metabolically simple forms of life. Today these bacteria are regarded as highly specialized bacteria capable of growth in a completely inorganic medium (Rittenberg, 1969; Kelley, 1971).

There is, however, some merit in the idea that inorganic sulfur metabolism may be a primitive physiological characteristic. Reduced inorganic compounds must have been abundant in the primitive anaerobic environment of the earth and as the supply of preformed organic materials approached exhaustion, these inorganic compounds represented a new source of chemical energy that could be exploited by these early organisms. Reduced sulfur compounds are attractive in this regard because of the anaerobic sulfur cycle, the ease with which reduced sulfur compounds can be oxidized and the existence of the only case of substrate-level conversion of chemical energy into ATP in inorganic metabolism. Although these ideas are highly speculative, they have indicated new areas for investigation which have already proved to be productive. The possible evolutionary significance of inorganic sulfur metabolism will be discussed in terms of the sulfur cycle, sulfur isotope data, physiology and biochemistry of sulfate reduction and the amino acid sequence of ferredoxin.

THE SULFUR CYCLE

The sulfur cycle (Fig. 1) is not only a schematic representation of the reactions involved in the transformation of sulfur compounds but also represents a simple system for the utilization of radiant energy and the partial support of life. For primitive forms of life to survive, it was necessary for them to develop systems capable of utilizing radiant energy and generating ATP or low potential electrons. As a candidate for the first system capable of utilizing radiant energy, the sulfur cycle offers a number of advantages. First, it is microbiologically the most simple inorganic cycle. The combination of a sulfate-reducing bacterium and a photosynthetic bacterium contain all the enzymes required for the cyclical transformation of sulfur compounds. Second, it is the only known inorganic transformation in which there occurs a substrate-phosphorylation enzyme. Third, sulfur cycles occur today. A series of lakes in North Africa produce elemental sulfur and the probable mechanism of this accumulation has been described (Butlin & Postgate, 1954). The lakes are fed by waters containing sulfate and in the lake the sulfate is reduced to sulfide by sulfate-reducing bacteria. Along the edge of the lakes, photosynthetic bacteria oxidize the sulfide to elemental sulfur and presumably supply the organic materials that support the reduction of sulfate. The accumulation of elemental sulfur is so extensive that as the lakes recede in the dry season the sulfur is harvested for commercial use. The photosynthetic bacteria can also oxidize elemental sulfur to sulfate and thus provide a sequence of cyclic reactions that can support life in light under anaerobic conditions. The accumulation of elemental sulfur in light from sulfate has been demonstrated in the

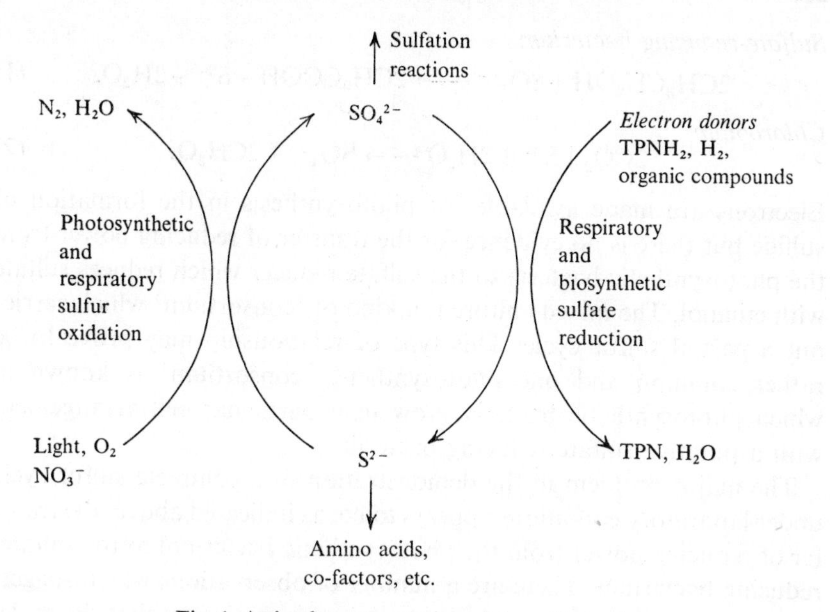

Fig. 1. A simple representation of the sulfur cycle.

laboratory (Butlin & Postgate, 1954) although the growth of such a mixed culture on CO_2 and sulfate has not been so clearly demonstrated. The major problem appears to be the transfer of reducing power generated during photosynthesis to the sulfate-reducing bacteria. If this transfer of reducing materials requires the death and breakdown of some of the photosynthetic bacteria, then it may only be possible to observe the sulfur cycle under complex natural conditions (Gemerden, 1967).

Recently an interesting case of the sulfur cycle in a mixed culture has been analyzed by Gray (1972). *Chloropseudomonas ethylica* has long been regarded as an interesting but anomalous type of photosynthetic bacterium which was capable of growth on ethanol. The cultural properties were constant but cultures were described as being pleomorphic as they contained several cell types. The organism has been extensively studied and a low potential *c*-type cytochrome isolated and sequenced (Ambler, 1971). Analysis of the sequence indicated that this cytochrome was closely related to cytochromes c_3 which are found in all species of *Desulfovibrio*. Gray has presented convincing evidence that *Chloropseudomonas ethylica* is a mixed culture composed of a *Chlorobium* and an unidentified sulfate-reducing bacterium. These results have been confirmed by Pfennig (personal communication). The basic reactions involved are as follows:

Sulfate-reducing bacterium

$$2CH_3CH_2OH + SO_4^{2-} \longrightarrow 2CH_3COOH + S^{2-} + 2H_2O. \quad (1)$$

Chlorobium

$$2CO_2 + S^{2-} + 2H_2O \xrightarrow{h\nu} SO_4^{2-} + 2CH_2O. \quad (2)$$

Electrons are made available for photosynthesis in the formation of sulfide but there is no evidence for the transfer of reducing power from the photosynthetic bacteria to the sulfate reducer which reduces sulfate with ethanol. The mixed culture is a kind of 'consortium' which carries out a partial sulfur cycle. This type of relationship may prove to be rather common and one photosynthetic 'consortium' is known in which photosynthetic bacteria grow in a particular cell arrangement with a putative sulfate-reducing bacterium.

The major problem in the demonstration of a complete sulfur cycle under laboratory conditions appears to be, as indicated above, the transfer of reducing power from the photosynthetic bacterium to the sulfate-reducing bacterium. There are a number of observations which suggest that such transfers do occur between certan bacteria and that the molecule actually transferred is molecular hydrogen. *Methanobacillus omelianskii* has been shown to be a mixed culture consisting of a methane forming bacterium (H organism) and an organism which oxidizes ethanol to acetate and H_2 (S organism) (Bryant *et al.* 1967). Although the methane bacterium grows well on H_2 and CO_2 the S organism grows poorly on ethanol even when the H_2 is continuously removed (Reddy, Bryant & Wolin, 1972). Both organisms grow well together on ethanol and methane formation is postulated to function as a highly efficient hydrogen trap which 'pulls' ethanol oxidation in the direction of acetate formation. Growth of the sulfate-reducing bacteria *Desulfovibrio vulgaris* and *D. desulfuricans* on lactate in the absence of the obligatory terminal electron acceptor, sulfate, has been observed when these organisms are cultured with the H organism isolated from *M. omelianskii*. Methane is formed presumably by the utilization of the electrons derived from both the oxidation of lactate and pyruvate (Bryant, 1969). From these and other observations concerning altered fermentations in mixed cultures, Wolin and Bryant have developed the concept of interspecies hydrogen transfer.

We have recently been concerned with the purification of hydrogenase from *Desulfovibrio gigas*. It was observed (Bell & LeGall, unpublished) that a considerable amount of hydrogenase could be removed from the cells merely by washing with phosphate buffer (LeGall, Mazza & Dragoni, 1965). Using procedures commonly employed to demonstrate

Table 1. *Evidence for the periplasmic localization of hydrogenase*

Fraction*	% protein or activity			
	Protein	Hydro-genase	Desulfo-viridin	APS reductase
Shock fluid	15	90	5	0
Lysate of shocked cell	85	10	95	100

* Cells were shocked by the method of Neu & Heppel (1965), hydrogenase was determined as described by LeGall *et al.* (1971), desulforividin by absorption at 630 nm and APS reductase by the method of Peck, Deacon & Davidson (1965).

the periplasmic location of enzymes (Neu & Heppel, 1965), it was possible to remove 90 % of the hydrogenase from *D. gigas* (Table 1). This result indicates that hydrogenase is located at the surface of the cell in the so-called periplasmic space. This unexpected observation suggests a new physiological role for hydrogenase in the efficient trapping of small amounts of hydrogen from its environment. It should be noted that hydrogen is not formed during normal growth of the sulfate-reducing bacteria although extracts contain relatively high levels of hydrogenase (LeGall *et al.* 1971) and that the sulfate-reducing bacteria are recognized as utilizing extremely low levels of hydrogen in the corrosion of iron (Postgate, 1959*a*). In the rumen fermentation, hydrogen appears to be an important intermediate in the formation of methane (Hungate, 1967). These results support the idea that molecular hydrogen is an important compound in interspecies electron transfer (Bryant, 1969) and suggest that anaerobic bacteria possess specific structural adaptations for symbiotic growth.

The evidence indicates that a simple sulfur cycle involving only two micro-organisms is feasible and that the organisms will be nutritionally linked through the transfer of inorganic sulfur compounds and molecular hydrogen (Fig. 2). This simple sulfur cycle is not a primary producing system as electrons must be derived from an electron donor other than water. In the presence of light and sulfide, only ATP or low potential electrons are generated (Tagawa & Arnon, 1962) and this type of system may represent an intermediate stage in the evolution of green plant photosynthesis. The establishment of a sulfur cycle was probably a vital stage in the evolution of bacteria. It liberated these organisms from dependency on the dwindling supply of fixed organic compounds for energy, but they probably still utilized fixed carbon for biosynthetic reactions. The diversity of organic compounds in the primitive seas must have fostered metabolic diversity among the existing organisms as competition for energy sources became more intense. This

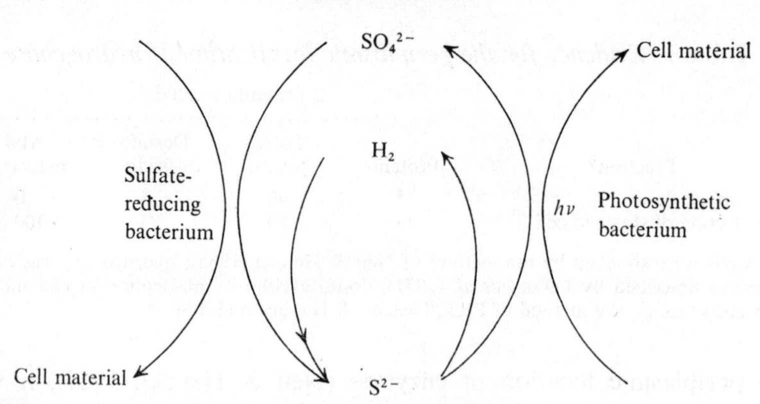

Fig. 2. The simple sulfur cycle.

simple sulfur cycle could have provided the first renewable energy source the existence of which would permit new types of micro-organisms to evolve. It is even possible that utilization of the sulfur cycle offered such a tremendous nutritional advantage that all existing bacteria are derived from organisms that participated in the cycle.

$^{32}S/^{34}S$ RATIOS

Naturally occurring sulfur consists of several isotopes, the major ones being ^{32}S and ^{34}S, but the ratio of these two isotopes varies in samples of naturally occurring sulfur. This has been established and studied by use of the mass spectrometer which determines $^{32}S/^{34}S$ ratios with a high degree of precision. The extent of the variation in the sulfur ratios is indicated in Table 2. The sulfur in meteors has a relatively constant $^{32}S/^{34}S$ ratio and is presumed to be representative of the natural abundance of sulfur isotopes in the universe (Kaplan & Hulston, 1966). The sulfur in the sulfate of sea water is enriched in ^{34}S but this value is subject to some variation, probably due to local conditions. Reduced sulfur compounds formed by geological phenomena, such as volcanic action, generally have about the same isotopic composition as meteoritic sulfur; there can, however, be some variations in these values. Other deposits of reduced sulfur compounds can be enriched in ^{32}S to values greater than that of either sea water or meteoritic sulfur. For example, the ^{32}S content of elemental sulfur in the sulfur domes of Louisiana and Texas is considerably higher than that found in the surrounding $CaSO_4$ from which the elemental sulfur was presumably formed (Thode, Wanless & Wallouch, 1954). It was postulated that sulfur deposits showing high $^{32}S/^{34}S$ ratios were of biogenic origin and

Table 2. $^{32}S/^{34}S$ *ratios in sulfur of various origins*

Source	$^{32}S/^{34}S$
Sea water	21.8
Geologically reduced sulfur	22.1–22.3
Meteoritic	22.22
Sulfate-reducing bacteria	22.8
Biologically reduced sulfur	up to 23.1

Table 3. *Time scale for the origin of various biological activities*

Years × 10^9	Phenomenon
5	Formation of the earth
4	Physiological temperatures
3	Respiratory sulfate reduction
2	Photosynthesis – green plant
1	Oxygen atmosphere – oxidative phosphorylation
Present	

this conclusion was supported by the observation that sulfur isotopes are fractionated during respiratory or dissimilatory sulfate reduction by the sulfate-reducing bacterium *D. desulfuricans* and increase in the $^{32}S/^{34}S$ ratio. Conditions for maximum fractionation include excess sulfate and a temperature suboptimal for growth. Other biological, chemical and physical phenomena which might effect this enrichment have been studied in detail (Thode, Kleerekoper & McElcheran, 1951; Kaplan & Rittenberg, 1964; Nakai & Jensen, 1964), but it appears that dissimilatory sulfate reduction is the only known process that produces the required enrichment in ^{32}S. It has been possible to measure $^{32}S/^{34}S$ ratios in sulfur isolated from various geological strata. The occurrence of the biological $^{32}S/^{34}S$ ratio should indicate that dissimilatory sulfate reduction was occurring at the time that the stratum from which the sulfur was obtained was deposited. However, the possibility always exists that the enriched sulfur was either produced or transported into the stratum later. Sulfur enriched in ^{32}S has been found in rocks formed 2–3.5 × 10^9 years ago (Ault & Kulp, 1959). This suggests that dissimilatory sulfate reduction and probably sulfate-reducing bacteria existed early in the history of life on earth. In Table 3 a time scale for the origin of various biological activities and dissimilatory sulfate reduction is shown. This suggests that dissimilatory sulfate reduction is very ancient. Perhaps sulfate-reducing bacteria have been stabilized in evolution by the mechanistic limitations of respiratory sulfate reduction and are 'living fossils'.

A consequence of the existence of dissimilatory sulfate reduction is the need for a large supply of sulfate, but large amounts of sulfate were probably not available in the anaerobic, reduced environment in which life is believed to have evolved. Hence a prerequisite for the existence of sulfate-reducing bacteria may have been organisms capable of growing anaerobically on reduced sulfur compounds and producing sulfate either by means of a primitive type of photosynthesis or by employing oxidized organic compounds as their terminal electron acceptors. The energy for growth would be derived from the single substrate phosphorylation that occurs in the oxidation of sulfide to sulfate. These ideas suggest that it may be of value to examine the physiology and biochemistry of the sulfate-reducing bacteria, *Clostridia* and photosynthetic bacteria with regard to evolutionary relationships.

MECHANISM OF SULFATE REDUCTION

Sulfate can satisfy both the oxidized and reduced sulfur requirements for the growth of most plants and bacteria (Roy & Trudinger, 1970). Physiologically, this reduction of sulfate is a biosynthetic pathway subject to regulation so that most organisms reduce only enough sulfur to satisfy their nutritional requirements. This type of sulfate reduction has been termed 'assimilatory sulfate reduction' and the form in which sulfate is reduced (Wilson, Asahi & Bandurski, 1961) appears to be 3′-phosphoadenylyl sulfate (PAPS) formed (Reactions 3 and 4) from ATP and sulfate by ATP sulfurylase and APS kinase (Robbins & Lipmann, 1958).

$$ATP + SO_4^{2-} \rightarrow APS + PP_i, \tag{3}$$

$$APS + ATP \rightarrow PAPS + ADP, \tag{4}$$

$$PAPS + NADPH + H^+ \rightarrow SO_3^{2-} + PAP + NADP. \tag{5}$$

PAPS is reduced in most bacteria and yeasts by the enzyme PAPS reductase, but in algae and plants there is evidence that APS is the form in which sulfate is reduced (Schmidt, 1972) although sulfotransferase reactions still require PAPS. Several properties of PAPS reductase should be noted. The reduction requires three proteins, one of which is probably thioredoxin. The reaction has not been shown to be reversible, formation of the enzyme is repressed by cysteine in bacteria, the prosthetic groups, if any, have not been established and the enzyme shows activity toward APS as well as PAPS.

Sulfate also functions as a terminal electron acceptor for the anaerobic respiration of the sulfate-reducing bacteria classified as *Desulfovibrio* or *Desulfotomaculum*. In this process, large amounts of sulfide are produced which can effect extensive pollution of the environment. This type of sulfate reduction is termed respiratory or dissimilatory sulfate reduction. The form in which sulfate is reduced is APS (Ishimoto, 1959; Peck, 1959) formed by ATP sulfurylase. Inorganic pyrophosphatase is probably required for the formation of significant amounts of APS. APS is directly reduced to AMP and sulfite by the enzyme, APS reductase (Reaction 6):

$$APS + 2e \rightarrow AMP + SO_3^{2-}. \tag{6}$$

APS reductase is an iron-containing flavoprotein of molecular weight 220000 (Peck, Deacon & Davidson, 1965). Extracts of the sulfate-reducing bacteria lack APS kinase and do not form, reduce or metabolize PAPS; however, they do have low levels of a biosynthetic-type sulfite reductase (Lee, LeGall & Peck, 1971). The reaction catalyzed by the APS reductase is reversible with O_2 or $Fe(CN)_6^{3-}$ as electron acceptor (Reaction 7) and results in the formation of a high-energy sulfate bond which can be exchanged for phosphate (Reaction 8) in the presence of the enzyme ADP sulfurylase (Robbins & Lipmann, 1958).

$$SO_3^{2-} + AMP \rightarrow APS + 2e, \tag{7}$$

$$APS + P_i \rightarrow ADP + SO_4^{2-}. \tag{8}$$

This is the only instance in the metabolism of inorganic compounds where there exists a substrate level phosphorylation. The enzymes appear to be involved in the oxidation of inorganic sulfur compounds in certain of the Thiobacilli (Peck, 1960*a*; Bowen, Happold & Taylor, 1966) and in the Thiorhodaceae and Chlorobacteriaceae (Thiele, 1968). APS reductase is constitutive (Wheldrake & Pasternak, 1965), appears to be of limited distribution and seems to be similar in different organisms (Table 4).

The reduction of sulfite in the sulfate-reducing bacteria appears to proceed by a related but different mechanism than that found in the biosynthetic pathway. The biosynthetic assimilatory sulfite reductase consists of two types. The first type is the sulfite reductase specific for reduced methyl viologen (and probably ferredoxin), which catalyzes the six electron reduction of sulfite to sulfide without the accumulation of intermediates. The prosthetic groups of the reductase are non-heme iron and an anomalous heme which has recently been identified (Murphy *et al.* 1973*b*). The reductase has been found mainly in plants (Asada, Tamura & Bandurski, 1969) and has a molecular weight of about 80000. The second type of sulfite reductase (Siegel, Murphy & Kamin, 1973) is able to utilize $NADPH_2$ as well as reduced methyl viologen as electron donor and is a high molecular weight complex consisting of

Table 4. *Comparison of adenylyl sulfate reductase from various sources*

Property	*Desulfovibrio vulgaris* (Peck *et al.* 1965)	*Thiobacillus denitrificans* (Bowen *et al.* 1966)	*Thiobacillus thioparus* (Lyric & Suzuki, 1970)	*Thiocapsa roseopersicina* (Trüper & Rogers, 1971)
Molecular weight	220000	n.d.	170000	180000
Sedimentation constant	9.2*S*	9.54*S*	n.d.	9.7*S*
FAD per molecule	1	0.7 per 10^5 g protein	1	1
NHI per molecule	10–12	5 per 10^5 g protein	8–10	4
Heme per molecule	0	0	0	2
Electron acceptor specificity				
$Fe(CN)_6^{3-}$ assay	+	+	+	+
pH optimum	7.5	7.2	7.4	8.0
Cytochrome *c* assay	+	n.d.	+	+
pH optimum	9.5	n.d.	9.5	9.0
O_2 assay	+	+	n.d.	n.d.

n.d.: not determined.

four sulfite reductase subunits and eight flavoprotein subunits containing FMN and FAD which are responsible for the utilization of $NADPH_2$ as electron donor (Siegel *et al.* 1971).

The dissimilatory pathway of sulfite reduction in the sulfate-reducing bacteria has been shown to be different from the assimilatory pathway in that it involves at least three distinct enzymes and free inorganic intermediates. The accumulation of thiosulfate during the reduction of sulfite was reported by Suh & Akagi (1969) in extracts of *D. vulgaris* and Kobayashi, Tachibana & Ishimoto (1969) demonstrated that trithionate and thiosulfate were sequentially accumulated during the reduction of sulfite by fractionated extracts of *D. vulgaris*. The following series of reactions was proposed for the reduction of sulfite to sulfide:

$$3HSO_3^{1-} + 3H^+ + 2e \rightarrow S_3O_6^{2-} + 3H_2O, \quad (9)$$

$$S_3O_6^{2-} + 2e \rightarrow S_2O_3^{2-} + SO_3^{2-}, \quad (10)$$

$$S_2O_3^{2-} + 2e \rightarrow S^{2-} + SO_3^{2-}. \quad (11)$$

Lee & Peck (1971) later described the purification of the trithionate forming enzyme termed bisulfite reductase (Reaction 9) and its identification as desulfoviridin, the green protein of taxonomic significance for the genus *Desulfovibrio* (Postgate, 1959*b*). The enzyme was called bisulfite reductase to distinguish it from the biosynthetic sulfite reductase. The presence of trithionate reductase (Reaction 10) in extracts of *D. vulgaris* was reported only by Kobayashi *et al.* (1969); however, the thiosulfate reductase (Reaction 11) has been purified and studied by Haschke & Campbell (1971).

The 'Norway Strain' of *Desulfovibrio desulfuricans* completely lacks desulfoviridin (Miller & Saleh, 1964) but retains the ability to reduce sulfate to sulfide at normal rates. Extracts catalyzed the formation of trithionate and it has been shown that this activity is due to a reddish-brown pigment, termed desulforubidin, which replaced but is completely analogous to desulfoviridin (Lee, Yi, LeGall & Peck, 1973).

The spore-forming sulfate-reducing bacteria, species of *Desulfotomaculum*, characteristically do not contain desulforviridin, but Trudinger (1970) has demonstrated the presence of an enzyme, termed P_{582}, which reduces sulfite to sulfide. Its absorption spectrum is similar to that of the biosynthetic sulfite reductase and the properties of the enzyme suggest that it contains a heme-like prosthetic group (Trudinger & Chambers, 1973). Akagi (personal communication) has recently shown that preparations of this enzyme do form trithionate and it is becoming clear that all three bisulfite reductases (desulfoviridin, desulforubidin and P_{582}) from the sulfate-reducing bacteria can form small but variable amounts of sulfide in addition to the major product, trithionate (Kobayashi, Takahashi & Ishimoto, 1972).

Recently, in a collaboration between the Duke and Georgia groups (Murphy *et al.* 1973*a*), it has been shown that all three bisulfite reductases are hemoproteins and that the hemes are identical to the novel heme, an octacarboxylic tetrahydroporphyrin of the isobacteriochlorin type, found in the biosynthetic sulfite reductase. The reason for the occurrence of these three different forms of bisulfite reductases and their relationship to each other are obscure, but one might speculate that groups of sulfate-reducing bacteria have arisen several times during evolution. The requirement for inorganic intermediates in respiratory sulfate reduction probably results from the bioenergetics of the process and will be considered in a later section.

The most striking similarities in the sulfur metabolism of the sulfate-reducing bacteria and the photosynthetic bacteria are the presence of APS reductase and ADP sulfurylase; however, it should be noted that sulfite and trithionate now appear to be implicated as intermediates in both the oxidative and reductive pathways (London & Rittenberg, 1964) and it should be interesting to examine in detail the sulfur metabolism of the Clostridia as they may be related to bacterial forms from which the photosynthetic bacteria and sulfate-reducing bacteria were derived.

BIOENERGETICS AND ELECTRON CARRIERS

The physiology and biochemistry of respiratory sulfate reduction indicate that oxidative phosphorylation is required for anaerobic growth with most substrates. As previously indicated (Reaction 6) the activation of sulfate requires ATP and, because of the pyrophosphate cleavage, two high energy phosphate bonds are utilized. As the bacteria can reduce sulfate with only molecular hydrogen present, it follows that ATP must be generated during the reduction of sulfate coupled with the oxidation of hydrogen. The occurrence of an oxidative phosphorylation can be demonstrated using whole cells and uncoupling agents such as dinitrophenol (DNP). DNP reversibly inhibits the reduction of sulfate, which requires ATP, but not the reduction of thiosulfate or sulfite which does not require ATP (Peck, 1960*b*). From a consideration of the probable reactions involved in the lactate–sulfate fermentation and the ethanol–sulfate fermentation (Reactions 12–15) it appears that oxidative phosphorylation is necessary for growth on these and probably other substrates.

$$2 \text{ ethanol} \rightarrow 2 \text{ acetaldehyde} + [4H\cdot], \qquad (12)$$

$$2 \text{ acetaldehyde} + 2P_i \rightarrow 2 \text{ acetyl phosphate} + [4H\cdot], \qquad (13)$$

$$2 \text{ acetyl phosphate} + AMP + 2H^+ \rightarrow 2 \text{ acetate} + ATP, \qquad (14)$$

$$SO_4^{2-} + ATP + [8H\cdot] \rightarrow S^{2-} + 2H_2O + AMP + 2P_i + 2H^+. \qquad (15)$$

Two high-energy phosphates are produced in the fermentation at the substrate level but two high-energy phosphates are consumed in the reduction of sulfate. The net yield of ATP from substrate phosphorylation is zero and it is necessary to assume oxidative phosphorylation to explain growth with these substrates. The presence of oxidative phosphorylation coupled to the oxidation of hydrogen with sulfite (Peck, 1966) and fumarate (Barton, LeGall & Peck, 1970) has been demonstrated in cell-free extracts. Lactate, but not pyridine nucleotide, can also function as an electron donor with fumarate.

The energy released in the oxidation of hydrogen by sulfate is about 40 kcal mol^{-1} sulfate. In terms of bioenergetics, the efficiency of the oxidative phosphorylation is probably less than 50% and if one assumes that the available energy is released equally in each of the four reductive steps, none of the reductive steps would provide sufficient energy to support an oxidative phosphorylation. The occurrence of intermediates in the reduction of sulfite may represent an adaptation to restrict the major energy-releasing reactions in sulfate reduction to

Table 5. *Some properties of electron carriers from the sulfate-reducing bacteria*

Electron carrier	Molecular weight	Prosthetic group	Electron acceptor for:	Electron donor for:	Reference
Cytochrome *b*	Unknown	Heme	Hydrogenase	Fumaric reductase	Hatchikian & LeGall, 1972
Cytochrome c_3	12000	Heme (4)	Hydrogenase	Sulfate reduction	Postgate, 1956
Cytochrome c_{553}	9000	Heme (1)	Formic dehydrogenase	Unknown	Hatchikian *et al.* 1971
Cytochrome cc_3	26000	Heme (8)	Unknown	Thiosulfate reduction	Hatchikian *et al.* 1971
Ferredoxin	6500	Non-heme iron (4)	Reduced cyto-chrome c_3	Sulfate reduction	LeGall & Dragoni, 1966
Flavodoxin	16000	FMN	Reduced cyto-chrome c_3	Sulfate reduction	LeGall & Hatchikian, 1967
Rubredoxin	6500	Non-heme iron	$NADH_2$	Cytochrome c_3	LeGall, 1968

two or possibly three reductive steps to permit phosphorylation coupled to electron transfer. This view tends to be supported by the fact that the E_0' of thiosulfate reduction is close to that of the hydrogen electrode and that the E_0' of the trithionate/thiosulfate couple is around +0.2 mV (M. Kamen, personal communication). Further, electron transfer may occur from specific electron donors, such as lactate, through specific electron transfer chains to specific electron acceptors such as trithionate reductase.

D. desulfuricans was the first non-photosynthetic anaerobe in which the presence of a cytochrome, cytochrome c_3, was demonstrated (Ishimoto, Koyama & Nagai, 1954; Postgate, 1956). Since that time, a number of small molecular weight electron carriers have been isolated from extracts of these organisms and studied. Some of the characteristics of these electron carriers are summarized in Table 5. They have all been shown to be active in electron transfer but, because most of the observations have been made with crude or impure preparations, it is not yet possible to assign discrete physiological roles to each carrier. In fact, the multiplicity of carriers has proved to be a hindrance in defining the roles of a specific carrier, and the occurrence of some of the carriers in the soluble protein may be an artifact of cell breakage or growth conditions. For example, one would expect that many of these carriers are involved in oxidative phosphorylation and should normally be located in the membranes or vesicle fraction of these bacteria. The function of the carriers has recently been discussed by LeGall & Postgate (1973).

As these electron carriers are of low molecular weight and their amino acid sequences relatively easy to determine, they appear to be ideal for defining evolutionary relationships. Cytochrome c_3 from a number of the sulfate-reducing bacteria has been sequenced but, apart from the heme binding sites, there is little homology with other types of cytochromes. It seemed at one time that *Chloropseudomonas ethylica* contained a low potential cytochrome of the c_3-type which showed structural relationships to cytochrome c_3 from the sulfate-reducing bacteria. However, it now appears that this cytochrome was derived from the sulfate-reducing bacterium that has been found in these cultures (Gray, Fowler, Nugent & Fuller, 1972).

Ferredoxins have been isolated from the photosynthetic bacteria *Chromatium*, *Chlorobium*, *Thiosulfatophilum* (Buchanan, Matsubara & Evans, 1969) *Rhodospirillum rubrum* (Shanmugan, Buchanan & Arnon, 1972) and the sulfate-reducing bacteria (Akagi, 1965; LeGall & Dragoni, 1966). Only the ferredoxins from *Chromatium* and *Desulfovibrio gigas* have been sequenced but the amino acid composition of the *Chlorobium* ferredoxin suggests that it is similar to the clostridial ferredoxins (Buchanan *et al.* 1969). Two types of ferredoxin, a plant type and a bacterial type, have been isolated from extracts of a light-grown photosynthetic bacterium, *Rhodospirillum rubrum* (Shanmugan *et al.* 1972). Two types of ferredoxin have also been isolated from *Desulfovibrio gigas* (J. LeGall, unpublished) again indicating close relationships between the sulfate-reducing and photosynthetic bacteria. The anomalous bacterial-type ferredoxin from *D. gigas* has been isolated and the amino acid sequence determined (Travis, Newman, LeGall & Peck, 1971). This ferredoxin contains six cysteine residues but only four non-heme irons and has areas with structural relationships to both plant- and bacterial-type ferredoxins.

SEQUENCE ANALYSIS OF FERREDOXIN FROM *DESULFOVIBRIO GIGAS*

Sequence analysis of the ferredoxins from the fermentative anaerobes has indicated that these proteins are clearly homologous and that there is symmetry between the two halves of the molecule, presumably the result of gene duplication (Eck & Dayhoff, 1966*a*). The plant-type ferredoxins are even more homologous but it has not been possible to determine with any degree of certainty the relationships of these ferredoxins to the bacterial types or whether the evolution of plant-type ferredoxins involved gene duplication.

Two ideas have been considered concerning the position of photosynthetic bacteria in the evolution of green plant photosynthesis. On the basis of physiological, biochemical and morphological considerations, some authors (Klein & Cronquist, 1967; Hall, Cammack & Rao, 1971) have suggested that the photosynthetic bacteria are direct ancestral forms in the evolution of green plant photosynthesis. This view is based on diverse but inadequate data and must be regarded as highly speculative. Gaffron (1962) has proposed that the photosynthetic bacteria represent a relatively unimportant evolutionary development in that they arose only after the development of cell types capable of green plant photosynthesis. This hypothesis was supported by previous analyses of the known amino acid sequences of a limited number of ferredoxins from each of the major groups. Three general conclusions (Matsubara, Jukes & Cantor, 1968) are summarized as follows: first, all ferredoxins appear to be distantly related; second, from the existing sequences, it is not possible to state whether the gene duplication evident in bacterial ferredoxins occurred before or after the evolutionary divergence of plants and bacteria; third, the divergence of plants and bacteria was an early evolutionary event. These considerations have resulted in a bifurcated representation of the evolutionary tree for the ferredoxin molecule, one branch leading to plants and the second branch, containing the doubling event, leading to bacteria. Utilizing the technique of Eck & Dayhoff (1966*b*) a phylogenetic tree (Fig. 3) has been drawn for the ferredoxin molecule which incorporates the information obtained from the ferredoxin of *D. gigas* (Eck, Travis & Peck, unpublished). The length of the branches is indicative of the number of base changes required to generate one sequence from another. This phylogenetic tree for ferredoxin indicates that the ferredoxins of green plants are directly descended from bacterial ferredoxin and accordingly that gene duplication was an ancestral event for all known ferredoxins. The length of the branches cannot be correlated directly with a time scale but the general topology of the evolutionary tree probably represents the actual evolutionary relationships between the fermentative anaerobes, photosynthetic bacteria and green plants. The use of amino acid sequences to demonstrate phylogenetic relationships can be justified by analogy with the analyses of the primary structures of cytochrome *c* which provided independent verification of the established evolutionary relationships of the eukaryotes from which the proteins were isolated.

The ferredoxins from green plants form a distinct group, as do the ferredoxins from the mesophilic fermentative anaerobes. The divergence

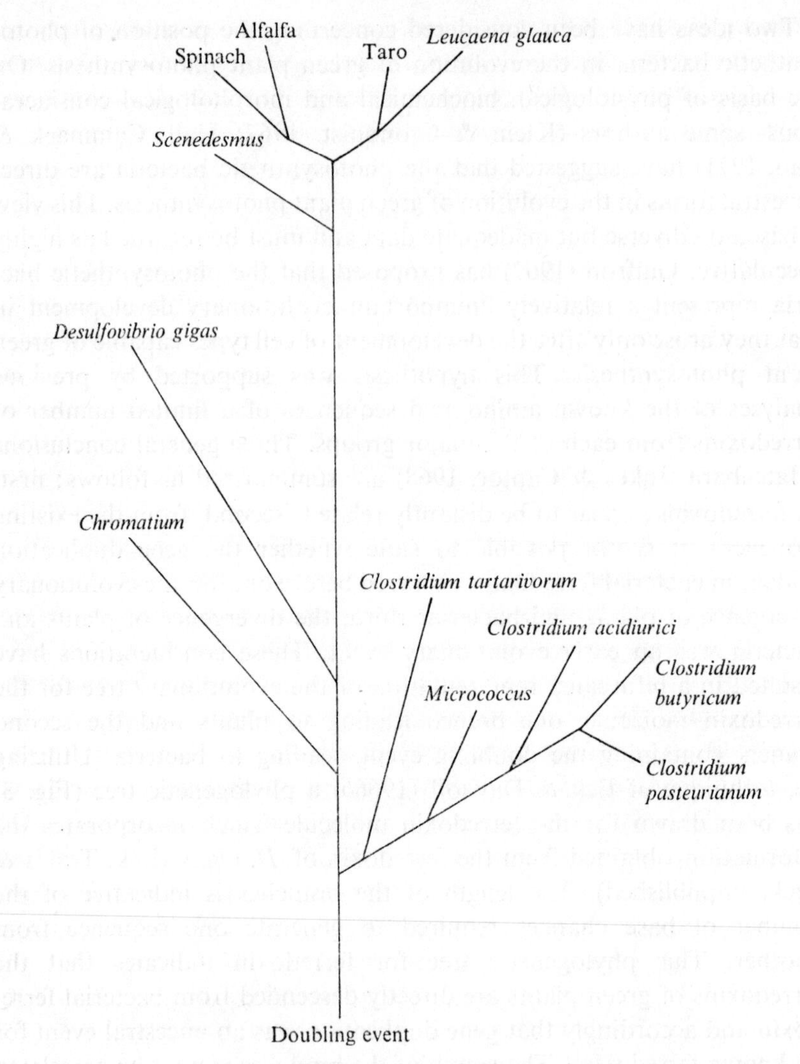

Fig. 3. Phylogenetic tree for ferredoxins.

of the ferredoxin from the thermophilic fermentative anaerobe *Clostridium tartarivorum* can be rationalized as a specific response to selection pressure for temperature stability and probably represents an evolutionary cul-de-sac. Evolutionary relationships between the sulfate-reducing bacteria and photosynthetic bacteria have been postulated from a consideration of their physiology and biochemistry.

The position of the ferredoxin from *D. gigas* indicates that the sulfate-reducing bacteria were not antecedents of photosynthetic bacteria but rather evolved from ancestral types which were photosynthetic bacteria. Although initially surprising, this evolutionary relationship is consistent with the idea that the accumulation of sulfate, the obligatory terminal electron acceptor for the sulfate-reducing bacteria, was the result of bacterial photosynthesis.

GENERAL CONSIDERATIONS

The sulfur isotope data indicate that the sulfur cycle was established during an early evolutionary period and the bioenergetics of the sulfur cycle suggest that it could have served as an intermediate stable stage for the utilization of radiant energy. These considerations also indicate that cytochromes and oxidative phosphorylation must have evolved under anaerobic conditions before the advent of the oxygen atmosphere. In accordance with these ideas, the sulfate-reducing bacteria and photosynthetic bacteria appear to be evolutionarily related, the sulfate-reducing bacteria having been derived from photosynthetic bacteria. More information is required to establish firmly these relationships and to demonstrate relationships between the Clostridia and photosynthetic bacteria. Little is known about the sulfur metabolism of either of these groups of bacteria but the presence of sulfite reduction in the Clostridia suggests that they are capable of synthesizing hemes.

Evolutionary considerations such as these are speculative but they can be useful in suggesting new directions for research. They may indicate, for example, what proteins from which organisms it might be useful to sequence, in which organisms the biochemistry of electron transfer should be studied to establish a comparative biochemistry of this topic, and the possible existence of new physiological types of micro-organisms that might be isolated and studied. The evolution of mechanisms for the utilization of radiant energy may also be revealed by such considerations and prove to be useful for the solution of modern energy problems.

SUMMARY

The sulfur cycle is not only a schematic representation of the reactions involved in the enzymatic transformations of organic and inorganic compounds of sulfur, but also conceptually represents a relatively simple biological system for the utilization of radiant energy. The sulfur cycle can involve a minimum of two types of organisms, sulfate-reducing and photosynthetic bacteria, as has recently been shown to occur with the consortium of two species formerly known as *Chloropseudomonas ethylica*. The sulfur cycle is not a primary producing system as reducing power for carbon dioxide fixation and biosynthesis must be supplied from the environment. During the early evolution of fermentative bacteria, the amount of fixed carbon must have dwindled rapidly and the competition for new sources of energy must have fostered a great metabolic diversity among existing organisms. Reduced inorganic sulfur compounds were abundant in the primitive environment and represented alternative sources of energy for the support of life. It is logical to suppose that these inorganic substrates would be utilized and have led to the development of a simple sulfur cycle. The establishment of an anaerobic sulfur cycle driven by light may have provided the micro-organisms involved with constant nutritional conditions and provided the first renewable energy source for life. This idea is discussed in terms of the biochemistry and physiology of the sulfate-reducing bacteria, the photosynthetic bacteria and the clostridia. An analysis of the amino acid sequence of a ferredoxin from a sulfate-reducing bacterium suggests that the sulfate-reducing bacteria and photosynthetic bacteria are sequentially related in evolution.

REFERENCES

AKAGI, J. M. (1965). The participation of a ferredoxin of *Clostridium nigrificans* in sulfite reduction. *Biochemical and Biophysical Research Communications*, **21**, 72–7.

AMBLER, R. P. (1971). The amino acid sequence of cytochrome *c*-551.5 (cytochrome c_7) from the green photosynthetic bacterium *Chloropseudomonas ethylica*. *FEBS Letters*, **18**, 351–3.

ASADA, K., TAMURA, G. & BANDURSKI, R. S. (1969). Methyl viologen-linked sulfite reductase from spinach leaves. *Journal of Biological Chemistry*, **244**, 4904–15.

AULT, W. V. & KULP, J. L. (1959). Isotopic geochemistry of sulfur. *Geochimica et Cosmochimica Acta*, **16**, 201–35.

BARTON, L. L., LEGALL, J. & PECK, H. D. JR. (1970). Phosphorylation coupled to oxidation of hydrogen with fumarate in extracts of the sulfate-reducing bacterium *Desulfovibrio gigas*. *Biochemical and Biophysical Research Communications*, **41**, 1036–42.

BOWEN, T. J., HAPPOLD, F. C. & TAYLOR, B. F. (1966). Studies on adenosine-5′-phosphosulphate reductase from *Thiobacillus denitrificans*. *Biochimica et Biophysica Acta*, **118**, 566–76.

BRYANT, M. P. (1969). Symbiotic association of certain ethanol and lactate fermenting bacteria. *Abstracts of the* 158*th Meeting of the American Chemical Society, Microbiology Section*, p. 18.

BRYANT, M. P., WOLIN, E. A., WOLIN, M. J. & WOLFE, R. S. (1967). *Methanobacillus omelianskii*, a symbiotic association of two species of bacteria. *Archiv für Mikrobiologie*, **59**, 20–31.

BUCHANAN, B. B., MATSUBARA, H. & EVANS, M. C. W. (1969). Ferredoxin from the photosynthetic bacterium *Chlorobium thiosulfatophilum*. A link to ferredoxins from non-photosynthetic bacteria. *Biochimica et Biophysica Acta*, **189**, 46–53.

BUTLIN, K. R. & POSTGATE, J. R. (1954). The microbiological formation of sulfur in Cyrenaican lakes. In *Biology of Deserts*, ed. J. L. Cloudsley-Thompson, pp. 112–22. London: Institute of Biology.

ECK, R. V. & DAYHOFF, M. O. (1966*a*). *Atlas of Protein Sequence and Structure.* Silver Spring, Maryland: National Biomedical Research Foundation.

ECK, R. V. & DAYHOFF, M. O. (1966*b*). Evolution of the structure of ferredoxin based on living relics of primitive amino acid sequence. *Science*, **152**, 363–6.

GAFFRON, H. (1962). On dating stages in photochemical evolution. In *Horizons in Biochemistry*, ed. M. Kasha & B. Pullman, pp. 59–89. New York: Academic Press.

GEMERDEN, H. V. (1967). On the bacteria sulfur cycle of inland waters. Thesis, University of Leiden.

GRAY, B. H. (1972). *Chloropseudomonas ethylica*: A symbiotic mixture of *Chlorobium limicola* and heterotrophs. Thesis, University of Tennessee.

GRAY, B. H., FOWLER, C. F., NUGENT, N. A. & FULLER, R. G. (1972). A reevaluation of the presence of low midpoint potential cytochrome 551.5 in the green photosynthetic bacterium *Chloropseudomonas ethylica. Biochemical and Biophysical Research Communications*, **47**, 322–7.

HALL, D. O., CAMMACK, R. & RAO, K. K. (1971). Role for ferredoxins in the origin of life and biological evolution. *Nature, London*, **233**, 136–8.

HASCHKE, R. H. & CAMPBELL, L. L. (1971). Thiosulfate reductase of *Desulfovibrio vulgaris. Journal of Bacteriology*, **106**, 603–7.

HATCHIKIAN, E. C., BRUSCHI, M., LEGALL, J. & DUBOURDIEU, M. (1971). Cristallization et propriétés d'un cytochrome intervenant dans la réduction du thiosulfate par *Desulfovibrio gigas. Bulletin de la Société Française de Physiologie Vegetale*, in Press.

HATCHIKIAN, E. C. & LEGALL, J. (1972). Evidence for the presence of a *b*-type cytochrome in the sulfate reducing bacterium, *Desulfovibrio gigas* and its role in the reduction of fumarate by molecular hydrogen. *Biochimica et Biophysica Acta*, **267**, 479–84.

HUNGATE, R. E. (1967). Hydrogen as an intermediate in the rumen fermentation. *Archiv für Mikrobiologie*, **59**, 158–64.

ISHIMOTO, M. (1959). Sulfate reduction in cell-free extracts of *Desulfovibrio. Journal of Biochemistry, Tokyo*, **46**, 105–6.

ISHIMOTO, M., KOYAMA, J. & NAGAI, Y. (1954). Biochemical studies on sulfate-reducing bacteria. IV. The cytochrome system of sulfate-reducing bacteria. *Journal of Biochemistry, Tokyo*, **41**, 763–70.

KAPLAN, J. R. & HULSTON, J. R. (1966). The isotopic abundance and content of sulfur in meteorites. *Geochimica et Cosmochimica Acta*, **30**, 479–96.

KAPLAN, J. R. & RITTENBERG, S. C. (1964). Microbiological fractionation of sulphur isotopes. *Journal of General Microbiology*, **34**, 195–212.

KELLY, D. P. (1971). Autotrophy: Concepts of lithotrophic bacteria and their organic metabolism. *Annual Review of Microbiology*, **25**, 177–210.

KLEIN, R. M. & CRONQUIST, A. (1967). A consideration of the evolutionary and taxonomic significance of some biochemical, micromorphological and physiological characters in the Thallophytes. *Quarterly Review of Biology*, **42**, 105–296.

KOBAYASHI, K., TACHIBANA, S. & ISHIMOTO, M. (1969). Intermediary formation of trithionate in sulfite reduction by a sulfate-reducing bacterium. *Journal of Biochemistry, Tokyo*, **65**, 155–7.

KOBAYASHI, K., TAKAHASHI, E. & ISHIMOTO, M. (1972). Biochemical studies on sulfate-reducing bacteria. XI. Purification and some properties of sulfite reductase, Desulfoviridin. *Journal of Biochemistry, Tokyo*, **72**, 879–87.

LEE, J. P., LEGALL, J. & PECK, H. D. JR. (1971). Purification of an assimilatory-type of sulfite reductase from *Desulfovibrio vulgaris*. *Federation Proceedings*, **30**, 1202.

LEE, J. P. & PECK, H. D. JR. (1971). Purification of the enzyme reducing bisulfite to trithionate from *Desulfovibrio gigas* and its identification as Desulfoviridin. *Biochemical and Biophysical Research Communications*, **45**, 583–9.

LEE, J. P., YI, C. S., LEGALL, J. & PECK, H. D. JR. (1973). Isolation of a new pigment, Desulforubidin, from *Desulfovibrio desulfuricans* (Norway Strain) and its role in sulfite reduction. *Journal of Bacteriology*, in Press.

LEGALL, J. (1968). Purification partielle et étude de la NAD: rubrédoxine oxydoreductase de *D. gigas*. *Annales de l'Institut Pasteur*, **114**, 109–15.

LEGALL, J., DERVARTANIAN, D. V., SPILKER, E., LEE, J. P. & PECK, H. D. JR. (1971). Evidence for the involvement of non-heme iron in the active site of hydrogenase from *Desulfovibrio vulgaris*. *Biochimica et Biophysica Acta*, **234**, 525–30.

LEGALL, J. & DRAGONI, N. (1966). Dependance of sulfite reduction on a crystallized ferredoxin from *Desulfovibrio gigas*. *Biochemical and Biophysical Research Communications*, **23**, 145–9.

LEGALL, J. & HATCHIKIAN, E. C. (1967). Purification et propriétés d'une flavoprotéine intervenant dans la réduction du sulfite. *Compte rendu des séances de l'Académie des Sciences*, **264**, 2580–3.

LEGALL, J., MAZZA, G. & DRAGONI, N. (1965). Le cytochrome c_3 de *Desulfovibrio gigas*. *Biochimica et Biophysica Acta*, **99**, 385–78.

LEGALL, J. & POSTGATE, J. R. (1973). The physiology of the sulfate-reducing bacteria. *Advances in Microbial Physiology*, in Press.

LONDON, J. & RITTENBERG, S. C. (1964). Path of sulfur in sulfide and thiosulfate oxidation by Thiobacilli. *Proceedings of the National Academy of Sciences, USA*, **52**, 1183–90.

LYRIC, R. M. & SUZUKI, I. (1970). Enzymes involved in the metabolism of thiosulfate by *Thiobacillus thioparus*. II. Properties of adenosine-5′-phosphosulfate reductase. *Canadian Journal of Biochemistry*, **48**, 344–54.

MATSUBARA, H., JUKES, T. H. & CANTOR, C. R. (1968). Structural and evolutionary relationships of ferredoxins. *Brookhaven Symposia in Biology*, **21**, 201–16.

MILLER, J. D. A. & SALEH, A. M. (1964). A sulphate-reducing bacterium containing cytochrome c_3 but lacking Desulfoviridin. *Journal of General Microbiology*, **37**, 419–23.

MURPHY, M. J., SIEGEL, L. M., KAMIN, H., DERVARTANIAN, D. V., LEE, J. P., LEGALL, J. & PECK, H. D. JR. (1973*a*). An iron tetrahydroporphyrin prosthetic group common to both assimilatory and dissimilatory sulfite reductases. *Biochemical and Biophysical Research Communications*, **54**, 82–8.

MURPHY, M. J., SIEGEL, L. M., KAMIN, H. & ROSENTHAL, D. (1973*b*). Reduced nicotinamide adenine dinucleotide phosphate–sulphite reductase of Enterobacteria. II. Identification of a new class of heme prosthetic group: an iron–tetrahydroporphyrin (isobacteriochlorin type) with eight carboxylic acid groups. *Journal of Biological Chemistry*, **248**, 2801–14.

NAKAI, N. & JENSEN, M. L. (1964). The kinetic isotope effect in the bacterial reduction and oxidation of sulfur. *Geochimica et Cosmochimica Acta*, **28**, 1893–912.

NEU, H. C. & HEPPEL, L. A. (1965). The release of enzymes from *Escherichia coli* by osmotic shock and during the formation of spheroplasts. *Journal of Biological Chemistry*, **240**, 3685–92.

PECK, H. D. JR. (1959). The ATP-dependent reduction of sulfate with hydrogen in extracts of *Desulfovibrio desulfuricans*. *Proceedings of the National Academy of Sciences, USA*, **45**, 701–8.

PECK, H. D. JR. (1960*a*). Adenosine 5′-phosphosulfate as an intermediate in the oxidation of thiosulfate by *Thiobacillus thioparus*. *Proceedings of the National Academy of Sciences, USA*, **46**, 1053–7.

PECK, H. D. JR. (1960*b*). Evidence for oxidative phosphorylation during the reduction of sulfate with hydrogen by *Desulfovibrio desulfuricans*. *Journal of Biological Chemistry*, **235**, 2734–8.

PECK, H. D. JR. (1966). Phosphorylation coupled with electron transfer in extracts of the sulfate reducing bacterium, *Desulfovibrio gigas*. *Biochemical and Biophysical Research Communications*, **22**, 112–18.

PECK, H. D. JR. (1967). Some evolutionary aspects of inorganic sulfur metabolism. In *Lectures on Theoretical and Applied Aspects of Modern Microbiology*, pp. 1–22. University Park: University of Maryland.

PECK, H. D. JR., DEACON, T. E. & DAVIDSON, J. T. (1965). Studies on adenosine 5′-phosphosulfate reductase from *Desulfovibrio desulfuricans* and *Thiobacillus thioparus*. I. The assay and purification. *Biochimica et Biophysica Acta*, **96**, 429–46.

POSTGATE, J. R. (1956). Cytochrome c_3 and desulphoviridin; pigments of the anaerobe *Desulphovibrio desulfuricans*. *Journal of General Microbiology*, **14**, 545–72.

POSTGATE, J. R. (1959*a*). Sulphate reduction by bacteria. *Annual Review of Microbiology*, **13**, 505–20.

POSTGATE, J. R. (1959*b*). A diagnostic reaction of *Desulphovibrio desulphuricans*. *Nature, London*, **183**, 481–2.

REDDY, C. A., BRYANT, M. P. & WOLIN, M. J. (1972). Characteristics of S organism isolated from *Methanobacillus omelianskii*. *Journal of Bacteriology*, **109**, 539–45.

RITTENBERG, S. C. (1969). The roles of exogenous organic matter in the physiology of chemolithotrophic bacteria. *Advances in Microbial Physiology*, **3**, 159–96.

ROBBINS, P. W. & LIPMANN, F. (1958). Separation of the two enzymatic phases in active sulfate synthesis. *Journal of Biological Chemistry*, **233**, 681–5.

ROY, A. B. & TRUDINGER, P. A. (1970). *The Biochemistry of Inorganic Compounds of Sulphur*. London: Cambridge University Press.

SCHMIDT, A. (1972). On the mechanism of photosynthetic sulfate reduction. An APS-sulfotransferase from Chlorella. *Archiv für Mikrobiologie*, **84**, 77–86.

SHANMUGAM, K. T., BUCHANAN, B. B. & ARNON, D. I. (1972). Ferredoxins in light and dark grown photosynthetic cells with special reference to *Rhodospirillum rubrum*. *Biochimica et Biophysica Acta*, **256**, 477–86.

SIEGEL, L. M., KAMIN, H., RUEGER, D. G., PRESSWOOD, R. P. & GIBSON, Q. H. (1971). An iron-free sulfite reductase flavoprotein from mutants of *Salmonella typhimurium*. In *Flavins and Flavoproteins*, ed. H. Kamin, pp. 523–54. Baltimore, University Park Press.

SIEGEL, L. M., MURPHY, M. J. & KAMIN, H. (1973). Reduced nicotinamide adenine dinucleotide phosphate–sulfite reductase of enterobacteria. I. The *Escherichia coli* hemoflavoprotein; molecular parameters and prosthetic groups. *Journal of Biological Chemistry*, **248**, 251–64.

SUH, B. & AKAGI, J. M. (1969). Formation of thiosulfate from sulfite by *Desulfovibrio vulgaris*. *Journal of Bacteriology*, **99**, 210–15.

TAGAWA, K. & ARNON, D. I. (1962). Ferredoxins as electron carriers in photosynthesis and in the biological production and consumption of hydrogen gas. *Nature, London*, **195**, 537–43.

THIELE, H. H. (1968). Sulfur metabolism in Thiorhodaceae. V. Enzymes of sulfur metabolism in *Thiocapsa floridana* and Chromatium species. *Antonie van Leeuwenhoek*, **34**, 350–6.

THODE, H. G., KLEEREKOPER, H. & MCELCHERAN, D. (1951). Isotope fractionation in the bacterial reduction of sulfate. *Research, London*, **4**, 581–2.

THODE, H. G., WANLESS, R. K. & WALLOUCH, R. (1954). The origin of native sulphur deposits from isotope fractionation studies. *Geochimica et Cosmochimica Acta*, **5**, 286–98.

TRAVIS, J., NEWMAN, D. J., LEGALL, J. & PECK, H. D. Jr. (1971). The amino acid sequence of ferredoxin from the sulfate-reducing bacterium *Desulfovibrio gigas*. *Biochemical and Biophysical Research Communications*, **45**, 452–8.

TRUDINGER, P. A. (1970). Carbon monoxide-reacting pigment from *Desulfotomaculum nigrificans* and its possible relevance to sulfite reduction. *Journal of Bacteriology*, **104**, 158–70.

TRUDINGER, P. A. & CHAMBERS, L. A. (1973). Reactions of P_{582} from *Desulfotomaculum nigrificans* with substrates, reducing agents and carbon monoxide. *Biochimica et Biophysica Acta*, **293**, 26–35.

TRÜPER, H. G. & PECK, H. D. JR. (1970). Formation of adenylyl sulfate in phototrophic bacteria. *Archiv für Mikrobiologie*, **73**, 125–42.

TRÜPER, H. G. & ROGERS, L. A. (1971). Purification and properties of adenylyl sulfate reductase from the phototrophic sulfur bacterium, *Thiocapsa roseopersicina*. *Journal of Bacteriology*, **108**, 1112–21.

WHELDRAKE, J. F. & PASTERNAK, C. A. (1965). The control of sulphate activation in bacteria. *Biochemical Journal*, **96**, 276–80.

WILSON, L. G., ASAHI, T. & BANDURSKI, R. S. (1961). Yeast sulfate-reducing system. I. Reduction of sulfate to sulfite. *Journal of Biological Chemistry*, **236**, 1822–9.

EVOLUTION WITHIN NITROGEN-FIXING SYSTEMS

J. R. POSTGATE

Agricultural Research Council Unit of Nitrogen Fixation, University of Sussex, Falmer, Brighton BN1 9QJ

Four aspects of the evolution of nitrogen-fixing systems will be discussed.

(1) The origin of the system.

(2) Its distribution among living organisms once it had originated.

(3) The evolution of physiological and anatomical devices to prevent interfering reactions with oxygen.

(4) The evolution of symbiotic systems.

As with nearly all discussions of microbial evolution, the contribution will involve speculation, because direct experimental data, analogous to those on the sulphur isotope fractionation relevant to the early history of sulphur bacteria, are non-existent. The speculation will draw on recent knowledge of the enzymology, physiology and biology of biological nitrogen fixation. This knowledge has been reviewed exhaustively by several authors in the last two years (Burris, 1971; Bergersen, 1971; Dalton & Mortenson, 1972; Benemann & Valentine, 1972) and has been the subject of two extensive monographs (Postgate, 1971*a*; Quispel, 1974) and a brief one (Postgate, 1972), so another account of the position is superfluous. A brief survey of some of the views on evolution to be elaborated here was given by Postgate (1972).

THE ORIGIN OF BIOLOGICAL NITROGEN FIXATION

Ability to fix nitrogen is restricted to prokaryotes and the property is associated with 'ancient' enzymes such as hydrogenase, ferredoxins, methane-producing and pyruvic phosphoroclastic systems (Imshenetskii, 1963; Peck, 1966/7). Therefore one should consider its status as a truly primitive enzymic system. On conventional views of the geochemical history of this planet (Bernal, 1967; Rutten, 1971) the planetary atmosphere was anaerobic before about 1.5×10^9 years ago and free ammonia was a normal constituent of the environment. Life had probably been established for $1.5–2.0 \times 10^9$ years already, so anaerobic prokaryotes would have had a long evolutionary history with seemingly abundant

Table 1. *Principal substrates reduced by nitrogenase*

The substrates below are reduced by nitrogenase preparations with the consumption of ATP. References may be found in the reviews cited in the text, Hardy & Knight (1967) and Hardy & Burns (1968). Lower alkyl acetylenes and alkyl cyanides are also reduced slowly. Formulae are written in forms which emphasise analogy to triply bonded dinitrogen.

Substrate	Formula	Product(s)
Dinitrogen	N_2	NH_3
Acetylene	$HC{\equiv}CH$	C_2H_4
Hydrogen cyanide	$HC{\equiv}N$	CH_4+NH_3
Acrylonitrile	$CH_2{=}CH{-}C{\equiv}N$	$CH_3CH{=}CH_2+NH_3$
Methyl isocyanide	$CH_3{-}\overset{+}{N}{\equiv}\overset{-}{C}$	$CH_3NH_2+CH_4$ (+C_2 and higher hydrocarbons)
Vinyl isocyanide	$CH_2{=}CH{-}\overset{+}{N}{\equiv}\overset{-}{C}$	include CH_4
Cyanogen*	$N{\equiv}C{-}C{\equiv}N$	include CH_4
Hydrogen azide	$H{-}\overset{-}{N}{-}\overset{+}{N}{\equiv}N$	N_2+NH_3
Nitrous oxide	$N{\equiv}\overset{+}{N}{-}\overset{-}{O}$	N_2+H_2O
Hydrogen thiocyanate†	$H{-}S{-}C{\equiv}N$	include CH_4
Methyl isothiocyanate†	$CH_3{-}\overset{+}{N}{\equiv}C{-}\overset{-}{S}$	include CH_4
Hydrogen ion	H^+	H_2

* M. Kelly, unpublished. † R. R. Eady, unpublished.

fixed nitrogen. These considerations led Silver & Postgate (1973) to ask whether the fixation of nitrogen could seriously be regarded as the original function of the enzyme system now called nitrogenase, since the enzyme would apparently confer no selective advantage on its possessors. Nitrogenase is known to reduce a variety of other small, triply bonded molecules such as acetylene, azide, cyanide, and isocyanide (see reviews cited earlier), some of which (e.g. cyanides, cyanogen and acetylene) are known to have been present in the anaerobic phase of this planet's existence, and it also, like hydrogenase, evolves hydrogen from water. A list of the classes of substrate reduced by nitrogenase is given in Table 1; for all these reactions, considerable amounts of ATP are consumed *in vitro* and probably also *in vivo* (Postgate, 1974); a typical stoichiometry is 15 mol ATP consumed for each N_2 molecule reduced to $2NH_3$. Silver & Postgate (1973) considered the following possible functions for an ancient nitrogenase system.

1. Anaerobic ATP generation as a terminal respiratory acceptor analogous to sulphate in the physiology of sulphate-reducing bacteria. This proposal was an extension of a suggestion by Parker & Scutt (1960) that nitrogen fixation could once have been a respiratory process.

2. A 'hydrogen escape valve' (Gray & Gest, 1965), providing a means of dissipating reducing power during anaerobic metabolism.

3. A process for rendering harmless potentially toxic substances such as cyanides or cyanogen.

They concluded that functions 1 and 2 were absurdly uneconomic in view of the high ATP consumption but that 3, a detoxification process, was at least as probable as the view that the enzyme fixed nitrogen.

Burns & Hardy (1973) took the view that nitrogenase was once widespread, but that it tended to be lost as this planet's atmosphere became oxidising. Such discussions overtly accept that the primitive environment contained adequate amounts of fixed nitrogen yet that nitrogenase is an 'ancient' enzyme, a view implicit in most early discussions of this problem (Imshenetskii, 1963; Burris, 1963; Peck, 1966/7), including Klein & Cronquist's (1967) monumental survey of microbial evolution. An alternative to the second viewpoint is that, despite appearances, nitrogenase is not primitive: that the appearance of nitrogen fixation among living things is a relatively recent evolutionary event. The reasons for regarding nitrogenase as primitive derive mainly from arguments by association, though De Ley (1968, p. 144) reached a similar conclusion from a consideration of G/C ratios of various free-living nitrogen-fixing bacteria. The prevalence of nitrogenase among organisms possessing 'ancient' enzymes such as hydrogenase and ferredoxins is one such association; another is the evidence that the aerobic genus *Azotobacter*, presumably more highly evolved than the anaerobes, possesses ferredoxins, flavodoxins and a form of hydrogenase, which appear to be physiological relics of a more primitive metabolism, and that these are intimately involved in the functioning of nitrogenase (see Burris, 1971). A third is that only prokaryotes, present-day representatives of the most primitive forms of terrestrial life, possess the enzyme.

One can counter such arguments by suggesting that such associations result from nitrogenase having such special chemical requirements, particularly with regard to the reduction step, that only organisms possessing the physiological vestiges of such 'ancient' enzymes were capable of developing a functional nitrogenase system. The enzyme is certainly a very special one, posing difficult mechanistic questions from a chemical point of view (Hardy & Knight, 1967; Chatt & Richards, 1971; Leigh, 1971).

A recent evolutionary emergence would at least be consistent with the known cross-reactivity of the component proteins of nitrogenase. Nitrogenase from all sources studied consists of two separable proteins (Burris, 1971; Postgate, 1971*b*), both containing non-haem iron and labile sulphur, the larger one containing molybdenum as well (Table 2). I shall call these protein 1 (the Mo–Fe protein) and protein 2 (the

Table 2. *Some properties of nitrogenase proteins from various bacteria*

Source*	Molecular weight	Subunits	Molecules per enzyme molecule: Mo	Fe	S^{2-}	Comments
Cp1	220000	2×59500+ 2×50700	2	24	24	Dalton & Mortenson (1972) and Mortenson, Zumft, Huang & Palmer (1973)
Cp2	55000	2×27500	–	4	4	
Av1	270000	2 types, about 40000	2	32–34	26–28	Burns, Holsten & Hardy (1970) and Hardy *et al.* (1971)
Av2	40000		–	...	...	
Kp1	218000	2×50000+ 2×60000	1	17	17	Eady *et al.* (1972)
Kp2	66800	2×34600	–	4	4	
Rj1	180000	...	1	9	...	Data of limited value as preparations were impure (Bergersen, 1971)
Rj2	51000	...	–	...	...	

* See Table 3 for code.

Table 3. *Cross-reactions of nitrogenase proteins from various bacteria*

+ means 80–100 % of activity of homologous cross (+); ± = about 50 %; *tr* = about 10 %; – = no activity; ... = no test recorded.

	Av1	Ac1	Kp1	Bp1	Cp1	Rr1	Mf1	Rj1	Dd1
Av2	+	...	+	–	–	...	...	+	...
Ac2	...	+	+	±	...	*tr*	±	...	...
Kp2	+	+	+	+	–	+	+	...	(±)
Bp2	+	*tr*	+	+	–	±	±	+	(±)
Cp2	–	...	–	+	+	–	–	–	...
Rr2	...	...	...	...	...	+	...	...	...
Mf2	...	+	+	+	...	+	+	...	...
Rj2	+	...	...	...	...	...	...	+	...

Code: Av is *Azotobacter vinelandii*, Ac is *A. chroococcum*, Kp is *Klebsiella pneumoniae*, Bp is *Bacillus polymyxa*, Cp is *Clostridium pasteurianum*, Rr is *Rhodospirillum rubrum*, Mf is *Mycobacterium flavum* 301, Rj is *Rhizobium japonicum* bacteroids, Dd is *Desulfovibrio desulfuricans*. Data collated from diverse sources; highly purified proteins 1 and/or 2 were not necessarily used. References given by Burris (1971) with, in addition, Biggins, Kelly & Postgate (1971) for Mf, Murphy & Koch (1971) for Rj. Data for Dd from T. H. Blackburn (unpublished).

Fe protein) and, when it is necessary to specify the organism of origin, use the code Kp1, Mf2 etc., as explained in Table 3 and used earlier (Postgate, 1971*b*; Eady, Smith, Cook & Postgate, 1972). Cross-reactivity is illustrated in Table 3 which extends a comparable table published by Burris (1971) and includes all data available to the writer early in 1973: Kp1, for example, complements Ac2 completely, giving a fully functional nitrogenase when preparations of the two proteins are mixed, but it does not cross-react with Cp2. Bp1, on the other hand, cross-reacts with Cp2 and also with Kp2, though it crosses only partially (i.e. gives a nitrogenase of low activity) with Ac2. Such cross-reactivity, where tested, extends to immunological activity: antiserum to Kp1 will also precipitate Av1, Bp1, Mf1 or Dd1 on Ouchterlony plates, though identity is not indicated (J. Postgate, unpublished); Dr W. J. Brill (personal communication) has observed comparable mutual cross-reactivity using antisera to Kp1 and Av1. These cross-reactions, indicating considerable structural uniformity, imply that there has been rather little evolution during the biological history of the enzyme, an implication entirely consistent with a relatively recent emergence. Structural uniformity, however, could equally imply that the chemical requirements for a functional nitrogenase are so rigid that little evolution has been possible at all, as with a much smaller protein, mitochondrial cytochrome *c*, which has persisted with little structural change for some 10^9 years (Fitch & Margoliash, 1970).

A useful datum which is not available but which would be helpful in resolving this problem would be geochemical information concerning the era when fixed nitrogen became a limiting nutrient as far as biological productivity is concerned (which it is over most of the world's land surface today); before that time, selection pressure in favour of the emergence of nitrogenase must have been slight and highly localised.

Setting aside the problems of when and why nitrogenase emerged, it is not likely that so complex an enzyme system developed fully formed; it more probably arose from pre-existing enzymes having other functions. The presence of non-haem iron, associated with labile sulphur, in both proteins 1 and 2 directs attention to the photosynthetic apparatus or the phosphoroclastic system, with their different classes of ferredoxins, as conceivable precursors and tends to exclude the rubredoxins and the haemoproteins. Only information on the sequence of the polypeptide chains in the subunits of proteins 2 and the two types of subunits of proteins 1 (Table 2) would permit fruitful speculation along these lines, but it is pertinent to point out that:

1. Hydrogenase, an invariable component of the phosphoroclastic system, seems usually to accompany nitrogenase (Wilson, 1958) though the character of the hydrogenase differs from organism to organism: Azotobacter and bacteroid hydrogenases are particulate and show different substrate specificity from soluble hydrogenases (Dixon, 1972). Mechanistically hydrogenase action involves hydride transfer reactions (Krasna & Rittenberg, 1954): it is also a non-haem iron protein (Nakos & Mortenson, 1971*a*, *b*; LeGall *et al.* 1971).

2. Apart from the quinol form of Azotobacter flavodoxin (Yates, 1972), the only effective donors to nitrogenase are free radical reductants (reduced ferredoxins, viologen dyes or $Na_2S_2O_4$), a class of compounds which also appears as a result of the generation of the primary photo-electron in photosynthesis.

3. The e.p.r. spectrum of Kp2, a protein with 4 Fe and 4 S atoms per molecule, recalls that of a plant ferredoxin (B. E. Smith, personal communication), so its two subunits (Eady *et al.* 1972) could be structurally of this class.

Thus both the phosphoroclastic and the photosynthetic apparatus today have properties and components which could be evolutionary precursors of nitrogenase.

Another system which could be related is suggested by the observation of Cheniae & Evans (1957) of a positive correlation between the effectiveness of fixation by soya bean nodules and the nitrate reductase activity of the bacteroids. Ketchum *et al.* (1970) showed that fragments of molybdenum-containing enzymes such as xanthine oxidase could restore nitrate reductase activity to a mutant of *Neurospora crassa* defective in ability to synthesise the Mo-containing component of its nitrate reductase. Nitrogenase proteins 1 could also provide the missing fragment (Nason *et al.* 1971). Evans & Russell (1971) discussed whether parts of the component proteins of nitrogenase and nitrate reductase might share a common biosynthetic pathway, a possibility which could account for nitrate inhibition of nodulation. From an evolutionary viewpoint, these observations provide a third class of possible precursors of nitrogenase: the molybdo-enzymes with their associated iron–sulphur protein components.

THE DISTRIBUTION OF NITROGEN-FIXING SYSTEMS

A classical approach to the evolution of metazoa has been taxonomic: organisms are arranged in groups on taxonomic grounds, intermediate forms are sought (real, fossilized or conceptual), phylogenetic patterns are observed and, often with a strong element of intuition, evolutionary hierarchies are proposed. Morphological restrictedness, allied to biochemical diversity and genetic mutability, make micro-organisms inappropriate subjects for such an approach. Imshenetskii (1963), discussing the presence of hydrogenase, an 'ancient' enzyme, in all nitrogen-fixing bacteria, took the reasonable view that the anaerobe *Clostridium pasteurianum* represented a more primitive type than the aerobe Azotobacter. He regarded the latter as related to the blue-green algae. Burris (1963) also regarded anaerobic nitrogen-fixing bacteria as primitive, pointing out that the photosynthetic bacteria, many of which fix nitrogen, formed a plausible evolutionary link to the nitrogen-fixing blue-green algae. Facultative anaerobes such as *Bacillus polymyxa*, which only fix nitrogen anaerobically, may represent transitional stages in the evolution of aerobic fixation.

The list of authenticated nitrogen-fixing systems has changed considerably in recent years. Simple inspection of it can suggest information bearing on the evolutionary problem, though one must bear in mind that it may change drastically again as more organisms are examined. Table 4 presents a list of nitrogen-fixing systems as it stood early in 1973. Organisms in an earlier list (Table 3 of Postgate, 1971*b*) have been excluded if their status as nitrogen fixers, as distinct from scavengers of traces of fixed nitrogen (Hill & Postgate, 1969), has been seriously challenged and the challenge not refuted. Thus the aerobic Pseudomonas, Azotomonas, Arthrobacter and Spirillum groups of bacteria are not accepted. Since 1970 the writer and Miss S. Hill have tested, with negative results, some 50 such 'ghosts' from various laboratories: mostly Gram-negative rods which form convincing glutinous colonies on nitrogen-free agar but which neither fix N_2 nor reduce acetylene, whether tested aerobically, anaerobically or at a low pO_2. All eukaryotes are now excluded, Dr J. R. Millbank or the writer having tested about a dozen more putative nitrogen-fixing yeasts and aspergilli, with negative results, since Millbank's (1969, 1970) evidence that putative nitrogen-fixing yeasts and Pullularia were also 'ghosts'. *Chloropseudomonas ethylica* disappears from the list because it proves to be a mixture of Chlorobium and Desulfovibrio (Gray, Fowler, Nugent & Fuller, 1972). Symbiotic systems are included only in outline; mycorrhiza is excluded because the reports did not rigidly exclude fixation by free-living prokaryotes. Reports of symbioses with whales and termites have also been excluded.

Inspection of Table 4 permits a number of comments.

1. The property of nitrogen fixation at present seems rigidly restricted to prokaryotes.

2. The property seems to be much more widespread among anaerobic than among aerobic heterotrophs. This is probably not true among phototrophs, although this impression may result from a widespread interest among scientists in blue-green algae as nitrogen fixers contrasting with a more limited interest in nitrogen fixation by phototrophic bacteria.

3. Among the facultative and obligately anaerobic heterotrophs, the

Table 4. *Annotated table of systems capable of biological nitrogen fixation*

Free-living microbes	*Comments*
Obligate aerobic bacteria:	
The family Azotobacteraceae (genera: *Azotobacter, Azomonas, Azotococcus, Beijerinckia, Derxia*)	All strains fix
Mycobacterium flavum 301, *M. roseo-album, M. azotabsorptum, Mycobacterium* sp. 571	Fedorov & Kalininskaya (1961); probably more closely related to Nocardia than Mycobacterium (Biggins & Postgate, 1971*a*)
Methylosinum trichosporium	Whittenbury *et al.* (1970); previously called *Pseudomonas methanitrificans* (Coty, 1967)
Thiobacillus ferro-oxidans	Mackintosh (1971). Acetylene test only. Requires confirmation
Aerobic phototrophs (blue-green algae):	
Family Nostocaceae	At least 9 genera, all heterocystous
Family Stigonemataceae	At least 5 genera, all heterocystous
Family Chroococcaceae	Gloeocapsa, possibly others } No heterocysts, fixation only at low pO_2 and low illumination. See Stewart (1971)
Family Oscillatoriaceae	Plectonema and others } No heterocysts, fixation only at low pO_2 and low illumination. See Stewart (1971)
Facultatively anaerobic bacteria:	*Only* fix when growing anaerobically
Klebsiella pneumoniae, K. rubiacearum	Fixation distributed haphazardly among Klebsiella (Mahl *et al.* 1965)
Bacillus polymyxa, B. macerans, possibly other bacilli	Fixation distributed haphazardly among bacilli (Grau & Wilson, 1962). See also Mishustin & Shil'nikova (1971)
Enterobacter aerogenes, E. cloacae	Line & Loutit (1971); Raju, Evans & Seidler (1972)
Escherichia intermedia, E. coli	Line & Loutit (1971). *E. coli* prepared by genetic manipulation (Dixon & Postgate, 1972)
Unidentified rod	'*Pseudomonas azotogensis* strain V', Hill & Postgate (1969)
Phototrophic bacteria:	Only fix anaerobically
The family Athiorhodaceae: *Rhodospirillum rubrum, Rhodopseudomonas palustris,* Rhodomicrobium	Lindstrom, Tove & Wilson (1950); one strain of each tested
The coloured sulphur bacteria: *Chlorobium thiosulfatophilum, Chromatium vinosum, C. minutissimum, Ectothiorhodospira shaposhnikovii*	N_2 fixation is rare among chlorobia (Zakhveteva, Malafeeva & Kondrat'eva, 1970)
Obligate anaerobes:	
Clostridium pasteurianum, C. butyricum, numerous other clostridia	See Mishustin & Shil'nikova (1971); not all strains of the named species fix readily
Methane bacteria?	*Methanobacillus omelianskii* fixes (Pine & Barker, 1964) but is a stable mixture of two types (Bryant *et al.* 1967). Which fixes is still unknown, despite persistent enquiries (R. S. Wolfe counsels patience)
Sulphate-reducing bacteria: *Desulfovibrio desulfuricans, D. vulgaris, D. gigas; Desulfotomaculum orientis, D. ruminis*	Not all strains or species of Desulfovibrio fix nitrogen. The thermophilic *Desulfotomaculum nigrificans* does not fix

Table 4 (*cont.*)

Symbiotic systems (examples only)	*Comments*
Algal association:	
Lichens (fungus and blue-green alga)	Association usually close
Some liverworts (thallus includes blue-green algae)	Association rather casual
Azolla (water fern and blue-green algae)	Association rather casual
Some Cycads (blue-green algae in coralloid roots)	Association rather casual
Leaf nodule systems:	
Psychotria (leaf nodules contain bacteria including *K. rubiaciarum*)	Syntrophism needs substantiation
Root nodule systems:	
Leguminous plants + Rhizobium	Near-obligate symbiosis: no fixation by rhizobia until transformed into bacteroids within nodule
Non-leguminous plants + Frankia	Endosymbiont named by Becking (1970) though never isolated
Root associations:	Casual, some host specificity
Paspalum (sand grass) + Azotobacter	Döbereiner, Day & Dart (1972)
Maize + *E. cloacae*	Raju, Evans & Siedler (1972)
Leaf associations:	Casual
Azotobacter + tropical plants	Ruinen (1956)
Douglas fir + unidentified bacteria	Jones (1970)
Animals:	
Ruminants	See Hobson *et al.* (1973); amounts fixed probably trivial
New Guinea natives	Bergersen & Hipsley (1970)

ability to fix nitrogen appears to be haphazardly distributed at the species level. This situation contrasts with that in the aerobic Azotobacteraceae, in which all genera and species fix nitrogen (as they must by definition) but seemingly related types which do not fix are not known.

4. Little can usefully be said about the evolutionary relationships of various taxa which include the nitrogen-fixing bacteria. The Azotobacteraceae are a group of related organisms which can be placed in an evolutionary series according to their oxygen tolerance (see below), but the question of their evolutionary neighbours is no clearer than it was half a century ago. The nitrogen-fixing 'mycobacteria' might bear some relationship to nocardias (Biggins & Postgate, 1971*a*) among which nitrogen fixation has been reported but not confirmed (Hill & Postgate, 1969). A hierarchy of blue-green algae based on oxygen sensitivities will be discussed later; their evolutionary relationship to the phototrophic bacteria is widely accepted but nothing can be added to

Burris's (1963) comments on this topic. The resemblance of some flexibacteria (e.g. Leucothrix) to a chlorogenous blue-green alga (Lewin & Lounsbery, 1969) suggests that nitrogen fixation ought to be sought within the flexibacteria. An evolutionary relationship between facultative bacilli and butyric clostridia is consistent with the appearance of nitrogen fixation within both groups; the occurrence of the property among facultative enterobacteria is also consistent with a presumed evolutionary interrelationship. There is no compelling reason to regard the two genera of sulphate-reducing bacteria, *Desulfovibrio* and *Desulfotomaculum*, as phylogenetically related, so the scattered appearance of nitrogenase within both groups adds nothing to what we already know, that nitrogen fixation is to be found in many heterotrophic bacteria (sulphate reducers, methane producers, clostridia) which seem primitive even by the standards of prokaryotes (Klein & Cronquist, 1967). Evolutionary relations among symbiotic systems will be discussed later.

If evolutionary pathways do not emerge from Table 4, a broader view can be taken of the fact that nitrogen fixation appears to be distributed in a rather scattered manner among the obligate and facultative anaerobes. Examples of this situation are the survey of Mahl, Wilson, Fife & Ewing (1965) who detected fixation by thirteen out of thirty-one strains of *Klebsiella pneumoniae* (but none by strains of *K. rhinosclermatis*, Serratia or Enterobacter). In a comparable survey of Klebsiella strains available from culture collections, Dr R. A. Dixon in the writer's laboratory detected, by means of the acetylene test, nitrogen fixation in seven out of sixteen strains of *K. aerogenes*, in none of five strains of *K. pneumoniae*, and in five of twenty-one Klebsiella strains not identified at species level. Fixation seems to be more common among the facultatively anaerobic bacilli than among the Klebsiellae: fifteen out of seventeen strains of *B. polymyxa* fixed nitrogen (Grau & Wilson, 1962) and so did twelve out of fifteen strains of *B. macerans* (Witz, Detroy & Wilson, 1967). In the genus *Desulfovibrio*, nitrogen fixation has been detected in three strains of *D. vulgaris*, the one known strain of *D. gigas*, and three out of four of *D. desulfuricans*, but not in the type strains of *D. salexigens* or *D. africanus* (Riederer-Henderson & Wilson, 1970; Postgate, 1970). More strains of Desulfovibrio clearly require examination, a point which applies with even more force to other anaerobes. But among the Klebsiellae the scattered distribution of nitrogen fixation is striking, and it is tempting to deduce that the genetic information has been fairly readily gained and lost during evolution. If so, it could well have existed as an extrachromosomal element, a plasmid, having properties analogous to a drug-resistance transfer factor. Some circumstantial evidence in favour of this view exists and may be itemized:

1. Like other genetic determinants, that for nitrogen fixation (*nif*) can be transferred intra-specifically by transduction (Streicher, Gurney & Valentine, 1971) or conjugation (Dixon & Postgate, 1971).

2. *Nif* has been transferred by inter-generic matings from *K. pneumoniae* to *Escherichia coli* and stable, nitrogen-fixing hybrids of *E. coli* have been obtained (Dixon & Postgate, 1972).

3. In such hybrids, regulation by ammonia repression occurs (Dixon & Postgate, 1972), suggesting that *nif* is a gene cluster with many of the properties of an operon. Streicher, Gurney & Valentine (1972) mentioned co-transductional analysis suggesting that *nif* is a gene cluster with all components close together in *K. pneumoniae*. Such a gene cluster is consistent with integration of a plasmid-borne *nif* operon during evolutionary history.

4. While there is no reason to believe that *K. pneumoniae nif* is other than integrated into the chromosome, evidence is accumulating in the writer's laboratory that, in *E. coli* hybrids, it is usually present as a plasmid (Cannon, Dixon, Postgate & Primrose, 1974*a*, *b*). Fig. 1 illustrates an example of a hybrid with *E. coli* K12 in which covalently closed circular DNA, absent from the original strain, is present when *nif* has been acquired and lost from a nif$^-$ segregant. Genetic evidence based on relative rates of transfer of drug-resistance determinants and *nif* support this view, though unequivocal assignment of *nif* to the plasmid has not been possible at the time of writing. In another hybrid, the stable *E. coli* strain C-M7, the *nif* genes appear to have become integrated into the *E. coli* chromosome.

Nif can thus be transduced, at least in part; it can be plasmid-borne, and can also become integrated into a new genus in the laboratory. The view that *nif* has become distributed among prokaryotes by genetic transfer among pre-existing strains would be one way of accounting for its scattered distribution among strains which are readily transducible and prone to conjugate (enteric bacteria) and perhaps among readily transformable types (bacilli). It would also account for the divergent G/C ratios of the nitrogen-fixing bacteria discussed by De Ley (1968, p. 144). It would be consistent with a late evolutionary emergence of the property, and with the fact that the property is still dispensable: not essential to the survival of many species which possess it. One would expect it to have become stable and universal as it became genetically integrated, particularly in highly evolved genera such as

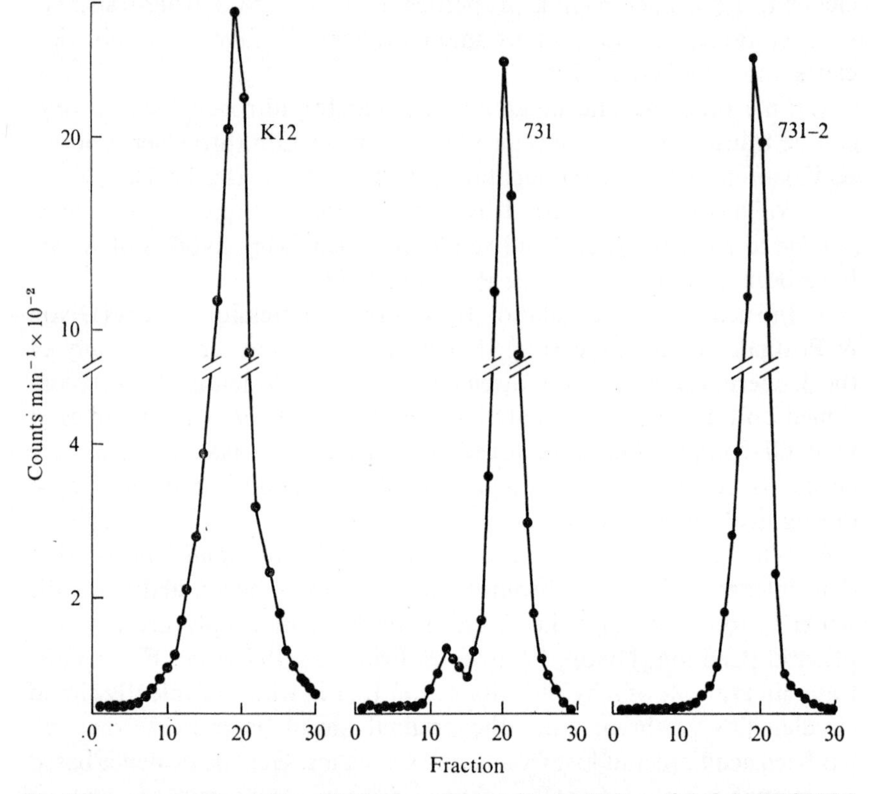

Fig. 1. Buoyant density profiles of DNA from *Escherichia coli* K12, a *nif*$^+$ hybrid (UNF-731) and a *nif*$^-$ segregant (UNF-7312). DNA was extracted from ^{3}H-thymidine-labelled bacteria with lysozyme, EDTA and a detergent, and the bulk of the chromosomal DNA sedimented with the debris. The supernatant ('cleared lysate') was treated with ethidium bromide to distinguish supercoiled (plasmid) from open and residual chromosomal DNA and centrifuged in a CsCl/gradient to equilibrium. The small peak of supercoiled DNA in UNF-731, lost from the segregant 7312 and absent from the original K12 strain, represents the presumptive *nif*-bearing plasmid. (Courtesy of Dr F. C. Cannon.)

Azotobacter. A corollary is that one would expect *nif* sometimes to be found in nature as a plasmid, perhaps being 'curable' by known procedures.

THE EVOLUTION OF OXYGEN-EXCLUDING MECHANISMS

The fact that nitrogenase proteins are irreversibly inactivated by exposure to air, even if they originate from aerobic organisms (Burris, 1971 and other review articles cited on p. 263) has led to the development of several physiological and anatomical stratagems to avoid exposure of the functioning enzyme to oxygen. The basic problem is that, under air, the saturation concentrations of dissolved oxygen (0.25 mM at 25 °C and 0.2 atm) and nitrogen (0.51 mM at 25 °C and 0.8 atm) in water are low but not very different. The two molecules are of very similar size and mobility, so the exclusion of one, oxygen, combined with the need for a reaction site of high affinity for the other, nitrogen, presents a physico-chemical problem to any biological system. The anaerobes and facultative anaerobes might be said to avoid this problem since they do not fix nitrogen when dissolved oxygen is present. There remains a problem of protection of the nitrogenase enzyme when the anaerobe is not growing but is surviving an oxygen stress (the obligate anaerobe *Clostridium pasteurianum* can be harvested in air and an active nitrogenase can then be extracted) but we know nothing of any protection mechanism (possibilities briefly discussed by Postgate, 1974) and can say nothing of its evolution.

Respiratory protection

As far as facultative anaerobes are concerned, Klucas (1972) showed that *K. pneumoniae*, normally regarded as only able to fix anaerobically, could synthesise nitrogenase in air provided its respiration was such as to keep the dissolved oxygen concentration very low. His findings are confirmed by independent experiments in the writer's laboratory by Miss S. Hill, who induced a chemostat culture of *K. pneumoniae* M5a1 to fix nitrogen vigorously under 0.4 atm O_2, sustaining a dense population in a medium containing small amounts of yeast extract. Such work enables one to envisage a physiological evolution in the direction of the respiratory protection of nitrogenase found in Azotobacter. The role of respiration as a protective process in this organism has been discussed extensively by the present writer and others (Dalton & Postgate, 1969; Drozd & Postgate, 1970; Postgate, 1971*a*; Hill, Drozd & Postgate, 1972; Senior, Beech, Ritchie & Dawes, 1972; Postgate, 1974; Lees & Postgate, 1973; Jones, Brice, Wright & Ackrell, 1973). Among the most persuasive pieces of evidence that respiration in Azotobacter protects active nitrogenase as well as performing its normal physiological function are the following:

1. Nitrogen fixation is most efficient when the oxygen supply limits respiration.

2. Oxygen stress, and adaptation to oxygen, lead to augmented respiration without a change in nitrogenase activity.

3. With intact bacteria, optimal nitrogenase activity is not necessarily shown at the pO_2 of air but is affected by the cultural history of the population. The optimal pO_2 for activity is greater the more the

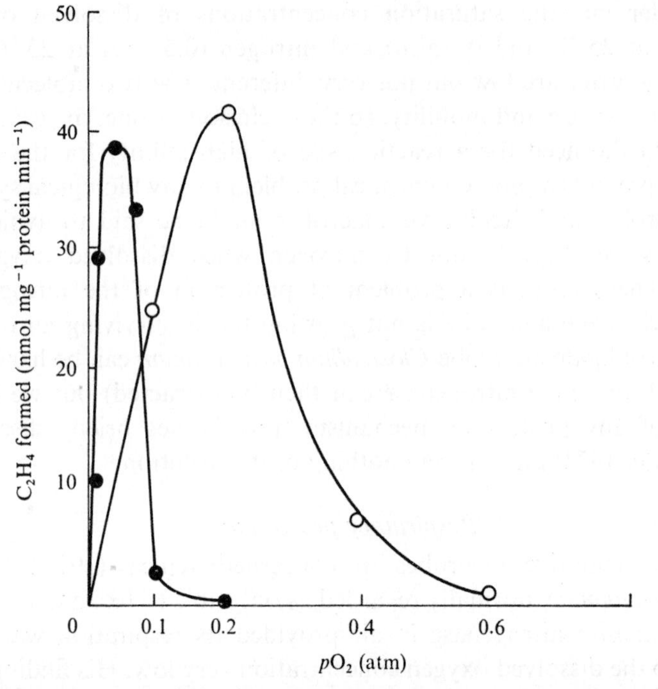

Fig. 2. Effect of pO_2 on acetylene-reducing activity of cultures of *Azotobacter chroococcum*. Whole chemostat cultures habituated to high or low pO_2 values were stopped and the acetylene-reducing activity tested at various pO_2 values. ●, Culture grown with 0.09 atm O_2; ○, culture grown with 0.55 atm O_2. (Courtesy of Dr J. W. Drozd.)

respiratory activity of the population, though the absolute activity at the optimal pO_2 is unchanged (Fig. 2).

4. By exposing populations to pure O_2, or by suddenly applying an oxygen stress to phosphate-limited populations, it is possible to overcome respiratory protection and also such other protective mechanisms as may exist. Specific damage to the nitrogenase enzyme, notably the very oxygen-sensitive protein 2 (Fe protein), is then observed.

If the oxygen-scavenging effect of respiration can protect functioning nitrogenase in Klebsiella, then the obligately aerobic Azotobacter must be regarded as more highly evolved organisms, by virtue of their high respiratory activity, their ability (usually) to adjust respiration to the ambient oxygen concentration (Lees & Postgate, 1973) and their branched electron transport pathway which avoids over-production of ATP during respiratory protection (Jones *et al.* 1973). The aerobe *Mycobacterium flavum* is, when fixing nitrogen, very sensitive to oxygen and normally unable to grow in air from small inocula unless traces of fixed nitrogen (as ammonium salt or yeast extract) are available

(Biggins & Postgate, 1969); since its respiratory activity (Q_{O_2} = 80 μl O_2 mg^{-1} organism h^{-1}, Biggins & Postgate, 1971*b*) is normally about 4 % that of an 'average' Azotobacter (Q_{O_2} *ca* 2000 μl O_2 mg^{-1} h^{-1}; values up to 4000 have been recorded), inability to conduct respiratory protection at least partly accounts for this behaviour. Although it is an obligate aerobe, and grows without any difficulty in air when provided with fixed nitrogen as ammonium ion, this organism behaves almost as a micro-aerophile when fixing nitrogen, thus representing a physiologically primitive type of aerobic fixer. A comparable physiological situation exists in the genus *Derxia*, a member of the Azotobacteraceae, in which Hill (1971) has shown that the characteristic colonial dimorphism on nitrogen-free agar plates in air can be interpreted in terms of sensitivity to oxygen greater than in the genus *Azotobacter*. Though respiratory coefficient data are not available for this organism, it is clearly more primitive in its oxygen-excluding mechanisms than Azotobacter, but more advanced than Klebsiella. One can thus envisage a physiological evolution of respiratory protection, starting with Clostridium which has none:

clostridium → klebsiella → *Mycobacterium flavum*
(also *Derxia gummosa*?) → azotobacter

Gum formation

Gum or mucoid matter is very characteristic of aerobic nitrogen-fixing bacteria, particularly among Azotobacteraceae such as Beijerinckia and Derxia. In part it arises from the nutritional status of such organisms: the growth rate of *Azotobacter chroococcum* is normally limited by the rate of nitrogen fixation so polysaccharide tends to be formed as in any N-limited population (see Hill *et al.* 1972). But gum could act as a buffer to oxygen stress by slowing diffusion of O_2 to the sensitive sites (Postgate, 1971*a*); the fact that it would delay diffusion of N_2 as well would be unimportant because the gross requirement of an organism for this element is an order of magnitude lower. However, nothing useful can be said on the evolution of gum formation except to note that it is more pronounced in the intermediate Derxia than in the more highly evolved Azotobacter.

Compartmentation

Another physiological stratagem for excluding oxygen is compartmentation: restriction of the nitrogenase to a location to which oxygen has little or no access. With molecules so similar in size, mobility and solubility as O_2 and N_2, this must be mechanistically difficult in biological systems and, in fact, such systems as can be cited are imperfect.

The nitrogenase of the most highly evolved aerobic genus, *Azotobacter*, can assume a form in which it is stable to air. This is the form in which crude nitrogenase is normally obtained when cell-free enzyme preparations are made from *A. vinelandii* (Bulen, Burns & LeComte, 1965) or *A. chroococcum* (Kelly, 1968) and such preparations are subcellular particles which sediment at about 150000*g* over 2 to 4 h. Comparable subcellular particles from *Mycobacterium flavum* are oxygen-sensitive (Biggins & Postgate, 1971*b*).

It follows that, in Azotobacter, the nitrogenase proteins can exist in some conformation in which their oxygen-sensitive sites are protected from damage by oxygen. The hypothesis that this ability to assume an oxygen-insensitive form *in vitro* represents a means of protecting the nitrogenase *in vivo* has been discussed by the present writer and colleagues several times. The essential features of the view are:

1. That conformational protection is regulated in the living organism.
2. That it is a 'back-up' protective process which operates only when the oxygen stress is such that respiratory protection is inadequate.
3. That the conformationally protected enzyme is inactive towards its substrates.

The evidence in favour of the view that regulation of conformational protection is a real part of the physiology of Azotobacter has recently been reviewed (Postgate, 1974); essentially it rests on the behaviour of N_2-limited chemostat cultures of *A. chroococcum* with regard to their rates of nitrogen fixation and on the ability of the organisms seemingly to 'switch off' and 'switch on' their nitrogenase activity much more rapidly in response to changes in oxygen supply than changes in respiration rate could account for. From an evolutionary point of view it is interesting that *M. flavum*, whose subcellular particles are not conformationally protected, can yield membranous fragments of disrupted organisms in which some of the nitrogenase is retained in an oxygen-tolerant form. An intracellular network of subcellular membranes is characteristic of nitrogen-fixing *A. vinelandii* (Oppenheim & Marcus, 1970) and *A. chroococcum* (Hill *et al.* 1972; Drozd, Tubb & Postgate, 1972) and is absent from ammonia-grown populations. The oxygen-tolerant subcellular particles of Azotobacter contain cytochromes and other membrane-associated materials; it is tempting to take the view that conformational protection represents an intracellular compartmentation of nitrogenase, in association with a membrane, in which a rapid shift to the oxygen-tolerant conformation can be brought about in response to extracellular environmental changes. If this hypothesis is correct, the internal membranous network of Azotobacter represents a relatively highly evolved oxygen-exclusion apparatus which would not be found in less evolved aerobes such as *M. flavum*, though conformational protection on a modest scale might still be possible at the normal peripheral membrane.

In terms of its oxygen relations, Azotobacter seems to be a highly evolved organism, and this is consistent with the universality of ability to fix nitrogen within the family. Also supporting this view is the fact that substrate-yield experiments indicate a lower consumption of ATP *in vivo* than in Klebsiella or Clostridium (see Hill *et al.* 1972; Postgate, 1974) despite the substrate wastage inherent in respiratory protection: such relative biochemical efficiency would be consistent with the property having a relatively long evolutionary history within the genus.

Compartmentation in phototrophs

Conformational protection in Azotobacter implies compartmentation of nitrogenase at a subcellular level. Compartmentation on a grosser scale is familiar in the heterocystous blue-green algae, which have a special problem with oxygen because it is a product of their photosynthesis, unlike the situation in nitrogen-fixing photosynthetic bacteria such as *Rhodospirillum rubrum*. There is good circumstantial evidence that, when such organisms grow at normal pO_2 values, the heterocysts are the specific sites of nitrogen fixation (see Stewart, 1971). At one time the heterocystous blue-green algae were thought to be the only nitrogen-fixing types, and the heterocyst the sole nitrogen-fixing organ. Now examples of both filamentous and unicellular nitrogen-fixing blue-green algae are known which form no heterocysts. It is a characteristic of most of these organisms that they show evidence of oxygen sensitivity: they only fix nitrogen at low pO_2 values or at low illuminations, when photosynthetic oxygen production is low. Evidence for processes analogous to respiratory protection in blue-green algae is available, such as oxygen-stimulated photorespiration in *Anabaena cylindrica* (Lex, Silvester & Stewart, 1972) and oxygen sensitivity in *A. flos-aquae* (Bone, 1971) influenced by nutritional status (Bone, 1972). Stewart (1971) recalled that the blue-green algae are present-day representatives of very primitive microbes and suggested that the micro-aerophily of certain species when fixing nitrogen might be a relic of adaptation to a primitive environment. This view presumes that the ancient organisms actually fixed nitrogen, and the arguments presented earlier against such a view in bacteria apply equally to the blue-green algae. Nevertheless, once the ability to fix nitrogen developed within the blue-green algae, a logical course of evolution can be seen, from micro-aerophilic fixation, possibly assisted by gummy capsules in certain Chroococcaceae, perhaps via respiratory protection alone, to respiratory protection supplemented by the specialised heterocyst. At the time of writing the problem is still unresolved whether, in addition to the oxygen-excluded process in the heterocysts, micro-aerophilic nitrogen fixation takes place in the vegetative filaments of heterocystous blue-green algae comparable to the micro-aerophilic fixation by the filamentous Plectonema. Evolutionary considerations would lead one to expect micro-aerophilic fixation in the vegetative filaments of at least some heterocystous species.

The root nodule as a compartment

The evolution of the nodule symbioses is discussed in the next section, but it is pertinent to mention here that the legume nodule is an oxygen-sensitive, aerobic nitrogen-fixing system in which ATP generation at the expense of oxygen reduction must be coupled to nitrogenase function. Recent evidence suggests that leghaemoglobin plays a part in this process (Bergersen, Turner & Appleby, 1973) and the fact that leghaemoglobin is substantially unoxygenated in functioning nodules indicates that the effective pO_2 inside the nodule is very low (Appleby, 1961). Thus the legume nodule may be regarded as another example of a compartment for nitrogen fixation to which access of oxygen is strictly limited.

Interference by hydrogen ions

Though this section is ostensibly concerned with interference with nitrogen fixation by oxygen, it is important to recall that the hydrogen ion also interferes with nitrogenase function. All nitrogenase preparations show an ATP-activated hydrogen-evolving reaction (see reviews cited on p. 263) even when a reducible substrate is present. This fact implies that, *in vitro*, a proportion of the electrons generated are donated to hydrogen ions. Thus nitrogen-fixing organisms which do not produce hydrogen in their normal metabolism, as Clostridium and anaerobically-grown Klebsiella or Bacillus do, may have developed mechanisms for excluding the hydrogen ion from the enzyme, or for re-using the hydrogen produced. Dixon (1972) has suggested that the particulate hydrogenases of Azotobacter and Rhizobium bacteroids might perform a recycling function. Nothing useful can yet be said concerning the evolution of such processes.

EVOLUTION OF SYMBIOTIC ASSOCIATIONS

Symbiotic associations, involving partial or complete interdependence, can normally be regarded as having evolved from more casual commensalism. The phyllosphere association first recognised by Ruinen (1956) may thus be the precursor of leaf nodule associations such as that observed between Psychotria, a tropical plant, and *Klebsiella rubiacearum* (Centifanto & Silver, 1964). However, the failure of Becking (1971) to obtain clear indications with $^{15}N_2$ and the acetylene test of symbiotic fixation in Psychotria, even when inoculated with *K. rubiacearum*, leaves the real status of this association doubtful. Associations between blue-green algae and plants ranging from cycads and liverworts, where the mutuality of the association is sometimes doubtful (Watanabe & Yamamoto, 1971, contrasting with Wittman, Bergersen & Kennedy, 1965), to lichens and the water fern, Azolla, where the mutual benefit of the association is undisputed (see Mishustin & Shil'nikova, 1971), also lead to no productive speculation on evolution at this stage, although the bearing of these associations on the 'age' of the fixation process will be discussed later.

The importance of associations between plants and free-living nitrogen-fixing bacteria in the rhizosphere is now being recognised. Probably the first clear example was Döbereiner's demonstration of a specific Azotobacter species associated with the rhizosphere of sand grasses in soils in which the principal nitrogen-fixing aerobes belonged to the genera Beijerinckia or Derxia (Döbereiner & Campelo, 1971). The rhizosphere of the grass *Paspalum notatum* appears to provide a particularly suitable environment for the growth of *Azotobacter paspali*, particularly with regard to oxygen tension, and the association could bring nearly 100 kg N ha^{-1} to soil, calculated from acetylene reduction rates (Döbereiner, Day & Dart, 1972, 1973). Comparable associations have been recognised in the rhizosphere of rice (Rinaudo, Balandreau & Dommergues, 1971), maize (Raju, Evans & Seidler, 1972) in which a nitrogen-fixing Enterobacter species is involved and Hedge Woundwort *Stachys sylvatica* (Harris & Dart, 1973). Association of Klebsiella and related bacteria with the rhizosphere of legumes was noted by Evans, Campbell & Hill (1972). The presence of root exudates, together with a low pO_2, must make the rhizosphere an ideal micro-environment for casual association of plant and free-living bacteria, from which more interdependent root associations probably evolved.

Rhizobium, the classical nitrogen-fixing symbiont, is not taxonomically related to any of the known nitrogen-fixing rhizosphere commensals; its nearest relative appears to be the sometimes phytopathogenic genus *Agrobacterium* (Gibbins & Gregory, 1972). It is also unrelated to the as yet unidentified symbiont in the nodules of non-leguminous plants which Becking (1970) has called *Frankia*, a genus with ten species. If his assignment of this organism to the Actinomycetales is justified, its nearest free-living relative would be the organism of doubtful taxonomic status called *Mycobacterium flavum* 301, this organism being closer to Nocardia than to Mycobacterium, Corynebacterium or Arthrobacter (Biggins & Postgate, 1971*a*).

Despite the fact that no evolutionary sequence can be seen in taxonomic terms, there seems to be a physiological evolution among the root associations in the direction of increasing interdependence between plant and microbe:

klebsiella → *Azotobacter paspali* → rhizobium → *Frankia.*

In the sequence above, Klebsiella shows casual syntrophism and *A. paspali* some host specificity; Rhizobium is more host-dependent in that it does not fix away from the plant. It can be cultured readily, without fixation, away from its host but the weight of evidence is that rhizobia do not form nitrogenase until they have become the pleomorphic 'bacteroids' within an effective nodule. Frankia appears even more host-dependent in that it has not yet been cultivated away from the plant, though it has been grown *in vitro* in alder tissue culture (Becking, 1965) and can obviously survive and presumably grow as a soil saprophyte, since the microbial symbiont is not seed-borne (Bond, 1963).

The only symbiotic system which has been the subject of detailed speculation regarding evolutionary origin is the rhizobium–legume interaction. Bond, MacConnell & McCullum (1956) regarded the nodule systems as ancient in botanical terms but almost modern to the microbial evolutionist, extending in time from the present to the earliest angiosperms. Norris (1956) discussed the evolution of the legume symbiosis at some length and collected much data on the evolution of leguminous plants. Parker (1957) regarded symbiotic systems as sophisticated developments from more casual root associations. Questioning some of Norris's conclusions about the soil and climatic conditions in which the symbiosis originated, Parker (1968) extended the discussion and concluded that the prototype symbiont, probably originating from a free-living organism, probably resembled most closely the slow-growing 'cowpea' type of organism rather than the faster-growing *R. meliloti* or *R. trifolii.* The leguminous plants are rather ancient angiosperms and species differ widely in their ability to nodulate (Parker quoted 90 % of the Mimosaceae and Papilionaceae as able to nodulate but only 23 % of the Caesalpiniaceae), which suggested that ability to nodulate was acquired after the Leguminales had become differentiated into its three families. This would place the emergence of nodulation later than the Cretaceous era (about 10^8 years ago) and raises the possibility that nodulation emerged more than once, from different prototype symbionts. It is a pity that nodules, being soft, have left no satisfactory fossil traces to indicate when they first appeared.

Dilworth (1969) reported experiments which showed, essentially, that two different species of plant, infected with the same strain of Rhizobium, formed nodules with chemically distinguishable leghaemoglobin molecules, a situation which must have involved the transcription and translation of information from the plant genome. Leghaemoglobin, though important physiologically in fixation by legume nodules, is not an essential component of the bacteroids' nitrogen-fixing system (Evans & Russell, 1971), so speculation on its evolution will not be pursued here, despite the fact that it raises interesting questions concerning the origin of such an 'animal-type' protein in plants (Appleby & Dilworth, 1973). However, the evidence that the plant can specify information which is relevant to nodule function, taken with the obligate nature of the nitrogen-fixing symbiosis, suggests strongly that the plant intervenes in the control of nitrogenase synthesis by bacteroids. Dilworth & Parker (1969) considered three possible ways:

1. The bacteroids may possess the whole *nif* gene complex, but repressed. The plant might produce an inducer or derepressor.
2. The *nif* gene complex may be held by the plant.
3. The *nif* genes may be held partly by the plant and partly by the bacteria.

They were inclined to exclude (1) on the principal ground that derepressed mutants would surely have appeared. Derepressed mutants of *Azotobacter vinelandii* have now been reported (Gordon & Brill, 1972). Possibility (2) was not favoured on the grounds that nitrogenase is now known to be formed within the bacteroid and that expression of *nif* by the plant without the involvement of bacteria would surely have developed. While accepting that (1) and (2) were not rigidly excluded, they favoured (3), suggesting that some of the *nif* genes were transferred to the bacteroid, perhaps as a plasmid, from the plant. For this to be logical, they assumed that the possession of *nif* had, at some time, been ecologically disadvantageous to Rhizobium, which thus gained selective advantage by depositing some *nif* genes with a host plant. No new information has emerged to allow elaboration or revision of these viewpoints, except the increasing evidence that informational molecules transfer from microbes to plants with even greater freedom than Dilworth & Parker (1969) had reason to believe (see Ledoux, 1971).

CODA

In a contribution so fraught with speculation it would be improper to give this final section a title as firm as 'conclusions'. One can dimly perceive certain directions the evolution of nitrogen fixation is likely to have taken, notably in reconciling so oxygen-sensitive a process with an oxidising environment, but no precise evolutionary sequences can yet be proposed. Even the question of the time at which fixation emerged is more complex than it seemed a decade ago: it could be truly ancient (older than or contemporary with Gunflint chert: $2\text{–}3 \times 10^9$ years old), it could be roughly co-incident with the transition to an oxidising atmosphere ($1\text{–}1.5 \times 10^9$ years old) or relatively recent, after the eukaryotes became well established (*ca* 5×10^8 years old). I am attracted by the view that the ability to fix nitrogen has appeared relatively recently on this planet, for reasons which may be summarised:

1. Absence of evidence for selective pressure in favour of fixation during the earlier aeons of terrestrial life (Delwiche, 1970).
2. Random distribution in certain bacterial genera and the existence of *nif* as a transmissible gene cluster in Klebsiella.
3. Lack of evolutionary divergence among nitrogenase proteins from various organisms.

On such a view, the association of nitrogen fixation with the primitive characters mentioned earlier (p. 263) might seem to be coincidental, but a logical connection can be found. Nitrogen fixation is a highly reductive process, requiring a potential lower than the hydrogen electrode (evident from the ATP-activated hydrogen evolution reaction which accompanies the reduction of other substrates *in vitro*, see Table 1) and also complete exclusion of oxygen. 'Primitive' organisms such as *Clostridium pasteurianum* or *Desulfovibrio* contain electron transport proteins (ferredoxins and flavodoxins) capable of operating in this range and may well have been the organisms best suited to 'receive' the ability to fix nitrogen when it emerged and became genetically transmissible. A ferredoxin has been found in nitrogen-fixing *Bacillus polymyxa* (Shethna, Stombaugh & Burris, 1971) and comparable electron transport factors probably exist in Klebsiella and other facultative anaerobes. Particularly relevant is the presence of ferredoxins and flavodoxins in the aerobic Azotobacter and their specific association with nitrogenase function (Burris, 1971): was Azotobacter obliged to retain the evolutionary relics of an anaerobic habit in order to reconcile nitrogen fixation with the aerobic way of life? These electron transport proteins are constitutive both in anaerobes and, when they occur, in aerobes: unlike the nitrogenase proteins, their synthesis is not repressed by fixed nitrogen as ammonia (though iron deficiency promotes flavodoxin formation in *C. pasteurianum*). Hence they presumably (though not necessarily) have physiological functions distinct from transporting electrons to nitrogenase. Possession of such electron transport factors may have been a prerequisite of expression of a late-appearing *nif* gene cluster in evolutionary history – a requirement which would account for the ready acceptance of *nif* by the photosynthetic organisms, particularly the blue-green algae, whose photometabolism already involved such low potential electron transport factors.

This view of the evolution of nitrogen fixation would predict that the traces of blue-green algae to be found in Precambrian cherts (Bernal, 1967; Rutten, 1971) are relics of organisms which had not yet 'learned' to fix nitrogen. It would carry the corollary that genera such as Azotobacter, in which nitrogen fixation is universal and associated with a complex, specialised physiology, are young in an evolutionary sense.

A late appearance of ability to fix nitrogen would also explain the restriction of the property to prokaryotes. Nitrogen fixation requires elaborate mechanisms to exclude oxygen and eukaryotes such as yeasts and plants are highly aerobic. This aerobic character must be a hindrance to the expression of *nif* in eukaryotes, yet it ought not to have excluded the property from eukaryotes for the whole evolutionary history of terrestrial life, since prokaryotes have developed diverse means of coping with oxygen. Moreover, the functioning mitochondrion in eukaryotes is a subcellular organelle in which all the cytochrome *b* is normally fully reduced, implying an ambient E_h value well below zero. So there seems to be no compelling physiological objection to the existence in eukaryotes of a comparable subcellular compartmentation to accommodate *nif* if appropriate selective pressure exists – as now it often does. But if *nif* appeared late, when eukaryotes were well established, and if it could only spread at first among anaerobic prokaryotes, then its restriction to prokaryotes becomes more logical, because only they are readily susceptible to genetic modification by conjugation, transduction and transformation. It may be that *nif* has been accepted by eukaryotes often, despite the genetic restriction processes of eukaryotes, but that the appropriate metabolic modifications for its expression have not yet been possible. In short, eukaryotes may already have been too genetically rigid and too aerobic by the time *nif* became available. Such a view promotes optimism in considering the possibility of transferring functional *nif* to plants, still only a highly speculative possibility despite excitement in the National Press during 1972.

Late emergence of *nif* would also account for another, if minor, paradox. The endophytic origin of the chloroplast as a plant organelle is now widely accepted. The close associations of nitrogen-fixing blue-green algae with plants ranging from fungi (in lichens) through thallophytes (liverworts), pteridophytes (Azolla) and cycads, even if plant–algal syntrophism has not always been clearly demonstrated, must surely have provided conditions favourable to the incorporation of the alga as a nitrogen-fixing organelle during evolutionary time. The fact that no such organelles have developed may mean that there is some incompatibility that we are unaware of between nitrogen fixation and existence within eukaryotic protoplasm. (A high dissolved ammonium ion level, repressive to nitrogenase synthesis, would be an example of such a condition; one which presumably operates to restrict nitrogen fixation by the bacteria in the rumens of ruminant animals.) On the other hand, it may be that evolutionary history, as far as *nif* is concerned, has been too brief for a stable partnership of this kind to emerge by normal natural selection – in which case, judicious assistance of Nature by the scientist, duly preceded by fashionable consideration of the ecological consequences, might be advantageous to many forms of terrestrial life, not least ourselves.

I am grateful to Dr T. H. Blackburn for valuable discussion and to Dr M. J. Dilworth for reading and criticising a draft of the manuscript; both are absolved from responsibility for the views expressed here.

REFERENCES

Appleby, C. A. (1961). The oxygen equilibrium of leghemoglobin. *Biochimica et Biophysica Acta*, **60**, 226–35.

Appleby, C. A. & Dilworth, M. J. (1973). Leghemoglobin and other nodule proteins. In section on biochemistry, ed. R. C. Burns, in *Dinitrogen Fixation: Chemistry, Biochemistry, Biology, Agronomy and Ecology*, ed. R. W. F. Hardy. New York: Wiley.

Becking, J. H. (1965). *In vitro* cultivation of alder root-nodule tissue containing the endophyte. *Nature, London*, **207**, 855–87.

Becking, J. H. (1970). *Frankeaceae* fam.nov. (Actinomycetales) with one new combination and six new species of the genus *Frankia* Brunchorst 1886, 174. *International Journal of Systematic Bacteriology*, **20**, 201–20.

BECKING, J. H. (1971). The physiological significance of the leaf nodules of Psychotria. In *Biological Nitrogen Fixation in Natural and Agricultural Habitats*, ed. T. A. Lie & E. G. Mulder. *Plant and Soil* (special volume), pp. 361–76.

BENEMANN, J. R. & VALENTINE, R. C. (1972). The pathways of nitrogen fixation. *Advances in Microbial Physiology*, **8**, 59–104.

BERGERSEN, F. J. (1971). Biochemistry of symbiotic nitrogen fixation in legumes. *Annual Review of Plant Physiology*, **22**, 121–40.

BERGERSEN, F. J. & HIPSLEY, E. H. (1970). Presence of N_2-fixing bacteria in the intestines of man and animals. *Journal of General Microbiology*, **60**, 61–6.

BERGERSEN, F. J., TURNER, G. L. & APPLEBY, C. A. (1973). Studies on the physiological role of leghaemoglobin in soya bean root nodules. *Biochimica et Biophysica Acta*, **292**, 271–82.

BERNAL, J. D. (1967). *The Origin of Life*. London: Weidenfeld & Nicolson.

BIGGINS, D. R., KELLY, M. & POSTGATE, J. R. (1971). Resolution of nitrogenase of *Mycobacterium flavum* 301 into two components and cross reaction with nitrogenase components from other bacteria. *European Journal of Biochemistry*, **20**, 140–3.

BIGGINS, D. R. & POSTGATE, J. R. (1969). Nitrogen fixation by cultures and cell-free extracts of *Mycobacterium flavum* 301. *Journal of General Microbiology*, **56**, 181–93.

BIGGINS, D. R. & POSTGATE, J. R. (1971*a*). Confusion in the taxonomy of a nitrogen-fixing bacterium currently classified as *Mycobacterium flavum* 301. *Journal of General Microbiology*, **65**, 119–23.

BIGGINS, D. R. & POSTGATE, J. R. (1971*b*). Nitrogen fixation by extracts of *Mycobacterium flavum* 301. *European Journal of Biochemistry*, **19**, 408–15.

BOND, G. (1963). The root nodules of non-leguminous angiosperms. In *Symbiotic Associations, Symposia of the Society for General Microbiology*, **13**, 72–91. Ed. P. S. Nutman & B. Mosse. London: Cambridge University Press.

BOND, G., MACCONNELL, J. T. & MCCULLUM, A. H. (1956). The nitrogen nutrition of *Hippophae rhamnoides* L. *Annals of Botany*, **20**, 501–12.

BONE, D. H. (1971). Kinetics of synthesis of nitrogenase in batch and continuous cultures of *Anabaena flos-aquae*. *Archiv für Mikrobiologie*, **80**, 242–51.

BONE, D. H. (1972). The influence of canavanine, oxygen and urea on the steady state levels of nitrogenase in *Anabaena flos-aquae*. *Archiv für Mikrobiologie*, **85**, 13–24.

BRYANT, M. P., WOLIN, E. A., WOLIN, M. J. & WOLFE, R. S. (1967). *Methanobacillus omelianskii*, a symbiotic association of two species of bacteria. *Archiv für Mikrobiologie*, **59**, 20–31.

BULEN, W. A., BURNS, R. C. & LECOMTE, J. R. (1965). Nitrogen fixation: hydrosulfite as electron donor with cell-free preparations of *Azotobacter vinelandii* and *Rhodospirillum rubrum*. *Proceedings of the National Academy of Sciences, USA*, **53**, 532–9.

BURNS, R. C. & HARDY, R. W. F. (1973). *Nitrogen fixation in bacteria and higher plants*, chapter 3. New York: Springer-Verlag.

BURNS, R. C., HOLSTEN, R. D. & HARDY, R. W. F. (1970). Isolation by crystallization of the Mo–Fe protein of *Azotobacter* nitrogenase. *Biochemical and Biophysical Research Communications*, **39**, 90–9.

BURRIS, R. H. (1963). In *Evolutionary Biochemistry*, ed. A. I. Oparin. *Proceedings of the 5th International Congress of Biochemistry, Moscow*, **3**, 173–7. New York: Macmillan.

BURRIS, R. H. (1971). In *The Chemistry and Biochemistry of Nitrogen Fixation*, pp. 106–60. Ed. J. R. Postgate. London: Plenum Press.

CANNON, F. C., DIXON, R. A., POSTGATE, J. R. & PRIMROSE, S. B. (1974*a*). Chromosomal integration of Klebsiella nitrogen fixation genes in *Escherichia coli*. *Journal of General Microbiology*, **80**, in press.

CANNON, F. C., DIXON, R. A., POSTGATE, J. R. & PRIMROSE, S. B. (1974*b*). Plasmids formed in nitrogen-fixing *Escherichia coli–Klebsiella pneumoniae* hybrids. *Journal of General Microbiology*, **80**, in press.

CENTIFANTO, Y. M. & SILVER, W. S. (1964). Leaf nodule symbiosis 1. Endophyte of *Psychotria bacteriophila*. *Journal of Bacteriology*, **88**, 776–81.

CHATT, J. & RICHARDS, T. L. (1971). In *The Chemistry and Biochemistry of Nitrogen Fixation*, ed. J. R. Postgate, pp. 57–103. London: Plenum Press.

CHENIAE, G. & EVANS, H. J. (1957). On the relation between nitrogen fixation and nodule nitrate reductase of soyabean root nodules. *Biochimica et Biophysica Acta*, **26**, 654–5.

COTY, V. (1967). Atmospheric nitrogen fixation by hydrocarbon-oxidizing bacteria. *Biotechnology and Bioengineering*, **9**, 25–38.

DALTON, H. & MORTENSON, L. E. (1972). Dinitrogen (N_2) fixation (with a biochemical emphasis). *Bacteriological Reviews*, **36**, 231–60.

DALTON, H. & POSTGATE, J. R. (1969). Effect of oxygen on growth of *Azotobacter chroococcum* in batch and continuous cultures. *Journal of General Microbiology*, **54**, 463–73.

DE LEY, J. (1968). Molecular biology and bacterial phylogeny. In *Evolutionary Biology*, ed. T. Dobzhansky, M. K. Hecht & W. C. Steere, vol. 2, pp. 103–56. New York: Appleton-Century-Crofts.

DELWICHE, C. C. (1970). The nitrogen cycle. *Scientific American*, **223**, 137–46.

DILWORTH, M. J. (1969). The plant as the genetic determinant of leghaemoglobin production in the legume root nodules. *Biochimica et Biophysica Acta*, **184**, 432–41.

DILWORTH, M. J. & PARKER, C. A. (1969). Development of the nitrogen-fixing system in legumes. *Journal of Theoretical Biology*, **25**, 208–15.

DIXON, R. A. & POSTGATE, J. R. (1971). Transfer of nitrogen fixation genes by conjugation in *Klebsiella pneumoniae*. *Nature, London*, **234**, 47–8.

DIXON, R. A. & POSTGATE, J. R. (1972). Genetic transfer of nitrogen fixation from *Klebsiella pneumoniae* to *Escherichia coli*. *Nature, London*, **237**, 102–3.

DIXON, R. O. D. (1972). Hydrogenase in legume root nodule bacteroids: occurrence and properties. *Archiv für Mikrobiologie*, **85**, 193–201.

DÖBEREINER, J. & CAMPELO, A. B. (1971). Non-symbiotic nitrogen-fixing bacteria in tropical soils. In *Biological Nitrogen Fixation in Natural and Agricultural Habitats*, ed. T. A. Lie & E. G. Mulder. *Plant and Soil* (special volume), pp. 457–70.

DÖBEREINER, J., DAY, J. M. & DART, P. J. (1972). Nitrogenase activity and oxygen sensitivity of the *Paspalum notatum–Azotobacter paspali* association. *Journal of General Microbiology*, **71**, 103–16.

DÖBEREINER, J., DAY, J. M. & DART, P. J. (1973). Rhizosphere associations between grasses and nitrogen-fixing bacteria: effect of O_2 on nitrogenase activity in the rhizosphere of *Paspalum notatum*. *Soil Biology and Biochemistry*, **5**, 157–9.

DROZD, J. W. & POSTGATE, J. R. (1970). Effects of oxygen on acetylene reduction, cytochrome content and respiratory activity of *Azotobacter chroococcum*. *Journal of General Microbiology*, **63**, 63–73.

DROZD, J. W., TUBB, R. S. & POSTGATE, J. R. (1972). A chemostat study of the effect of fixed nitrogen sources on nitrogen fixation, membranes and free amino-acids in *Azotobacter chroococcum*. *Journal of General Microbiology*, **73**, 221–32.

EADY, R. R., SMITH, B. E., COOK, K. A. & POSTGATE, J. R. (1972). Nitrogenase of *Klebsiella pneumoniae*. *Biochemical Journal*, **128**, 655–75.

EVANS, H. J., CAMPBELL, N. E. R. & HILL, S. (1972). Asymbiotic nitrogen-fixing bacteria from the surfaces of nodules and roots of legumes. *Canadian Journal of Microbiology*, **18**, 13–21.

EVANS, H. J. & RUSSELL, S. A. (1971). In *The Chemistry and Biochemistry of Nitrogen Fixation*, ed. J. R. Postgate, pp. 191–244. London: Plenum Press.

FEDOROV, M. V. & KALININSKAYA, T. (1961). A new species of nitrogen-fixing *Mycobacterium* and its physiological peculiarities. *Mikrobiologiya*, **30**, 9–14.

FITCH, W. M. & MARGOLIASH, E. (1970). The evolution of cytochromes. In *Evolutionary Biology*, vol. 4, ed. T. Dobzhansky, M. K. Hecht & W. C. Steere, pp. 6–36. New York: Appleton-Century-Crofts.

GIBBINS, A. M. & GREGORY, K. F. (1972). Relatedness among *Rhizobium* and *Agrobacterium* species determined by three methods of nucleic acid hybridization. *Journal of Bacteriology*, **111**, 129–41.

GORDON, J. K. & BRILL, W. J. (1972). Mutants that produce nitrogenase in the presence of ammonia. *Proceedings of the National Academy of Sciences, USA*, **69**, 3501–3.

GRAU, F. H. & WILSON, P. W. (1942). Physiology of nitrogen fixation by *Bacillus polymyxa*. *Journal of Bacteriology*, **83**, 490–6.

GRAY, B. H., FOWLER, C. F., NUGENT, N. A. & FULLER, R. C. (1972). Re-evaluation of the presence of a midpoint potential cytochrome 551–S in the green photosynthetic bacterium *Chloropseudomonas ethylica*. *Biochemical and Biophysical Research Communications*, **47**, 322–7.

GRAY, C. T. & GEST, H. (1965). Biological formation of molecular hydrogen. *Science*, **148**, 186–92.

HARDY, R. W. F. & BURNS, R. C. (1967). Biological nitrogen fixation. *Annual Reviews of Biochemistry*, **37**, 331–58.

HARDY, R. W. F., BURNS, R. C., HEBERT, R. R., HOLSTEN, R. D. & JACKSON, E. K. (1971). Biological nitrogen fixation: a key to World protein. In *Biological Nitrogen Fixation in Natural and Agricultural Habitats*, ed. T. A. Lie & E. J. Mulder. *Plant and Soil* (special volume), pp. 561–90.

HARDY, R. W. F. & KNIGHT, E. J. (1967). Biochemistry and postulated mechanisms of nitrogen fixation. In *Progress in Phytochemistry*, ed. L. Reinhold & Y. Liwschitz, pp. 407–89. New York: Wiley.

HARRIS, D. & DART, P. J. (1973). Nitrogenase activity in the rhizosphere of *Stachys sylvatica* and some other dicotyledonous plants. *Soil Biology and Biochemistry*, **5**, 277–9.

HILL, S. (1971). Influence of oxygen concentration on the colony type of *Derxia gummosa* grown on nitrogen-free media. *Journal of General Microbiology*, **67**, 77–83.

HILL, S., DROZD, J. W. & POSTGATE, J. R. (1972). Environmental effects on the growth of nitrogen-fixing bacteria. *Journal of Applied Chemistry and Biotechnology*, **22**, 541–58.

HILL, S. & POSTGATE, J. R. (1969). Failure of putative nitrogen-fixing bacteria to fix nitrogen. *Journal of General Microbiology*, **58**, 277–85.

HOBSON, P. M., SUMMERS, R., POSTGATE, J. R. & WARE, D. A. (1973). Nitrogen fixation in the rumen of a living sheep. *Journal of General Microbiology*, **77**, 225–6.

IMSHENETSKII, A. A. (1963). In *Evolutionary Biochemistry*, ed. A. I. Oparin. *Proceedings of the 5th International Congress of Biochemistry, Moscow*, **3**, 139–48. New York: Macmillan.

JONES, C. W., BRICE, J. M., WRIGHT, V. & ACKRELL, B. A. C. (1973). Respiratory protection of nitrogenase in *Azotobacter vinelandii*. *FEBS Letters*, **29**, 77–81.

JONES, K. (1970). Nitrogen fixation in the phyllosphere of the Douglas fir, *Pseudotsuga douglasii*. *Annals of Botany*, **34**, 239–44.

KELLY, M. (1968). The kinetics of the reduction of isocyanides, acetylenes and the cyanide ion by nitrogenase preparation from *Azotobacter chroococcum* and the effects of inhibitors. *Biochemical Journal*, **107**, 1–6.

KETCHUM, P. A., COMBIER, H. Y., FRAZIER, W. A., MADANSKY, C. & NASON, A. (1970). In *vitro* assembly of *Neurospora* assimilatory nitrate reductase from protein subunits of a *Neurospora* mutant and the xanthine-oxidizing or aldehyde oxidase systems of higher animals. *Proceedings of the National Academy of Sciences, USA*, **66**, 1016–23.

KLEIN, R. M. & CRONQUIST, A. (1967). A consideration of the evolutionary and taxonomic significance of some biochemical micromorphological and physiological characters in the thallophytes. *Quarterly Review of Biology*, **42**, 108–296.

KLUCAS, R. (1972). Nitrogen fixation by *Klebsiella* grown in the presence of oxygen. *Canadian Journal of Bacteriology*, **18**, 1845–50.

KRASNA, A. I. & RITTENBERG, D. (1954). The mechanism of action of the enzyme hydrogenase. *Journal of the American Chemical Society*, **76**, 3015–20.

LEDOUX, L. (1971). *Information Molecules in Biological Systems*. Amsterdam: North-Holland.

LEES, H. & POSTGATE, J. R. (1973). The behaviour of *Azotobacter chroococcum* in oxygen- and phosphate-limited chemostat culture. *Journal of General Microbiology*, **74**, 161–6.

LEGALL, J., DERVARTANIAN, D. V., SPILKER, E., JIN-PO-LEE & PECK, H. D. (1971). Evidence for the involvement of non-heme iron in the active site of hydrogenase from *Desulfovibrio vulgaris*. *Biochimica et Biophysica Acta*, **234**, 525–30.

LEIGH, G. J. (1971). In *The Chemistry and Biochemistry of Nitrogen Fixation*, ed. J. R. Postgate, pp. 19–56. London: Plenum Press.

LEWIN, R. A. & LOUNSBERY, D. M. (1969). Isolation, cultivation and characterization of Flexibacteria. *Journal of General Microbiology*, **58**, 145–70.

LEX, M., SILVESTER, W. B. & STEWART, W. D. P. (1972). Photorespiration and nitrogenase activity in the blue-green alga, *Anabaena cylindrica*. *Proceedings of the Royal Society* B, **180**, 87–102.

LINDSTROM, E. S., TOVE, S. R. & WILSON, P. W. (1950). Nitrogen fixation by the green and purple sulfur bacteria. *Science*, **112**, 197–9.

LINE, M. A. & LOUTIT, M. W. (1971). Non-symbiotic nitrogen-fixing organisms from some New Zealand tussock-grassland soils. *Journal of General Microbiology*, **66**, 309–18.

MACKINTOSH, M. E. (1971). Nitrogen fixation by *Thiobacillus ferrooxidans* species. *Journal of General Microbiology*, **66**, i–ii.

MAHL, M. C., WILSON, P. W., FIFE, M. R. & EWING, W. K. (1965). Nitrogen fixation by members of the tribe *Klebsiellae*. *Journal of Bacteriology*, **89**, 1482–5.

MILLBANK, J. R. (1969). Nitrogen fixation in moulds and yeasts – a reappraisal. *Archiv für Mikrobiologie*, **68**, 32–9.

MILLBANK, J. R. (1970). The effect of conditions of low oxygen tension on the assay of nitrogenase in moulds and yeasts using the acetylene reduction technique. *Archiv für Mikrobiologie*, **72**, 375–7.

MISHUSTIN, E. N. & SHIL'NIKOVA, V. K. (1971). *Biological Fixation of Atmospheric Nitrogen*. London: Macmillan.

MORTENSON, L. E., ZUMFT, W. G., HUANG, T. C. & PALMER, G. (1973). The structure and function of nitrogenase of *Clostridium pasteurianum*: electron paramagnetic studies. *Biochemical Society Transactions*, **1**, 35–7.

MURPHY, P. M. & KOCH, B. L. (1971). Compatability of the components of nitrogenase from soybean nodules and free-living nitrogen-fixing bacteria. *Biochimica et Biophysica Acta*, **253**, 295–7.

NAKOS, G. & MORTENSON, L. E. (1971*a*). Structural properties of hydrogenase from *Clostridium pasteurianum* W5. *Biochemistry*, **10**, 2442–9.

NAKOS, G. & MORTENSON, L. E. (1971*b*). Purification and properties of hydrogenase, an iron–sulfur protein, from *Clostridium pasteurianum* W5. *Biochimica et Biophysica Acta*, **227**, 576–83.

NASON, A., LEE, K., PAN, S., KETCHUM, P. A., LAMBERTI, A. & DE VRIES, J. (1971). *In vitro* formation of assimilatory reduced nicotinamide adenine dinucleotide phosphate: nitrate reductase from a *Neurospora* mutant and a component of molybdenum enzymes. *Proceedings of the National Academy of Sciences, USA*, **68**, 3242–6.

NORRIS, D. O. (1956). Legumes and the rhizobium symbioses. *Empire Journal of Experimental Agriculture*, **24**, 247–70.

OPPENHEIM, J. & MARCUS, L. (1970). Correlation of ultrastructure in *Azotobacter vinelandii* with nitrogen source for growth. *Journal of Bacteriology*, **101**, 286–91.

PARKER, C. A. (1957). Evolution of nitrogen-fixing symbiosis in higher plants. *Nature, London*, **179**, 593–4.

PARKER, C. A. (1968). On the evolution of symbiosis in legumes. *Festskrift til Hans Lauritz Jensen*, pp. 107–16. Lamvig, Denmark: Gradgaard Nielsens.

PARKER, C. A. & SCUTT, P. B. (1960). The effect of oxygen on nitrogen fixation by *Azotobacter*. *Biochimica et Biophysica Acta*, **38**, 230–8.

PECK, H. D. (1966/7). Some evolutionary aspects of inorganic sulphur metabolism. *Lectures on Theoretical and Applied Aspects of Modern Microbiology*. University Park: University of Maryland Press.

PINE, M. J. & BARKER, H. A. (1964). Studies in the methane bacteria. XI. Fixation of atmospheric nitrogen by *Methanobacterium omelianskii*. *Journal of Bacteriology*, **68**, 589–91.

POSTGATE, J. R. (1970). Nitrogen fixation by sporulating sulphate-reducing bacteria including rumen strains. *Journal of General Microbiology*, **63**, 137–9.

POSTGATE, J. R. (1971*a*). *The Chemistry and Biochemistry of Nitrogen Fixation*. London: Plenum Press.

POSTGATE, J. R. (1971*b*). Relevant aspects of the physiological chemistry of nitrogen fixation. In *Microbes and Biological Productivity*, *Symposia of the Society for General Microbiology*, **21**, 297–307. Ed. D. E. Hughes & A. H. Rose. London: Cambridge University Press.

POSTGATE, J. R. (1972). *Biological Nitrogen Fixation*. Watford: Merrow Publishing Co.

POSTGATE, J. R. (1974). Prerequisites of biological nitrogen fixation in free-living organisms. In *Biological Nitrogen Fixation*, ed. A. Quispel. Amsterdam: North-Holland.

QUISPEL, A. (1974). *Biological Nitrogen Fixation*. Amsterdam: North-Holland.

RAJU, P. N., EVANS, H. J. & SEIDLER, R. J. (1972). An asymbiotic nitrogen-fixing bacterium from the root environment of corn. *Proceedings of the National Academy of Sciences, USA*, **69**, 3474–8.

RIEDERER-HENDERSON, M. & WILSON, P. W. (1970). Nitrogen fixation by sulphate-reducing bacteria. *Journal of General Microbiology*, **61**, 27–31.

RINAUDO, G., BALANDREAU, J. & DOMMERGUES, Y. (1971). Algal and bacterial non-symbiotic nitrogen fixation in paddy soils. In *Biological Nitrogen Fixation in Natural and Agricultural Habitats*, ed. T. A. Lie & E. G. Mulder. *Plant and Soil* (special volume), pp. 471–9.

RUTTEN, M. G. (1971). *The Origin of Life by Natural Causes*. Amsterdam: Elsevier.

RUINEN, J. (1956). The phyllosphere. III. Nitrogen fixation in the phyllosphere. *Plant and Soil*, **22**, 375–94.

SENIOR, P. J., BEECH, G. A., RITCHIE, G. A. F. & DAWES, E. A. (1972). The role of oxygen limitation in the formation of poly-β-hydroxybutyrate during batch and continuous culture of *Azotobacter beijerinckii*. *Biochemical Journal*, **128**, 1193–201.

SHETHNA, Y. I., STOMBAUGH, N. A. & BURRIS, R. H. (1971). Ferredoxin from *Bacillus polymyxa*. *Biochemical and Biophysical Research Communications*, **42**, 1108–16.

SILVER, W. S. & POSTGATE, J. R. (1973). Evolution of asymbiotic nitrogen fixation. *Journal of Theoretical Biology*, **40**, 1–10.

STEWART, W. D. P. (1971). Physiological studies on nitrogen-fixing blue-green algae. In *Biological Nitrogen Fixation in Natural and Agricultural Habitats*, ed. T. A. Lie & E. G. Mulder. *Plant and Soil* (special volume), pp. 377–91.

STREICHER, S., GURNEY, E. & VALENTINE, R. C. (1971). Transduction of the nitrogen-fixation genes in *Klebsiella pneumoniae*. *Proceedings of the National Academy of Sciences, USA*, **68**, 1174–7.

STREICHER, S., GURNEY, E. & VALENTINE, R. C. (1972). The nitrogen-fixation genes. *Nature, London*, **239**, 495–9.

WATANABE, A. & YAMAMOTO, M. (1971). Algal nitrogen fixation in the tropics. In *Biological Nitrogen Fixation in Natural and Agricultural Habitats*, ed. T. A. Lie & E. G. Mulder. *Plant and Soil* (special volume), pp. 403–13.

WHITTENBURY, R., PHILLIPS, K. C. & WILKINSON, J. F. (1970). Enrichment, isolation and some properties of methane-utilizing bacteria. *Journal of General Microbiology*, **61**, 205–18.

WILSON, P. W. (1958). Asymbiotic nitrogen fixation. In *Encyclopedia of Plant Physiology*, ed. W. Ruhland, vol. 8, pp. 9–47. Berlin: Springer-Verlag.

WITTMAN, W., BERGERSEN, F. J. & KENNEDY, G. S. (1965). The coralloid roots of *Macrozamia communis* L. Johnson. *Australian Journal of Biological Science*, **18**, 1129–42.

WITZ, D. F., DETROY, R. W. & WILSON, P. W. (1967). Nitrogen fixation by growing cells and cell-free extracts of the Bacillaceae. *Archiv für Mikrobiologie*, **55**, 369–81.

YATES, M. G. (1972). Electron transport to nitrogenase in *Azotobacter chroococcum*: azotobacter flavodoxin hydroquinone as an electron donor. *FEBS Letters*, **27**, 63–7.

ZAKHVATEVA, B. V., MALAFEEVA, I. V. & KONDRAT'EVA, E. N. (1970). Nitrogen fixation capacity of photosynthesizing bacteria. *Mikrobiologiya*, **39**, 761–6.

EVOLUTION IN VIRUSES

W. K. JOKLIK

Department of Microbiology and Immunology, Duke University Medical Center, Durham, North Carolina 27710, *USA*

INTRODUCTION

In attempting to define the evolutionary history of viruses, scientists are confronted with a problem similar to that facing a tracker who finds a series of scattered ill-defined footprints: were they all made by the animal he is tracking, and can anything be deduced from them concerning the animal's habits? So with viruses: where do they come from, and where are they going? These are questions which are difficult to answer, since there is no palaeontological evidence to guide us.

Three main possibilities have been considered in speculations concerning the origin of viruses and their relationship to other forms of life. First, it is conceivable that viruses are the surviving descendants of primitive precellular forms of life, which, when higher forms became established, failed to survive except in the protected environment provided by the adoption of a strictly parasitic form of life. Second, viruses have been held to represent the degenerate descendants of larger pathogenic micro-organisms. The development of parasitism in many forms is a frequently occurring event in evolution. Loss of the ability to synthesize substances that cannot pass across cell membranes would, of necessity, lead to intracellular parasitism, which could ultimately proceed so far as to reduce the genetic apparatus of the parasite to a small fraction of its former self. The third possibility is that viruses took their origin from subcellular elements which escaped from normal restraints. There is no lack of organelles which might have served as precursors: mitochondria, chloroplasts, kinetoplasts, plasmids, chromosomes, genes, RNA–protein complexes, and others.

The evidence in favor of each of these hypotheses has been considered repeatedly, most recently by Luria & Darnell (1967), and will not be reviewed again here: suffice it to say that it currently appears likely that viruses originated from subcellular organelles. The main purpose of this review is not to consider how viruses originated, but how they evolved. More specifically, we will address ourselves to providing answers to questions such as:

1. Are viruses likely to have arisen as a unique event in some primordial cell, or could viruses have arisen repeatedly in different, higher cells?

2. Is there evidence that any of the viruses extant today are linear descendants of each other?

3. Are larger viruses likely to be ancestors of smaller viruses?

4. Are viruses of eukaryotes descendants of viruses of prokaryotes?

5. Are those viruses of higher plants, insects and vertebrates that share a common morphology likely to have had a common origin?

6. What are the likely evolutionary histories of viruses possessing double-stranded RNA and single-stranded DNA?

7. Do viruses possess features that can be regarded as evidence of progression along an evolutionary pathway?

8. Are viruses that are capable of integrating their genomes into that of the host cell precursors of viruses that cannot?

The answers to these and other questions cannot be discerned by looking into the past, nor does virus evolution proceed sufficiently rapidly today for any insight to be gained by direct experimentation; rather one must turn to comparative virology, that is, to an examination of the nature of viruses and of their interactions with their hosts. We will therefore first review in some detail the two properties of viruses that are most likely to provide the most definitive indicators of common evolutionary history, namely the structure of the viral genome and the structure of the viral capsid. We will then consider both of these in relation to viral specificity and host range, and finally apply the knowledge currently available to providing answers to the questions listed above.

THE STRUCTURE OF VIRAL GENOMES

Type of nucleic acid

Until recently, viruses seemed to belong to only two classes: those that contained a single molecule of double-stranded DNA in their virions, such as the poxviruses, herpesviruses, adenoviruses, papovaviruses, certain insect viruses and many of the bacteriophages, and those that contained a single molecule of single-stranded RNA, such as the picornaviruses, togaviruses, many plant viruses and some bacteriophages. Recently, however, numerous other forms of virion nucleic acids have come to light. First, certain bacteriophages, the mammalian parvoviruses and the insect densonucleosis virus (Kurstak, 1972) contain single-stranded rather than double-stranded DNA. In the case of the bacteriophages, which include two groups of viruses, one with icosahedral, the other with filamentous (helical) nucleocapsids, the DNA is circular and plus-stranded (the same polarity as the messenger RNA); in the case of the parvoviruses the DNA is linear and also plus-stranded in all cases except AAV (adeno-associated virus), which consists of equal amounts of two kinds of morphologically identical particles, one with plus-stranded, the other with minus-stranded DNA. This is also true for densonucleosis virus.

Second, a group of viruses has been characterized which, although containing RNA, and RNA capable of acting as mRNA at that, nevertheless do not replicate it like other RNA viruses. Instead, these viruses, which include primarily the RNA tumor viruses, appear to possess a mechanism for transcribing their RNA into double-stranded DNA which then becomes integrated into the host cell genome, and progeny virion RNA seems to be formed by transcription of this integrated DNA. The genomes of these viruses would thus be DNA integrated into the host genome and never excised, while their virion RNA would be the mRNA corresponding to this integrated DNA.

Third, there are several groups of viruses which contain not plus-stranded, but minus-stranded RNA. These RNAs cannot act as messenger RNA in the infected cell; they must first be transcribed into RNA of the opposite polarity, which is accomplished by an RNA polymerase present within the virion. The viruses which conform to this pattern are the mammalian myxoviruses and rhabdoviruses, a plant virus group of which sowthistle yellow virus, potato yellow dwarf virus, lettuce necrotic yellows virus and wheat striate mosaic virus are the best known representations, and perhaps tomato spotted wilt virus. As will be seen below, these viruses share not only possession of minus-stranded RNA, but also morphological features.

Finally, several viruses contain genomes consisting of double-stranded rather than single-stranded RNA; and some of these viruses have also been shown to contain an RNA polymerase so that their genomes may be transcribed into messenger RNA. These viruses belong to two groups: the reoviridae, a family which includes vertebrate, insect and plant viruses, and a group of viruses found in *Neurospora*, *Penicillium* and *Aspergillus* species.

Segmentation

Many viral genomes possess a structure which was unsuspected until recently but which is now known to be shared by genomes which are otherwise quite unrelated: this is segmentation. Among the viruses which possess segmented genomes are:

(i) *The RNA tumor viruses*

Their RNA is a 70 S hydrogen-bonded complex of several subunits which fall into three classes (Robinson, Pitkanen & Rubin, 1965;

Duesberg, 1968; Faras *et al.* 1973): (1) three to four large subunits each of which comprises some 10000 nucleotides; (2) ten to twenty subunits which are similar in size to tRNAs, only dissociate at temperatures exceeding 80 °C and apparently serve as primers for the transcription of the large subunits into DNA; and (3) additional small subunits ranging from 4 to 7 S (70 to 200 nucleotides) which dissociate at lower temperatures and are therefore less firmly hydrogen-bonded. Their function is not known. RNA tumor virions also contain free molecules very similar to or identical with those of class 3. Among this population of free molecules, as well as among those in classes 2 and 3, are some which are tRNAs as indicated by their ability to accept amino acids (Erikson & Erikson, 1970). Their presence raises fascinating questions concerning their evolutionary history (see below).

The polypeptide coding information of RNA tumor virus RNA presumably resides exclusively in the large subunits. There is some doubt concerning how many types of these subunits each virion contains. It has proved impossible so far to resolve the 3–4 subunits in sarcoma viruses (transforming viruses) either in density gradients or in polyacrylamide gels; nor has it been possible to resolve the corresponding subunits in leukemia viruses (non-transforming viruses) which however are significantly smaller than those in sarcoma viruses (Duesberg & Vogt, 1973). Not only therefore are the large subunits of RNA tumor viruses very similar in size, but there is also genetic evidence compatible with the view that they are all identical; and it has been suggested that RNA tumor viruses may in fact be polyploid (Vogt, 1973). However, it has also been argued that a single subunit may not contain sufficient genetic information for both virus replication and cell transformation, and preliminary estimates of the genetic complexity of RNA tumor virus RNA by reassociation kinetics analysis have suggested a size closer to 30000 than to 10000 nucleotides (Bishop *et al.* 1973).

Nothing is known as yet concerning the physical state of the double-stranded virus-specific DNA which is postulated to be incorporated into the host genome: whether it corresponds to only one kind of RNA subunit or to several, and if the latter, whether the corresponding segments are linked in tandem, whether they are separated by spacers, or whether they are widely distributed throughout the genome.

(ii) *The Visna virus group*

Their RNA appears to have a structure rather similar to that of the RNA tumor viruses.

(iii) *The influenza viruses*

They contain either five or six segments of single-stranded RNA, each encased within its own helical nucleocapsid.

(iv) *The reoviridae*

They contain either ten or twelve segments of double-stranded RNA (Shatkin, Sipe & Loh, 1968). It is with them that the clearest insight has been gained concerning the nature of the genetic information encoded in each segment; such insight is, of course, a prerequisite for understanding why the evolution of all these viruses has proceeded along the pathway of segmentation. As pointed out above, the nature of the genetic information encoded in the large subunits of RNA tumor virus genomes is not yet known; and while certain influenza RNA segments appear to be just the right size for coding some of the influenza capsid polypeptides, there is as yet no experimental proof that they actually do so. In the case of reovirus, however, it has been shown that all ten segments of double-stranded RNA are completely transcribed into plus-stranded RNA molecules which are translated both *in vitro* and *in vivo* into all the capsid polypeptides and into two non-capsid polypeptides; eight of the ten possible polypeptides have been definitely identified, and the remaining two have been noticed repeatedly. It seems clear therefore that the reovirus genome is segmented into ten discrete cistrons (or chromosomes) (Joklik, 1973).

(v) *Certain fungal viruses*

Virus-like particles have been isolated from certain species of *Penicillium* and *Aspergillus* which contain several, generally three, size classes of double-stranded RNA molecules; strangely enough, some also contain single-stranded RNA (Bozarth, Wood & Mandelbrot, 1971).

(vi) *Certain plant viruses*

At least three groups of plant viruses contain segmented genomes: those of which alfalfa mosaic virus is the prototype (four segments), those of which tobacco rattle virus is the prototype (two segments, the smaller being variable in size, depending on the strain), and a group of more or less closely related viruses of which tobacco ringspot virus, cowpea mosaic virus, brome mosaic virus, pea enation mosaic virus and cowpea chlorotic mottle mosaic virus are the prototypes (two or three segments) (van Kammen, 1972).

Viruses of groups v and vi are of great interest since their genome segments are not encapsidated together but individually; these viruses therefore consist of several different types of virus particles, all with the same capsid polypeptide composition, but each containing a different genome segment. Production of viral progeny is therefore contingent upon simultaneous infection with at least one of each type of particle.

The principal advantage gained from genome segmentation is the ease with which genomes of new constitution can arise; for this can then proceed not merely by recombination involving breakage and reformation of covalent bonds, but by very simple genome segment reassortment. A virus which seems to have taken advantage of this mechanism is human influenza virus which has been shown to give rise to new virus strains by exchanging some of its own genome segments with those of equine, porcine, and avian strains; and one of the theories of the origin of the new virus strains responsible for the recurrent human pandemics of influenza is that they arise via genome segment reassortment between human and animal virus strains (Laver & Webster, 1973; Skehel, this volume).

The viruses which possess individually encapsidated genome segments pose an interesting problem in evolution. On the one hand it is conceivable that all segments were originally encapsidated in one large capsid composed of ancestral versions of the present-day capsid polypeptides (which permitted the construction of larger capsids); on the other hand it is also conceivable that the present-day multiple particle systems evolved from two or more different viruses each of which gradually discarded a portion of its own genome while taking advantage of the corresponding portion of another virus which performed more efficiently. The controlling factors in either case would most probably reside in the specificities involved in encapsidation. Unfortunately very little is yet known concerning them, in spite of the fascinating work of Klug and his colleagues with TMV. There is no question that RNA–protein interactions are involved; but numerous examples of phenotypic mixing and phenotypic masking are known which suggest that only limited regions of absolute specificity are required for those interactions between nucleic acids and proteins which are required for encapsidation, and that similarity of overall size and shape may sometimes be sufficient.

The size of viral genomes

The sizes of viral nucleic acids cannot fail but be of great evolutionary significance, for it is inconceivable that they should not have changed greatly since the emergence of each virus as a separate entity.

The size of viral nucleic acids varies enormously, over a 100-fold range and even more if the recently discovered viroids are considered. The largest genomes are those of poxviruses; their molecular weight is 160–200 million, enough to code for 80000–100000 amino acids (160–200 proteins of 500 amino acids). Poxviruses are also the most structurally complex viruses; some thirty to forty proteins have been identified within their virions (Sarov & Joklik, 1972), but there may be many more, since they comprise some 3×10^7 amino acids, that is, about 10^5 protein molecules, and present techniques do not permit detection of viral capsid polypeptides present to the extent of less than 0.1 % of the total (still 100 protein molecules per poxvirus particle). It is therefore not possible to estimate just how complex the poxvirus particle really is, and what proportion of the poxvirus genome codes for structural polypeptides and for the apparatus necessary to assemble them into virions. Structurally poxviruses possess features both of smaller viruses as well as of organisms more complex than viruses. The inner portion of the poxvirus particle, the core, which possesses a coat made up of regularly arranged subunits, is constructed much like other viruses; but the outer regions are so complex that it has repeatedly been suggested that if any viruses represent the reductive form of some larger and more complex intracellular parasite, it would be the poxviruses (see below).

(i) *Economizing viral genetic information by having capsid polypeptides assume regulatory functions*

Consideration of the evolutionary significance of viral genome size is not as profitable at the upper end of the scale as at the lower; for here it is fascinating to see how viruses have adapted to shortage of genetic information storage capacity. There is, for example, the problem of using the minimum amount of genetic information for ensuring the orderly progression of the viral growth cycle. Recent work has provided extensive insight into how this problem is handled by small RNA phages like Qβ and is beginning to provide clues concerning how mammalian picornaviruses cope with it. In both cases parental RNA must serve not only as the messenger for protein synthesis, but also as the template for the synthesis of the minus strand; further, the viruses presumably need far more capsid polypeptide molecules than polypeptides that serve enzymic or regulatory functions. The problems are therefore to ensure that parental RNA can be traversed both in the 5′- to 3′-direction by ribosomes and in the 3′- to 5′-direction by replicase, and to prevent the wasteful synthesis of large amounts of unneeded polypeptide molecules. In the case of Qβ the genome is polycistronic: it is divided into three cistrons (from the 5′-end, the *A*-protein, the coat protein, and the replicase cistron), each with its own ribosome attachment site. Its secondary structure is such that by far the most efficient of these sites is that at the beginning of the coat protein cistron, which lies in the middle. Attachment of ribosomes there renders the site at the beginning of the replicase cistron also accessible for ribosome binding, so that during the initial stages of infection coat protein and replicase (i.e. the phage-specified replicase component) are formed. The entire replicase then binds not only to the 3′-end of the RNA but also to the ribosome attachment site of the coat protein cistron, thereby preventing ribosomes from binding there; and it then transcribes the viral genome once ribosomes (which travel in the opposite direction to the replicase) have been cleared. Formation of progeny genomes then ensues, and during the later stages of infection cycle these are translated principally into coat protein, which itself inhibits ribosome attachment at the *A*-protein and replicase cistron ribosome attachment sites (Kolakofsky & Weissmann, 1971; Weber *et al.* 1972). Qβ therefore solves the two problems by using the polypeptides coded by it, not only for structural and catalytic but also for regulatory purposes, that is, by assigning them dual functions.

Encephalomyocarditis virus (EMC), poliovirus and the Coxsackie viruses, the prototypes of the mammalian picornaviruses, handle the problem differently. There appears to be only one ribosome attachment site (Öberg & Shatkin, 1972), and the entire RNA is translated into one very large polypeptide, the polyprotein, with a molecular weight of 250000–300000, which can be cleaved in nine locations by as yet unidentified proteases (Summers & Maizel, 1968; Jacobson & Baltimore, 1968; Butterworth, Hall, Stoltzfus & Rueckert, 1971). The polyprotein is not cleaved immediately at all these sites, but cleavage proceeds in a defined sequence and results finally in the formation of the four capsid polypeptides, the replicase, and several polypeptides the function of which is not yet known. It is not clear yet how the problem of causing capsid polypeptides to be formed in greatest amounts is solved, or indeed, whether it is solved at all, since there is evidence that the polyprotein is formed even at late stages of the infection cycle. However it may be significant that the code for the capsid polypeptides is located adjacent to the 5′-terminus of the viral genome, and certain intermediate cleavage products of the polyprotein, one of which has been suggested as a regulator of viral RNA–ribosome affinity (the 'equestron', Cooper, Steiner-Pryor & Wright, 1973), may also serve as termination factors at specific locations during the later stages of the infection cycle, as the polypeptides coded by the more distal regions of the RNA are no longer required.

The assumption of regulatory functions by viral capsid polypeptides is not the only example of such polypeptides possessing dual functions. Another example is provided by the assumption of catalytic functions, as in the case of the various virion RNA polymerases. This is discussed further below.

(ii) *Economizing viral genetic information by utilizing host proteins*

Another way in which viruses have responded to limited genetic information storage capacity is by using host proteins for their own purposes. Perhaps the best known example is that of the Qβ replicase. This enzyme is a large molecule which consists of four different polypeptide chains, only one of which is virus-coded; the other three are elongation factor *T*s, elongation factor *Tu* and a third host-coded protein, factor *i*, which is responsible for replicase binding to the coat protein cistron ribosome attachment site, and the function of which in uninfected cells has not yet been identified (Blumenthal, Landers & Weber, 1972; Groner *et al.* 1972).

Another example is provided by the protease(s) used to cleave the picornavirus polyprotein. If these enzymes are cellular proteases, and all indications so far are that they are, then they and the polyprotein must obviously be exquisitely attuned to each other. The polyprotein contains nine bonds which can be cleaved; but these bonds are not all cleaved simultaneously but sequentially, as if they became accessible to the enzymes one by one in some well defined sequence. This would be understandable if the bonds in question were of a type uniquely susceptible to hydrolysis, and if they became exposed in turn in different breakdown products. There is a precedent for this type of sequential exposure of susceptible bonds: it is known that the three polypeptides which together make up the outer capsid shell of reovirus are degraded by chymotrypsin in a strictly sequential manner, each being degraded via several intermediates that possess a sufficiently long half-life for them to be characterized with respect to size (Joklik, 1972). Further, it is of interest that when reovirus infects cells, its outer capsid shell is partially degraded by cellular enzymes via a pathway which is exactly the same, as far as one can tell, as that catalyzed by chymotrypsin *in vitro*.

The highly specialized cleavage of picornavirus polyprotein by cellular enzymes has an interesting parallel in the highly specialized cleavage of T7 messenger RNA in infected *E. coli*. The early mRNA region of

T7 DNA is transcribed into one RNA chain which is subsequently cleaved by a host enzyme, an endonuclease termed the sizing factor, into five segments which correspond to the individual monocistronic mRNAs (Dunn & Studier, 1973). Once again the use of a host enzyme for such a highly specific task indicates an extraordinarily high degree of adaptation of the virus to its host.

Yet a further example is provided by the papovaviruses. It is clear that the replication of their DNA proceeds via a complex series of reactions which includes (i) initiation of DNA synthesis at a specific site, (ii) bidirectional DNA replication, (iii) a discontinuous mechanism of chain growth with a possible requirement for two polymerases, one required for the synthesis of 4 S fragments and the second involved in filling in the gaps in growing chains, and (iv) a mechanism which allows for nicking and sealing of parental strands to permit DNA replication and yet preserve a covalently closed structure in the template strand of the replicating molecules. Yet the genetic information capacity of SV40 DNA is so limited that the only function involved in viral DNA replication which is coded by the viral genome appears to be that required for initiation of DNA synthesis. It has been suggested that this single function may have been preserved in a way that permits the viral genome to compete effectively with cellular DNA for the DNA replication machinery (Salzman & Thoren, 1973).

(iii) *Economizing viral genetic information by discarding information that can be supplied by other viruses*

The third consequence of reducing genome size may be the development of defectiveness and the resulting necessity to rely on helper viruses for the exercise of essential functions. Numerous examples of this have been discovered. The best known defective viruses are the mammalian sarcoma viruses, adeno-associated virus (AAV) and the plant satellite viruses. The mammalian sarcoma viruses can transform cells, but can almost never multiply unless they are simultaneously infected with leukemia viruses; the function that the leukemia viruses supply appears to be the synthesis of the envelope. Adeno-associated viruses, on the other hand, have been restricted to multiplying in one particular type of host cell, namely one simultaneously infected with an adenovirus. The nature of the essential function which is supplied by adenovirus is not known with certainty: however, it is clear that cells abortively infected with adenoviruses can become permissive for AAV, and present indications are that it is an early adenovirus function which is required, possibly stimulation of host DNA synthesis. Another example is provided by the tobacco necrosis satellite virus, the nucleic acid of which contains the code for its coat protein but not for a replicase; it can therefore not multiply on its own, but only in cells infected with tobacco necrosis virus. The possible evolutionary relationships between these defective viruses and their helpers will be discussed below.

The ultimate in genetic information loss appears to have resulted in viral forms which have lost their ability to synthesize capsid protein and therefore exist only in the form of naked nucleic acid. These are the viroids, which consist solely of RNA molecules some 200 nucleotides long, which are probably incapable of coding for polypeptide (Diener, 1972). The nature of the enzyme which replicates them is presently a mystery. Viroids have so far been implicated as the infectious agents of potato spindle tuber, citrus exocortis and chrysanthemum stunt disease. There is some evidence that the first two are caused by the same viroid.

(iv) *The advantages and disadvantages of discarding genetic information*

Before leaving the subject of discarding genetic information, it should be pointed out that a variant with a reduced genome size will only become the dominant population component if it has a survival advantage over the larger parent. The smaller genome may have an advantage because it takes less time to replicate and be translated, uses fewer precursors and less energy, and provides less opportunity for miscoding both during transcription and translation; but this may not be sufficient to offset the loss of the function which is eliminated. To take some examples: the T-even bacteriophages code for numerous enzymes which are also present in every cell which they infect; yet the code for these enzymes has been conserved throughout their evolutionary history, presumably because phages which can code for them have a survival advantage over those that do not. Another example is provided by the enzyme thymidine kinase which is coded by herpesviruses and poxviruses. The function of thymidine kinase in mammalian cells is not known; thymidine is not on the normal pathway of TTP biosynthesis. Since its level increases markedly at the beginning of the S phase, it is generally assumed that the enzyme has some regulatory function. The amount of virus-coded thymidine kinase formed under conditions of single infection (which is the normal mode of infection under natural conditions) must be extremely small compared with the amount present in the host cell; furthermore, it is known that mutants of vaccinia virus and herpesvirus unable to code for thymidine kinase (TK^- mutants) can grow well in cellular mutants which are unable to synthesize thymidine kinase (TK^- cells) (Dubbs & Kit, 1964*a*, *b*). Thymidine kinase is thus apparently not at all essential for vaccinia virus and herpesvirus multiplication; yet the viruses code for it. The most plausible explanation is that the yield of TK^+ virus is very slightly larger than that of TK^- virus; sufficiently larger, that is, for TK^+ virus to continue to remain the dominant population component.

Whereas there is little doubt that amino acid coding information has been repeatedly discarded from viral genomes during the course of evolution, it is less certain that portions of viral genomes which do not code for amino acids can be discarded; for they seem to be much more essential. One example is provided by the genome of adenoviruses. Adenovirus DNA exists in virions in the form of a linear molecule which is not circularly permuted. It is however terminally redundant to the extent of some 2–4 %, and the redundancy is of a remarkable kind, since the direction of the repeated sequence is different at the two ends. The result is that on denaturation and renaturation adenovirus DNA forms circles with short double-stranded tails (Wolfson & Dressler, 1972). The purpose of this odd sort of reverse redundancy is not known; however, the fact that it has been conserved clearly indicates that without it the virus could not survive. A second example is provided by the RNA-containing bacteriophages f2, MS2 and R17. The RNA of these phages, as pointed out above, consists of three cistrons, each of which is preceded by a region some 50–150 nucleotides long. These regions are generally assumed to possess regulatory functions such as controlling affinity for ribosomes and frequency of translation; indeed, as pointed out above, the regions preceding the coat protein and replicase cistrons can apparently combine with both coat protein and replicase which greatly decreases their ability to combine with ribosomes and therefore the frequency with which these cistrons are translated. The RNA of this family of phages has now been examined with respect to base sequence: far more variation in sequence has so far been detected in the coding portions of these RNAs than in the non-translated regions preceding them (Robertson & Jeppesen, 1972). It seems that the information in them is essential in almost every detail in order for these phage to be able to replicate.

Sequences in viral genomes which can combine with amino acids

Finally, attention should be drawn to the possible significance of the presence in virions of RNA capable of combining with amino acids. It has already been pointed out that at least some of the primer molecules for DNA synthesis in RNA tumor viruses have amino acid-acceptor activity and that these molecules appear to be so tightly hydrogen-bonded to the larger subunits that small regions of the latter can be regarded as essentially being their anti-strands. It is not yet known what codes for these primer molecules. If they are host-coded tRNAs, it would be a remarkable example of viruses using small host-specified RNA molecules for their own purposes. The advantage of using host molecules in this instance may not be so much the conservation of genetic information as avoidance of the necessity of carrying both the code and the anticode of these primer molecules in the same RNA molecule. As for the apparent ability of at least some primer molecules to accept amino acids, this property may be irrelevant because the reason for using tRNAs may be their abundance in the cell. However, this does not explain why other 4 S RNA molecules are also associated with the large subunits of RNA tumor virus RNA, molecules which are hydrogen-bonded less firmly (since they melt off at lower temperatures) than those that prime; nor does it explan why still more 4 S RNA molecules which are not associated with the 70 S RNA complex at all are also present in RNA tumor virions. Neither of these two classes of molecules are random populations of cellular tRNA molecules, but both seem to be selected by the virus.

The second example of virion RNA molecules capable of combining with amino acids occurs among the plant viruses. TYMV RNA can be charged with valine (Yot *et al.* 1970) and TMV RNA with histidine (Öberg & Philipson, 1972); further TMV RNA is cleaved by a tobacco endonuclease into a series of fragments about 55 nucleotides long, which can be charged with serine and methionine (Sela, 1972). At least one of these RNA–AA complexes (TYMV–RNA–valine) can donate its amino acid in protein synthesizing reactions, possibly after conversion to a 4.5 S fragment (Haenni, Prochiantz, Bernard & Chapeville, 1973). The significance of viral genomes possessing 3′-termini capable of acting as amino acid acceptors and donors, and of cleavage products of viral genomes behaving similarly, is not clear at present; but this ability is obviously relevant to the consideration of their evolutionary history.

THE STRUCTURE OF VIRAL CAPSIDS

Capsid structure is another fundamental virus property. As we shall see below, there are many viruses of vertebrates, insects and plants, and even of protozoa and fungi, which exhibit no cross hybridization whatsoever among their genomes, no serological cross-reactions, and entirely different effects on their hosts ranging from lytic to transforming or abortive infection, very severe to scarcely detectable cytopathogenicity, and almost universally fatal disease to inapparent infection; yet their morphology is not infinitely variable, but always (or almost always) conforms closely to one of a very limited number of patterns. Capsid structure, the fitting together of capsid components, is thus, together with nucleic acid structure, a viral property that has been strongly conserved during evolution.

The capsids in which viral genomes are enclosed differ enormously at first glance, yet they conform essentially to two basic patterns: in one, the nucleic acid is extended, in the other it is condensed. In the former the nucleic acid is surrounded by protein molecules arranged helically so as to yield structures with a single rotational axis (helical nucleocapsids); these either exist in the free state or they are coiled more or less tightly inside envelopes. In the other pattern the condensed nucleic acid forms the central portion of a quasispherical nucleocapsid which consists of a shell of protein molecules clustered into groups of two to ten. These clusters form morphological units, termed capsomers, which can be seen with the electron microscope, vary in size and shape from virus to virus, and are arranged extremely precisely according to icosahedral patterns characterized by 5:3:2 rotational symmetry. No other symmetry pattern has been observed among viruses.

These two basic patterns are clearly those which have evolved as being the most efficient for transferring nucleic acids from cell to cell. At one time it was thought that the protein components of virus particles served solely a structural function, but clearly this is not the case. Virus particles often contain proteins which have some other specific function (apart from the regulatory functions during virus replication which have already been referred to above). These proteins fall into two classes: those which are also obviously structural components, and those which are not. The distinction between these two classes is not always clear, but it nevertheless appears worthwhile making. Examples of proteins in the latter category may be the various nucleases found in vaccinia virus and RNA tumor virus particles, and the ligase present in the latter. As for examples of proteins in the first class, there are several which are worth considering in some detail, for they are likely to have arisen during the course of the evolution of viruses in response to some peculiarity of their structure or of their interaction with cells. First, some structural polypeptides are also RNA polymerases. The clearest example of this type is provided by the transcriptase of reoviridae; others that may be in this category are the RNA polymerases of myxoviruses, rhabdoviruses, and poxviruses. Second, some structural polypeptides are hemagglutinins. On enveloped viruses the hemagglutinins are generally present in the form of spikes made up of glycoproteins which are anchored in the lipid bilayer of the viral envelope. However, it is by no means necessary that hemagglutinins be glycoproteins, since many icosahedral viruses which contain no glycoprotein also hemagglutinate. The significance of the ability to hemagglutinate is not clear. One hypothesis is that viruses agglutinate red blood cells because the receptor sites on them are the same as or similar to the acceptor sites on susceptible cells, and that ability to hemagglutinate therefore indicates strong interaction with cellular receptors. One group of viruses, the myxoviruses, possesses not only a hemagglutinin but also an enzyme, the neuraminidase, which removes terminal neuraminic acid groups from polysaccharides. The reason why only this group of viruses possesses a neuraminidase may be that myxoviruses have a predeliction for multiplying in cells of the respiratory tract which produce large amounts of mucopolysaccharides, and the enzyme may be necessary to ensure efficient liberation of these viruses. Possession of the enzyme would thus be a remarkable example of adaptation.

The second example is even more remarkable; it is the extraordinarily complex mechanisms that many bacteriophages have evolved for injecting their DNA into host cells. In its most highly developed form this mechanism consists of some ten to twenty polypeptides with specialized functions: polypeptides which form fibers used for establishing specific contact with the bacterial surface; spikes which anchor the phage tail down firmly; tail plates to which the spikes are attached; tail sheaths capable of contracting when activated by an enzyme system; tail cores which are caused to penetrate through the bacterial cell wall when the sheath contracts; and tail collars through which tail cores are attached to phage heads. The elaboration of this mechanism often occupies up to one-half of the total information content of the phage genome. Not all phages have evolved a mechanism of such complexity; but all phages except the very smallest ones have evolved tails, and while these often lack the contractile sheaths, they almost always possess fibers, knobs or some other appendage designed to make specific contact with their host and facilitate injection of their DNA.

HOST RANGE

Host range often provides clues concerning evolutionary paths and relationships; there are many examples of viruses with strikingly similar and characteristic morphology infecting hosts belonging to widely different phyla of the animal and plant kingdoms. The best example is provided by the group comprising wound tumor virus, rice dwarf virus and maize rough dwarf virus which multiply in both plants and insects, cytoplasmic polyhedrosis virus which multiplies only in insects, the orbiviruses (prototypes Bluetongue virus, African horse sickness virus and Colorado tick fever virus) which multiply in both insects and vertebrates, and the vertebrate reoviruses. All these viruses resemble each other closely in morphology; they all contain ten or twelve segments of double-stranded RNA of similar size; they are composed of polypeptides of similar size; and they all contain a transcriptase capable of transcribing their double-stranded RNA into single-stranded RNA. All these viruses are obviously closely related and they have in fact been grouped together as the family Reoviridae.

Other examples of this nature are known, although they have not yet been examined in such detail. Examples are:

(1) Viruses that consist of single-stranded RNA with molecular weights ranging from one to three million encapsidated in capsids of roughly the same size and morphology have long been known to infect bacteria, plants, invertebrates and vertebrates.

(2) Poxviruses have long been thought to be confined to vertebrates; but virus particles with typical poxvirus morphology are now being found in insects.

(3) Another group of DNA viruses which replicate in the cytoplasm, the polyhedral cytoplasmic deoxyviruses, which includes the mammalian African swine fever virus, frog virus 3, lymphocystis virus of fish and the insect *Tipula iridescens* virus.

(4) Viruses of fungi have recently been described which possess the typical morphology of herpesviruses. Their morphogenetic pathway appears to resemble that of mammalian herpesviruses.

(5) A group of plant viruses the prototypes of which are lettuce necrotic yellows virus, sowthistle yellow vein virus, wheat striate mosaic virus and potato yellow dwarf virus, closely resembles the vertebrate rhabdoviruses. Most, if not all, can also multiply in insects (leafhoppers). The Drosophila sigma virus also has rhabdovirus morphology.

(6) Like the viruses of the previous group, tomato spotted wilt virus and carrot mottle virus possess unit membranes probably derived from their host cells; however they are spherical (diameter about 80 nm) and resemble vertebrate myxoviruses rather than rhabdoviruses.

The significance of the fact that cells of widely differing families are infected with viruses obviously possessing common ancestors will be discussed below.

Whereas some viruses, such as the AAV discussed above, have an extraordinarily narrow host range, others have a host range which is very wide. Numerous examples are known of viruses capable of infecting plants and insects on the one hand or insects and vertebrates on the other. Not only can all these viruses infect cells which present widely differing surfaces, but they can also obviously use the synthetic apparatus of very widely differing host cells for their own replication. They are an excellent argument for the universality of the genetic code.

Between the viruses with such very wide host ranges and those with extremely narrow ones are the many viruses which infect only cells of one particular species with any reasonable degree of efficiency. There are numerous examples, the most familiar being the various strains of human viruses, some of which will also infect primates and monkeys, while others will not even infect these.

Carrying this type of analysis further at a different level, very little is known yet concerning what makes cells permissive or non-permissive to viral infection. As a rule, most of the cells in animals are non-permissive for any given virus; most viruses have a predeliction for multiplying in, say, cells of the respiratory tract, or the gastro-intestinal tract, or the liver, or the cells of the epidermal layer, and so on. This organ specificity no doubt represents the result of adaptation, that is, the selection of the most successful variants.

THE EVOLUTION OF VIRUSES

The purpose of the discussion so far has been to provide the background information that is necessary to enable us to gain some insight into the origin and evolutionary history of viruses. We will address ourselves specifically to providing answers to the following questions:

Are there any viruses which might be descended from regressed parasites rather than from cellular genetic elements that acquired autonomy?

The arguments in favor of these two opposing views of virus origin have been examined frequently in the past (Green, 1935; Burnet, 1945; Laidlaw, 1938; Waterson, 1963), the most recent as well as the most complete discussion being that by Luria & Darnell (1967). According to the first view, viruses represent the degenerate descendants of larger pathogenic micro-organisms. This theory, first proposed by Green and then by Laidlaw, enjoyed great vogue, but is not in favor today, since viruses possess no features which could reasonably be regarded as vestiges of cellular organization. Rather, viral capsids are morphogenetically analogous to cellular organelles like microtubules, actin filaments, bacterial flagella, etc., all of which are composed of regularly repeating protein subunits and arise by self-assembly. This applies even to viruses which themselves possess envelopes derived from plasma cell membranes: for the essential part of all such viruses is a nucleocapsid composed of nucleic acid and a sheath made up of repeated protein subunits, while the envelope, consisting of a lipid bilayer and one or a very few virus-coded protein species, is obviously used only to facilitate exit from and entry into cells. There is only one exception; and that concerns the poxviruses. Poxviruses, the largest of all viruses, are the only ones that can reasonably be considered as being descended from a larger cellular ancestor. Even there the structure of the core is difficult to reconcile with this view; however, the outer regions of the virus appear to be so complex, and the virus contains so many protein components (at least 30–35 different species are represented by more than 100 molecules each) that it is difficult to see how the genetic information for so complex a structure ever became associated. By the same token, it may well be that poxviruses are the youngest viruses still in the process of regressive evolution. The fact that they are exclusively animal viruses, with insects the lowest order host, is in accord with this hypothesis.

Did viruses arise repeatedly and independently, or could all viruses have a common ancestor? Is there any evidence that any viruses extant today are linear descendants of each other? Are larger viruses likely to be the precursors of smaller viruses?

There are altogether about thirty structural patterns of virus morphology; what is the likelihood that some have evolved from others? And that the smallest viruses are descendants of larger ones? Or is it possible that viral genomes have increased as well as decreased in size in the course of evolution? Various nucleic acid-containing cellular organelles have been considered as possible viral relatives or ancestors. Thus for DNA-containing viruses organelles such as protozoan kinetoplasts or centrioles or plasmids, that is, autonomous or semi-autonomous DNA-containing genetic elements, have been considered, not to mention chromosomes and genes, as well as mitochondria and chloroplasts (the vogue of which has recently diminished because of their possible cellular origin).

For RNA viruses the list of possible candidates is shorter; in particular no cellular RNA structure is yet known which is capable of replicating its own RNA. There appear to be two alternatives for the origin of RNA viruses: (1) they arose in the same way as DNA viruses at a time when primordial cells possessed autonomous genetic elements containing RNA;* (2) they are descendants of RNA-containing structures which acquired an RNA-replicating mechanism. Such structures might have resembled the ribonucleoprotein complexes which appear to be the form in which mRNA is transported from the nucleus to the cytoplasm. It is interesting in this connection that whereas the DNA of all DNA viruses replicates by the same mechanism (which appears to be identical with that which replicates cell DNA), there is considerable variation in the mechanisms by which the RNA of RNA viruses replicates (although the basic mechanism in all cases involves alternate transcription of plus and minus strands). As for the source of the RNA, this could clearly have been either cellular or viral (that is, mRNA of DNA viruses). It is difficult to decide which of these two alternatives is the more likely, unless even in primordial cells there was a generic difference between host and viral mRNAs.

Although RNA viruses could clearly have derived from DNA viruses and, indeed, now that the existence of RNA-dependent DNA polymerases has been demonstrated, DNA viruses could also be derived from RNA viruses, let us assume for the moment that DNA and RNA viruses arose independently. What then is the chance that there were just two original events, one for DNA and the other for RNA viruses, and that all other viruses are descended from these two ancestors? An evolutionary history of this nature has two primary prerequisites: mechanisms for changing genome size and mechanisms for changing virion structure. Mechanisms for changing genome size are readily apparent; genome size can decrease by deletion and increase by gene duplication, accretion of extra segments from related or unrelated genomes with and without joining by ligases, integration into the host genome followed by abnormal excision, or similar mechanisms. Changes in either direction of genome (and cistron) size could therefore certainly have occurred. However, the problems relating to changes in capsid structure are more serious; in particular, the structure of icosahedral capsids is fixed very precisely and it is very difficult to conceive of a way in which one type could evolve into another. For example, the loss of the adenovirus penton polypeptide, or even a diminution in its size, could certainly not be tolerated by the adenovirus capsid. Indeed, and this is

* It is conceivable that cells today still possess the ability to replicate RNA, but that this system has not yet been detected: after all, transcription of RNA into DNA was only recognized three years ago. An example of a possible cellular RNA replicating system that is sometimes cited in this context is the metagon of protozoa (Sonneborn, 1965).

the most telling argument against the notion that all viruses are derived from one or two precursors, there are today no *intermediates* between the thirty-odd morphological virus groups; this is in sharp contrast to the infinite gradation and variability among cells and organisms.

Are viruses of eukaryotes descendants of viruses of prokaryotes? We conclude therefore that the various virus groups arose as independent events. Several questions then arise. For example, are all the viruses *within each group* likely to be the result of one original event? This question concerns primarily the small icosahedral RNA viruses with diameters of about 25 nm and genome molecular weights of 1–2 million daltons, to which the MS2 and Qβ group of bacteriophages, many of the plant virus such as turnip yellow mosaic virus, brome mosaic virus, tomato bushy stunt virus and others, and the insect and mammalian picornaviruses belong. It is difficult to answer this question. The best answer is probably that the plant, insect and vertebrate viruses of this series resemble each other sufficiently closely to make a common origin not unlikely; indeed, as will be discussed below, it is quite likely that viruses have repeatedly crossed species barriers from the plant to the animal kingdom and vice versa. The small RNA phages, however, seem quite different from these in their molecular biology; and while it is not inconceivable that they share a common ancestor with them, it nevertheless appears more likely that they had an independent origin. The uncertainty surrounding the extent of the relationship between the small bacterial RNA viruses and the small plant, insect and vertebrate RNA viruses highlights one of the principal difficulties in tracing the evolutionary history of viruses: this is the enormous gap in our knowledge concerning the viruses of the cells of the simplest eukaryotes. No doubt many of these cells do have viruses associated with them, but practically nothing is known concerning them. It is, however, striking that almost all the viruses of fungi and yeasts which have so far been described are quite unique: most of them are small viruses containing double-stranded RNA segmented into no more than three segments which may be individually encapsidated, while some appear to comprise particles which contain both double- and/or single-stranded RNA. These viruses do not form obvious links between the viruses of bacteria and the viruses of the higher eukaryotes. It is, however, extremely interesting that viruses with the typical morphology of herpesviruses have recently been reported in fungi (Kazama & Schornstein, 1973). Further studies on the viruses of primitive eukaryotes should certainly prove very exciting.

Are viruses of higher plants, insects and vertebrates, which share a common morphology, likely to have had a common origin?

As already alluded to, there are many examples of viruses with strikingly similar morphology, nucleic acid structure and even enzyme complement that infect either plants, insects or vertebrates. Specific examples are as follows. The various viruses grouped together as reoviridae, with mammalian, avian, insect and plant hosts, differ considerably in polypeptide constitution and are unrelated serologically or as judged by nucleic acid hybridization; but they do display highly characteristic morphologic features such as the core spikes which are not seen on any other group of viruses, they all contain genomes consisting of 10–12 very similarly sized double-stranded RNA segments, and they all contain a DS → SS RNA polymerase. These features are sufficiently distinctive to render it extremely unlikely that these viruses arose independently.

There is also a group of viruses with the morphology of mammalian rhabdoviruses the various members of which infect vertebrates, insects and/or plants, and like the mammalian rhabdoviruses, at least some of the plant rhabdoviruses have been shown to possess the enzyme capable of transcribing single-stranded RNA into strands of the opposite polarity. Presumably therefore, this group of viruses has a common evolutionary history. Further examples are the vertebrate myxoviruses and tomato spotted wilt virus, the poxviruses that infect insects and those that infect vertebrates, the various polyhedral cytoplasmic deoxyviruses, and others.

We may conclude then that in all probability most of the major virus groups as currently recognized had an independent origin, and that all viruses within each group had a common origin. How can one envisage that, say, all reoviridae, all rhabdoviridae, all polyhedral cytoplasmic deoxyviruses, and so on, had a common origin, when each now has members which infect mammals and vertebrates in general, insects and also plants? The common factor would appear to be the insect: for viruses are known which can multiply in both insects and vertebrates on the one hand, and in insects and plants on the other. The group ancestor could therefore either have originated in insects and then evolved into variants capable of multiplying in vertebrates and plants; or it may have originated in either a plant or a vertebrate and then been transferred one to the other via an insect vector.

What are the likely evolutionary histories of viruses containing double-stranded RNA and single-stranded DNA?

The origins and evolutionary histories of both of these groups which occupy unique niches in the hierarchy of viruses are obscure. Some years ago, when it was discovered that the replication of single-stranded RNA proceeded via a (partially) double-stranded intermediate form, it seemed possible that just as there are viruses which contain plus-stranded RNA and others which contain minus-stranded RNA, there might be some in which the double-stranded replicative form of RNA is encapsidated into virions. However, this is clearly not the case. Progeny double-stranded RNA of reoviridae is formed by a unique mechanism: parental genomes are not uncoated (another unique feature of this group of viruses, for the genomes of all other viruses are uncoated following infection), but are transcribed in a fully conservative manner into plus-stranded RNA. Later in the infection cycle this plus-stranded RNA is then transcribed into minus strands which remain associated with their template, thereby generating progeny double-stranded RNA. While this mechanism clearly has features in common with that by which single-stranded viral RNA replicates, it nevertheless possesses a sufficient number of unique features to suggest that if double- and single-stranded RNA-containing viruses do share a common evolutionary history, this was a very long time ago.

The single-stranded DNA-containing viruses pose a similar problem. It is unlikely that they evolved from double-stranded DNA-containing viruses, since the replication of single- and double-stranded DNA seems to proceed by fundamentally different mechanisms: double-stranded DNA replicates via the simultaneous discontinuous synthesis of both strands, while single-stranded DNA replicates via a mechanism formally analogous to that by which single-stranded RNA replicates, namely by the sequential continuous synthesis of plus and minus strands, with each strand serving alternately as the template for the synthesis of the other. It therefore seems likely that single-stranded DNA viruses originated independently of those containing double-stranded DNA. Until recently the principal difficulty with this hypothesis was that single-stranded DNA was not known to occur at all in cells, but it has recently been found that extensive regions of cellular DNA do in fact exist in the single-stranded state (Amalric, Bernard & Simard, 1973), and it is conceivable that some of this material became autonomous in a manner similar to that postulated for single-stranded RNA.

This brings us to the question of whether the various single-stranded DNA-containing viruses could have had a common origin. Bacterial, insect and mammalian viruses of this type are known; the genomes of the bacterial ones are covalently closed circles, those of the animal ones are linear molecules. Each group consists of viruses which differ markedly and possibly fundamentally among themselves. Thus the capsids of some single-stranded DNA-containing bacteriophages are icosahedral, of others helical; the DNA replication of some involves RNA, that of others does not; and some of the mammalian viruses contain only the plus strand, while adeno-associated virus and densonucleosis virus contain either the plus or the minus strand in different particles. The single-stranded DNA-containing viruses are therefore a heterogeneous group; but additional data concerning their growth cycles are necessary before speculation concerning their derivation from a common ancestor can become worthwhile.

Do viruses possess features that can be regarded as evidence of progression along an evolutionary pathway?

The principal argument against the hypothesis that some virus groups evolved from others is the total lack of viral forms intermediate between those within the various morphological groups. There is, however, one group of viruses which possess an organ that is clearly in the process of evolution: that is the tail of bacteriophages.

There are four major groups of bacteriophages which most probably originated independently: those that contain single-stranded RNA; those that contain single-stranded DNA; those that possess a capsid which contains lipid; and those (the majority) that possess tails. Phage tails are the only viral feature which exhibits a more or less complete gradation in morphology and functional complexity, and they therefore appear to be the products of an evolutionary pathway.

Phage tails vary greatly in complexity. There are simple tails which consist merely of long non-contractile or very short squat tubes and which lack, as far as one can tell, special organs of attachment; non-contractile tails with a single fiber at their tip which attaches to pili and allows the phage to slide along them to their base where the DNA is injected; non-contractile tails with knobs at their distal ends; tails with contractile sheaths and base plates and spikes; and contractile tails with base plates and spikes to which fibers are attached at their ends, or along their length, or where they join the head.

As pointed out above, the information required for the morphopoeisis of the most complex tails requires over one-half of the genome of many phages. Phages with tails of any particular degree of complexity are not confined to any particular group of bacteria: most bacterial species seem to be hosts to phages of all degrees of complexity. There seems little doubt that all tailed phages had a common origin and that the more complex tails have evolved from the simpler ones. The fact that the genes involved in tail formation are strongly clustered in the phage genome – clustered in such a way as to reflect the level of gene expression, with genes the polypeptides of which are only needed in very few copies (less than five), an intermediate number of copies (6–24), and many copies (over 100) being grouped together – and thus appear to form an integrated genetic system, is in accord with this idea (King & Laemmli, 1973). The major question concerning the evolutionary history of this morphogenetic system is the nature of the selective advantage conferred by progressively increasing complexity: for phages with very simple tails seem just as successful as those with the most complex ones.

Are viruses that are capable of integrating their genomes into that of the host cell precursors of viruses that cannot?

No discussion of the origin and evolution of viruses would be complete without a consideration of the evolutionary significance of the ability of viral genomes to become inserted into host cell genomes. Viruses of almost all DNA virus groups have developed this ability; among them are numerous types of bacterial viruses and all families of mammalian DNA viruses except the poxviruses (which cannot integrate their genomes because they do not enter the nuclei of their host cells). In all cases ability to become integrated requires special information. Viral genomes that can become integrated can therefore be divided into two parts, one which codes for the functions necessary for the vegetative multiplication of the virus, and another which codes for the functions necessary for integration, the maintenance of the integrated state, and the prevention of the full expression of the viral genome in the integrated state. The genetic machinery necessary for exercising the latter three classes of functions, that is, for the establishment and management of the lysogenic state, has been studied in greatest detail in the case of bacteriophage lambda. Phages of this type, which can either multiply vegetatively or become lysogenic, originated presumably as lytic phages. It is clear that they can evolve still further in several directions. For example, if they lose the genetic information necessary for excision or for some function essential for viral development, they become permanent additions to the host genome. Such integrated viral genomes, known as cryptic prophages in bacterial systems and conjectured to be of potentially crucial importance in human cancer, can only be detected by genetic recombination with non-defective viral genomes. Alternately, host-cell genetic information can be incorporated into such viral genomes. If the virus has an icosahedral capsid, so that there is a limit to the amount of genetic information that can be encapsidated in it, then any accretion of host genetic material must be balanced by a corresponding reduction in the amount of phage genetic material; this leads to the formation of defective transducing virus particles which are well known in the case of phage, and which probably also exist in the case of animal viruses, where they have not yet been described, presumably because of the absence of means of detecting them. If, on the other hand, the capsid is not limited with respect to space, either because extra space is available or because it is expandable, as is the case for enveloped viruses, host genetic information can be incorporated into the viral genome without viral information having to be deleted; such viruses are not defective, and they can therefore still multiply.

Numerous vertebrate viruses do indeed contain genetic information for changing the growth and social behavior of their host cells, that is, for transforming them. Among these viruses are herpesviruses, adenoviruses, papovaviruses and RNA tumor viruses (and perhaps also the Visna group of viruses), all of which contain not only the information for their own multiplication, but also that for transforming their host cells. The fact that these two sets of information are independent is indicated by the existence of mutants which are deficient in either one or the other. Currently much attention is focused on one class of these viruses, namely the mammalian sarcoma viruses, which appear to be defective in their ability to multiply in the absence of a helper. As for the original source of the information necessary for transformation, the best that one can do at this time is to postulate that it may originally have been host genetic information. The main points are that a virion's ability to integrate its genome into that of the host cell is almost certainly a highly specialized acquired function representing an example of extreme adaptation, and that viruses with this ability are therefore very unlikely to have been the precursors or ancestors of lytic viruses.

The detection of relationships by nearest neighbor base analysis

A possible way of assessing relationships among the various virus groups and therefore, hopefully, of detecting common evolutionary histories, is by means of nearest-neighbor nucleic acid base analysis, as suggested by Subak-Sharpe (1967, 1969, this volume). This gives a measure of the deviation from random expectation of the frequencies of the sixteen nucleic acid base doublets, which in turn permits comparison of the general design of different nucleic acids. Certainly one may expect that viral genomes which have evolved in similar conditions within the same or closely related environments would possess similar designs, and resemblance of general design might indeed suggest evolutionary relationship. Interestingly enough, some of the major groups of animal viruses yield two types of patterns according to this criterion. First, a group composed of small viruses, namely the picornaviruses, the parvoviruses and the papovaviruses (that is, viruses with single-stranded RNA as well as single- and double-stranded DNA) display patterns which resemble in considerable detail both each other and that of mammalian cell DNA. Second, another group which comprises the larger DNA-containing viruses, namely the adenoviruses, the herpesviruses and the poxviruses, exhibits designs which are much closer to random and quite different from that of cellular DNA, nor do they strikingly resemble each other. It is not clear how to interpret this information; does it suggest, for example, that all small mammalian viruses have evolved from the same information source in mammalian cells or their not too distant ancestors, while the larger adenoviruses, herpesviruses, and poxviruses originated in earlier cells? Or should the data be interpreted as signifying that the smaller the viral genome the more readily does it adapt to its cellular environment? The conclusion that all small viruses, irrespective of the chemical nature of their genome, had a common origin is unlikely to be true; the fact that the larger viruses are not closely related, on the other hand, is in accord with the conclusion reached above.

REFERENCES

AMALRIC, F., BERNARD, S. & SIMARD, R. (1973). Detection of single-stranded DNA in the nucleolus. *Nature, New Biology*, **243**, 38–41.

BISHOP, J. N., CHUN, TSAN-DENG, FARAS, A. J., GOODMAN, H. M., LEVINSON, W. E., TAYLOR, J. M. & VARMUS, H. E. (1973). Transcription of the Rous sarcoma virus genome by RNA-directed DNA polymerase. In *Proceedings of the ICN–UCLA Symposium in Molecular Biology 'Virus Research'*, ed. C. F. Fox & W. S. Robinson. New York: Academic Press.

BLUMENTHAL, T., LANDERS, T. A. & WEBER, K. (1972). Bacteriophage Qβ replicase contains the protein biosynthesis elongation factors EF Tu and EF Ts. *Proceedings of the National Academy of Sciences, USA*, **69**, 1313–17.

BOZARTH, R. F., WOOD, H. A. & MANDELBROT, A. (1971). The *Penicillium stoloniferum* virus complex: two similar double-stranded RNA virus-like particles in a single cell. *Virology*, **45**, 516–23.

BURNET, F. M. (1945). *Virus as Organism.* Cambridge, Mass.: Harvard University Press.

BUTTERWORTH, B. E., HALL, L., STOLTZFUS, C. M. & RUECKERT, R. R. (1971). Virus-specific proteins synthesized in encephalomyocarditis virus-infected HeLa cells. *Proceedings of the National Academy of Sciences, USA*, **68**, 3083–7.

COOPER, P. D., STEINER-PRYOR, A. & WRIGHT, P. J. (1973). A proposed regulator for poliovirus: the equestron. *Intervirology*, **1**, 1–10.

DIENER, T. O. (1972). Viroids. *Advances in Virus Research*, **17**, 295–313.

DUBBS, D. R. & KIT, S. (1964*a*). Isolation and properties of vaccinia mutants deficient in thymidine kinase inducing activity. *Virology*, **22**, 214–26.

DUBBS, D. R. & KIT, S. (1964*b*). Mutant strains of herpes simplex deficient in thymidine kinase-inducing activity. *Virology*, **22**, 493–502.

DUESBERG, P. H. (1968). Physical properties of Rous sarcoma virus RNA. *Proceedings of the National Academy of Sciences, USA*, **60**, 1511–18.

DUESBERG, P. H. & VOGT, P. K. (1973). RNA species obtained from clonal lines of avian sarcoma and from avian leukosis viruses. *Virology*, **54**, 207–19.

DUNN, J. J. & STUDIER, F. W. (1973). T7 early RNAs are generated by site-specific cleavages. *Proceedings of the National Academy of Sciences, USA*, **70**, 1559–63.

ERIKSON, E. & ERIKSON, R. L. (1970). Isolation of amino acid acceptor RNA from purified avian myeloblastosis virus. *Journal of Molecular Biology*, **52**, 387–90.

FARAS, A. J., GARAPIN, A. C., LEVINSON, W. E., BISHOP, J. M. & GOODMAN, H. M. (1973). Characterization of the low molecular weight RNAs associated with the 70 S RNA of Rous sarcoma virus. *Journal of Virology*, **12**, in Press.

GREEN, R. G. (1935). On the nature of filtrable viruses. *Science*, **82**, 443–5.

GRONER, Y., SCHEPS, R., KAMEN, R., KOLAKOFSKY, D. & REVEL, M. (1972). Host subunit of Qβ replicase is translation control factor *i*. *Nature, New Biology*, **239**, 19–20.

HAENNI, A. L., PROCHIANTZ, A., BERNARD, O. & CHAPEVILLE, E. F. (1973). TYMV valyl-RNA as an amino-acid donor in protein biosynthesis. *Nature, New Biology*, **41**, 166–8.

JACOBSON, M. F. & BALTIMORE, D. (1968). Polypeptide cleavages in the formation of poliovirus proteins. *Proceedings of the National Academy of Sciences, USA*, **61**, 77–84.

JOKLIK, W. K. (1972). Studies on the effect of chymotrypsin on reovirions. *Virology*, **49**, 700–50.

JOKLIK, W. K. (1973). The transcription and translation of reovirus RNA. In *Proceedings of the ICN–UCLA Symposium in Molecular Biology 'Virus Research'*, ed. C. F. Fox & W. S. Robinson. New York: Academic Press.

KAZAMA, F. Y. & SCHORNSTEIN, K. L. (1973). Ultrastructure of a fungus herpes-type virus. *Virology*, **52**, 478–87.

KING, J. & LAEMMLI, U. K. (1973). Bacteriophage T4 tail assembly: structural proteins and their genetic identification. *Journal of Molecular Biology*, **75**, 315–37.

KOLAKOFSKY, D. & WEISSMANN, C. (1971). Possible mechanism for transition of viral RNA from polysome to replication complex. *Nature, New Biology*, **231**, 42–6.

KURSTAK, E. (1972). Small DNA densonucleosis virus (DNV). *Advances in Virus Research*, **17**, 207–41.

LAIDLAW, P. T. (1938). *Virus Diseases and Viruses*. London: Cambridge University Press.

LAVER, W. G. & WEBSTER, R. G. (1973). Studies on the origin of pandemic influenza. III. Evidence implicating duck and equine influenza viruses as possible progenitors of the Hong Kong strain of human influenza. *Virology*, **51**, 383–91.

LURIA, S. E. & DARNELL, J. E. JR. (1967). *General Virology*, 2nd edition. New York, London and Sydney: Wiley.

ÖBERG, B. & PHILIPSON, L. (1972). Binding of histidine to tobacco mosaic virus RNA. *Biochemical and Biophysical Research Communication*, **48**, 927–32.

ÖBERG, B. F. & SHATKIN, A. J. (1972). Initiation of picornavirus protein synthesis in ascites cell extracts. *Proceedings of the National Academy of Sciences, USA*, **69**, 3589–93.

ROBERTSON, H. D. & JEPPESEN, P. G. N. (1972). Extent of variation in three related bacteriophage RNA molecules. *Journal of Molecular Biology*, **68**, 417–28.

ROBINSON, W. S., PITKANEN, A. & RUBIN, H. (1965). The nucleic acid of Rous sarcoma virus: purification of the virus and isolation of the nucleic acid. *Proceedings of the National Academy of Sciences, USA*, **54**, 137–44.

SALZMAN, N. P. & THOREN, M. M. (1973). Inhibition in the joining of DNA intermediates to growing Simian virus 40 chains. *Journal of Virology*, **11**, 721–30.

SAROV, I. & JOKLIK, W. K. (1972). Studies on the nature and location of the capsid polypeptides of vaccinia virions. *Virology*, **50**, 579–92.

SELA, I. (1972). Tobacco enzyme-cleaved fragments of TMV–RNA specifically accepting serine and methionine. *Virology*, **49**, 90–4.

SHATKIN, A. J., SIPE, J. D. & LOH, P. C. (1968). Separation of the ten reovirus genome segments by polyacrylamide gel electrophoresis. *Journal of Virology*, **2**, 986–91.

SONNEBORN, T. M. (1965). The metagon: RNA and cytoplasmic inheritance. *American Naturalist*, **99**, 279–307.

SUBAK-SHARPE, J. H. (1967). Doublet patterns and evolution of viruses. *British Medical Bulletin*, **23**, 161–8.

SUBAK-SHARPE, J. H. (1969). The doublet pattern of the nucleic acid of oncogenic and non-oncogenic viruses and its relationship to that of mammalian DNA. In *Proceedings of the 8th Canadian Cancer Research Conference*, ed. J. F. Morgan, pp. 242–60. Oxford: Pergamon Press.

SUMMERS, D. F. & MAIZEL, J. V. JR. (1968). Evidence for a large precursor protein in poliovirus synthesis. *Proceedings of the National Academy of Sciences, USA*, **59**, 966–71.

VAN KAMMEN, A. (1972). Plant viruses with divided genomes. *Annual Review of Phytopathology*, **10**, 125–63.

VOGT, P. K. (1973). Genome of avian RNA tumor viruses: a discussion of four models. In *Proceedings of the 4th Lepetit Colloquium on 'Possible Episomes in Eukaryotes'*, ed. L. G. Silvestri. Amsterdam: North-Holland (in Press).

WATERSON, A. P. (1963). The origin and evolution of viruses. I. Virus origins: degenerate bacteria or vagrant genes. *New Scientist*, **18**, 200–2.

WEBER, H., BILLETER, M. A., KAHANE, S., WEISSMANN, C., HINDLEY, J. & PORTER, A. (1972). Molecular basis for repressor activity of Qβ replicase. *Nature, New Biology*, **237**, 166–70.

WOLFSON, L. & DRESSLER, D. (1972). Adenovirus-2 DNA contains an inverted terminal repetition. *Proceedings of the National Academy of Sciences, USA*, **69**, 3054–7.

YOT, P., PINCK, M., HAENNI, A.-L., PURANTON, H. M. & CHAPEVILLE, F. (1970). Valine-specific tRNA-like structure in turnip yellow mosaic virus RNA. *Proceedings of the National Academy of Sciences, USA*, **67**, 1345–52.

THE ORIGIN OF PANDEMIC INFLUENZA VIRUSES

J. J. SKEHEL
National Institute for Medical Research,
Mill Hill, London NW7 1AA

The outstanding characteristic of the influenza viruses is their ability to cause one of the most important pandemic diseases affecting mankind. Since the first isolation of an influenza virus from humans in 1933 (Smith, Andrewes & Laidlaw, 1933) three major pandemics have been recorded, and each of these was caused by the introduction of antigenically different viruses of unknown origin into a non-immune population. As a result of these experiences, extensive studies have been undertaken to identify the viruses responsible, to compare their properties and to trace their evolution and the results of these efforts will be briefly reviewed and discussed.

There are three types of influenza virus designated A, B and C which are distinguished by immunological criteria (Pereira, 1969). Viruses of all three types infect man but only type A viruses have been isolated from other species. In addition only type A viruses have caused pandemic disease and consequently only the properties of these will be considered. Viruses within this type are further divided into subtypes on the basis of their immunological properties, and the individual viruses within these subtypes are commonly referred to as virus strains.

All influenza type A viruses consist of the viral genome in association with several types of protein, enclosed by a bilayer of lipid molecules from which glycoprotein spikes project (Plate 1*a*). The chemical and immunological properties of the proteins which are located inside the lipid envelope are similar for all viruses of this type. Thus, the individual internal proteins are invariant in relative abundance and molecular weight from strain to strain (Brand, 1972); the two proteins present in largest amount have similar amino acid sequences as judged by analyses of their tryptic peptides (Laver, 1964); and they appear to have similar immunological properties since antisera prepared against either of the two major internal proteins from any type A virus react in complement fixation or immunodiffusion assays with the corresponding protein from any other virus of this type (Brand, 1972). Because of these similarities the small differences which may occur between the internal proteins of the viruses of different strains, although they may be related to differences in certain biological properties of the viruses, are not considered to be directly involved in the selection of the new strains of virus which cause pandemics. These proteins will not, therefore, be considered further.

The properties of the envelope glycoproteins, on the other hand, differ from strain to strain and on each occasion the viruses which caused pandemic disease were characterized by marked differences in the properties of their glycoprotein components when these were compared with those of previously recognized viruses. For this reason the major portion of this chapter will be concerned with the properties of these molecules. It will be divided into four sections concerning:

(*a*) a description of those properties of the virus genome relevant to the theories concerning the origins of influenza;

(*b*) a review of the basic properties of the envelope glycoproteins;

(*c*) an account of the recorded variation in the immunological properties of influenza viruses; and

(*d*) a discussion of the theories proposed to explain its mechanism.

THE VIRAL GENOME

The earliest studies on the nature of the genome involved analyses of the progeny viruses which resulted from the infection of mice with a mixture of two different influenza viruses. In these and subsequent experiments recombination frequencies of up to 50 % were recorded between certain gene markers (Burnet & Lind, 1949; Hirst, 1962). These results were interpreted as indicating that the viral genome was made up of several pieces of nucleic acid (Burnet, 1956; Hirst, 1962). The results of more recent direct analyses are compatible with this conclusion and indicate that the genome of influenza viruses is made up of seven species of single-stranded RNA of total molecular weight approximately 4.5×10^6 daltons (Pons & Hirst, 1968; Duesberg, 1968; Bishop, Obijeski & Simpson, 1971; Skehel, 1971). The results of fingerprint analyses and estimations of the number and nature of 5′-terminal nucleotides and 3′-terminal nucleosides in the total genome (Horst *et al.* 1972; Lewandowski, Content & Leppla, 1971; Young & Content, 1971) indicate that these seven species of RNA are not generated as artefacts of the procedures used for their extraction and in studies of viral RNA synthesis in infected cells they have not been found to be covalently associated at any stage during virus replication. Indeed, the results of these studies and of analyses of protein synthesis in infected cells (Skehel,

1972) strongly suggest that during replication, transcription, and translation the seven distinct segments of the viral genome and their transcripts function independently as seven separate genes. Thus each of the nucleic acid species appears to contain the information for the synthesis of a single protein during virus replication.

THE GLYCOPROTEINS OF THE VIRUS

The influenza virus particle contains six different types of protein (Compans, Klenk, Caliguiri & Choppin, 1970; Schulze, 1970; Skehel & Schild, 1971). As stated above, these can be conveniently divided into two groups: (*a*) those associated with the virus RNA genome and located inside the lipid envelope, and (*b*) the envelope glycoproteins. The latter group consists of two of the six types of virus protein and these are readily distinguished by their biological, structural and immunological properties. One is a haemagglutinin, the other a neuraminidase.

The haemagglutinin, so-called because of its ability to agglutinate chicken erythrocytes (Hirst, 1942), is responsible for attaching virus particles to cells during infection. It accounts for approximately 25 % of the virus protein and is a rod-shaped molecule about 14 nm long with an apparently triangular cross-section (Plate 1*b*). It has a molecular weight of approximately 210000 daltons and is made up of two types of glycopolypeptide which are linked together in the molecule by disulphide bonds (Skehel & Schild, 1971). The apparent molecular weights of these two glycopolypeptide components varies with the strain of virus but the sum of their molecular weights is always approximately 80000 daltons. Considering the shape of the molecule, its molecular weight and the molecular weights of the polypeptide subunits, the haemagglutinin contains either two or more probably three polypeptides of each type. Recent analyses of protein synthesis in influenza virus infected cells have shown that the two glycopolypeptide components are formed by proteolytic cleavage of a glycosylated precursor polypeptide of molecular weight 80000 daltons and as a consequence the synthesis of them both appears to be directed by a single influenza virus gene (Skehel, 1972). In this connection also the observations mentioned above concerning the variation in size of the glycopolypeptide components depending on the particular strain of virus, are considered to indicate that the position of proteolytic cleavage which generates the two glycopolypeptides may vary for the haemagglutinin precursors of different viruses.

The function of the other envelope glycoprotein, the neuraminidase, is not known. During virus replication it is inserted along with the haemagglutinin into the plasma membrane of the infected cell (A. J. Hay, in preparation) and is responsible for removing sialic acid residues from the membrane glycoproteins at certain sites (Klenk, Compans & Choppin, 1970) but the significance of this action is not presently clear. The molecular weight of the enzyme is approximately 240000 daltons and it is made up of four identical glycopolypeptide subunits of molecular weight 60000 daltons (Wrigley, Skehel, Charlwood & Brand, 1973). These are held together by disulphide bonds and form a molecule which consists of a narrow stalk 10 nm in length extending from the centre of one face of a box-like structure of dimensions 8 nm × 8 nm × 4 nm (Plate 1*c*). The free end of the stalk of the isolated protein is associated in the virus particle with the lipid bilayer.

Before considering the results of comparative studies of the immunological properties of the glycoproteins from different strains of virus it is convenient to note the following points.

(1) Antibodies which neutralize virus infectivity are specifically directed against the haemagglutinin. Although antibodies which react with the neuraminidase appear to affect the spread of virus from infected to uninfected cells, they do not affect the infectivity of virus incubated with them *in vitro* (Seto & Rott, 1966; Webster & Laver, 1967).

(2) The two glycoproteins are not covariant. Thus certain influenza viruses may have haemagglutinins with markedly different immunological properties but the properties of their neuraminidase components may be very similar and vice versa.

(3) The carbohydrate side chains of both glycoproteins are synthesized under the action of the glycosyl transferases of the infected cells. The composition and structure of these side chains, therefore, are at least in part determined by the host cell and constitute in viruses grown in certain cells host-specific antigenic determinants in addition to the virus-specified determinants located in the polypeptide chain. These host-specific determinants are frequently referred to as the host antigen (Haukenes, Harboe & Mortensson-Egnund, 1965; Laver & Webster, 1966).

VARIATION IN INFLUENZA VIRUSES

The comparative immunology of influenza has to a large extent been considered mainly in epidemiological investigations. In these, the most common test system used to determine the relationship between viruses has involved the inhibition of haemagglutination by antiviral sera. In more recent studies similar assays of the inhibition of neuraminidase

Table 1.

	Antiserum											
Antigen	PR	Weiss	Cam	FM	FW	FLW	Den	Jap/ 305	Jap/ 170	Alb	Tok	Aichi
Reciprocal haemagglutination inhibition test results with human influenza virus type A strains, 1934–68												
A/PR/8/34	*804*	226	14	10	28	10	—	—	—	—	—	—
A/Weiss/43	113	*509*	640	80	28	80	—	—	—	—	—	—
A1/Cam/46	10	40	*2036*	160	113	40	—	—	—	—	—	—
A1/FM/1/47	—	14	905	*640*	269	28	10	—	—	—	—	—
A/FW/1/50	—	—	28	34	*380*	—	—	—	—	—	—	—
A1/FLW/1/52	—	—	20	20	48	*640*	17	—	—	—	—	—
A1/Denver/1/57	20	—	—	—	17	80	*226*	—	10	40	—	—
A2/Japan/305/57	—	—	—	—	—	—	—	*320*	320	160	28	28
A2/Japan/170/62	—	—	—	—	—	—	—	320	*1018*	452	160	56
A2/Albany/3/64	—	—	—	—	—	—	—	160	452	*2036*	160	40
A2/Tokyo/3/67	—	—	—	—	—	—	—	40	56	160	*640*	10
A2/Aichi/2/68	—	—	—	—	—	—	—	—	10	14	320	*254*
Reciprocal neuraminidase inhibition test results with human influenza virus type A strains, 1934–68.												
A/PR/8/34	*1660*	80	118	69	—	26	—	—	—	—	—	—
A/Weiss/43	631	*525*	1350	270	155	32	59	—	—	—	—	—
A1/Cam/46	148	—	*2460*	502	252	—	252	—	—	—	—	—
A1/FM/1/47	24	28	2520	*708*	332	60	142	—	—	—	—	—
A/FW/1/50	—	—	28	—	*525*	152	44	—	—	—	—	—
A1/FLW/1/52	—	26	76	35	525	*955*	302	—	—	—	—	—
A1/Denver/1/57	—	—	22	—	252	124	*447*	—	—	—	—	—
A2/Japan/305/57	—	—	—	—	—	—	—	*4080*	8520	6030	692	1660
A2/Japan/170/62	—	—	—	—	—	—	—	1420	*5250*	9550	1420	390
A2/Albany/3/64	—	—	—	—	—	—	—	2240	3170	*5020*	1870	759
A2/Tokyo/3/67	—	—	—	—	—	—	—	339	302	1000	*2820*	1590
A2/Aichi/2/68	—	—	—	—	—	—	—	266	258	708	2630	*944*

The tests employed infected allantoic fluid as immunogen and antigen, and chicken antisera. No entry indicates less than 50 % inhibition using undiluted antisera. The titres indicate the reciprocal of the dilution of antisera at which haemagglutination or neuraminidase activity was inhibited by 50 %. Data from Dowdle *et al.* (1974).

activity by these sera have also been employed. The results of these studies which are summarized in the examples given in Table 1 (Dowdle, Coleman & Gregg, 1974) indicate the following sequence of variation in the antigenic properties of influenza viruses since the time of the first isolation of the virus from humans in 1933 until 1969. From 1933 to 1946 the properties of both the haemagglutinin and neuraminidase components of all virus strains which were examined were similar. Thus, although these viruses were sufficiently distinct one from the other to cause epidemics of disease throughout this period, for example in 1936–7, in 1940–1 and in 1943, they were nevertheless sufficiently related in serological tests to be grouped together and have been termed the A0 strains. In 1946 and 1947 viruses were isolated which showed only limited cross-reactions with the A0 viruses. In addition, although in experimental animals and in man these viruses induced the production of antibodies which combined with the A0 viruses, vaccination with the A0 strains did not stimulate an antibody response against the 1947 strains (Francis, 1952). As a consequence of this the 1947 viruses were classified as A1 influenza and related viruses continued to be isolated until 1957. As in the previous period of A0 virus prevalence, from 1947 until 1957 gradual progressive changes in the immunological properties of the A1 viruses were recorded and for example in 1950–1 and in 1953, these changes were accompanied by epidemics of disease.

The data presented in Table 1 show that direct or indirect serological relationships can be established between either the haemagglutinins or the neuraminidases of the A0 and the A1 viruses. In 1957, however, viruses were isolated in Asia which contained glycoproteins which were both basically unrelated immunologically to those of previously recognized strains (Table 1). These viruses formed another subtype of influenza termed A2, were responsible for the pandemic of Asian influenza in 1957 and subsequently continued to cause epidemics until 1968. It was again noted as with the A1 and A0 strains that even though the A2 viruses were markedly different from the A1 viruses, antisera prepared against certain A2 strains reacted with members of the A1 subtype (Jensen, Dunn & Robinson, 1958). These and other results suggested that viruses isolated late in the A1 period were more closely related to the A2 viruses than were other A1 and A0 viruses. During the period from 1957 to 1968 the course of the variation in the immunological properties of the glycoproteins of the A2 viruses was carefully traced and, until 1965, little change was noted from the properties of the initial isolates of 1957. In the viruses recovered from 1966 to 1968, however, considerable antigenic heterogeneity was detected and it appeared, therefore, that approximately eight years of relative stability were followed by a period of enhanced variation. In 1968 viruses were isolated during an epidemic in Hong Kong which clearly differed in the properties of their haemagglutinin components from the A2 viruses, although the properties of their neuraminidases were similar (Dowdle, Coleman, Hall & Knez, 1969). These viruses caused the pandemic of 1968–9 and closely related viruses continue to cause epidemics at this time. Because of the similarities between the Hong Kong and the A2 viruses which, from results such as those shown in Table 1 appeared to be more closely related antigenically than the A1 and the A2 viruses, the Hong Kong viruses were not initially considered as a separate subtype. A recently revised system of nomenclature for influenza viruses

(WHO, 1972) has taken into account the marked differences between the haemagglutinin components of the viruses, however, and in this the Hong Kong viruses are classified as H3N2 and the Asian A2 strains as H2N2 where H and N stand for haemagglutinin and neuraminidase respectively. For completion A0 viruses are now designated H0N1 and A1 viruses A1N1. This classification scheme more conveniently describes the similarities and differences between the pandemic strains and clearly indicates that the influenza viruses isolated since 1933 are divided into four subtypes on the basis of the properties of their haemagglutinins, and that during this time only two markedly different neuraminidase antigens have been recorded.

Two general observations may also be considered at this stage which will be referred to again later.

(1) The appearance of a different strain of virus results in the elimination of the previously recognized viruses. In particular, viruses of different subtypes are not simultaneously present in the population.

(2) The relationships between different viruses have been extended to viruses which occurred before 1933 by examination of the antibodies in sera collected from elderly people (Mulder & Masurel, 1958; Masurel, 1969). In addition, there are many observations of common characteristics for the glycoproteins of the viruses of man and viruses which infect animals and birds (WHO, 1972).

All of the results which assess the relatedness of different viruses are clearly important to the epidemiologist in his surveillance studies, particularly from the point of determining the appropriate time at which to recommend changes in the composition of influenza vaccines. They are not necessarily so useful, however, in demonstrating the fine degrees of relationship which may exist between certain viruses, e.g. between A1 strains and A2 strains, and which are necessary for an understanding of the evolution of influenza. As stated above they involve in the main the results of sub-type-specific complement fixation tests and more generally of haemagglutination inhibition or neuraminidase inhibition assays which are specific for the two-envelope glycoproteins. These tests are, however, subject to certain limitations. In addition to the usual inaccuracies and sources of variation which must be considered in any assay system involving antigen–antibody interactions such as the heterogeneity of the population of induced antibodies, the variation in the average affinity for antigen of the antibody molecules obtained at different times after immunization, and the species of animal from which the antibodies are derived, in the influenza virus system there are two other main sources of complication.

(1) Depending on the source of the viruses, antisera produced against them may contain antibodies specific for the host antigenic determinants which reside in the carbohydrate chains of the virus glycoproteins. The combination of such antibodies with the glycoproteins can clearly affect the results of both haemagglutination inhibition and neuraminidase inhibition assays. Moreover, since the glycoproteins of all viruses produced in a particular host system will contain similar carbohydrate side chains, they will all induce the synthesis of similar antibodies against these determinants. Consequently cross-reactions may be observed which are not a reflection of similarities in the amino acid sequence of the virus glycoproteins. However, although the carbohydrate groups of viruses grown in hens' eggs (which are the most commonly used host system) are immunogenic, this is not the case for those grown in some other host systems, e.g. calf kidney cells in tissue culture or duck eggs. It is, therefore, possible to circumvent this difficulty. Similarly the problem can be overcome by using egg-grown viruses as immunogens and using exclusively chicken sera in the tests.

(2) Although the haemagglutinin and neuraminidase are completely different glycoproteins, they are so arranged in the lipid envelope of the virus that antibodies specifically directed against either molecule may, as a result of associating with the appropriate glycoprotein, affect the biological activity of the other component by steric hindrance (Schulman & Kilbourne, 1969). Thus, for example, when comparing the immunological properties of two viruses which have a similar neuraminidase but different haemagglutinins, antisera produced against either virus will combine with both through the neuraminidase components. As a result of this combination the haemagglutinating capacity of the viruses may be reduced and they may appear to be related in haemagglutination inhibition assays. This consideration is important when comparing the A0 with the A1 viruses and the 1957 A2 strains with those isolated after the 1968 pandemic, since their neuraminidases were closely related antigenically. This complication has been avoided in some comparative studies by using only monospecific antisera prepared against the purified glycoproteins. Thus, in the above example, the use of antisera prepared against the purified haemagglutinin instead of intact virus particles removes the difficulty of interpretation. However, a note of caution concerning the use of isolated proteins as immunogens should be made, since it is possible that structural modifications may occur following their extraction from virus particles which may selectively affect different immunogenic determinants on the molecules. Another procedure which can be employed to overcome this difficulty

involves the preparation of recombinant viruses (by procedures which will be mentioned later) in which the characteristics of the glycoprotein components can be predetermined by an appropriate choice of the parent viruses. Considering again the above example, therefore, a recombinant virus can be selected for use as the immunogen which contains the haemagglutinin under study but a different neuraminidase component from the virus with which it is to be compared.

Unfortunately, in many studies of the differences in properties of the glycoproteins of different viruses, these sources of complication have not been ruled out. However, in several recent investigatons both recombinant viruses of the type mentioned above (Schulman & Kilbourne, 1969; Dowdle, Marine, Coleman & Knez, 1972) and purified glycoproteins (Webster & Laver, 1972) have been used as immunogens in suitable animal systems and the results of these studies have been interpreted to support at least one theory of the origin of different influenza viruses.

THEORIES CONCERNING THE ORIGIN OF INFLUENZA VIRUSES

As a consequence of the pattern of progressive change in the immunological properties of the virus glycoproteins which was noted in epidemiological studies, the term antigenic drift has been used to denote the process of change which operates in the periods between pandemics (Burnet, 1960). The more distinct differences observed between viruses which caused pandemics occur as a result of 'antigenic shift'.

It is generally considered that the gradual changes which are observed in the properties of the glycoprotein components of viruses isolated between pandemics, result from the selection under immunological pressure of virus mutants which arise spontaneously. As a consequence of the immune status of the population at that time such mutants, which may have either haemagglutinin or neuraminidase components with different properties, have a growth advantage. This situation is not unique to influenza A viruses. It is also observed with influenza B viruses (Pereira, 1969) and for example with the viruses of Foot-and-Mouth Disease (Brooksby, 1958).

The proposed mechanism of immunological selection is consistent with the observed pattern of antigenic change following the introduction of a new subtype of virus. Thus, for example, the sequence of changes in the properties of the A2 (H2N2) viruses between 1957 and 1968 which were described above is interpreted to indicate that, as more members of the population developed antibodies against the pandemic strain, variation in the immunological characteristics of the glycoproteins of subsequently isolated viruses became more obvious. Support for this mechanism of antigenic drift also comes from laboratory experiments in which it has been reported on several occasions (Archetti & Horsfall, 1950; Isaacs & Edney, 1950) that mutant viruses with significantly different immunological properties can be selected by passage of a virus, for example, in the presence of specific antibody of low affinity or in partially immune animals (Hamre, Loosli & Gerber, 1958). In one of these investigations (Laver & Webster, 1968) a combination of chemical and immunological techniques was used to compare the properties of the haemagglutinin components of the initial virus and selected mutants. The results indicated that, although the mutants had significantly different antigenic properties from the initial virus, the maps of the tryptic peptides derived from their haemagglutinins were very similar. It was therefore suggested that the drift in the antigenic properties of these viruses was not accompanied by major changes in the amino acid sequence of the haemagglutinin. Similar conclusions have also been made from the results of experiments in which the amino acid sequences of haemagglutinins from immunologically different A2 (H2N2) viruses were compared by this method (Laver & Webster, 1972).

It is reasonably clear then that the available data on both the chemical and the immunological properties of the virus glycoproteins are consistent with this view of the origin of influenza viruses during periods of antigenic drift.

There is somewhat less agreement concerning the processes involved in the generation of those viruses which cause pandemics. In the main two theories have been proposed. In the first it is suggested that antigenic shift may be simply an extension of the processes described above for antigenic drift: the second directly or indirectly implicates the influenza viruses of other species in the process.

The suggestion that antigenic shift is an extension of antigenic drift is based solely on determinations of the immunological properties of the virus glycoproteins. It rests on the observations that, depending on the test system, the immunogen and the antibody preparations employed, direct or indirect antigenic relationships can be established between essentially any two influenza viruses. Thus, as noted previously, although in 1948, 1957 and 1968, viruses were isolated which had markedly different antigenic properties from those of the previous subtype, in all three cases antigenic relatedness between the new viruses and their predecessors could be established by one test or another. The significance of these estimates of relationship has, however, been questioned. Particularly in the case of the 1968 Hong Kong virus (H3N2) and viruses of the previous Asian A2 subtype (H2N2), the complication that the viruses contained similar neuraminidase components has been stressed by several authors (Webster & Laver, 1972; Schulman & Kilbourne, 1969). Although antisera prepared against the Hong Kong virus appeared to react in haemagglutination inhibition tests with certain of the Asian viruses, using either recombinant viruses containing the Hong Kong haemagglutinin and an unrelated neuraminidase or purified Hong Kong haemagglutinin as immunogens, similar indications of antigenic relationship were not observed in several studies. On the other hand, in other investigations in which antisera prepared against such recombinants were also used, limited antigenic cross-reaction could be detected (Dowdle *et al.* 1972). No doubt these differences will eventually be resolved, but at present they serve to emphasize the uncertainty which exists concerning the detailed immunological relationships betwen viruses of different subtypes.

Although cross-reactions between the Asian and the Hong Kong viruses may be complicated by the presence of similar neuraminidase glycoproteins in the viruses of both subtypes, this is not the case for the A1 and the A2 viruses. In the viruses isolated in 1957 both the haemagglutinin and the neuraminidase components appeared to be quite distinct antigenically from those of the A1 viruses. Even so, asymmetric relationships between the A2 and A1 viruses have been reproducibly observed which, moreover, allow the conclusion that the A1 viruses of 1956 were more closely related to the Asian strains than earlier isolates of the A1 subtype. Thus, for example, antisera prepared against the 1957 Asian strains also react with the 1956 A1 viruses but not vice versa. There appear to be reasonable grounds in this case, therefore, to propose that the antigenic shift which resulted in the appearance of the A2 viruses could have involved the selection of mutants with haemagglutinins which, although they contained different antigenic determinants from viruses of the A1 subtype, also contained sequences in common with them.

The explanation of this process which is most often given is essentially a modification of the theory proposed by Francis in the early 1950s (see Francis & Maassab, 1965). In this theory (Fazekas de St Groth, 1970) it is suggested that the structure of the virus glycoproteins is sufficiently fluid to allow the exposure or development of different antigenic sites on the haemagglutinins and neuraminidases of different viruses. In cases where marked differences between the immunological properties of certain viruses are observed, it is suggested that these may be a reflection of mutations which affect the amino acid sequence at sites in the glycoprotein molecules distinct from those involved at that particular time in the process of antigenic drift. It is also suggested that as a consequence of these mutations the structure of the glycoproteins in the region of the antigenic determinants is modified in such a way that the binding of neutralizing antibodies is prevented and, therefore, the mutants have an extreme growth advantage. It is not proposed that in this process new viruses are selected from derivatives of the ultimate viruses which resulted from the previous process of antigenic drift. What is suggested is that at some stage during the development of the subtype, mutations occurred at a distinct site which nevertheless had an effect on the binding of antibodies to the determinant characteristic of that subtype.

One of the consequences of such a scheme might be that before the appearance of a pandemic strain viruses would be isolated which had immunological properties in common with viruses of both the old and the new subtypes. The identification of such 'bridging strains' has been reported (Fazekas de St Groth, 1969). Several viruses isolated in Australia and New Zealand in 1967 were proposed to have characteristics in common with both the A2 Asian viruses and the Hong Kong virus of 1968 and these were considered to be 'bridging strains' between the A2 (H2N2) and the Hong Kong (H3N2) subtypes. As yet, detailed immunological analyses of both the neuraminidases and the haemagglutinins of these viruses have not been reported, nor has antigenic shift been demonstrated in laboratory experiments similar to those described previously for simulated antigenic drift. However, until conclusive evidence to the contrary is presented, this theory remains as a plausible explanation for the mechanism of development of influenza viruses with distinct immunological properties.

In addition, the theory is supported to some extent by the suggestions that influenza viruses may re-cycle (Davenport, Minuse, Hennessy & Francis, 1969). These are based on the observations that sera from people born before 1878 and between then and 1892 contain antibodies which react with the A2 Asian viruses (H2N2) and the Hong Kong viruses (H3N2) respectively. It is, therefore, proposed that the influenza viruses which caused epidemics late in the nineteenth century had similar immunological properties, at least as far as their haemagglutinins were concerned, to those of the more recent pandemic strains. In line with this theory of antigenic shift, therefore, it is possible that this phenomenon of reappearance of viruses with similar immunological characteristics to previously identified strains is a reflection of a restricted number of structures in which distinct antigenic determinants can be presented on the virus glycoprotein molecules which would still allow the proteins to fulfil their functions in virus replication and hence ensure that the viruses which contain the glycoproteins in these forms would be viable.

Whatever the merits of this theory the extension of it must await more chemical evidence on the nature of the changes in the sequence of the glycoproteins which result in the marked differences in their immunological properties, since at present the data which are available do not appear to support it.

This evidence is obtained from comparative analyses of the tryptic peptides derived from the haemagglutinins of different viruses and most effort has been concentrated on the haemagglutinins from the Asian A2 (H2N2) and the Hong Kong influenza (H3N2) viruses (Laver & Webster, 1972). Amino acid analyses of purified haemagglutinins indicate that approximately seventy peptides should be released by tryptic digestion. This mixture of peptides is difficult to analyse and because of this, advantage has been taken of the fact that the haemagglutinin subunit is made up of two types of glycopolypeptide of molecular weight approximately 55000 and 25000 daltons, as described above. Analyses of the tryptic peptides of either of these separated glycopolypeptides, and in particular the smaller, is relatively simple (Laver, 1971), and the separation patterns obtained are considerably easier to compare. As a consequence, many of the reported analyses are of the separate haemagglutinin polypeptides.

The results of these analyses clearly indicate that both the large and the small haemagglutinin polypeptides of viruses from different subtypes have distinctly different amino acid sequences. On the other hand, it is also clear that the polypeptides obtained from viruses of the same subtype yield almost identical tryptic peptides and are, therefore, very similar in amino acid sequence. From these results it appears that marked differences in the immunological properties of the virus glycoproteins are accompanied by gross changes in their amino acid sequence and it has, therefore, been persuasively argued that a theory which suggests that a limited number of point mutations in the haemagglutinin gene could give rise to haemagglutinins with such different properties is unlikely to be correct.

However, although it is experimentally convenient to analyse the large and small haemagglutinin glycopolypeptide components separately, it is not necessarily valid to draw conclusions with regard to the amino acid sequence of the complete protein from the results of such analyses. This is principally because the apparent molecular weights of the two haemagglutinin components are not invariant from strain to strain. In fact, in the viruses used in the analyses referred to above, although the haemagglutinin glycopolypeptides of all A2 Asian (H2N2) viruses examined appear to be identical with molecular weights of 50000 and 33000 daltons, the haemagglutinins of all the Hong Kong (H3N2) viruses consist of glycopolypeptides of different molecular weights, namely 58000 and 25000. As discussed previously, it is suggested that these differences are due to differences in the position of cleavage of the haemagglutinin precursor glycopolypeptide in virus-infected cells and clearly if this is the case it is not correct to assume, for example when the smaller glycopolypeptides of two different viruses are considered, that equivalent portions of the different haemagglutinins are being compared. The positions of cleavage of the haemagglutinin precursors of different viruses probably reflect conformational differences in these molecules. This suggestion is also supported by results which indicate that the extent of glycosylation of the smaller haemagglutinin glycopeptide differs for viruses of different subtypes (Hayman, Skehel & Crumpton, 1973), and also by observations of the differing stability of the native haemagglutinins to digestion with the protease bromelain (Brand & Skehel, 1972), and to denaturation in the presence of ionic detergents (Laver, 1964). Until the significance of these different properties is appreciated, therefore, it is not possible to state categorically that the amino acid sequence of the different haemagglutinin molecules is markedly different.

The suggestion that the influenza viruses which infect man are derived from other species has been considered for many years. For example, even before human influenza viruses were first isolated, a virus obtained from pigs had been proposed as the causative agent of the catastrophic pandemic of 1918 (Shope, 1936) and although such connections have not been proven, on each occasion of an influenza pandemic the possibility that animal viruses were involved has been considered. These considerations are based on the results of many studies which indicate that the glycoprotein components, in particular the neuraminidases, of certain animal and avian viruses have similar immunological properties to viruses which caused epidemics in humans (WHO, 1972). Thus, for example, the neuraminidase of the A0 (H0H1) and the A1 (H1N1) subtypes is related to those of viruses isolated from pigs and from ducks; the neuraminidase of the A2 Asian (H2N2) and the Hong Kong (H3N2) viruses is similar to the neuraminidase of a virus from turkeys; and the haemagglutinin of the Hong Kong viruses is related to those of viruses isolated from horses and from ducks. In some cases it is not possible to decide whether or not the animals, particularly those of domesticated species, where infected before an epidemic in humans or after it, but it should be noted that in the last example above the equine and duck viruses were isolated in 1963, long before the Hong Kong virus of 1968.

These observations of immunological relatedness have recently been supported by comparative chemical analyses of the glycoproteins of the animal and the human viruses. It has been clearly shown that the pattern of tryptic peptides from certain animal and human viruses are very similar (Laver & Webster, 1973). Thus, for example, the smaller haemagglutinin polypeptide isolated from the above mentioned duck and equine viruses appear by this criterion to have essentially the same amino acid sequences as the equivalent polypeptide from Hong Kong virus.

Clearly then this evidence and the immunological data indicate that similar glycoproteins to those of viruses which cause pandemics in humans are present in animal influenza viruses, and because of the suddenness of the emergence of new epidemic strains and of the marked chemical and immunological differences between their glycoproteins and those of their immediate predecessors it has appeared logical to propose that the epidemics did not occur as the result of a selection of mutants of human viruses but rather were caused by distinct viruses related in some way to those of animals.

The nature of this relationship is not completely understood, but in the main two mechanisms which could be involved in the production of such viruses have been proposed. In the first it might be assumed that those viruses isolated from humans which are immunologically related to recognized animal viruses were simply host-range mutants of the latter which had acquired the capability for growth in man. There is at present no evidence that this process has been involved in the generation of viruses which caused epidemic disease in humans, but the reverse situation involving the infection of animals by viruses from humans has certainly been demonstrated (Kundin & Easterday, 1972; Paniker & Nair, 1970). There is, therefore, no reason to suppose that the spread of infection from animals to man could not occur (Kasel & Couch, 1969). It is also evident that this process and the suggestions made in the direct mutation theory discussed above are not mutually exclusive, since it seems possible that the so-called bridging strains could develop equally well in animals as in man and it has in fact been proposed that animal hosts are involved in the recycling of influenza viruses and may serve as a reservoir of potential epidemic strains (Davenport *et al.* 1969).

The second mechanism which has been proposed is probably the more widely favoured at the present time. In this it is suggested that the viruses which cause pandemics in humans are recombinant viruses which result from the simultaneous infection of a suitable host with viruses from both human and animal sources (see Webster, 1972). The recombinants which result from such a mixed infection would include viruses which had the capacity to infect humans but which also contained glycoprotein components derived from the animal virus parent. Such viruses would be presented with a non-immune human population and would, therefore, have an extreme growth advantage.

This proposition is supported by a considerable amount of circumstantial evidence and is primarily based on the observations described previously on the structure of the influenza virus genome and the concomitant ability of influenza viruses to recombine.

Since the early experiments of Burnet & Lind (1949) and of Hirst (1962) it has been clear that the progeny viruses which are obtained from mixed infections with different influenza viruses are made up of a mixture of viruses with different characteristics with regard to antigenic composition, growth capacity, ability to agglutinate red cells of different species, virulence in mice, etc. (Burnet, 1959; McCahon & Schild, 1971). The frequency and ease with which these hybrid viruses were obtained led to the suggestion that the genome of influenza viruses may be segmented and that different virus genes may be located on distinct segments, and this has been amply supported by subsequent analyses of the genome as discussed above. Considering these observations on the structure of the virus genome it appears probable, therefore, that the vast majority of recombinant viruses are formed by a process of reassortment of the genome segments produced during the simultaneous replication of the parent viruses in a mixed infection. The basis of this theory then is that similar recombination events occur during mixed infections in nature which may involve the infection of animals with a virus of human origin at a time when they are also infected with an animal virus or the infection of humans already suffering from influenza with an animal virus. Again, neither of these situations has been observed directly under natural conditions, but the passage of influenza viruses from species to species, the mixed infection of animals with viruses obtained from those of other species and from their own, and the characteristics of the recombinant viruses obtained from such infections, have been thoroughly established.

Much of the relevant data concerning the latter two phenomena particularly with regard to the immunological properties of the glycoprotein components of the recombinant viruses have been obtained from experiments with pigs and turkeys which were recently reported (Webster, 1972; Webster, Campbell & Granoff, 1973).

In this investigation two sorts of experiments were done. In the first, pigs were inoculated with a mixture of swine influenza virus and fowl plague virus and turkeys were infected with a mixture of fowl plague virus and turkey influenza. From both experimental systems a mixture of viruses was recovered which contained all of the possible combinations of envelope glycoproteins. Thus, from the pigs the resulting mixture contained both parental viruses, and recombinants containing either the haemagglutinin from the swine virus and the neuraminidase of fowl plague virus or the haemagglutinin from fowl plague and the neuraminidase from the swine virus. The corresponding recombinants were isolated in the experiment with turkeys. From these results it was clear that recombination can occur *in vivo* following a mixed infection.

The second type of experiment carried the process even nearer to the natural situation, since instead of inoculating animals with a mixture of viruses infection was allowed to spread through a flock of turkeys following the introduction of birds infected with either turkey influenza or fowl plague virus. Again, in this experiment recombinant viruses were isolated, although only one of the possible recombinants, namely that containing fowl plague haemagglutinin and turkey virus neuraminidase, was identified. The results of this type of experiment, therefore, indicate that hybrid viruses can be produced under conditions which closely parallel those in nature and indicate the feasibility of the proposition that antigenically distinct viruses which result from a natural recombination event, for example between a human and a swine virus, may be responsible for causing pandemic disease in man.

The preference which has been expressed for a scheme involving recombination over the suggestion that host-range mutants of animal viruses may be directly transferred to humans is mainly based on the observations that both the Hong Kong (H3N2) and the A1 (H1N1) viruses contained neuraminidase glycoproteins which were very similar to those of the previous pandemic virus. Because of these observations it has been considered more likely that recombinations involving transfer of only the gene coding for the haemagglutinin occurred rather than the direct infection of humans with a virus from another species which contained a similar neuraminidase to those of the human viruses of the previous subtypes.

These then are the two main theories currently advanced to explain the origin of those influenza viruses which cause pandemics. They are not necessarily mutually exclusive but clearly the alternative mechanisms of variation involving either mutation or gene reassortment during mixed infection are quite distinct. The former based on immunological data more readily accounts for certain of the demonstrated relationships between viruses of different subtypes and for the elimination from the population of other viruses with the appearance of a new strain. It is, however, certainly not supported at present by the results of chemical analyses of the virus glycoproteins. It might also be considered that the direct mutation theory gives insufficient emphasis to the peculiar properties of the segmented single-stranded RNA genome of influenza viruses and it is precisely these characteristics upon which the alternative theory depends. It may be noted here that these properties are not unique to the genomes of type A influenza viruses; those of type B influenza viruses have similar characteristics. In fact these observations, together with those that influenza B viruses have not been isolated from hosts other than man (Pereira, 1969), and that they do not appear to undergo such extreme variations as those resulting from antigenic 'shift' in the influenza A viruses, have been cited as additional evidence in favour of the recombination theory.

Because of this and the other consistencies mentioned previously this theory has gained widespread acceptance, but it should be stated that at present neither theory can explain all the observations on the variation of influenza viruses and neither theory is supported by enough immunological or chemical evidence to exclude the other. This is mainly because the immunological evidence is equivocal and the chemical data are insufficiently comprehensive to be completely convincing.

The variation in the immunological properties of influenza viruses, therefore, and the origin of the new viruses which cause pandemics pose important questions for future investigations in immunochemistry and clearly the experiments of Laver & Webster represent an interesting first approach to these problems.

REFERENCES

ARCHETTI, I. & HORSFALL, F. L. JR. (1950). Persistent antigenic variation of influenza A viruses after incomplete neutralization *in ovo* with heterologous immune serum. *Journal of Experimental Medicine*, **92**, 441–662.

BISHOP, D. H. L., OBIJESKI, J. F. & SIMPSON, R. W. (1971). Transcription of the influenza ribonucleic acid genome by a virion polymerase. II. Nature of the *in vitro* polymerase product. *Journal of Virology*, **8**, 74–80.

BRAND, C. M. (1972). Studies on the structural components of influenza viruses. Ph.D. thesis, Council for National Academic Awards.

BRAND, C. M. & SKEHEL, J. J. (1972). Crystalline antigen from the influenza virus envelope. *Nature, New Biology*, **238**, 145–7.

BROOKSBY, J. B. (1958). The virus of foot and mouth disease. *Advances in Virus Research*, **5**, 1–37.

BURNET, F. M. (1956). Structure of influenza virus. *Science*, **123**, 1101–4.

BURNET, F. M. (1959). Genetic interactions between animal viruses. In *The Viruses*, ed. F. M. Burnet & W. M. Stanley, vol. 3, Animal Viruses, pp. 275–306. New York: Academic Press.

BURNET, F. M. (1960). *Principles of Animal Virology*, 2nd edition, p. 398. New York: Academic Press.

BURNET, F. M. & LIND, P. E. (1949). Recombination of characters between two influenza virus strains. *Australian Journal of Science*, **12**, 109–10.

COMPANS, R. W., KLENK, H. D., CALIGUIRI, L. A. & CHOPPIN, P. W. (1970). Influenza virus proteins. I. Analysis of polypeptides of the virion and identification of spike glycoproteins. *Virology*, **42**, 880–9.

DAVENPORT, F. M., MINUSE, E., HENNESSY, A. V. & FRANCIS, T. JR. (1969). Interpretations of influenza antibody patterns of man. *Bulletin of the World Health Organization*, **41**, 453–60.

DOWDLE, W. R., COLEMAN, M. T. & GREGG, M. B. (1974). Natural History of influenza type A in the United States, 1957–1972. *Progress in Medical Virology*, **16**, in Press.

DOWDLE, W. R., COLEMAN, M. T., HALL, E. C. & KNEZ, V. (1969). Properties of the Hong Kong influenza virus. 2. Antigenic relationship of the Hong Kong virus haemagglutinin to that of other influenza A viruses. *Bulletin of the World Health Organization*, **41**, 419–24.

DOWDLE, W. R., MARINE, W. M., COLEMAN, M. T. & KNEZ, V. (1972). Haemagglutinin relationships of Hong Kong (H3) and Asian (H2) influenza strains delineated by antigen specific recombinants. *Journal of General Virology*, **16**, 127–34.

DUESBERG, P. H. (1968). The RNAs of influenza virus. *Proceedings of the National Academy of Sciences, USA*, **59**, 930–7.

FAZEKAS DE ST GROTH, S. (1969). New criteria for the selection of influenza vaccine strains. *Bulletin of the World Health Organization*, **41**, 651–7.

FAZEKAS DE ST GROTH, S. (1970). Evolution and hierarchy of influenza viruses. *Archives of Environmental Health*, **21**, 293–303.

FRANCIS, T. JR. (1952). Significance of antigenic variation of influenza viruses in relation to vaccination in man. *Federation Proceedings*, **11**, 808–12.

FRANCIS, T. JR. (1959). Serological Variation. In *The Viruses*, ed. F. M. Burnet & W. M. Stanley, vol. 3, Animal Viruses, pp. 251–73. New York: Academic Press.

FRANCIS, T. JR. & MAASSAB, H. F. (1965). Influenza viruses. In *Viral and Rickettsial Infections of Man*, 4th edition, ed. F. L. Horsfall, Jr. & I. Tamm, pp. 689–740. Philadelphia: Lippincott.

HAMRE, D., LOOSLI, C. G. & GERBER, P. (1958). Antigenic variants of influenza A virus (PR8 strain). III. Serological relationships of a line of variants derived in sequence in mice given homologous vaccine. *Journal of Experimental Medicine*, **107**, 829–44.

HAUKENES, G., HARBOE, A. & MORTENSSON-EGNUND, K. (1965). A uronic and sialic acid free chick allantoic mucopolysaccharide sulphate which combines with influenza virus HI-antibody to host material. I. Purification of the substrate. *Acta Pathalogica et Microbiologica Scandinavica*, **64**, 534–42.

HAYMAN, M. J., SKEHEL, J. J. & CRUMPTON, M. J. (1973). Purification of virus glycoproteins by affinity chromatography using Lens culinaris phytohaemagglutinin. *FEBS Letters*, **29**, 185–8.

HIRST, G. K. (1942). The quantitative determination of influenza virus and antibodies by means of red cell agglutination. *Journal of Experimental Medicine*, **75**, 47–64.

HIRST, G. K. (1962). Genetic recombination with Newcastle disease virus, polioviruses and influenza. *Cold Spring Harbor Symposia on Quantitative Biology*, **27**, 303–9.

HORST, J., CONTENT, J., MANDELES, S., FRAENKEL-CONRAT, H. & DUESBERG, P. (1972). Distinct oligonucleotide patterns of distinct influenza virus RNA's. *Journal of Molecular Biology*, **69**, 209–16.

ISAACS, A. & EDNEY, M. (1950). Variation in laboratory stocks of influenza viruses: genetic aspects of variations. *British Journal of Experimental Pathology*, **31**, 209–16.

JENSEN, K. E., DUNN, F. L. & ROBINSON, R. Q. (1958). Influenza 1957. A variant and the pandemic. *Progress in Medical Virology*, **1**, 165–209.

KASEL, J. A. & COUCH, R. B. (1969). Experimental infection in man and horses with influenza A viruses. *Bulletin of the World Health Organization*, **41**, 447–52.

KLENK, H. D., COMPANS, R. W. & CHOPPIN, P. W. (1970). An electron microscopic study of the presence or absence of neuraminic acid in enveloped viruses. *Virology*, **42**, 1158–62.

KUNDIN, W. D. & EASTERDAY, B. C. (1972). Hong Kong influenza infection in swine: experimental and field observations. *Bulletin of the World Health Organization*, **47**, 489–91.

LAVER, W. G. (1964). Structural studies on the protein subunits from three strains of influenza virus. *Journal of Molecular Biology*, **9**, 109–24.

LAVER, W. G. (1971). Separation of two polypeptide chains from the haemagglutinin subunit of influenza virus. *Virology*, **45**, 275–88.

LAVER, W. G. & WEBSTER, R. G. (1966). The structure of influenza viruses. III. Chemical studies of the host antigen. *Virology*, **30**, 104–15.

LAVER, W. G. & WEBSTER, R. G. (1968). Selection of antigenic mutants of influenza viruses. Isolation and peptide mapping of their haemagglutinin proteins. *Virology*, **34**, 193–202.

LAVER, W. G. & WEBSTER, R. G. (1972). Studies on the origin of pandemic influenza. II. Peptide maps of the light and heavy polypeptide chains from the haemagglutinin subunits of A2 influenza viruses isolated before and after the appearance of Hong Kong influenza. *Virology*, **48**, 445–55.

LAVER, W. G. & WEBSTER, R. G. (1973). Studies on the origin of pandemic influenza. III. Evidence implicating duck and equine influenza viruses as possible progenitors of the Hong Kong strain of human influenza. *Virology*, **51**, 383–91.

LEWANDOWSKI, L. J., CONTENT, J. & LEPPLA, S. H. (1971). Characterization of the subunit structure of the ribonucleic acid genome of influenza virus. *Journal of Virology*, **8**, 701–7.

MCCAHON, D. & SCHILD, G. C. (1971). An investigation of some factors affecting cross-reactivation between influenza A viruses. *Journal of General Virology*, **12**, 207–19.

MASUREL, N. (1969). Serological characteristics of a 'new' serotype of influenza A virus: the Hong Kong strain. *Bulletin of the World Health Organization*, **41**, 461–8.

MULDER, J. & MASUREL, N. (1958). Pre-epidemic antibody against 1957 strain of Asiatic influenza in serum of older people living in the Netherlands. *Lancet*, i, 810–14.

PANIKER, C. K. & NAIR, C. M. (1970). Infection with A2 Hong Kong influenza virus in domestic cats. *Bulletin of the World Health Organization*, **43**, 859–62.

PEREIRA, H. G. (1969). Influenza: Antigenic spectrum. *Progress in Medical Virology*, **11**, 46–79.

PONS, M. W. & HIRST, G. K. (1968). Polyacrylamide gel electrophoresis of influenza virus RNA. *Virology*, **34**, 385–8.

SCHULMAN, J. L. & KILBOURNE, E. D. (1969). Independent variation in nature of haemagglutinin and neuraminidase antigens of influenza virus: Distinctiveness of haemagglutinin antigen of Hong Kong/68 virus. *Proceedings of the National Academy of Sciences, USA*, **63**, 326–33.

SCHULZE, I. T. (1970). The structure of influenza virus. I. The polypeptides of the virion. *Virology*, **42**, 890–904.

SETO, J. T. & ROTT, R. (1966). Functional significance of sialidase during influenza virus multiplication. *Virology*, **30**, 731–7.

SHOPE, R. E. (1936). The incidence of neutralizing antibodies for swine influenza virus in the sera of human beings of different ages. *Journal of Experimental Medicine*, **63**, 655–65.

SIMPSON, R. W. & HIRST, G. K. (1961). Genetic recombination among influenza viruses. I. Cross reactivation of plaque forming capacity as a method for selecting recombinants from the progeny of crosses between influenza A strains. *Virology*, **15**, 436–51.

SKEHEL, J. J. (1971). Estimations of the molecular weight of the influenza virus genome. *Journal of General Virology*, **11**, 103–9.

SKEHEL, J. J. (1972). Polypeptide synthesis in influenza virus infected cells. *Virology*, **49**, 23–36.

SKEHEL, J. J. & SCHILD, G. C. (1971). The polypeptide composition of influenza A viruses. *Virology*, **44**, 396–408.

SMITH, W., ANDREWES, C. H. & LAIDLAW, P. P. (1933). A virus obtained from influenza patients. *Lancet*, ii, 66.

WEBSTER, R. G. (1972). On the origin of pandemic influenza viruses. *Current Topics in Microbiology and Immunology*, **59**, 75–105.

WEBSTER, R. G., CAMPBELL, C. H. & GRANOFF, A. (1973). The 'in vivo' production of 'New' influenza viruses. III. Isolation of recombinant influenza viruses under simulated conditions of natural transmission. *Virology*, **51**, 149–62.

WEBSTER, R. G. & LAVER, W. G. (1967). Preparation and properties of antibody directed specifically against the neuraminidase of influenza virus. *Journal of Immunology*, **99**, 49–55.

WEBSTER, R. G. & LAVER, W. G. (1972). Studies on the origin of pandemic influenza. I. Antigenic analysis of A2 influenza viruses isolated before and after the appearance of Hong Kong influenza using antisera to the isolated haemagglutinin subunits. *Virology*, **48**, 433–44.

WHO (1971). A revised system of nomenclature for influenza viruses. *Bulletin of the World Health Organization*, **45**, 119–23.

WHO (1972). Influenza in animals. *Bulletin of the World Health Organization*, **47**, 439–542.

WRIGLEY, N. G., SKEHEL, J. J., CHARLWOOD, P. A. & BRAND, C. M. (1973). The size and shape of influenza virus neuraminidase. *Virology*, **51**, 525–9.

YOUNG, R. J. & CONTENT, J. (1971). 5′-Terminus of influenza virus RNA. *Nature, New Biology*, **230**, 140–2.

EXPLANATION OF PLATE

Electron micrographs of:

(*a*) Influenza virus particles showing the glycoprotein surface projections.
(*b*) Haemagglutinin subunits, and
(*c*) neuraminidase subunits purified from viruses disrupted by detergent treatment.

All three specimens were negatively stained using sodium silicotungstate at pH 7.0. Scale bar refers to all three micrographs.

Kindly supplied by N. G. Wrigley.

PLATE 1

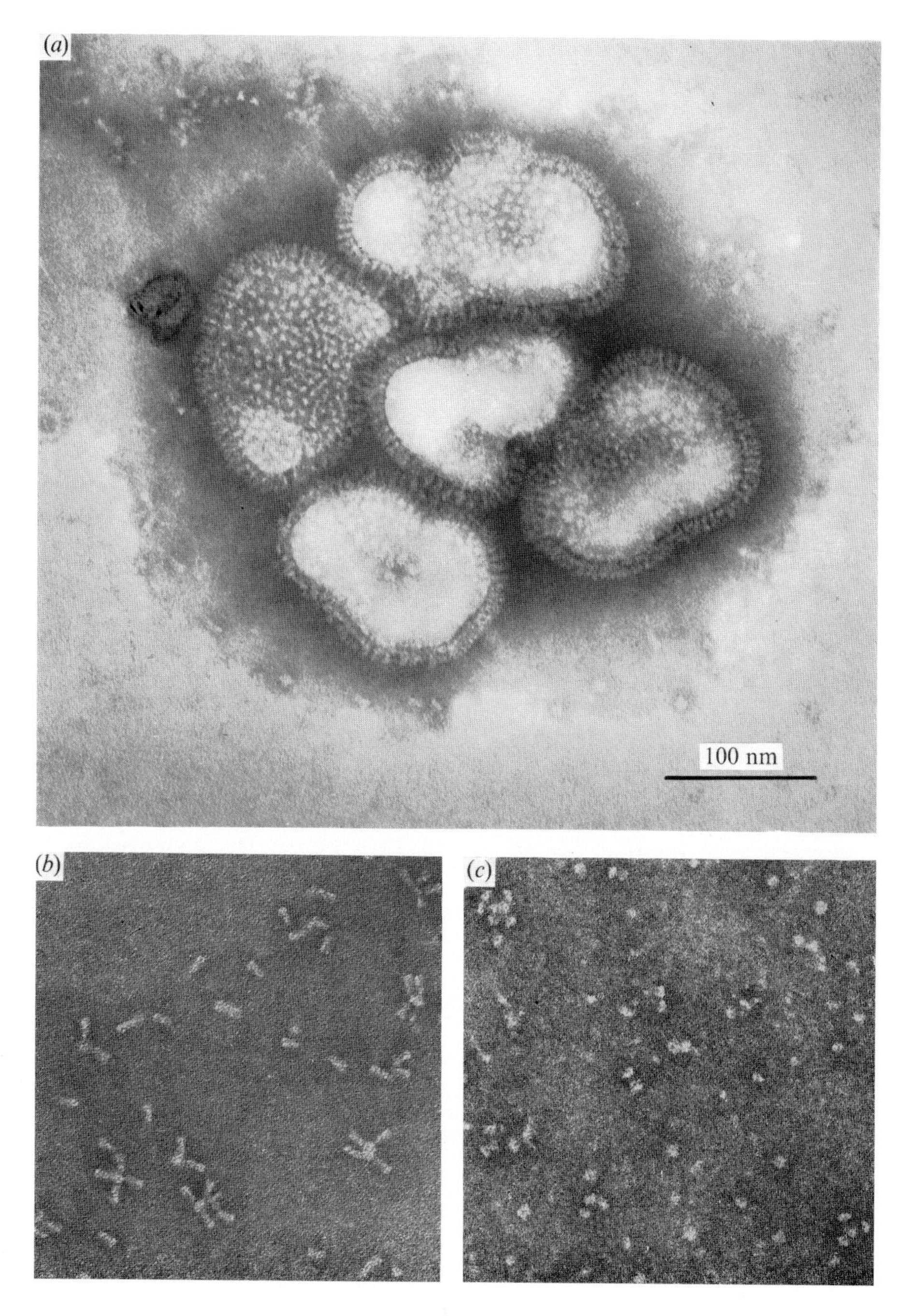

(*Facing p.* 342)

THE EVOLUTIONARY ORIGIN AND SPECIATION OF THE GENUS *TRYPANOSOMA*

J. R. BAKER

Medical Research Council Biochemical Parasitology Unit, Molteno Institute, Downing Street, Cambridge CB2 3EE

INTRODUCTION

Since I have already covered historical views on the evolution of trypanosomes and their relatives (Baker, 1963, 1965) I shall here state them briefly and bring them up to date (April, 1973). I shall then summarise my own views on the possible major lines of evolution within the genus and concentrate on one group, the Salivaria, which appear to be undergoing active evolutionary development.

Trypanosoma Gruby, 1843 is a genus of uniflagellate parasitic protozoa belonging to the class Zoomastigophorea of the subphylum Sarcomastigophora (Table 1). Within this class the trypanosomes (a vernacular synonym for *Trypanosoma*) and some related organisms are characterised by possessing a single usually extensive mitochondrion; the DNA of this organelle is concentrated into a large mass, the kinetoplast, which is about 1 μm in diameter and readily visible by light microscopy when appropriately stained. The kinetoplast is spatially, though not as far as is known functionally, associated with the flagellum, hence its name, given before its true nature was recognised. These organisms are grouped in the order Kinetoplastida. *Trypanosoma* itself, and seven closely related genera, constitute the suborder Trypanosomatina with a single family, the Trypanosomatidae. All members of this suborder are parastic. Three genera, including *Trypanosoma*, have two distinct hosts, one vertebrate and one, the vector, a blood-sucking invertebrate, in which they undergo complicated life cycles with often striking morphological and physiological transformations which are not fully understood (Newton, Cross & Baker, 1973). Certain species of *Trypanosoma* and *Leishmania* parasitise man or his domestic mammals and may be important pathogens. In their vertebrate hosts, trypanosomes inhabit the blood and tissue fluids, though members of one subgenus (*Schizotrypanum*) also have an intracellular stage. In their invertebrate vectors most species live in the gut lumen and complete their development (i.e. transform into 'metacyclic' stages which are capable of infecting the vertebrate host) in the hindgut; they

Table 1. *Classification of the family Trypanosomatidae**

Phylum Protozoa
  Subphylum Sarcomastigophora
    Class Zoomastigophorea
      Order Kinetoplastida
        Suborder Trypanosomatina
          Family Trypanosomatidae
            Genera:
              Herpetomonas, *Leptomonas*, *Crithidia*, *Blastocrithidia* } Monoxenous parasites of invertebrates
              Phytomonas – Dixenous parasite of invertebrates and plants
              Endotrypanum, *Leishmania*, *Trypanosoma* } Dixenous parasites of invertebrates and vertebrates
              Subgenera:
                Schizotrypanum (intracellular amastigotes in vertebrate host)
                Megatrypanum, *Herpetosoma*, *Duttonella*, *Nannomonas*, *Pycnomonas*, *Trypanozoon* } No intracellular stage in vertebrate host
                (*Duttonella*, *Nannomonas*, *Pycnomonas*, *Trypanozoon* } 'Salivarian' trypanosomes)
                Species†
                  brucei brucei
                  brucei gambiense, *brucei rhodesiense* } infective to man
                  evansi evansi
                  evansi equiperdum
                  equinum

* This table is by no means a complete classification of the family, more detail being included for those groups which are discussed more fully in the text.

† The assignment of rank to the minor taxa comprising the subgenus *Trypanozoon* is almost a matter of personal choice (see pp. 352–5); this is mine.

are then transferred to the vertebrate host either by the latter's ingestion of infected invertebrates or their faeces, or by faecal contamination of lesions such as the puncture caused by the invertebrate when it feeds. In a small but important group of species, the salivarian trypanosomes of mammals, which include pathogens of man and domestic animals, the infective forms develop in the proboscis or salivary glands of the vector, a tsetse fly, *Glossina* sp., and are reintroduced to the vertebrate host by inoculation when the vector feeds. A few other species of *Trypanosoma*, including some parasitising aquatic vertebrates and leeches (Hirudinea) share this characteristic, as do two other genera within the family. The reverse transfer, from vertebrate to invertebrate, is always accomplished via the latter's mouthparts as it feeds on the blood of an infected vertebrate. Fig. 1 shows diagrammatically the main features of a trypanosome's somatic organisation and some of the

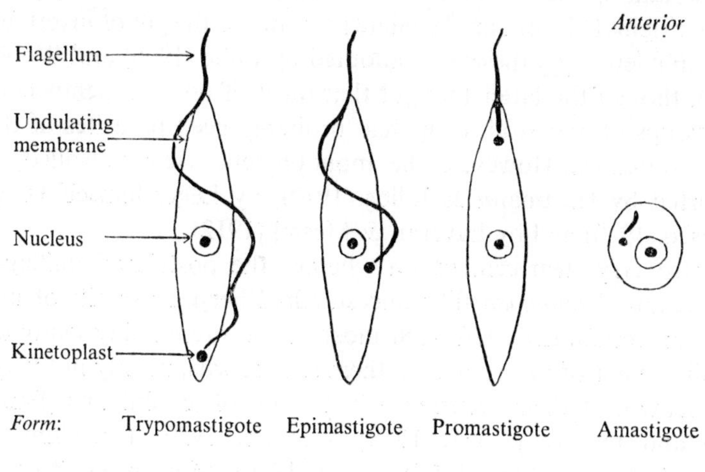

Fig. 1. Diagrammatic representation of gross structure of a trypanosome and some of the forms which may occur during its life cycle (based on Hoare & Wallace, 1966).

major modifications which this may undergo during the life cycle. Further information on trypanosomes in general, and those of mammals in particular, can be obtained from the monograph by Hoare (1972).

EVOLUTIONARY ORIGIN OF THE GENUS

Species of Trypanosomatidae with the promastigote body form inhabiting the gut of a variety of invertebrates, especially insects, which are the only hosts in their monoxenous life cycle, occur widely (Wallace, 1966); they are transmitted by means of encysted amastigotes in their host's faeces and constitute the genus *Leptomonas*. The promastigote somatic organisation is basically similar to that of the vast group of free-living, chlorophyll-bearing flagellates of the class Phytomastigophorea such as *Euglena* (Kudo, 1966) which may be among the most primitive of eukaryotic organisms. It is therefore generally assumed that the dixenous trypanosomes, with their more complex life cycles and trypomastigote body form, were derived from forms similar to the present-day *Leptomonas* spp., a second host being incorporated into the life cycle when some of the original invertebrate hosts adopted haematophagy. However, certain authors have doubted this. Léger (1904*a*) suggested that trypanosomes had evolved from a biflagellate kinetoplastid flagellate (*Cryptobia*) inhabiting the blood of fish, through the loss of the anterior flagellum. He later (Léger, 1904*b*) suggested that only certain species of *Trypanosoma* had evolved in this way, others having originated from uniflagellate parasites of the gut of invertebrates. This diphyletic hypothesis was adopted by Lühe (1906) and Woodcock (1906), though the latter thought that most, if not all, members of the two groups of ancestral forms had probably been parasites of the gut of invertebrates. However, the monophyletic view prevailed, being supported by Hartmann & Jollos (1910), by Léger himself (Léger & Duboscq, 1910) and by Laveran & Mesnil (1912).

Controversy then centred on whether the postulated uniflagellate, monoxenous *Leptomonas*-like ancestor had been a parasite of invertebrates or vertebrates. Although most of the present-day leptomonads inhabit the gut of invertebrates, there are records, mostly not very well documented, of their occurrence in the gut of amphibians (Fantham, 1922) and reptiles (Léger, 1918; Wenyon, 1920; Franchini, 1921; Strong, 1924; Fantham, 1926; Bayon, 1926). Minchin (1909) was the first to propose that the ancestral leptomonads were parasites of the gut of vertebrates rather than invertebrates. He based his suggestion on the fact that invertebrate hosts of trypanosomes may be insects or leeches (or, as is now known, mites); in his words, 'the constant trypanosome host is the vertebrate; the inconstant host is the invertebrate'. However, this argument seems to bear little weight when it is remembered that many trypanosomatids have no vertebrate host at all. It is difficult to explain the widespread occurrence of trypanosomatids in insects which are not haematophagous by this hypothesis. Minchin (1912) himself changed his views and later wrote that the ancestor of *Trypanosoma* was 'originally a parasite of the invertebrate alone' (Minchin, 1914). Minchin's earlier view was supported by Mesnil (1918), but little more was heard of it until Lavier (1943) and Dias (1948) resurrected it. The contrary view, that *Trypanosoma* evolved from monoxenous uniflagellate promastigote parasites of the gut of invertebrates, has prevailed among most parasitologists (Wenyon, 1926; Grassé, 1952; Manwell, 1961; Hoare, 1972) – rightly, in my view. In addition to the numerous records of such parasites in insects (Wallace, 1966), there are a few reports of their presence in 'lower' invertebrates such as nematodes (Bütschli, 1878; Chatton, 1924; Goodey & Triffit, 1927), molluscs (Porter, 1914; de Mello, 1921; Fantham, 1925) and rotifers (Chatton, cited by Grassé, 1952, p. 611), which support the idea that the family Trypanosomatidae was orginally parasitic in very early invertebrates, before blood-sucking developed as a means of nourishment and probably before there were any vertebrates with blood to suck. If it is accepted that trypanosomatids were parasites of invertebrates long before the adoption of haematophagy, and indeed before the evolutionary divergence of arthropods and annelids from their presumed common ancestor (Buchsbaum, 1951) before the Cambrian period began some 500×10^6 years ago (Oakley & Muir-Wood, 1949), then the fact that some members of the three major groups of haematophagous invertebrates (Annelida, Acarina, Insecta) act as vectors of trypanosomes, which caused Minchin (1908) to doubt the origin of the genus from parasites of invertebrates, becomes readily explicable.

DEVELOPMENT OF THE TRYPOMASTIGOTE FORM

This form (Fig. 1) is found only in the genus *Trypanosoma* and usually occurs only in those stages of the life cycle which inhabit the blood and tissue fluids of the vertebrate host. There are, however, two exceptions to this. One is the metacyclic stage which develops at the end of that part of the life cycle which is undergone in the invertebrate vector, and which is the stage capable of infecting the vertebrate host. This stage is trypomastigote and is pre-adapted physiologically and morphologically for life in the vertebrate (Vickerman, 1971). The other exception is in the salivarian trypanosomes (see below), in which the trypomastigote form is retained during an initial period of multiplication in the vector insect (the 'procyclic' trypomastigote; Newton, Cross & Baker, 1973). These forms can perhaps be regarded as an intercalation into the life cycle of an accessory multiplicative phase before the morphological transformation which usually occurs at this stage of the cycle (Newton, Cross & Baker, 1973), possibly to increase the parasites' chances of survival in a vector, the great majority of individuals of which appear to be resistant to infection by the parasite. Thus the trypomastigote form appears to be primarily related to life in the vertebrate's bloodstream or tissue fluid; the intracellular stages of those species of *Trypanosoma* which have such a stage in their vertebrate hosts are not trypomastigote. It is tempting therefore to suggest that this form is, at least in part, a mechanical adaptation to life in a viscous but liquid environment. The significant feature of the trypomastigote, from this point of view, may be the presence over most or all of the length of the cell of an undulating membrane – a wavy fin-like expansion of the pellicle produced by the undulations of the flagellum which is attached at intervals to the pellicle by 'maculae adherentes' resembling desmosomes (Vickerman, 1969); such an expansion might well increase the trypanosome's motility in the plasma – and, perhaps equally importantly, serve more efficiently to stir and thus to renew the surrounding medium.

SPECIATION WITHIN THE GENUS

Speciation of trypanosomes of vertebrates other than mammals

In general, *Trypanosoma* seems to have evolved with its vertebrate host groups. Presumably the earliest representatives of the genus were those introduced by leeches and early haematophagous insects to aquatic vertebrates (fish) and the early terrestrial vertebrates (amphibia), which are thought to have evolved before or during the Devonian period, about 300×10^6 years ago (Oakley & Muir-Wood, 1949). It is not necessarily true that all present-day species of *Trypanosoma* were introduced to their vertebrate hosts by their current vectors. It is entirely possible that *Trypanosoma grayi* of the African crocodile was first introduced to that reptile or its evolutionary progenitors by a leech and only subsequently became adapted to its present vector, *Glossina palpalis*, when the latter adopted the apparently risky habit of feeding inside the crocodile's mouth (Hoare, 1929). Having been introduced to each evolving group of vertebrates – either anew from an invertebrate or more probably by evolving together with the new vertebrate group and its associated blood-sucking invertebrates as part of a dynamic tryptych or 'ephemeral triangle' – the trypanosomes then presumably continued to evolve to fill each ecological niche (i.e. each new species of vertebrate) available to them in that group. However, this apparently rather simple situation may not apply to the trypanosomes of mammals, which are dealt with below.

Speciation of stercorarian trypanosomes of mammals

The term stercorarian (Hoare, 1964) is used in contradistinction to salivarian to distinguish those species of *Trypanosoma*, the majority, which complete their development in the invertebrate host in the hindgut; the salivarian species do so in the vector's foregut (proboscis) or salivary glands. By definition, the two terms were restricted to species of *Trypanosoma* which parasitise mammals. Originally they were introduced as the names of two taxa of undetermined rank between genus and subgenus, but it is perhaps better to use them only as descriptive adjectives.

Within the stercorarian species parasitising mammals two major groupings can be recognised: those (the subgenus *Schizotrypanum*) which have an intracellular phase in their vertebrate host and those (all other subgenera) which do not. This dichotomy may represent an early division of the family into what I have previously termed two 'stocks' (Baker, 1965; Hoare, 1948), originally referred to as 'leptomonad' and 'crithidial' stocks respectively but following the terminology introduced by Hoare & Wallace (1966) better called the promastigote and epimastigote stocks (Fig. 1). Since my earlier review, increased knowledge of the relationship between the flagellar apparatus and the rest of the trypanosomatid cell resulting from electron microscope studies (Wallace, 1966; Hoare & Wallace, 1966) has suggested that the genera *Herpetomonas* and *Crithidia* may represent a third, minor evolutionary line. The genus *Endotrypanum* and the subgenus *Schizotrypanum* (of *Trypanosoma*) occupy equivocal positions. Only two species of the former are known (Shaw, 1969); both inhabit the erythrocytes of their vertebrate hosts (sloths), one (*E. schaudinni*) as epimastigotes and one (*E. monterogeii*) as trypomastigotes. In their probable vectors (phlebotomine Diptera or sandflies), however, they develop as promastigotes (Shaw, 1964). The subgenus *Schizotrypanum* lives in its mammalian hosts mainly as trypomastigotes and amastigotes, though transition forms between these two stages may resemble pro- and epimastigotes. In its vectors (reduviid Hemiptera) it develops mainly as epi- and trypomastigotes. Some of the more 'primitive' species of *Trypanosoma*, such as *T. diemyctyli* of amphibians, also produce promastigotes during development in their vectors (Hirudinea or leeches) as shown by Barrow (1953). This last exception may perhaps be explained as a retention in ontogeny of an early stage in phylogeny, the presumed ancestral monoxenous promastigote of invertebrates, which has been omitted from the life cycle of more recently evolved species of the genus. But the development undergone by *Endotrypanum* in its vector, and the dependence upon amastigote forms shown by members of the subgenus *Schizotrypanum* (amastigotes are its only reproductive forms in the vertebrate host), may indicate that these two groups are probably more closely allied to the promastigote stock than the epimastigote stock, though possessing some of the characters of both (Shaw, 1966; Cameron, 1956). These ideas are incorporated in the proposed phylogenetic 'tree' shown in Fig. 2 (p. 359).

Origin and speciation of the salivarian trypanosomes of mammals

This relatively homogeneous group (Table 1) is in many ways atypical of the genus. (1) All its members (except a few which have subsequently adopted other modes of transmission) have as their vector hosts Diptera of the genus *Glossina* and, as these flies are restricted to tropical Africa, the salivarian trypanosomes are similarly limited with certain exceptions; see pp. 355–7 below. (2) All complete their development in the proboscis or salivary glands of their vector, instead of its hindgut, and are transmitted via its proboscis – a feature which they share with only a few other species. *T. rangeli* is the only other species of trypanosome parasitising mammals which develops in the salivary glands of its vector (a reduviid Hemipteran) and is transmitted via the vector's proboscis. Possibly one species parasitising birds develops similarly in mosquitoes, but the evidence is slight (Nair & David, 1956). *Trypanosoma platydactyli* of *Tarentola mauretanica* (a gecko) develops in the mid- and foregut only of the sandfly *Phlebotomus parroti* and not in the hindgut, but transmission is apparently by the gecko's eating a sandfly rather than via the latter's proboscis (Adler & Theodor, 1935). However, all the trypanosomes of aquatic vertebrates and leeches (Hirudinea), whose life cycles have been adequately studied, develop in the leeches' mid- and foregut and not in the hindgut; they are transmitted to the vertebrate host either by the latter's ingestion of an infected leech or via the proboscis when the leech feeds (Brumpt, 1906*a*, *b*, *c*, 1914; Robertson, 1908, 1912; Nöller, 1913; Barrow, 1953; Qadri, 1962): this fact will be discussed later in connection with the evolutionary origin of the salivarian trypanosomes. (3) All the salivarian species reproduce in their vertebrate hosts by division of trypomastigotes in the circulating blood and tissue fluids – though possibly as other forms also (Ormerod & Venkatesan, 1971*a*, *b*) – whereas most if not all other species divide as epi- or amastigotes and usually not in the peripheral blood. (4) Perhaps related to this, the salivarian trypanosomes tend to be more pathogenic to their vertebrate hosts. This may be related to their ability to evade the host's immune response by changing their surface antigens (Vickerman, 1971). (5) Three of the salivarian subgenera (not *Duttonella*) differ from other trypanosomes by commencing their multiplication in the invertebrate as trypomastigotes instead of, as probably occurs with all other species of *Trypanosoma*, first transforming into epimastigotes or sometimes promastigotes. (6) The salivarian subgenera have physiological peculiarities also; the respiratory metabolism of some at least of the forms in the circulating blood of the vertebrate host does not depend on mitochondrial enzymes and a cytochrome chain but on an extramitochondrial aerobic system, which may be present in other species and stages but which is not their major oxidative pathway. In the vector, however, the parasites, except perhaps for the final infective trypomastigote stages, appear to respire at least in part by more conventional intramitochondrial pathways (Vickerman, 1971; Newton, Cross & Baker, 1973).

All these peculiarities suggest a common origin for the salivarian trypanosomes. Furthermore, none of these species is efficient at infecting its vector. The highest rate obtained for a salivarian subgenus

(*Duttonella*) is about 90 % under optimal experimental conditions (Desowitz, 1957), but a more usual figure is about 20 % (Hoare, 1949). The rates for other salivarian subgenera are less – with *Trypanozoon* 45 % has been achieved by introducing parasites *per anum* (Wijers & McDonald, 1961) and 40 % as a result of heating the pupae of the vectors (Burtt, 1946), but the 'normal' experimental maximum is 21 % (Burtt, 1946). Usually, however, rates of about 1 % are common. *Nannomonas* is intermediate in this respect. This lack of susceptibility of the invertebrate host is generally assumed to indicate a relatively recent association between parasite and vector. Added weight is given to this idea by the fact that *Duttonella*, which has the highest infection rate, develops only in the vector's proboscis; *Nannomonas* develops in its midgut and proboscis; and *Trypanozoon* (and *Pycnomonas*), with the lowest infection rates, develop in the midgut and salivary glands. The inference is that the parasites have only fairly recently evolved the ability to infect *Glossina*, with successive adaptations leading to progressively deeper invasion of the insect. Hoare (1948) suggested that the ancestors of the salivarian species were stercorarian parasites of African mammals which relatively recently acquired the ability to survive and develop in the tsetse flies (*Glossina* spp.) into which, in the course of the flies' ingestion of blood from parasitaemic mammals, they must have been repeatedly introduced. The first subgenus to result from this adaptation was *Duttonella*; further adaptation, to life in the insect's midgut, led to the evolution of the subgenus *Nannomonas*; and finally, the sophisticated adaptation involving completion of the life cycle in the salivary glands led to the evolution of the subgenera *Pycnomonas* and *Trypanozoon*, the extremely low vector infection rates characteristic of the two latter subgenera being a mark of the recentness of this development.

An alternative hypothesis (Baker, 1963) is that the salivarian species might be derived directly from those leech-transmitted parasites of aquatic vertebrates which develop in either or both the fore- and midgut of their vector and are transmitted to their vertebrate hosts by injection through the proboscis when the vector feeds. Thus the proboscis-mediated transmission of the salivarian trypanosomes would be a primitive and not a secondary feature, which they retained on being carried by their evolving vertebrate hosts from water to land and hence forced to adapt to a different vector. This, however, implies that the salivarian trypanosomes are not a recently evolved group but a very ancient one and as pointed out above the evidence is against this, at least with regard to their relationship with their vector.

A third hypothesis has been proposed by Woo (1970), who had shown that the salivarian species *T. brucei* could infect turtles which had been acclimatised to a temperature of 35–37 °C, for at least twelve days and that the tropical crocodile *Caiman sclerops* could be infected for at least twelve weeks without temperature acclimatisation (Woo & Soltys, 1969). He therefore suggested that the salivarian trypanosomes might have evolved relatively recently from species which infected amphibious reptiles and were previously transmitted by leeches, the opportunity to adopt *Glossina* species as vectors being presented when these insects fed upon reptiles basking on the land, as they are known to do on *Crocodilus niloticus*. Thus this plausible hypothesis combines features of both of the previous ones: the relative newness of the association between parasite and vector suggested by Hoare (1948) and the retention rather than re-acquisition of final development in the vector's foregut or salivary glands, and transmission via its proboscis emphasised by Baker (1963). *T. platydactyli* (see p. 350 above) may have evolved similarly.

Thus Woo's (1970) proposal takes account of the evidence in favour of both the previous 'rival' hypotheses, and avoids the arguments against each. It is perhaps the closest approximation to the truth which has so far been suggested, and I have adopted it in the phylogenetic 'tree' proposed in Fig. 2.

Speciation within the subgenus Trypanozoon

The subgenus *Trypanozoon* (Table 1) includes *T. brucei gambiense* and *T. brucei rhodesiense*, the subspecies (or species or races – see Hoare, 1972, pp. 483–4) causing sleeping sickness in man. All its other species are pathogens of domestic animals.

A great deal of speculation has centred on the evolution of this subgenus, perhaps partly because of its importance to man, but also because it seems to be in the midst of a period of evolutionary radiation which, though frustrating to the taxonomist who likes clear-cut divisions between taxa, has resulted in some interesting relationships between the various constituent populations of the subgenus which are perhaps on the way to becoming fully fledged species. The subgenus can conveniently be considered in two groups – those which undergo cyclical development (i.e. multiply and undergo successive morphological transformations) in their vectors (*Glossina* spp.), and those which do not.

Species and subspecies developing cyclically in Glossina

This group includes *T. brucei brucei*, *T. brucei gambiense* and *T. brucei rhodesiense*. All three are morphologically indistinguishable; in the vertebrate's blood they are characteristically pleomorphic, existing as long slender trypomastigotes (multiplicative forms), short stumpy trypomastigotes (almost certainly forms which are infective to the vector), and forms intermediate between these two extremes (Plate 1). They differ, however, in the diseases that they produce. *T. b. brucei* does not infect man but, like all the group, can infect a wide range of mammals from mice to monkeys; it generally produces relatively high, though fluctuating, parasitaemias in laboratory rats and mice and is resistant *in vivo* to the organic arsenical drug tryparsamide. *T. b. gambiense* and *T. b. rhodesiense* infect a similar range of hosts, but they can also infect man; *T. b. rhodesiense* resembles *T. b. brucei* in its virulence to rats and mice and resistance to tryparsamide, whereas *T. b. gambiense* is much less virulent to man as well as to laboratory animals and is sensitive *in vivo* to tryparsamide. There are also epidemiological and geographical differences between *T. b. rhodesiense* and *T. b. gambiense* (Baker, 1974). The difference in virulence is not clear-cut and strains of intermediate virulence exist; moreover, virulence and other characters also change markedly after prolonged maintenance of strains under artificial conditions in laboratories. However, the reality of the distinction between these subspecies has been supported by the recent work of Newton & Burnett (1972) which has revealed small but constant differences between the buoyant densities in caesium chloride gradients of the various DNA components from the three subspecies, presumably reflecting differences in their nucleotide compositions, and in the numbers of components separable by this technique.

Their close similarity makes it almost certain that the three subspecies had a common evolutionary ancestor. As *T. b. brucei* is the most widespread geographically, and as, apart from man, it has the same wide spectrum of susceptible hosts as do the other two subspecies, it seems reasonable to conclude that the common ancestor resembled the present-day *T. b. brucei*, i.e. that the latter is the least changed descendant of the common ancestral species. It seems equally reasonable to assume that the two subspecies capable of infecting man arose from this ancestral form by acquiring that ability, possibly by a single-gene mutation, since the character involved seems to be a relatively simple one. What is in doubt is whether the two subspecies evolved independently from the ancestral species, or whether one is a later modification of the other. There are two reasons, neither conclusive, for thinking that *T. b. gambiense* may be the species that acquired the ability to infect man earlier: (1) records of human trypanosomiasis are known from much earlier times in West Africa, where at present only *T. b. gambiense* occurs, and clinically the disease described seems to resemble the chronic type characteristically produced by this subspecies; and (2) the lower virulence of *T. b. gambiense* for man is thought to indicate a longer period of adaptation of the parasite to its host. Certainly *T. b. gambiense* appears to have become more dependent on man as a host, since it has no known non-human mammalian reservoir host, and hence any adaptation leading to an increase in the parasitaemic life of an infected man would, by increasing the length of time for which he would be available as a source of infection for *Glossina*, benefit the parasite. *T. b. rhodesiense* was first recorded, from the Zambezi valley in Rhodesia, only at the beginning of this century and there is evidence (Ormerod, 1961, 1967) of its progressive northward spread since then, culminating in its recent appearance in Ethiopia (Hutchinson, 1971). There are many possible explanations for the lack of earlier records of this parasite (Duggan, 1970), but the evidence is persuasive that it is in evolutionary terms an extremely recent development. Also, it is by no means fully adapted to man, to whom, especially in areas where it has recently appeared, it may be very virulent, and it appears to depend upon wild or domestic ungulate hosts, in which it produces a much less acute infection, for its survival between epidemic episodes (Baker, 1974). Thus, if these arguments be accepted, one is left with the alternative hypotheses of the development of *T. b. rhodesiense* from *T. b. gambiense* or its independent origin from *T. b. brucei*. Most authors have preferred the former idea, pointing out that virulent strains, clinically resembling *T. b. rhodesiense*, occasionally occur in areas where normally only 'typical' *T. b. gambiense* is known. Willett (1956) suggested that the increase of virulence necessary to produce what is now known as *T. b. rhodesiense* resulted from the introduction of *T. b. gambiense* to a new geographical area (SE Africa) and hence to a new species of vector, *G. pallidipes* rather than *G. palpalis*, it being postulated that the injection of a larger number of metacyclic trypomastigotes by *G. pallidipes* increased the parasite's virulence to mammals. Ashcroft (1963) suggested that in an area such as the east African savannah, in which the diminished contact between man and *Glossina* and increased contact between wild mammals and *Glossina* forced the parasite to depend on wild ungulates for its maintenance, only strains virulent enough to maintain in ungulates parasitaemias adequate to infect *Glossina* would survive. Hence selection pressure would operate in the reverse direction from that applied in the west African situation of close man–fly contact – i.e. towards an increase of virulence, limited however, as rapid death of the ungulates would not be conducive to the parasites' survival. Thus, if virulence to man and to ungulates are related, the increased virulence to man of *T. b. rhodesiense* would be an adventitious concomitant of its ecologically induced dependence on a transmission cycle involving ungulates and *Glossina*. Man can be regarded as only a facultative host of *T. b. rhodesiense*, not an obligate one, as he is of *T. b. gambiense*. This hypothesis was accepted by Willett (1963), who combined it with his earlier (Willett, 1956) suggestion of the movement of *T. b. gambiense* south-eastwards across Africa to explain the origin of *T. b. rhodesiense*. Ormerod (1961), however, while accepting Ashcroft's (1963) epidemiological and epizootiological views, considered that *T. b. rhodesiense* arose in the Zambezi river basin, so far removed geographically from any known focus of *T. b. gambiense* that it almost certainly must have originated from *T. b. brucei*. This idea necessarily implies separate evolutionary origins of the two man-infecting subspecies from the ancestral species. As Hoare (1972, p. 536) has written, 'the correctness of either of these hypotheses cannot be assessed with certainty'.

Species and subspecies which do not develop cyclically in a vector

This group includes three species or subspecies, *T. evansi evansi*, *T. evansi equiperdum* and *T. equinum* (Table 1). Morphologically they resemble the slender or intermediate haematozoic forms (i.e. those inhabiting the blood of the vertebrate host) of *T. brucei* sspp.; only rarely do they show the characteristic pleomorphism of that species. *T. equinum* differs, however, in lacking a normal kinetoplast (Plate 1). None of these parasites undergoes cyclical development in a vector, though *T. e. evansi* and *T. equinum* are transmitted non-cyclically by various biting Diptera (e.g. *Stomoxys*, Tabanidae), in whose probosces they may remain viable long enough to be transferred to another host if the insect is disturbed while in the course of a meal on the blood of a parasitaemic host. *T. e. equiperdum* has no vector at all (see below). *T. e. evansi* is widely distributed throughout the tropical and warmer subtropical regions of all continents except North America and Australasia, where it has been eliminated; it is a pathogen of camels, equines, other domestic ungulates and dogs and has been recorded from the Indian elephant. *T. e. equiperdum* is now restricted to South America, having been eliminated from Europe, India and North America, and is a pathogen of equines. It lives almost exclusively in the tissue fluid of oedematous patches in the skin of these animals, especially on the genitalia, and is transmitted venereally when such patches are abraded during coitus. *T. equinum* is known mainly from South America, where it is a potentially pathogenic parasite of equines and other domestic ungulates, but has also been reported as 'akinetoplastic strains of *T. evansi*' from the Sudan and Nigeria (Hoare & Bennett, 1937; Killick-Kendrick, 1964) and may be expected to occur wherever *T. e. evansi* does (see below). Hoare (1972, pp. 484, 555) regards *T. equinum* as a synonym of *T. evansi*, from which it probably arose or arises by a mutation involving the loss of the kinetoplast – the condition of 'akinetoplasty' or, better, 'dyskinetoplasty'. Vickerman (1970) has shown that the kinetoplast membrane (= mitochondrial membrane) and some part of its content are in fact retained (see below). As *T. equinum* is therefore separable from *T. evansi* by a constant and definite morphological feature – dyskinetoplasty – it seems better to regard it as a separate species. The proposed scheme for evolutionary development of this group of trypanosomes, which will be summarised below, is almost entirely due to Hoare (1972).

It is postulated that camels, which have for centuries been used as beasts of burden on the caravan routes which extend from North Africa southwards into the tropical regions of that continent on both the east and west sides, must have sometimes become infected with *T. b. brucei* at the southern end of their journeys, where they would temporarily have come into contact with *Glossina*. After returning north to tsetse-free areas, transmission of the trypanosomes could have occurred non-cyclically via Tabanidae, it having been shown that *T. brucei* can be transmitted thus. Selection pressure would then operate on the trypanosomes to increase the parasitaemia in the camels' blood, thereby increasing their chances of non-cyclical transmission. This could occur by an increase in the proportion of long slender multiplicative trypomastigotes present in the population. At the same time, the short stumpy trypomastigotes which rarely if ever divide in the vertebrate but are almost certainly adapted to continuing cyclical development in the vector, would become not only redundant, since there was no longer any suitable vector for cyclical development, but actually disadvantageous, since they form in the vertebrate a non-reproducing and therefore useless section of the population. It is therefore easy to understand how such selection could lead to the evolution of a trypanosome strain in which the stumpy forms were suppressed and all individuals were of the multiplicative type. Concomitant with this loss of pleomorphism would be an increased population growth rate and the loss of the ability to infect tsetse flies, as is seen in *T. e. evansi.* The relative newness of this development is indicated by the fact that certain strains of *T. e. evansi* spontaneously or under experimental manipulation revert to the production of stumpy forms and therefore to sporadic pleomorphism. An exactly similar adaptive process occurs in strains of *T. brucei* sspp. maintained in laboratories by repeated syringe passage from rodent to rodent; they become highly virulent in the rodent host and lose both their pleomorphism and their ability to infect *Glossina.* The disease produced by *T. e. evansi* in camels is however chronic, which would have facilitated the wide spread of *T. e. evansi* by carriage in camels over the extensive caravan routes throughout north Africa and into Asia. The catholic feeding habits of tabanids doubtless spread the parasite to other domestic ungulates and dogs, which carried it to Australia and to North and South America; probably the Spanish Conquistadores introduced infected horses to the last-named continent. In South America, *T. evansi* infects vampire bats which transmit it from ungulate to ungulate – possibly a unique example of a trypanosome parasitising two species of mammal in a dixenous life cycle (Hoare, 1965).

The evolution of the dyskinetoplastic species *T. equinum* is explicable in similar terms. The DNA of the kinetoplast represents the trypanosome's aggregated intramitochondrial DNA and although its function is not fully understood, it is likely to be concerned with synthesis and control of the mitochondrial enzyme systems and therefore, presumably, with the 'switching on' of these systems which occurs when haematozoic trypomastigotes of *T. brucei* sspp. are introduced into the midgut of their insect vector or into a culture tube, in which their development mimics the initial phase of development in the vector. As Newton, Cross & Baker (1973) put it, the kinetoplast DNA 'may contain genetic information required for the synthesis of some or all of the oxidative enzymes present in culture forms [and presumably procyclic trypomastigotes in the vector's midgut] but absent from blood stream trypomastigotes of this group [*T. brucei* sspp.]'. In species such as *T. e. evansi,* in which cyclical development in a vector no longer occurs, there is presumably no need for the mechanism which synthesises these enzymes and 'switches on' the mitochondrion. Thus, if this is the kinetoplast's major or only function, it would have become redundant in such species. Therefore, if a mutation leading to the deletion of the 'kinetoplast gene' were to occur, the resulting dyskinetoplastic individuals would be at no selective disadvantage and could therefore theoretically become established in the parent population. Either by chance, for example natural transmission of one or a very few organisms to a new host, or by possessing some advantageous feature, the dyskinetoplastic individuals could replace the parent population; this is presumably how *T. equinum* evolved from a parent *T. e. evansi* strain. Presumably it is a polyphyletic species, having appeared in various areas of the world at various times, doubtless including the present, but becoming established only in certain unknown circumstances. Since drugs such as acriflavine selectively inhibit the replication of kinetoplast DNA, their use can lead to the artificial production of *T. equinum* in the laboratory (Newton, Cross & Baker, 1973). It is likely that the dyskinetoplasty mutation (if it is a mutation) occurs relatively frequently in all species of the subgenus *Trypanozoon*, since about 1% of haematozoic individuals of *T. brucei* sspp. are dyskinetoplastic, at least in laboratory rodents. However, under natural conditions involving cyclical transmission by tsetse flies, such mutant individuals would be filtered out of the population by their inability to 'switch on' their mitochondrial enzyme system and so develop in the vector. *T. evansi equiperdum* presumably also evolved from *T. e. evansi*, by adaptation to life in tissue fluids rather than blood, a tendency shown by all species of *Trypanozoon* to a lesser extent, and by adopting as its vector the penis of its host, the horse.

CONCLUSIONS

Evolutionary trends

The views outlined in this paper have been summarised in the dendrogram of Fig. 2. They are necessarily very speculative. However, at least one trend is clear, that manifested particularly by the subgenus *Trypanozoon*, probably the most recent subgenus of *Trypanosoma* to evolve and possibly still in the 'exponential phase' of its evolution. The trend is the tendency towards emancipation from the invertebrate host, which on the view accepted in this paper is the ancestral host of the entire family Trypanosomatidae. Commencing their evolution, therefore, as monoxenous parasites of the intestine of some early group of invertebrates ancestral to the present-day annelids and arthropods, the trypanosomatids exploited the opportunity offered by their hosts' adoption of phytophagy or haematophagy to include a second host in their life cycle, resulting in the development of the genera *Phytomonas*, *Leishmania*, *Endotrypanum* and *Trypanosoma*. The last-named genus may have been merely a 'passenger' in its earlier vertebrate hosts, undergoing little or no multiplication therein; it is possible that this

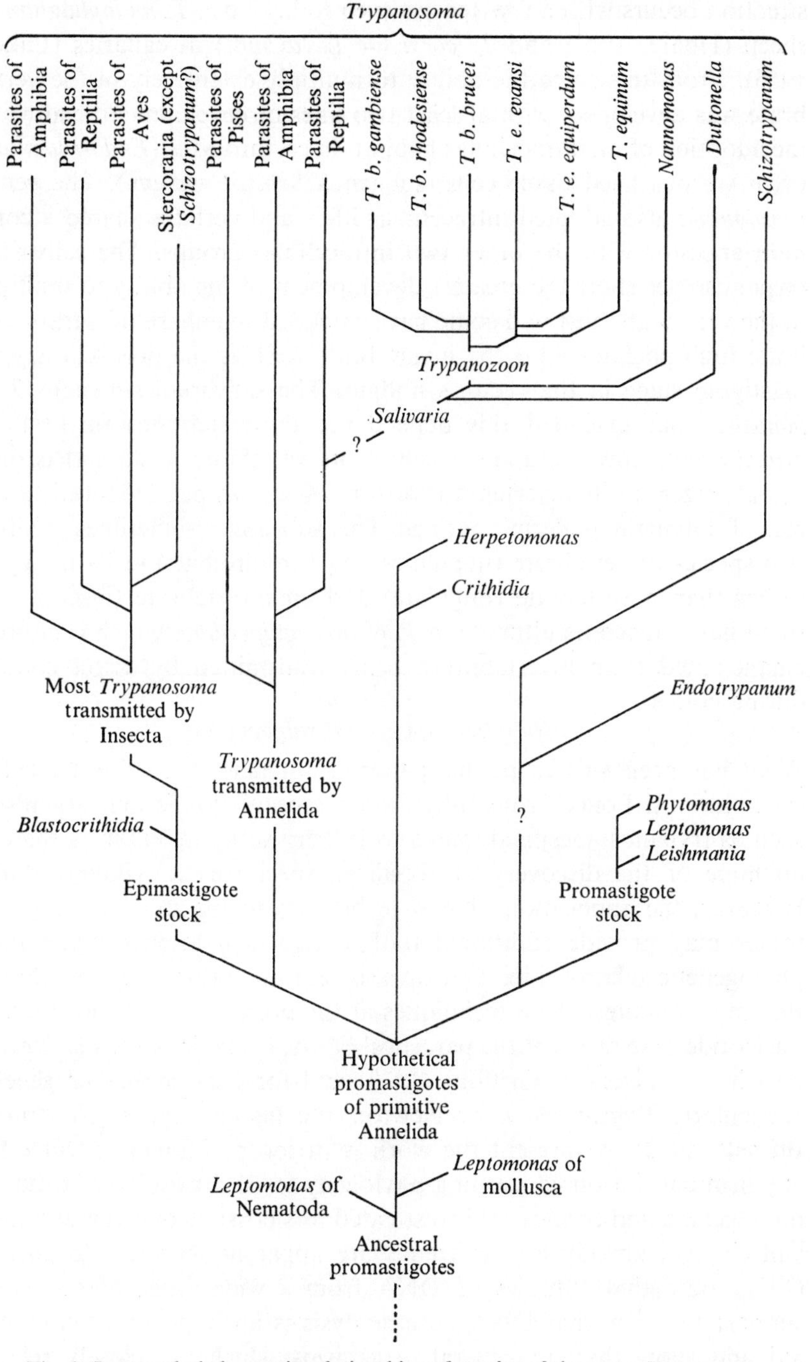

Fig. 2. Proposed phylogenetic relationships of species of the genus *Trypanosoma*, including some allied genera.

situation occurs with a few species even today – e.g. *T. melophagium* in sheep (Hoare, 1923) and *T. corvi* (= *T. 'avium'*) in canaries (Baker, 1956). From this stage the ability to multiply extensively in the vertebrate was developed, with at least two lines of development leading to the adoption of an intracellular habitat, in erythrocytes (*Endotrypanum*) or in various fixed tissue cells (subgenus *Schizotrypanum*). The genus *Leishmania* also adopted intracellular life – and perhaps shared a common ancestor with the other two intracellular groups. The salivarian trypanosomes show the greatest development of the ability to multiply in the vertebrate host and some have extended the phase of intra-vertebrate multiplication into the invertebrate host as the procyclic trypomastigote stage in the vector's midgut. The salivarian subgenus *Trypanozoon* has extended this dependence on vertebrates furthest. *T. evansi evansi* now multiplies solely in its vertebrate hosts and is only a 'passenger' in its invertebrate vector, which has been reduced to the role of a flying hypodermic syringe. This subspecies sometimes involves two species of vertebrate (ungulates and vampire bats) in its life cycle rather than a vertebrate (ungulate) and an invertebrate (insect). The trend has reached its ultimate in *T. evansi equiperdum*, which is entirely emancipated from invertebrates, being transmitted by direct contact during coitus.

New light on an old subject?

What has been written in this paper is not new; the subject itself is even older. As I once wrote, 'phylogenetic speculation [about organisms such as trypanosomes] is doomed to be frustrating in so far as there is no hope of the discovery of fossilized specimens...' (Baker, 1963). However, the application of new techniques to the study of trypanosomes may provide additional tools with which to investigate their phylogenetic relationships and thus, by extrapolation, their evolution. Foremost amongst these techniques at the present time is the study of nucleotide base ratios of the parasites' DNA, being done for *Leishmania* by Chance, Peters & Griffiths (1973) and for that and other genera, particularly *Trypanosoma*, by Newton and his colleagues (Newton & Burnett, 1972). At present the work is in too preliminary a stage for any firm conclusion concerning phylogeny to be drawn from it, but as more genera and species are investigated it is possible that this situation will change; already certain trends are apparent. Newton & Burnett (1972) state that 'Studies of DNA from a wide range of cell types support the view that DNA base analysis is an important taxonomic aid and show that, in general, organisms which are closely related genetically or by the criteria of numerical taxonomy have DNA base compositions which are similar...There is a progressive decrease in density from values corresponding to a high G+C [guanine plus cytosine] content in the kinetoplast DNA of *Crithidia*, through intermediate values in the stercoraria to densities corresponding to a high A+T [adenine plus thymine] content in the kinetoplast DNA of *brucei*-group trypanosomes [i.e. the subgenus *Trypanozoon*].' They go on: 'it is tempting to speculate whether the observed progressive increase in A+T content of kinetoplast DNA reflects a gradual evolutionary development...from monogenetic flagellates of insects through the stercorarian parasites to the *brucei*-group of salivaria'. Thus, very broadly, the results so far obtained by this promising technique at least do not seem to contradict the general outline of the evolution of the family Trypanosomatidae deduced from other facts and summarised in Fig. 2 of this paper.

Another technique currently being applied to the Trypanosomatidae is the hybridisation of kinetoplast DNA with complementary RNA as a means of studying base sequence homologies between different species and subspecies (Newton, Steinert & Borst, 1973; Steinert *et al.* 1973). There seem good grounds for hope that these and other sophisticated methods will eventually shed more light on the relationships and thus perhaps the phylogeny of the family.

SUMMARY

Trypanosoma species are uniflagellate parasitic protozoa possessing an unusually large mass of intramitochondrial DNA, the kinetoplast, and inhabiting blood, tissue fluids and, in some instances the cells, of all classes of vertebrates. Most species undergo part of their life cycle in the gut, and sometimes other organs of a second, invertebrate host. A few are pathogens of man or domestic animals.

The evolutionary ancestors of the genus are generally agreed to have been parasitic flagellates whose life cycle involved only a single host. These ancestors may have been biflagellate parasites of the blood of fish, or uniflagellate forms in the gut of either vertebrates or more probably invertebrates, the inclusion of a second host following the adoption of haematophagy by some invertebrates.

Speciation within the genus generally followed the evolution of the vertebrates. However, the salivarian trypanosomes which are transmitted to their vertebrate (mammalian) hosts via the mouthparts of the invertebrate may be direct descendants of an early group parasitising amphibians and hirudineans, or they may have evolved recently either from other species parasitising mammals or from species inhabiting reptiles. The last possibility seems the most probable.

A trend towards a secondarily simplified life cycle involving only a single, vertebrate, host is shown by the salivarian subgenus *Trypanozoon*. It may be related to development of a cyanide-insensitive component of oxidative metabolism and suppression of the usually dominant cytochrome-mediated pathway.

REFERENCES

ADLER, S. & THEODOR, O. (1935). Investigations on Mediterranean Kala-azar. X. A note on *Trypanosoma platydactyli* and *Leishmania tarentolae*. *Proceedings of the Royal Society* B, **116**, 543–4.

ASHCROFT, M. T. (1963). Some biological aspects of the epidemiology of sleeping sickness. *Journal of Tropical Medicine and Hygiene*, **66**, 133–6.

BAKER, J. R. (1956). Studies on *Trypanosoma avium* Danilewsky, 1885. III. Life cycle in vertebrate and invertebrate hosts. *Parasitology*, **46**, 335–52.

BAKER, J. R. (1963). Speculations on the evolution of the family Trypanosomatidae Doflein, 1901. *Experimental Parasitology*, **13**, 219–33.

BAKER, J. R. (1965). The evolution of parasitic protozoa. In *Evolution of Parasites. Symposia of the British Society for Parasitology*, **3**, 1–27. Ed. A. E. R. Taylor. Oxford: Blackwell Scientific Publications.

BAKER, J. R. (1974). The epidemiology of African sleeping sickness. In *Trypanosomiasis and Leishmaniasis: with special reference to Chagas' disease, Ciba Foundation Symposium*, **20**, in press. Ed. K. Elliot & M. O'Connor. Amsterdam: Associated Scientific Publishers.

BARROW, J. H. JR. (1953). The biology of *Trypanosoma diemyctyli* (Tobey). I. *Trypanosoma diemyctyli* in the leech, *Batrachobdella picta* (Verrill). *Transactions of the American Microscopical Society*, **72**, 197–216.

BAYON, H. P. (1926). *Herpetomonas* in the cloaca of African chamaeleons. *Parasitology*, **18**, 361–2.

BRUMPT, E. (1906*a*). Sur quelques espèces nouvelles de trypanosomes parasites des poissons d'eau douce; leur mode d'évolution. *Compte rendu hebdomadaire des séances et mémoires de la société de biologie*, **60**, 160–2.

BRUMPT, E. (1906*b*). Mode de transmission et évolution des trypanosomes des poissons. Description de quelques espèces de trypanoplasmes des poissons d'eau douce. Trypanosome d'un crapaud africain. *Compte rendu hebdomadaire des séances et mémoires de la société de biologie*, **60**, 162–4.

BRUMPT, E. (1906*c*). Rôle pathogène et mode de transmission du *Trypanosoma inopinatum* Ed. et Ét. Sergent. Mode d'inoculation d'autres trypanosomes. *Compte rendu hebdomadaire des séances et mémoires de la société de biologie*, **61**, 167–9.

BRUMPT, E. (1914). Le xénodiagnostic. Application au diagnostic de quelques infections parasitaires et en particulier à la trypanosomose de Chagas. *Bulletin de la société de pathologie exotique*, **7**, 706–10.

BUCHSBAUM, R. (1951). *Animals without Backbones*, vol. 2. Harmondsworth, Middlesex: Penguin Books. (Reprint in two volumes of the edition of 1948.)

BURTT, E. (1946). Incubation of tsetse pupae: increased transmission-rate of *Trypanosoma rhodesiense* in *Glossina morsitans*. *Annals of Tropical Medicine and Parasitology*, **40**, 18–28.

BÜTSCHLI, O. (1878). Beiträge zur Kenntniss der Flagellaten und einiger verwandten Organismen. *Zeitschrift für Wissenschaftliche Zoologie*, **30**, 205–81.

CAMERON, T. W. M. (1956). *Parasites and Parasitism*. London: Methuen.

CHANCE, M. L., PETERS, W. & GRIFFITHS, H. W. (1973). A comparative study of DNA in the genus *Leishmania*. *Transactions of the Royal Society of Tropical Medicine and Hygiene*, **67**, 24–5.

CHATTON, E. (1924). Sur un *Leptomonas* d'un nématode marin et la question de l'origine des trypanosomides. *Compte rendu hebdomadaire des séances et mémoires de la société de biologie*, **90**, 780–3.

DESOWITZ, R. S. (1957). Suramin complexes. II. Prophylactic activity against *Trypanosoma vivax* in cattle. *Annals of Tropical Medicine and Parasitology*, **51**, 457–63.

DIAS, E. (1948). [Abstract of discussion following paper by Hoare (1948).] *Proceedings of the 4th International Congresses on Tropical Medicine and Malaria*, **2**, 1117.

DUGGAN, A. J. (1970). An historical perspective. In *The African Trypanosomiases*, ed. H. W. Mulligan & W. H. Potts, pp. xli–lxxxviii. London: Allen & Unwin.

FANTHAM, H. B. (1922). Some parasitic Protozoa found in South Africa. V. *South African Journal of Science*, **19**, 332–9.

FANTHAM, H. B. (1925). Some parasitic Protozoa found in South Africa. VIII. *South African Journal of Science*, **22**, 346–54.

FANTHAM, H. B. (1926). Some parasitic Protozoa found in South Africa. IX. *South African Journal of Science*, **23**, 560–70.

FRANCHINI, G. (1921). Sur les flagellés intestinaux du type *Herpetomonas* du *Chamaeleon vulgaris* et leur culture, et sur les flagellés du type *Herpetomonas* de *Chalcides* (*Gongylus*) *ocellatus* et *Tarentola mauritanica*. *Bulletin de la société de pathologie exotique*, **14**, 641–5.

GOODEY, T. & TRIFFIT, M. J. (1927). On the presence of flagellates in the intestine of the nematode *Diplogaster longicauda*. *Protozoology* **3**, 47–58.

GRASSÉ, P.-P. (1952). Ordre des Trypanosomides. In *Traité de Zoologie*, ed. P.-P. Grassé, vol. 1 (1), pp. 602–88. Paris: Masson.

HARTMANN, M. & JOLLOS, V. (1910). Die Flagellatenordnung 'Binucleata'. Phylogenische Entwicklung und systematische Einteilung der Blutprotozoen. *Archiv für Protistenkunde*, **19**, 81–106.

HOARE, C. A. (1923). An experimental study of the sheep trypanosome (*T. melophagium* Flu, 1908), and its transmission by the sheep-ked (*Melophagus ovinus* L.). *Parasitology*, **15**, 365–424.

HOARE, C. A. (1929). Studies on *Trypanosoma grayi*. 2. Experimental transmission to the crocodile. *Transactions of the Royal Society of Tropical Medicine and Hygiene*, **23**, 39–56.

HOARE, C. A. (1948). The relationship of the haemoflagellates. *Proceedings of the 4th International Congresses on Tropical Medicine and Malaria*, **2**, 1110–16.

HOARE, C. A. (1949). *Handbook of Medical Protozoology*. London: Baillière, Tindall & Cox.

HOARE, C. A. (1964). Morphological and taxonomic studies on mammalian trypanosomes. X. Revision of the systematics. *Journal of Protozoology*, **11**, 200–7.

HOARE, C. A. (1965). Vampire bats as vectors and hosts of equine and bovine trypanosomes. *Acta tropica*, **22**, 204–16.

HOARE, C. A. (1972). *The Trypanosomes of Mammals*. Oxford: Blackwell Scientific Publications.

HOARE, C. A. & BENNETT, S. C. J. (1937). Morphological and taxonomic studies on mammalian trypanosomes. III. Spontaneous occurrence of strains of *Trypanosoma evansi* devoid of the kinetonucleus. *Parasitology*, **29**, 43–56.

HOARE, C. A. & WALLACE, F. G. (1966). Developmental stages of trypanosomatid flagellates: a new terminology. *Nature, London*, **212**, 1385–6.

HUTCHINSON, M. P. (1971). Human trypanosomiasis in Ethiopia. *Ethiopian Medical Journal*, **9**, 3–69.

KILLICK-KENDRICK, R. (1964). The apparent loss of the kinetoplast of *Trypanosoma evansi* after treatment of an experimentally infected horse with Berenil. *Annals of Tropical Medicine and Parasitology*, **58**, 481–90.

KUDO, R. R. (1966). *Protozoology*, 5th edition. Springfield, Illinois: Thomas.

LAVERAN, A. & MESNIL, F. (1912). *Trypanosomes et trypanosomiases*, 2nd edition. Paris: Masson.

LAVIER, G. (1943). L'évolution de la morphologie dans le genre *Trypanosoma. Annales de parasitologie humaine et comparée*, **19**, 168–97.

LÉGER, L. (1904*a*). Sur la structure et les affinités des Trypanoplasmes. *Compte rendu hebdomadaire des séances de l'academie des sciences*, **138**, 856–9.

LÉGER, L. (1904*b*). Sur les affinités de l'*Herpetomonas subulata* et la phylogène des trypanosomes. *Compte rendu hebdomadaire des séances et mémoires de la société de biologie*, **56**, 615–17.

LÉGER, L. & DUBOSCQ, O. (1910). *Selenococcidium intermedium* Lég. et Dub. et la systematique des Sporozoaires. *Archives de zoologie expérimentale et générale*, **5**, 187–238.

LÉGER, M. (1918). Infection sanguine par *Leptomonas* chez un saurien. *Compte rendu hebdomadaire des séances et mémoires de la société de biologie*, **81**, 772–4.

LÜHE, M. (1906). Die im Blute Schmarotzenden Protozoen und ihre nächsten Verwandten. In *Handbuch der Tropenkrankheiten*, 1st edition, ed. C. Mense, vol. 3, pp. 69–268. Leipzig: J. A. Barth.

MANWELL, R. D. (1961). *Introduction to Protozoology*. New York: St Martin's Press.

DE MELLO, F. (1921). Protozoaires parasites du *Pachelabra moesta* Reeve. *Compte rendu hebdomadaires des séances et mémoires de la société de biologie*, **84**, 241–2.

MESNIL, F. (1918). [Review of paper by C. França entitled 'Quelques considerations sur la classification des hématozoaires'.] *Bulletin de l'Institut Pasteur*, **16**, 536–7.

MINCHIN, E. A. (1908). Investigations on the development of trypanosomes in tsetse-flies and other Diptera. *Quarterly Journal of Microscopical Science*, **52**, 159–260.

MINCHIN, E. A. (1912). *An Introduction to the Study of the Protozoa*. London: Edward Arnold.

MINCHIN, E. A. (1914). The development of trypanosomes in the invertebrate host. Paper read at meeting of British Association for the Advancement of Science. Reported by Ashworth, J. H. (1914). Zoology at the British Association. *Nature, London*, **94**, 405–10.

NAIR, C. P. & DAVID, A. (1956). Observations on a natural (cryptic) infection of trypanosomes in sparrows (*Passer domesticus* Linnaeus). Part II. Attempts at experimental transmission and determination of the sites of infection in birds and mosquitoes. *Indian Journal of Malariology*, **10**, 137–47.

NEWTON, B. A. & BURNETT, J. K. (1972). DNA of Kinetoplastidae: a comparative study. In *Comparative Biochemistry of Parasites*, ed. H. Van den Bossche, pp. 185–98. New York: Academic Press.

NEWTON, B. A., CROSS, G. A. M. & BAKER, J. R. (1973). Differentiation in Trypanosomatidae. In *Microbial Differentiation. Symposia of the Society for General Microbiology*, **23**, 339–73. Ed. J. M. Ashworth & J. E. Smith. London: Cambridge University Press.

NEWTON, B. A., STEINERT, M. & BORST, P. (1973). Differentiation of haemoflagellate species by hybridization of complementary RNA with kinetoplast DNA.

Transactions of the Royal Society of Tropical Medicine and Hygiene, **67**, 259–60.

NÖLLER, W. (1913). Die Blutprotozoen des Wasserfrosches und ihre Übertragung. *Archiv für Protistenkunde*, **31**, 169–240.

OAKLEY, K. P. & MUIR-WOOD, H. M. (1949). *The Succession of Life through Geological Time*, 2nd edition. London: British Museum (Natural History).

ORMEROD, W. E. (1961). The epidemic spread of Rhodesian sleeping sickness, 1908–1960. *Transactions of the Royal Society of Tropical Medicine and Hygiene*, **55**, 525–38.

ORMEROD, W. E. (1967). Taxonomy of the sleeping sickness trypanosomes. *Journal of Parasitology*, **53**, 824–30.

ORMEROD, W. E. & VENKATESAN, S. (1971*a*). The occult visceral phase of mammalian trypanosomes with special reference to the lifecycle of *Trypanosoma (Trypanozoon) brucei*. *Transactions of the Royal Society of Tropical Medicine and Hygiene*, **65**, 722–35.

ORMEROD, W. E. & VENKATESAN, S. (1971*b*). An amastigote phase of the sleeping sickness trypanosome. *Transactions of the Royal Society of Tropical Medicine and Hygiene*, **65**, 736–41.

PORTER, A. (1914). The morphology and biology of *Herpetomonas patellae*, n.sp., parasitic in the limpet, *Patella vulgata*, together with remarks on the pathogenic significance of certain flagellates found in invertebrates. *Parasitology*, **7**, 322–9.

QADRI, S. S. (1962). An experimental study of the life cycle of *Trypanosoma danilewskyi* in the leech, *Hemiclepsis marginata*. *Journal of Protozoology*, **9**, 254–8.

ROBERTSON, M. (1908). A preliminary note on Haematozoa from some Ceylon reptiles. *Spolia Zeylanica, Ceylon*, **5**, 178–85.

ROBERTSON, M. (1912). Transmission of flagellates living in the blood of certain freshwater fishes. *Philosophical Transactions of the Royal Society*, B, **202**, 29–50.

SHAW, J. J. (1964). A possible vector of *Endotrypanum schaudinni* of the sloth *Choloepus hoffmani*, in Panama. *Nature, London*, **201**, 417–18.

SHAW, J. J. (1966). The relationship of *Endotrypanum* to other members of the Trypanosomatidae, and its possible bearing upon the evolution of certain haemoflagellates of the New World. *Proceedings of the 1st International Congress on Parasitology*, pp. 332–3. Oxford: Pergamon Press. [Abstract only.]

SHAW, J. J. (1969). *The Haemoflagellates of Sloths* (London School of Hygiene & Tropical Medicine Memoir 13). London: H. K. Lewis.

STEINERT, M., VAN ASSEL, S., BORST, P., MOL, J. N. M., KLEISEN, C. M. & NEWTON, B. A. (1973). Specific detection of kinetoplast DNA in cytological preparations of trypanosomes by hybridization with complementary RNA. *Experimental Cell Research*, **76**, 175–85.

STRONG, R. P. (1924). Investigations upon flagellate infections. *American Journal of Tropical Medicine*, **4**, 345–85.

VICKERMAN, K. (1969). On the surface coat and flagellar adhesion in trypanosomes. *Journal of Cell Science*, **5**, 163–94.

VICKERMAN, K. (1970). Ultrastructure of *Trypanosoma* and relation to function. In *The African Trypanosomiases*, ed. H. W. Mulligan & W. H. Potts, pp. 60–6. London: Allen & Unwin.

VICKERMAN, K. (1971). Morphological and physiological considerations of extracellular blood protozoa. In *Ecology and Physiology of Parasites*, ed. A. M. Fallis, pp. 58–89. Toronto: University Press.

WALLACE, F. G. (1966). The trypanosomatid parasites of insects and arachnids. *Experimental Parasitology*, **18**, 124–93.

WENYON, C. M. (1920). Observations on the intestinal protozoa of three Egyptian lizards, with a note on a cell-invading fungus. *Parasitology*, **12**, 350–6.

Wenyon, C. M. (1926). *Protozoology*, vol. 1. London: Baillière, Tindall & Cox.

Wijers, D. J. B. & McDonald, W. A. (1961). Anal feeding as a method of infecting tsetse flies with *Trypanosoma gambiense*. II. The results. *Annals of Tropical Medicine and Parasitology*, **55**, 46–8.

Willett, K. C. (1956). The problem of *Trypanosoma rhodesiense*, its history and distribution, and its relationships to *T. gambiense* and *T. brucei*. *East African Medical Journal*, **33**, 473–9.

Willett, K. C. (1963). Some principles of the epidemiology of human trypanosomiasis in Africa. *Bulletin of the World Health Organization*, **28**, 645–52.

Woo, P. T. K. (1970). Origin of mammalian trypanosomes which develop in the anterior-station of blood-sucking arthropods. *Nature, London*, **228**, 1059–62.

Woo, P. T. K. & Soltys, M. A. (1969). The experimental infection of reptiles with *Trypanosoma brucei*. *Annals of Tropical Medicine and Parasitology*, **63**, 35–8.

Woodcock, H. M. (1906). The haemoflagellates: a review of present knowledge relating to the trypanosomes and allied forms. *Quarterly Journal of Microscopical Science*, **50**, 151–231; 233–331.

EXPLANATION OF PLATE

Photomicrographs of haematozoic trypomastigotes of *Trypanosoma* (*Trypanozoon*) spp.

(*a*)–(*d*) *T. brucei* (Giemsa's stain): (*a*) long slender form; (*b*) long slender form undergoing division (note paired kinetoplasts and flagella, and elongated nucleus); (*c*) intermediate form; (*d*) short stumpy form. (*e*) *T. e. evansi* (note dividing organism with paired kinetoplasts and nuclei. Feulgen's stain); (*f*) *T. equinum* (= dyskinetoplastic strain of *T. evansi evansi*; note absence of visible kinetoplast. Feulgen's stain).

PLATE 1

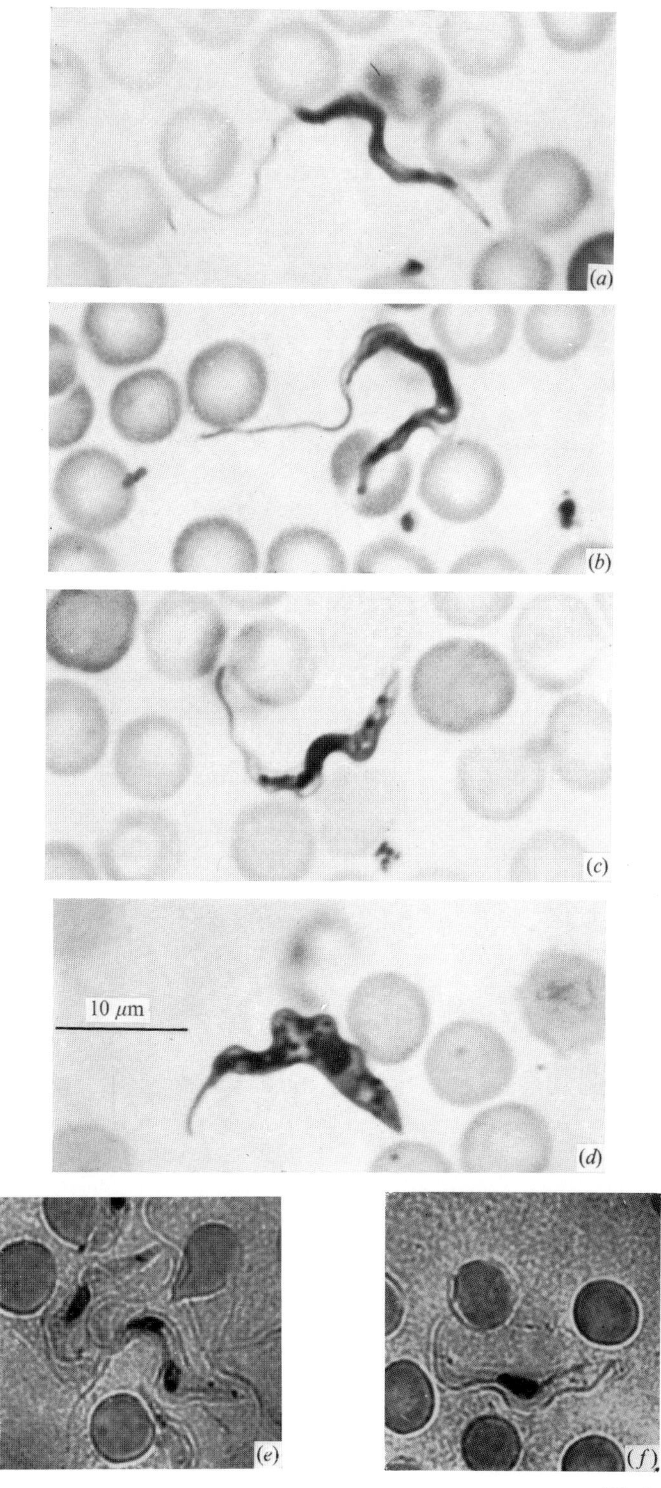

(Facing p. 366)

MICRO-ORGANISMS AND PLANTS: THE EVOLUTION OF PARASITISM AND MUTUALISM

D. H. LEWIS

Department of Botany, The University of Sheffield, Sheffield S10 2TN

INTRODUCTION

Under natural conditions, associations of varying intimacy exist between autotrophs and autotrophs, autotrophs and heterotrophs and heterotrophs and heterotrophs. From a conceptual analysis of such diverse organismic interactions, Heise & Starr (unpublished typescript, 1972: see Starr & Chatterjee, 1972) have proposed a universally applicable classification for them. This consists of a set of interacting continua based on three criteria which they term (*a*) *locational–occupational*, indicating where the organisms are located and how they are 'gaining a living' in relation to one another, (*b*) *valuational*, which concerns whether benefit or harm arises from the association, and (*c*) *degree of dependency* – whether association is necessary or contingent. The segments of these continua, conceptually identical with those of Heise & Starr, but with some nomenclatural changes,* are as follows: locational–occupational: *commensalism, symbiosis, predation*; valuational: *mutualism, neutralism, antagonism*; degree of dependency: *obligate, facultative*. This paper will largely be restricted to a consideration of symbiosis, i.e. associations of usually dissimilar organisms which exhibit permanent, or at least prolonged, intimate contact (Lewis, 1973*a*). Commensalism and predation will not receive more than passing comment. The principal concern will be the evolutionary relationships between the intersecting segments of the valuational and degree of dependency continua. Such a consideration involves an analysis both of the nutritional interactions between symbionts and of the relationship between symbiotic and free-living existence.

* M. P. Starr (personal communication) has substantially modified the original scheme of Heise & Starr, both in terminology and by inclusion of further continua. Details will be published in 1975.

SYMBIOTIC NUTRITION

Autotrophy

Although autotrophic symbionts are potentially capable of synthesizing both organic carbon and organic nitrogen compounds from inorganic sources, this potential may not be fully realized in some situations. For example, blue-green algae in the root nodules of cycads may not receive sufficient light to photosynthesize and may therefore be ecologically heterotrophic for carbon. Similarly, other potential autotrophs e.g. higher plants with N-fixing nodules, or sheathing or vesicular–arbuscular mycorrhizas, although capable of synthesizing organic nitrogen compounds from nitrate or ammonium ions, nevertheless derive their nitrogen from their associated heterotroph in organic form (Pate, 1968; Lewis, 1973*b*). A further major difference between the autotrophic nutrition of free-living and symbiotic species concerns the degree of retention of synthesized compounds. Although secretion of small amounts of fixed carbon from free-living algae or from roots and leaves of higher plants is a normal feature of metabolism, the proportion of total photosynthate which is released by autotrophic symbionts to their heterotrophic associates is very high (Smith, Muscatine & Lewis, 1969). These three aspects of symbiotic autotrophs – carbon heterotrophy, nitrogen heterotrophy and enhanced release of metabolites – are facultative features only, involving alteration of particular activities in the symbiotic state. 'Normal' behaviour can be resumed if the symbiotic associate is experimentally eliminated. These nutritional features are considered elsewhere (Smith *et al.* 1969; Smith, 1974).

Heterotrophy

Of greater significance for the thesis of this paper are the differences in heterotrophic behaviour within diverse symbiotic associations and the relationship between symbiotic and free-living heterotrophy. Before restricting discussion to green plants and chemoheterotrophic micro-organisms (fungi and bacteria, including actinomycetes), a classification of nutritional categories which is comparable to the continua of Heise & Starr and which is applicable to all chemoheterotrophic organisms, plant, animal and microbial, is presented. The relevance of the evolutionary schemes, which are proposed for nutritional interactions between plants and micro-organisms, to the evolution of similar interactions involving animals may then be gauged.

Classifications of the nutritional behaviour of fungi have been bedevilled by use of the criterion of culturability, one which has not

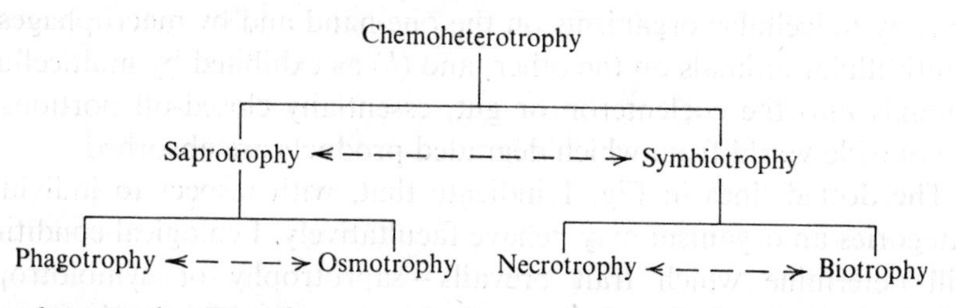

Chemoheterotrophy: Dependence on organic compounds as both energy and principal carbon sources.
Symbiotrophy: Derivation of organic compounds directly from another organism by virtue of permanent, intimate contact.
Saprotrophy: Derivation of organic compounds from non-living environment.
Biotrophy: Derivation of organic compounds from living cells of symbiotic host.
Necrotrophy: Derivation of organic compounds from killed cells of symbiotic host.
Phagotrophy: Assimilation of organic compounds in particulate form.
Osmotrophy: Assimilation of organic compounds in dissolved form.

The dashed lines indicate that particular organisms may behave facultatively in either of the arrowed modes.

Fig. 1. Patterns of chemoheterotrophic nutrition.

significantly entered similar considerations for bacterial behaviour. Since unculturability is a negative character and the division between the culturable and the uncultured will change as information increases and techniques improve, it does not form a sound basis for classification (Lewis, 1973*a*). Alexander (1971) pointed out that microbiologists with special interests in particular microbial groups used conflicting terminologies to describe nutritional categories and that it was not possible to 'satisfy contending microbiologists with acute semantic sensitivities'. The classification that is proposed here combines features of those used by Alexander (1971) and Stanier, Doudoroff & Adelberg (1971) for micro-organisms in general and that suggested for fungi by Lewis (1973*a*). This scheme (Fig. 1) uses symbiotic in a broad context to include both antagonistic (parasitic) and mutualistic associations (i.e. the complete spectrum of the valuational continuum of Heise & Starr) and is applicable to animals as well as to microbial chemoheterotrophs.

As well as osmotrophy, the typical bacterial and fungal mode of nutrition, saprotrophy can encompass phagotrophy, which involves ingestion of particulate matter and is typical of animals. In phagotrophy the digestive and assimilatory processes kill ingested organisms and tissues, degrade complex molecules and absorb products in a manner comparable to osmotrophy. It should be noted that two kinds of ingestion may be recognized within phagotrophy: (*a*) at the cellular level by unicellular organisms on the one hand and by macrophages of multicellular animals on the other, and (*b*) as exhibited by multicellular animals into the coelenteron or gut, essentially closed-off portions of the outside world from which degraded products are absorbed.

The dotted lines in Fig. 1 indicate that, with respect to individual categories an organism may behave facultatively. Ecological conditions will determine which trait prevails – saprotrophy or symbiotrophy (i.e. free-living or symbiotic), osmotrophy or phagotrophy, necrotrophy or biotrophy. Thus, it can be seen that nutritional considerations can be readily integrated into the Heise & Starr analysis by the erection of a fourth nutritional continuum; that shown in Fig. 1.* This differs from those considered by them in that it is immediately necessary to divide the primary continuum (saprotrophy and symbiotrophy) into subsegments. It should, however, be noted that the second dichotomy does not produce four equivalent categories. The split into osmotrophy and phagotrophy concerns mechanisms of assimilating nutrients. The split into biotrophy and necrotrophy concerns source of nutrients also. In practice, symbiotrophic heterotrophs usually obtain their nutrients in an osmotrophic manner.

Carbon sources of symbiotic and predatory chemoheterotrophs

This section examines the possible range of nutritional interactions between symbiotic or predatory chemoheterotrophs and their carbon sources in order to set the stage, with specific examples, for the discussion of evolutionary pathways which follows. As in the previous section, animals as carbon recipients will be briefly considered so that the overall pattern of interactions may be borne in mind when microbial associations are considered in detail.

As well as acting as carbon recipients, some heterotrophs may also function as carbon donors to other heterotrophs. This is illustrated in Fig. 2, which represents associations in a chequer-board of carbon donors (sources) and recipients. In most cases, particularly where the donor is photosynthetic, it can be assumed that a major form by which carbon moves from donor to recipient is as carbohydrate. However, there is also experimental evidence for movement of such compounds as alanine from photosynthetic algae (Smith, 1974) and other amino acids from higher plants (Scott, 1972). In some associations as noted above, carbon may also move from a heterotroph to an autotroph as

* Among his modifications to the Heise & Starr scheme (see footnote on p. 367), Starr has, independently, included a necrotrophy/biotrophy nutritional continuum (M. P. Starr, personal communication).

Carbon recipient	Carbon source				
	Algae**	Higher plants†	Bacteria‡	Fungi§	Animals
Algae**	△ ○ ✱	○ ✱	△	△ ○ ✱	○ ✱
Higher plants†	○ ✱	○	**○ ✱**	□ **○ ✱**	△
Bacteria‡	□ ○ ✱	**□ ○ ✱**	□	□	□ ○ ✱
Fungi§	□ ○ ✱	**□ ○ ✱**	△	□ △ ○	□ △ ○ ✱
Animals	△ ○ ✱	△ ○ ✱	△	△ ○ ✱	□ △ ○ ✱

Key to symbols:

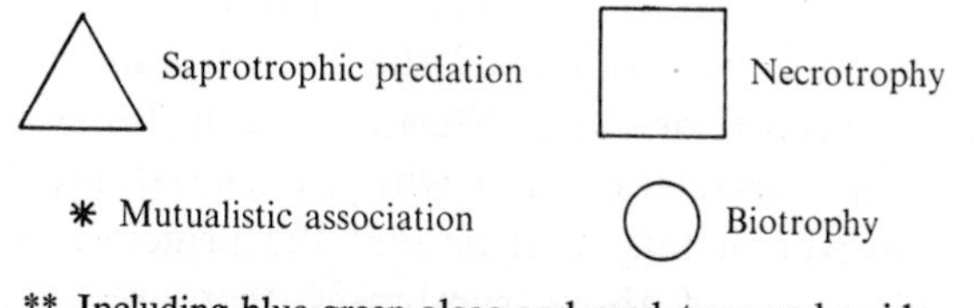

** Including blue-green algae and predatory euglenoids.
† Bryophyta, Pteridophyta and Spermatophyta.
‡ Including Actinomycetes.
§ Including slime moulds.

Associations with symbols in heavy lines are considered in more detail in Table 1.

Fig. 2. Nutritional interactions between predatory or symbiotic chemoheterotrophs and their carbon sources.

amino acids or amides. Undoubtedly, compounds of other elements pass between symbionts, such as phosphorus in mycorrhizas (Harley, 1969), and when more data are available it would be instructive to construct chequer-boards similar to Fig. 2 based on movement of particular compounds.

Table 1 classifies in more detail the associations in heavy type in Fig. 2 according to the nutritional class of the heterotroph and valuational class of the outcome of the symbiosis. Supposedly neutral associations such as symptomless infections of higher plants by fungi (Harley, 1950) are not included. In order to compose a similar table for algae, each algal phylum (e.g. Chlorophyta, Rhodophyta) would need to be considered separately along with the many associations that exist between two algae as well as between algae and other organisms. Fritsch (1935, 1952) provides a general survey of algal associations, Droop (1963) discusses those between algae and invertebrates, Joubert & Rijkenberg (1971) those between green algae and higher plants and Kohlmeyer & Kohlmeyer (1972) the apparently symptomless associations of some brown algae and fungi. The algal symbioses most thoroughly studied are lichens – mutualistic associations of fungi with either green or blue-green algae or in a few cases, both (Smith, 1962, 1963, 1974; Smith *et al.* 1969; Ahmadjian, 1967; Hale, 1970). Lesser known associations, such as those between species of red algae, are now also being investigated (Evans, Callow & Callow, 1973.)

Further discussion will be restricted largely to symbioses in Table 1 together with lichens. Of the other associations in Fig. 2, animals as either donors or recipients will not be considered further nor will the predatory activities of slime moulds, classified for simplicity as fungi. Algae and higher plants as recipients will only receive passing comment. Although the physiology of the necrotrophic bacterial genus *Bdellovibrio* has been reviewed by Starr & Huang (1972), little is known of the nutritional interrelationships within bacterial infections of algae and fungi. However, it may well be that the principles which emerge from a consideration of those symbioses that are to be discussed further are also applicable to these.

Concerning the final association in Fig. 2, fungi and fungi, there has been a resurgence of interest and research into their biology largely due to the efforts of H. L. Barnett and co-workers. Both necrotrophic and biotrophic mycoparasites exist (Barnett, 1963, 1964) and since several biotrophic species of both contact and haustorial kind have been cultured axenically (e.g. Kurtzman, 1968; Gain & Barnett, 1970; Barnett, 1970; Calderone & Barnett, 1972; Barker & Barnett, 1973)

Table 1. *Symbiotic associations between higher plants and heterotrophic micro-organisms*

Heterotrophs	Nutritional class	Valuational class	Types of association*	References
Bacteria	Necrotrophic	Parasitic	Many plant diseases (*Erwinia*, *Pseudomonas*, *Xanthomonas*, *Corynebacterium*)	Dowson, 1957; Starr, 1959; Starr & Chatterjee, 1972; Stapp, 1961; Wood, 1967; Wheeler, 1969
	Biotrophic	Parasitic	A few plant diseases, e.g. Crown gall (*Agrobacterium tumefaciens*); leafy gall (*Corynebacterium fascians*); olive knot (*Pseudomonas savastanoi*)	Stapp, 1961; Wood, 1967; Wheeler, 1969
		Mutualistic	Legume root nodules (*Rhizobium*). Leaf nodules of *Ardisia*, *Psychotria*, etc. (*Klebsiella*)	Nutman, 1963; Stewart, 1966; Lange, 1966
Actinomycetes	Necrotrophic	Parasitic	A few plant diseases, e.g. potato scab (*Streptomyces scabies*)	Waksman, 1959
	Biotrophic	Parasitic	?	
		Mutualistic	Non-legume root nodules, e.g. of *Alnus*, *Myrica*, *Casuarina*, etc.	Bond, 1963, 1967; Stewart, 1966; Becking, 1970
Fungi	Necrotrophic	Parasitic	Many plant diseases, e.g. soft rots, vascular wilts, damping off, leaf spots, anthracnose,' cankers and scabs	Wood, 1967; Wheeler, 1969; Webster, 1970
	Biotrophic	Parasitic	Many plant diseases, e.g. rusts, smuts, powdery and downy mildews, club root, potato blight, apple scab, ergot and choke of grasses	Wood, 1967; Wheeler, 1969; Webster, 1970
		Mutualistic	Sheathing mycorrhizas (*Boletus*, *Rhizopogon*, *Amanita*, *Russula* etc.). Vesicular–arbuscular mycorrhizas (*Endogone*)	Harley, 1969; Mosse, 1963; Nicolson, 1967

* Heterotroph in parentheses.

there is considerable scope for nutritional studies *in vitro*; techniques that have been used for other symbioses (Smith *et al.* 1969) are also applicable. Electron microscopy is elucidating the nature of the haustorial apparatus in such species as *Piptocephalis* (Armentrout & Wilson, 1969; Manocha & Lee, 1971).

CHARACTERISTICS OF BIOTROPHY

In a consideration of the physiological interaction in symbioses involving then uncultured fungi, Brian (1967) delimited and discussed six characteristic features – intracellular penetration, minimal tissue damage, highly developed physiological specialization and relatively restricted host range, morphologial disturbances in the host plant, 'green islands' and nuclear disturbances. Lewis (1973*a*) pointed out that these were not only features of physiologically obligate parasites but several were also essential attributes of all biotrophic infections of higher plants and algae by fungi. Consultation of the references giving descriptions of the biotrophic infections by bacteria and actinomycetes cited in Table 1 reveals that the features listed by Brian are also characteristic of these infections, whether parasitic or mutualistic.

Two characteristics are particularly relevant to the evolution of the biotrophic habit – minimal tissue damage, and morphological disturbances in the host plant. An understanding of the first requires consideration of the factors controlling enzyme systems which degrade tissues (Wood, 1967; Albersheim, Jones & English, 1969; Starr & Chatterjee, 1972). The second feature has also been reviewed by Wood (1967) for parasitic infections. To his compilation of morphological disturbances may be added the characteristic nodules of the mutualistic bacterial and actinomycete associations of Table 1, the modified root structure of sheathing mycorrhizas and the lichen thallus. In the last case, the morphological change is most evident in the fungus. Anatomical differences following vesicular–arbuscular mycorrhizal infection have been investigated (Daft & Okusanya, 1973) and are quantitative rather than qualitative in nature. The induced abnormal developments require active metabolism by the infected host cells, which therefore become important 'sinks' for the hosts' photosynthetic products. Since both plant growth and translocation of metabolites are under hormonal control (Wareing & Phillips, 1970), an understanding of hormonal changes and their consequences following infection is an essential prerequisite for understanding the biotrophic habit.

Before the detailed consideration of the possible origins of biotrophy, it is essential to emphasize that the biotrophic habit is of great antiquity. If the existence of fossil vesicular–arbuscular mycorrhizal infections of *Asteroxylon* and *Rhynia* is accepted (Harley, 1969), the basic features of the evolutionary processes to be discussed had already been selected for and become established soon after plants emerged from the sea and colonized the land 400 million years ago, and if the symbiotic origin of the organelles of eukaryotic cells (Margulis, 1970) is also accepted, the biotrophic habit occurred in the ancestors of all modern eukaryotic cells and was established over 1000 million years ago.

EVOLUTIONARY RELATIONSHIPS BETWEEN NUTRITIONAL CATEGORIES

From a consideration of the degree of dependency continuum of Heise & Starr (i.e. ecologically obligate or facultative symbiosis) and the nutritional continuum restricted to saprotrophy, necrotrophy and biotrophy, five categories of fungal behaviour were delimited by Lewis (1973*a*). These are shown in Table 2 with descriptions modified to encompass chemoheterotrophic bacteria and animals. The bracketed abbreviations in Table 2 are based on the recognition (*a*) that obligate and facultative refer to ecological and not cultural behaviour and (*b*) that facultative saprotrophy is merely a form of facultative necrotrophy or biotrophy within groups 2 or 4.

It would require a plant pathologist, an animal symbiologist and perhaps more ecological information to allot *particular* bacterial and animal symbionts into appropriate categories (see Buddenhagen, 1965; Crosse, 1968; Read, 1970). However, with reference to Table 1, with the exception of *Agrobacterium*, *Corynebacterium fascians* and *Pseudomonas savastanoi* which, with *Rhizobium*, are facultative biotrophs, bacterial infections of plants fall into groups 2 and 3. The nitrogen-fixing endophytes of non-legume nodules belong to group 5. In the schemes for the evolution of nutritional behaviour to be proposed, it is held that organisms of group 1 are, nutritionally, the most primitive and versatile and that those of group 5 are the most advanced and specialized.

In agreement with Brian's (1967) assessment of physiologically obligate parasitic fungi, most obligate biotrophs have highly developed physiological specialization and relatively restricted host range and so combine a present fitness for their ecological niche with loss of adaptability. This loss of adaptability, discussed by McNew (1960) and Garrett (1970), results in several well marked differences between the unspecialized parasites of group 2 and the specialized species of group 5.

Table 2. *Groups of chemoheterotrophs based on the degree of dependency and nutritional continua* (*after Lewis*, 1973*a*)

Group 1	*Ecologically obligately free-living saprotrophs* (*obligate saprotrophs*) Free-living organisms with no capacity for parasitic or mutualistic symbiosis.
Group 2	*Ecologically facultative symbionts, whose nutrition is necrotrophic in the parasitic mode, but otherwise saprotrophic* (*facultative necrotrophs*) Potentially symbiotic organisms ranging from those that are normally free-living but can adopt a parasitic existence to those that are normally parasitic but can have a limited free-living existence other than as gametes, cysts or spores.
Group 3	*Ecologically obligately symbiotic necrotrophs* (*obligate necrotrophs*) Specialized parasites whose free-living existence other than as gametes, cysts or spores is restricted to mere survival in dead, infected host tissue (i.e. the extreme form of group 2).
Group 4	*Ecologically facultative symbionts, whose nutrition is biotrophic in the symbiotic mode, but otherwise saprotrophic* (*facultative biotrophs*) Biotrophic symbionts with some capacity for a free-living existence other than as gametes, cysts or spores under natural conditions.
Group 5	*Ecologically obligately symbiotic biotrophs* (*obligate biotrophs*) Parasitic and mutualistic symbionts with no capacity for free-living existence other than as gametes, cysts or spores, or as laboratory cultures (i.e. the extreme form of group 4).

These include a change from high to low competitive saprophytic ability, a diminution and eventual disappearance of the capacity to destroy cells by production of cytolytic enzymes and a progression from indiscriminate general invasion of cells to restricted establishment inter- or intracellularly. Intracellular penetration may be via haustoria, i.e. with the bulk of the organism outside cells, or complete as with some rusts and chytrids, *Plasmodiophora* and mutualistic bacteria and actinomycetes. Also, the maximum potential growth rates of the specialized heterotroph are commonly much reduced and in the most specialized associations a host response to the symbiont occurs that is governed by the symbiont. Overall, as indicated by Garrett, a change occurs from *physical* to *economic* damage.

If the proposition that biotrophy is evolutionarily derived from saprotrophy is accepted, clearly this can occur directly or via necrotrophy. An understanding of the latter possibility necessitates some insight into the origin of necrotrophy. With reference to the nutritional groups in Table 2, several possible evolutionary pathways may be delimited. Groups 1 → 2 and groups 1 → 2 → 3 represent routes to necrotrophic behaviour of increasing specialization. Groups 1 → 4 and groups 1 → 4 → 5 are similar routes directly from primitive saprotrophy to advanced biotrophy. Finally, groups 1 → 2 → 5 is a pathway to biotrophy via necrotrophy. There appear to be few chemoheterotrophs in group 4, possibly indicating that few opportunities exist for an organism to indulge in the alternative nutritional modes this category represents. This perhaps further suggests that biotrophy has arisen via necrotrophy more often than directly from saprotrophy. It is essential to emphasize 'more often' since there can be no doubt that one kind of nutritional behaviour has arisen from another on numerous occasions.

ORIGIN OF NECROTROPHY

Necrotrophic nutrition involves the exploitation of cells killed by the heterotroph within an at least initially alive host. Two overlapping kinds of necrotrophic attack of higher plants have been recognized, the first in which massive enzymic destruction of host cells occurs (e.g. soft rots, vascular wilts and damping-off diseases) and the second in which death of tissues is possibly induced more by toxin production, e.g. cankers and scabs (Wood, 1967). In the latter, tissue destruction may occur secondarily. Since many fungal and bacterial obligate saprotrophs can produce cellulolytic, hemicellulolytic, pectinolytic and lignin-oxidizing enzymes, it is not difficult to foresee the possibility of the evolution of facultative necrotrophs from them, i.e. groups 1 → 2. This is particularly easy to envisage for the commensal saprotrophs that specifically inhabit the rhizosphere and phyllosphere. This hypothesis has been evaluated for both bacterial and fungal infections by Yarwood (1956), McNew (1960), Buddenhagen (1965), Buddenhagen & Kelman (1964), Crosse (1968), Garrett (1970) and Gray & Williams (1971). Garrett, in particular, discusses the relationship of the obligate necrotrophs of group 3 and the facultative necrotrophs of group 2, in doing so presenting the case for the progression 1 → 2 → 3.

ORIGIN OF BIOTROPHY

Origin of biotrophy directly from saprotrophy

It would seem that some insight into the direct origin of biotrophy from saprotrophy might be gained from a consideration of the facultative biotrophs of group 4. As originally delimited by Lewis (1973*a*) with respect to fungi, only a few lichen and sheathing mycorrhizal fungi could be segregated into this group although, as noted above, both the parasitic and mutualistic biotrophic bacteria can also be classified here. However, as both *Rhizobium* and *Pseudomonas savastanoi* retain some necrotrophic features (Thornton, 1930; Stapp, 1961), it is more likely that the current, more biotrophic, status of their symbioses with higher plants arose via necrotrophy than directly from saprotrophy.

With regard to lichens, many are tolerant of such environmental stresses as prolonged drought, frequent wetting and drying, extremes of temperature and nutrient deficiency (Smith, 1962). An essential feature of all lichens is the sustained loss of photosynthetic products from alga to fungus (Smith *et al.* 1969). The precise mechanisms responsible for this nutrient transfer are not understood but certainly depend on intimate physical contact between the symbionts and result in selective loss of glucose from blue-green algae and, dependent on the genus in the case of green algae, selective loss of the sugar alcohols, erythritol, ribitol or sorbitol (Smith, 1974). Smith *et al.* (1969) pointed out the similarity between some sooty mould fungi of the phyllosphere and lichen fungi in both their natural carbon sources and tolerance of environmental stress. It is therefore not surprising that some sooty moulds are occasionally lichenized. It is not difficult to envisage the development of permanent lichenized associations from such casual associations as these as well as from the lichenized algal covers discussed by Quispel (1942, 1943) and Ahmadjian (1967).

As with lichens, only a few species of the many fungi which form sheathing mycorrhizas are known to exist also as free-living mycelia. Such species include *Boletus bovinus*, *B. subtomentosus*, *B. scaber*, *Scleroderma aurantium*, *Laccaria amethystina* and *Cenococcum graniforme* (Meyer, 1966; Harley, 1969). The occurrence of such saprotrophic behaviour does not, however, by itself indicate direct development of biotrophy since some strains of at least one of these species, *B. subtomentosus*, produce cytolytic and lignin-oxidizing enzymes in culture (Lindeberg, 1948; Lundeberg, 1970). Like *Rhizobium* mentioned above, therefore, they are at least potentially necrotrophic. As will be emphasized in the next section, it is the fine enzymic control of such potential necrotrophy which most likely gives rise to biotrophy. This is in agreement with Harley's (1969) summary of the views of others who have speculated on the evolution of the sheathing mycorrhizal condition, 'mycorrhizal fungi have evolved from aggressive parasites by progressive selection of non-lethal varieties so that the symbiotic stage becomes indefinitely prolonged'. Harley here used 'symbiotic' in the same sense as 'mutualistic' is used in this article. Harley follows up this summary by stating that 'We are dealing, then, in ectotrophic mycorrhiza with a special case where, on a given root, one species of fungus in particular is so affected by local conditions as to become dominant in the root-surface zone'. and adding, 'This is, I believe, a statement of fact and does not mean to suggest any evolutionary origin.'

Origin of biotrophy via necrotrophy

Lewis (1973*a*) suggested that a change from necrotrophic to biotrophic nutrition of symbiotic fungi could come about by the following sequence of events.

(1) Catabolite repression of the degradative enzymes of necrotrophs limits their cytolytic activity.

(2) Concurrently with (1), localized production of plant hormones, either by the heterotroph or by the host under the influence of the heterotroph, and their movement from the relatively undamaged tissue results in changes of patterns of translocation of photosynthetic products within the host. In this way, infected areas become net importers of organic carbon.

(3) Continued provision of simple metabolites, consequent upon the changed patterns of translocation, maintains catabolite repression while serving the needs of the now biotrophic symbiont.

(4) Genetic loss of the capacity to produce cytolytic enzymes occurs in some species, resulting in obligate biotrophy owing to reduced competitive saprophytic ability.

The evidence for this hypothesis was largely derived from a consideration of fungal infections. The extent to which information from experiments with and analyses of phytopathogenic and mutualistic bacteria and actinomycetes and the infections they cause can be integrated with the fungal work to support the hypothesis is examined below.

Necrotrophic pathogens in general produce copious amounts of cytolytic enzymes *in vitro* as well as *in vivo*. The voluminous literature concerning this aspect of physiological plant pathology has been well reviewed. The theme of that of Albersheim *et al.* (1969) is especially pertinent to the first event of the above sequence since they state their hypothesis as follows, 'The control of polysaccharide-degrading enzyme production is, in our opinion, the key to the role played by saccharides in a plant's resistance to microbial disease.' Their paper therefore considers in detail what is known about the control (e.g. induction and repression) of individual enzymes during the initiation of infection. Not only do they point out that 'small changes in the composition, structure, or accessibility of the host's carbohydrates can alter qualitatively and quantitatively the array of polysaccharide-degrading enzymes secreted by an invading pathogen', but also that 'free sugars as well as polysaccharide components can participate in the regulation of polysaccharide-degrading enzyme production'. Of particular relevance to the thesis of this paper is the relationship between disease development and studies of catabolite repression of cytolytic enzymes of *Pyrenochaeta terrestris* and *Fusarium oxysporum in vivo* and *in vitro* (Horton & Keen, 1966*a*, *b*; Keen & Horton, 1966; Patil & Dimond, 1968; Biehn & Dimond, 1971). In these species, either or both cellulases and pectinases were shown to be subject to catabolite repression by simple sugars *in vitro*. Experimental manipulation of host tissues which increased sugar levels decreased enzyme production *in vivo* and reduced severity of symptoms. The converse was also true. Catabolite repression of degradative enzymes by sugars also occurs in some phytopathogenic bacteria (Starr & Chatterjee, 1972) and has been reviewed in general by Paigen & Williams (1970).

The close correlation between production of degradative enzymes and pathogenic virulence has also been stressed (Albersheim *et al.* 1969; Starr & Chatterjee, 1972) and it is significant that phytopathogenic bacteria lose virulence when cultured in media rich in carbohydrate (Stapp, 1961). The ecological parallel to continued culture on sugar-rich media in the laboratory is continued infection of plants with high sugar levels in their tissues. It is striking that most necrotrophic infections develop best in plants low in sugar, whereas biotrophs are dependent on tissues with abundant free sugar (Horsfall & Dimond, 1957; Lewis, 1973*a*).

Before discussing the manner by which tissues rich in sugar can be maintained in the face of microbial infection (i.e. events (2) and (3) above), it is pertinent to note how far biotrophic pathogens have indeed lost the capacity to produce cytolytic enzymes. Since *indiscriminate* cell wall destruction does not occur during biotrophic infection, the most significant place where *localized* degradation could take place would be during early infection and, in the case of fungi, haustorium formation. Although, formerly, this was considered to be an entirely mechanical process and may still be so in some cases, e.g. infection by *Plasmodiophora* (Aist & Williams, 1971), there is now increasing evidence from electron microscope and histochemical studies that localized production of 'cutinases' and cellulases occurs in powdery and downy mildews (Akai, Kunoh & Fukutomi, 1968; Kunoh & Akai, 1969; McKeen, Smith & Bhattacharya, 1969; Edwards & Allen, 1970; Sargent, Tommerup & Ingram, 1973). Edwards & Allen, studying *Erysiphe graminis*, state that the enzyme production appears to be confined to the tip of the infection peg and to be active only to distances of 0.1 μm. These fungi have yet to be grown axenically so that their *in vitro* enzymic potential has still to be examined, although cell-wall splitting enzymes of dormant and germinating urediospores of the rust *Puccinia graminis* were described by Van Sumere, Van Sumere-De Preter & Ledingham (1957). The biotrophic fungi that have been most extensively studied *in vitro* are those that form sheathing mycorrhizas. Lengthy catalogues of what particular species will and will not grow on are available (Harley, 1969; Palmer & Hacskaylo, 1970; Lundeberg, 1970). Although some species or strains will grow on cell wall components, including cellulose, there has been no systematic study of the control of potentially cytolytic enzymes. Melin (1948, 1953) and Norkrans (1950) did, however, hint at what would be now termed catabolite repression when they suggested that mycorrhizal fungi capable of utilizing cellulose only produce cellulase when soluble sugars in the root are depleted.

In order to shed more light on this subject, two lines of investigation should prove fruitful. Firstly, study of mechanisms controlling production of degrading enzymes by such fungi as *Elsinoe wisconsinensis* (Mason & Backus, 1969), *Phytophthora infestans* and *Rhynchosporium secalis* would be of interest. The first two biotrophs are closely related to wholly necrotrophic species and in the last two the biotrophic phase is followed by one of necrosis. Olutiola & Ayres (1973) have described the production of several cellulases by *R. secalis in vitro*. Similar studies with the biotrophic bacteria of Table 1 are also needed. The second potentially fruitful approach is comparable to the studies with necrotrophs above, where sugar levels in infected tissues were experimentally modified. If catabolite repression in biotrophic infections is maintained by high sugar levels, artificial lowering of these should de-repress cytolytic enzymes with consequent cell damage. This was demonstrated for leguminous root nodules by Thornton (1930) who described three conditions under which an effectively N-fixing, mutualistic strain of *Rhizobium* became actively parasitic; (*a*) in boron-deficient plants, (*b*) following transfer to the dark, and (*c*) in old nodules. Thornton attributed all three effects to diminished supply of carbohydrate to the nodule. Other bacterial as well as fungal infections need to be examined in this manner.

Continued supply of carbohydrate is thus the *sine qua non* of biotrophy (Allen, 1954). Altered patterns of translocation of photosynthetic products in hosts following infection by biotrophic parasites is now a well established phenomenon. In the case of aerial infections, these take the form of decreased export from and increased import to infected leaves. Such studies have been made for rusts (*Puccinia* and *Uromyces* spp.), smuts (*Ustilago nuda*), powdery mildews (*Erysiphe graminis*), downy mildews (*Pseudoperonospora cubensis*) and such species as

Plasmodiophora brassicae, Phytophthora infestans, Epichloe typhina and *Venturia inaequalis*. In sheathing mycorrhizas, increased import to roots occurs (Durbin, 1967; Smith *et al.* 1969; Scott, 1972; Lewis, 1973*a*; Hignett & Kirkham, 1967; Keen & Williams, 1969; Perl, Cohen & Rotem, 1972). In contrast to these effects by biotrophic fungi, neither necrotrophic bacteria nor fungi can induce import to diseased areas, although reduced export has been recorded (Shaw & Samborski, 1956; Lewis, 1973*a*).

The study of changes in patterns of translocation induced by biotrophic fungi has been facilitated by the fact that movement of carbohydrate to the sites of fungal infection is essentially one-way, although within sites there may be interchange of other metabolites between host and fungus. Carbon from the host commonly accumulates in specifically fungal products. In ascomycetes and basidiomycetes, these include acyclic sugar alcohols, trehalose and glycogen (Smith *et al.* 1969). Except in *Plasmodiophora*, where trehalose also occurs (Keen & Williams, 1969), comparable specific fungal products which accumulate have not been identified in 'lower' fungi, but are likely to be lipids.

Although there have been no comparable studies with biotrophic bacterial parasites, where it seems probable that essentially one-way movement will again predominate, there has been much work on the fate of photosynthetic products in nitrogen-fixing legume and non-legume nodules where carbon traffic is two-way between host and endophyte. Nearly all this work, however, has been in relation to nitrogen fixation and there appear to have been no *comparative* studies of translocation in nodulated and non-nodulated plants with respect to nutrition of the endophyte.

Minchin & Pate (1973) have shown that, in nodulated *Pisum sativum* during the 21–30-day period after sowing, 32 % of the carbon photosynthetically fixed by the plant is translocated to the nodules where, of this, only 18 % is consumed for nodule, including bacterial, growth. Respiration accounts for 38 % and the remaining 46 % is returned to the shoot as amino acids. Other studies demonstrate that movement of nutrients to nodules shows diurnal variation, being maximal close to noon both in legumes (Hardy, Holsten, Jackson & Burns, 1968) and in *Alnus* and *Myrica* (Wheeler, 1969, 1971) and that photosynthetically produced sugars are essential for the fixation process (Bach, Magee & Burris, 1958; Pate, 1962, 1968; Wheeler, 1971). These data suggest that maximal rates of fixation occur when recently produced photosynthates are translocated to the nodule. The presence of combined nitrogen inhibits nodulation and nitrogen fixation in both legumes and non-legumes (Stewart, 1966), and Small & Leonard (1969) have demonstrated that nitrate inhibits movement of photosynthetic products to nodules.

Movement of nutrients to the actual microbial endophytes of nodules has received less attention, perhaps because of difficulties owing to their small size and the fact that interchange of metabolites is occurring. Nevertheless, Lawrie & Wheeler (1973), by a micro-autoradiographic study of sections of nodules of *P. sativum*, frozen to retain soluble material, have shown that, particularly in the light, translocated photosynthetic products specifically accumulate within bacterial cells. Greatest accumulation of recent photosynthate occurs in cells partially filled with bacteroids and not those completely filled, although the latter are the major sites of nitrogen-fixing activity. It appears that in the cells filled with largely non-growing bacteroids there is a rapid flux of photosynthates *through* actively N-fixing cells without accumulation and that accumulation only occurs in the growing cells.

It is clear from this discussion of the effects of biotrophic micro-organisms on carbon movement in higher plants that sites of infection, whether parasitic or mutualistic, are also sites of accumulation of photosynthetic products, an accumulation which is only transient in the case of the root-nodulated nitrogen-fixing symbioses. The biotrophic movement of carbon from algae to symbiotic heterotrophs, both fungal and invertebrate, has been reviewed (Smith *et al.* 1969; Smith, 1974). The continued import of soluble sugars may bring about catabolite repression of degradative enzymes of the heterotrophs.

The mechanisms by which alterations to patterns of translocation occur are probably hormonal but exact details of how microbially produced hormones or microbially induced alterations in hormone production are effective must await an understanding of the role of hormones in non-infected higher plants. The implication of hormones in fungally induced changes in patterns of translocation has been discussed by Thrower (1965), Brian (1967), Durbin (1967), Smith *et al.* (1969), Scott (1972) and Lewis (1973*a*).

Although many micro-organisms – saprotrophs, necrotrophs and biotrophs – produce auxins, gibberellins and cytokinins (Gruen, 1959; Sequeira, 1963; Wood, 1967), it is only in biotrophic infections that the minimal tissue damaged involved permits not only ready transport of these hormones away from sites of infection but also the enhanced translocation of nutrients to these sites. As hormones also control morphogenesis in plants and induced abnormal morphology is a characteristic of biotrophic infections (Brian 1967; Wood, 1967; Yarwood,

1967), it is not surprising that the production of hormones in general and (since they seem to be involved in nutrient mobilization and cell division) cytokinins in particular has been investigated in infected tissues and by the organisms responsible in culture. Since the fungi of the most intensively studied diseases involving biotrophs have either not yet been cultured or, as in the case of rusts, only recently been cultured (Scott, 1972), most hormonal work with fungi has concerned changes in infected versus uninfected tissues (see reviews quoted at the end of the last paragraph). Thimann & Sachs (1966) discuss the role of cytokinins in infections by *Corynebacterium fascians*, Wood (1967) that of auxins in infections by *Agrobacterium tumefaciens*, Dullaart (1970*a*, *b*) that of auxins in legume and non-legume nodules and Dekhuijzen & Overeem (1971) that of cytokinins in infections by *Plasmodiophora*. The recent studies of Butcher, El Tigani & Ingram (personal communication) have indicated that in infections of *Brassica rapa* (Cruciferae) by *Plasmodiophora brassicae* the pathogen induces the conversion of the host metabolite glucobrassicin into indoleacetonitrile (an active auxin) which is then involved in club-root hypertrophy. It is significant that the physiological disorder of tobacco, false broomrape, which results in white succulent outgrowths first appearing as tumour-like growths on the roots is caused by increased cytokinin to auxin ratios in the roots (Hamilton, Lowe & Skoog, 1972). Clearly, abnormal growth and the attendant alterations in patterns of translocation can be brought about by induced hormonal imbalance.

Where biotrophic symbionts are culturable, hormone production *in vivo* has also been studied. For example, mycorrhizal fungi produce auxins and cytokinins (Moser, 1959; Ulrich, 1960; Miller, 1967, 1971; Laloue & Hall, 1973), and one or both of these classes of hormone are synthesized in culture by the parasitic bacterial symbionts, *Agrobacterium tumefaciens* and *Corynebacterium fascians*, and the mutualistic bacterial symbionts, *Rhizobium* spp. and *Chromobacterium lividum*, the endophyte of the leaf nodules of *Ardisia* spp. (Thimann & Sachs, 1966; Klämbt, Thies & Skoog, 1966; Wood, 1967; Dullaart, 1970*a*; Phillips & Torrey, 1972; Rodrigues Pereira, Houwen, Deurenberg-Vos & Pey, 1972). For *Corynebacterium fascians* Thimann & Sachs note, 'when kinetin is applied to plants, it causes the local accumulation of various nutrients. An accumulation of this sort must be a nutritional advantage to the parasite, and this is probably the reason, in an evolutionary sense, for its forming a cytokinin.'

Although catabolite repression and its maintenance can therefore contribute to the suppression of necrotrophic nutrition in favour of biotrophy, it remains to be seen how far stage 4 (see p. 379) of the process has progressed. Genetic loss of the capacity to produce any cytolytic enzymes at all is theoretically possible, although unproved. It certainly has not occurred in some biotrophs, including some of the unspecialized, as yet uncultured species, as shown by the role of enzymes in initial infections by powdery and downy mildews. The assertion of Albersheim *et al.* (1969) 'that every plant pathogen examined has shown ability to produce such enzymes' may, however, require modification. Much more work is needed concerning the effects of experimentally interrupting supplies of carbohydrate to sites of infection on maintenance of biotrophy.

BIOTROPHY AND MUTUALISTIC SYMBIOSIS

Little distinction has been made above between parasitic and mutualistic symbiosis. This has been deliberate since from the point of view of nutrition of the heterotrophic member there *is* little distinction. The essential features of mutualistic symbiosis involving photosynthetic organisms have been stated by Scott (1969) and discussed by Lewis (1973*a*, *b*). Fig. 3 illustrates the salient features which distinguish mutualistic from parasitic associations, the essential one being the increased ecological amplitude which dual associations have over the partners in isolation. This is brought about as a result of the two-way movement of nutrients which is only possible in a sustained manner in biotrophic interactions. As discussed above and elsewhere (Lewis, 1973*a*), biotrophy can change to necrotrophy depending on environmental conditions, so that mutualistic associations can also become parasitic.

Lewis (1973*b*) distinguished between two kinds of mutualistic interaction: those where, with reference to potentially deficient nutrient elements, one partner could *avoid* or *tolerate* the rigours of the environment by virtue of particular attributes of the other. With chemoheterotrophs, carbon deficiency is avoided by exploitation of the photosynthetic capacity of their partners. With nitrogen-fixing associations, nitrogen deficiency is avoided by the fungal component of lichens and by the higher plants with nodules. In both these cases, exploitation of gaseous reserves compensates for deficiencies in terrestrial and aquatic environments. With respect to mineral deficiencies, mutualistic symbioses, particularly sheathing and vesicular–arbuscular mycorrhizas and lichens, tolerate nutrient stress by efficient exploitation of the habitat (Smith, 1962; Baylis, 1972). In all cases, 'standard' nutrient cycles are short-circuited. This is particularly evident in the case of carbon and

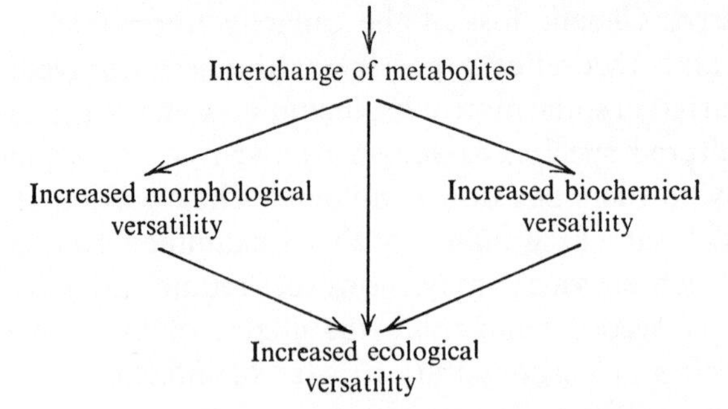

Fig. 3. Characteristics of mutualistic symbioses between autotrophs and heterotrophs (after Scott, 1969).

nitrogen where in the donor partners the incorporation of the element is not primarily into polymers but into simple, soluble substances which pass to the recipient. Mutualistic symbioses therefore utilize their resources efficiently by reduced expenditure of energy in the donors on synthesis of macromolecules and reduced expenditure of energy in the recipient on synthesis of degradative enzymes. Under conditions of nutrient stress, natural selection favours mutualistic symbiosis. Those we see today are the evolutionary equivalent of the ones, which under nutrient stress over a thousand million years ago, resulted in the eukaryotic cell (Margulis, 1970).

SUMMARY

The universally applicable classification of associations between organisms into three interacting continua proposed by Heise & Starr has been extended to include a fourth, nutritional, continuum. This recognizes that chemoheterotrophs, including animals, may derive nutrients in the free-living state (*saprotrophic*) or following intimate contact with other organisms (*symbiotrophic*). From the point of view of mechanisms of assimilating nutrients, saprotrophs may be *osmotrophic* or *phagotrophic* and, from the point of view of source of nutrients, symbiotrophs may be *biotrophic* or *necrotrophic*. By combining classifications based on ecological and nutritional criteria, five categories of microbial behaviour can be delimited: *obligate saprotrophs*, *facultative necrotrophs*, *obligate necrotrophs*, *facultative biotrophs* and *obligate biotrophs*. Although only illustrated for associations of photosynthetic organisms with bacteria, actinomycetes or fungi, these are also applicable to associations of diverse eukaryotes such as animals and algae.

These categories can form the basis of several series of evolutionarily related patterns of behaviour. This is illustrated with regard to the origins of necrotrophy and biotrophy directly from saprotrophy, and biotrophy from saprotrophy via necrotrophy. The last involves the interplay of alterations to patterns of translocation produced by microbially induced changes in hormonal balance in infected plants and catabolite repression of degradative enzymes of the chemoheterotrophs. The relevance of this hypothesis, originally formulated for fungal behaviour, to prokaryotic nutrition is demonstrated by a consideration of the nutrition of both parasitic and mutualistic nodule-forming bacteria and actinomycetes.

The evolution within biotrophy from parasitism to mutualism depends on the two-way transfer of nutrients between symbionts and the capacity of the symbioses to avoid or tolerate nutrient stresses in the environment. Any change from biotrophic to necrotrophic nutrition results in the breakdown of mutualism. Such a change is under environmental influence and depends on control of catabolite repression of degradative enzymes.

I am grateful to Dr P. M. Holligan, Dr C. T. Wheeler and Mr A. Howard for helpful advice and to Dr D. N. Butcher for permission to quote unpublished work.

REFERENCES

AHMADJIAN, V. (1967). *The Lichen Symbiosis.* Waltham, Mass.: Blaisdell.

AIST, J. R. & WILLIAMS, P. H. (1971). The cytology and kinetics of cabbage root hair penetration by *Plasmodiophora brassicae. Canadian Journal of Botany*, **49**, 2023–34.

AKAI, S., KUNOH, H. & FUKOTOMI, M. (1968). Histochemical changes of the epidermal cell wall of barley leaves, infected by *Erysiphe graminis hordei. Mycopathologica et Mycologia Applicata*, **35**, 175–80.

ALBERSHEIM, P., JONES, T. M. & ENGLISH, P. D. (1969). Biochemistry of the cell wall in relation to infective processes. *Annual Review of Phytopathology*, **7**, 171–94.

ALEXANDER, M. (1971). *Microbial Ecology.* New York: Wiley.

ALLEN, P. J. (1954). Physiological aspects of fungus diseases of plants. *Annual Review of Plant Physiology*, **3**, 225–48.

ARMENTROUT, V. N. & WILSON, C. L. (1969). Haustorium–host interaction during mycoparasitism of *Mycotypha microspora* by *Piptocephalis virginiana. Phytopathology*, **59**, 897–905.

BACH, M. K., MAGEE, W. E. & BURRIS, R. H. (1958). Translocation of photosynthetic products to soy bean nodules and their role in nitrogen fixation. *Plant Physiology*, **33**, 118–24.

BARKER, S. M. & BARNETT, H. L. (1973). Nitrogen and vitamin requirements for axenic growth of the haustorial mycoparasite, *Dispira cornuta. Mycologia*, **65**, 21–7.

BARNETT, H. L. (1963). The nature of mycoparasitism by fungi. *Annual Review of Microbiology*, **17**, 1–14.

BARNETT, H. L. (1964). Mycoparasitism. *Mycologia*, **56**, 1–19.

BARNETT, H. L. (1970). Nutritional requirements for axenic growth of some haustorial mycoparasites. *Mycologia*, **62**, 750–60.

BAYLIS, G. T. S. (1972). Fungi, phosphorus and the evolution of root systems. *Search*, **3**, 257–8.

BECKING, J. H. (1970). Plant–endophyte symbiosis in non-leguminous plants. *Plant and Soil*, **32**, 611–54.

BIEHN, W. L. & DIMOND, A. E. (1971). Effect of galactose on polygalacturonase production and pathogenesis by *Fusarium oxysporum* f.sp. *lycopersici*. *Phytopathology*, **61**, 242–3.

BOND, E. (1963). The root-nodules of non-leguminous angiosperms. In *Symbiotic Associations, Symposia of the Society for General Microbiology*, **13**, 72–91. Ed. P. S. Nutman & B. Mosse. London: Cambridge University Press.

BOND, G. (1967). Fixation of nitrogen by higher plants other than legumes. *Annual Review of Plant Physiology*, **18**, 107–26.

BRIAN, P. W. (1967). Obligate parasitism in fungi. *Proceedings of the Royal Society*, B, **168**, 101–18.

BUDDENHAGEN, I. W. (1965). The relation of plant pathogenic bacteria to the soil. In *Ecology of Soil-Borne Plant Pathogens*, ed. K. F. Baker & W. C. Snyder, pp. 269–84. London: John Murray.

BUDDENHAGEN, I. & KELMAN, A. (1964). Biological and physiological aspects of bacterial wilts caused by *Pseudomonas solanacearum*. *Annual Review of Phytopathology*, **2**, 203–30.

CALDERONE, R. A. & BARNETT, H. L. (1972). Axenic growth and nutrition of *Gonatobotryum fuscum*. *Mycologia*, **64**, 153–60.

CROSSE, J. E. (1968). Plant pathogenic bacteria in soil. In *The Ecology of Soil Bacteria*, ed. T. R. G. Gray & D. Parkinson, pp. 552–72. Liverpool: Liverpool University Press.

DAFT, M. J. & OKUSANYA, B. O. (1973). Effect of *Endogone* mycorrhiza on plant growth. VI. Influence of infection on the anatomy and reproductive development in four hosts. *New Phytologist*, **72**, 1333–9.

DEKHUIJZEN, H. M. & OVEREEM, J. C. (1971). The role of cytokinins in clubroot formation. *Physiological Plant Pathology*, **1**, 151–61.

DOWSON, W. J. (1957). *Plant Diseases Due to Bacteria*. London: Cambridge University Press.

DROOP, M. R. (1963). Algae and Invertebrates. In *Symbiotic Associations, Symposia of the Society for General Microbiology*, **13**, 171–99. Ed. P. S. Nutman & B. Mosse. London: Cambridge University Press.

DULLAART, J. (1970*a*). The auxin content of root nodules and roots of *Alnus glutinosa* (L.) Vill. *Journal of Experimental Botany*, **21**, 975–84.

DULLAART, J. (1970*b*). The bioproduction of indole-3-acetic acid and related compounds in root nodules and roots of *Lupinus luteus* L. and by its rhizobial symbionts. *Acta Botanica Neerlandica*, **19**, 573–615.

DURBIN, R. D. (1967). Obligate parasites: effect on the movement of solutes and water. In *The Dynamic Role of Molecular Constituents in Plant–Parasite Interaction*, ed. C. J. Mirocha & I. Uritani, pp. 80–99. St Paul, Minnesota: The American Phytopathological Society.

EDWARDS, H. H. & ALLEN, P. J. (1970). A fine-structure study of the primary infection process during infection of barley by *Erysiphe graminis* f.sp. *hordei*. *Phytopathology*, **60**, 1504–9.

EVANS, L. V., CALLOW, J. A. & CALLOW, M. E. (1973) Structural and physiological studies on the parasitic red alga *Holmsella* . *New Phytologist*, **72**, 393–402

FRITSCH, F. E. (1935). *The Structure and Reproduction of the Algae*, vol. 1. London: Cambridge University Press.

FRITSCH, F. E. (1952). *The Structure and Reproduction of the Algae*, vol. 2. London: Cambridge University Press.

GAIN, R. E. & BARNETT, H. L. (1970). Parasitism and axenic growth of the mycoparasite, *Gonatorhodiella highlei. Mycologia*, **62**, 1122–9.

GARRETT, S. D. (1970). *Pathogenic Root Infecting Fungi.* London: Cambridge University Press.

GRAY, T. R. G. & WILLIAMS, S. T. (1971). *Soil Micro-organisms.* Edinburgh: Oliver & Boyd.

GRUEN, H. E. (1959). Auxins and fungi. *Annual Review of Plant Physiology*, **10**, 405–40.

HALE, M. E. (1970). *The Biology of Lichens.* London: Edward Arnold.

HAMILTON, J. L., LOWE, R. H. & SKOOG, F. (1972). False broomrape: a physiological disorder caused by growth regulator imbalance. *Plant Physiology*, **50**, 303–4.

HARDY, R. W. F., HOLSTEN, R. D., JACKSON, E. K. & BURNS, R. C. (1968). The acetylene–ethylene assay for nitrogen fixation: laboratory and field evaluation. *Plant Physiology*, **43**, 1185–207.

HARLEY, J. L. (1950). Recent progress in the study of endotrophic mycorrhiza. *New Phytologist*, **49**, 213–47.

HARLEY, J. L. (1969). *The Biology of Mycorrhiza*, 2nd edition. London: Leonard Hill.

HIGNETT, R. C. & KIRKHAM, D. S. (1967). The role of extracellular melanoproteins of *Venturia inaequalis* in host susceptibility. *Journal of General Microbiology*, **48**, 269–75.

HORSFALL, J. G. & DIMOND, A. E. (1957). Interactions of tissue sugar, growth substances and disease susceptibility. *Zeitschrift für Pflanzenkrankheiten, Pflanzenpathologie und Pflanzenschutz*, **64**, 415–21.

HORTON, J. C. & KEEN, N. T. (1966*a*). Regulation of induced cellulase synthesis in *Pyrenochaeta terrestris* Gorenz *et al.* by utilisable carbon compounds. *Canadian Journal of Microbiology*, **12**, 209–20.

HORTON, J. C. & KEEN, N. T. (1966*b*). Sugar repression of endopolygalacturonase and cellulase synthesis during pathogenesis by *Pyrenochaeta terrestris* as a resistance mechanism in onion pink rot. *Phytopathology*, **56**, 908–16.

JOUBERT, J. J. & RIJKENBERG, F. H. J. (1971). Parasitic green algae. *Annual Review of Phytopathology*, **9**, 45–64.

KEEN, N. T. & HORTON, J. C. (1966). Induction and repression of endopolygalacturonase synthesis by *Pyrenochaeta terrestris. Canadian Journal of Microbiology*, **12**, 443–53.

KEEN, N. T. & WILLIAMS, P. H. (1969). Translocation of sugars into infected cabbage tissues during clubroot development. *Plant Physiology*, **44**, 748–54.

KLÄMBT, D., THIES, G. & SKOOG, F. (1966). Isolation of cytokinins from *Corynebacterium fascians. Proceedings of the National Academy of Sciences, USA*, **56**, 52–9.

KOHLMEYER, J. & KOHLMEYER, E. (1972). Is *Ascophyllum nodosum* lichenized? *Botanica marina*, **15**, 109–22.

KUNOH, H. & AKAI, S. (1969). Histochemical observation of the halo on the epidermal cell wall of barley leaves attached by *Erysiphe graminis hordei. Mycopathologica et Mycologia Applicata*, **37**, 113–18.

KURTZMAN, C. P. (1968). Parasitism and axenic growth of *Dispira cornuta. Mycologia*, **60**, 915–23.

LALOUE, M. & HALL, R. H. (1973). Cytokinins in *Rhizopogon roseolus.* Secretion of *N*[9-(β-D-Ribofuranosyl-9H) purin-6-yl carbamoyl] threonine in the culture medium. *Plant Physiology*, **51**, 559–62.

LANGE, R. T. (1966). Bacterial symbiosis with plants. In *Symbiosis*, ed. S. M. Henry, vol. 1, pp. 99–170. New York: Academic Press.

LAWRIE, A. C. & WHEELER, C. T. (1973). The supply of photosynthetic assimilates to nodules of *Pisum sativum* L. in relation to the fixation of nitrogen. *New Phytologist*, **72**, 1341–8.

LEWIS, D. H. (1973*a*). Concepts in fungal nutrition and the origin of biotrophy. *Biological Reviews*, **48**, 261–78.

LEWIS, D. H. (1973*b*). The relevance of symbiosis to taxonomy and ecology, with particular reference to mutualistic symbioses and the exploitation of marginal habitats. In *Taxonomy and Ecology*, ed. V. H. Heywood, pp. 151–72. London: Academic Press.

LINDEBERG, G. (1948). On the occurrence of polyphenol oxidases in soil-inhabiting Basidiomycetes. *Physiologia Plantarum*, **1**, 196–205.

LUNDEBERG, G. (1970). Utilization of various nitrogen sources, in particular bound soil nitrogen, by mycorrhizal fungi. *Studia Forestalia Suecica*, **79**, 1–95.

MCKEEN, W. E., SMITH, R. & BHATTACHARYA, P. K. (1969). Alteration of the host wall surrounding the infection peg of powdery mildew fungi. *Canadian Journal of Botany*, **47**, 701–6.

MCNEW, G. L. (1960). The nature, origin and evolution of parasitism. In *Plant Pathology – an Advanced Treatise*, ed. J. G. Horsfall & A. E. Dimond, vol. 2, pp. 19–69. New York: Academic Press.

MANOCHA, M. S. & LEE, Y. K. (1971). Host–parasite relations in mycoparasite. I. Fine structure of host, parasite and their interface. *Canadian Journal of Botany*, **49**, 1677–82.

MARGULIS, L. (1970). *Origin of Eucaryotic Cells*. New Haven: Yale University Press.

MASON, D. L. & BACKUS, M. P. (1969). Host–parasite relations in spot anthracnose of *Desmodium*. *Mycologia*, **61**, 1124–41.

MELIN, E. (1948). Recent advances in the study of tree mycorrhiza. *Transactions of the British Mycological Society*, **30**, 92–9.

MELIN, E. (1953). Physiology of mycorrhizal relations in plants. *Annual Review of Plant Physiology*, **4**, 325–46.

MEYER, F. H. (1966). Mycorrhiza and other plant symbioses. In *Symbiosis*, ed. S. M. Henry, vol. 1, pp. 171–255. New York: Academic Press.

MILLER, C. O. (1967). Zeatin and zeatin riboside from a mycorrhizal fungus. *Science*, **157**, 1055–7.

MILLER, C. O. (1971). Cytokinin production by mycorrhizal fungi. In *Mycorrhizae, Proceedings of the* 1*st North American Conference on Mycorrhizae*, pp. 168–74. Washington: US Government Printing Office.

MINCHIN, F. R. & PATE, J. S. (1973). The carbon balance of a legume and the functional economy of its root nodules. *Journal of Experimental Botany*, **24**, 259–71.

MOSER, M. (1959). Beiträge zur Kenntnis der Wuchsstoffbeziehungen im Bereich ectotrophen Mykorrhizen. *Archiv für Mikrobiologie*, **34**, 251–69.

MOSSE, B. (1963). Vesicular–arbuscular mycorrhiza: an extreme form of fungal adaptation. In *Symbiotic Associations*, *Symposia of the Society for General Microbiology*, **13**, 146–70. Ed. P. S. Nutman & B. Mosse. London: Cambridge University Press.

NICOLSON, T. H. (1967). Vesicular–arbuscular mycorrhiza – a universal plant symbiosis. *Science Progress*, **55**, 561–81.

NORKRANS, B. (1950). Studies in growth and cellulolytic enzymes of *Tricholoma* with special reference to mycorrhiza formation. *Symbolae botanicae upsaliensis*, **11**, 1–126.

NUTMAN, P. S. (1963). Factors influencing the balance of mutual advantage in legume symbiosis. In *Symbiotic Associations, Symposia of the Society for General Microbiology*, **13**, 51–71. Ed. P. S. Nutman & B. Mosse. London: Cambridge University Press.

OLUTIOLA, P. O. & AYRES, P. G. (1973). A cellulase complex in culture filtrates of *Rhynchosporium secalis* (barley leaf blotch). *Transactions of the British Mycological Society*, **60**, 273–82

PAIGEN, K. & WILLIAMS, B. (1970). Catabolite repression and other control mechanisms in carbohydrate utilization. *Advances in Microbial Physiology*, **4**, 251–324.

PALMER, J. G. & HACSKAYLO, E. (1970). Ectomycorrhizal fungi in pure culture. I. Growth on single carbon sources. *Physiologia Plantarum*, **23**, 1187–97.

PATE, J. S. (1962). Root exudation studies on the exchange of ^{14}C-labelled organic substances between the roots and shoot of the nodulated legume. *Plant and Soil*, **17**, 333–56.

PATE, J. S. (1968). Physiological aspects of inorganic and intermediate nitrogen metabolism (with special reference to the legume, *Pisum sativum* L.). In *Recent Advances of Nitrogen Metabolism in Plants*, ed. E. J. Hewitt & L. V. Cutting, pp. 214–40. London: Academic Press.

PATIL, S. S. & DIMOND, A. E. (1968). Repression of polygalacturonase synthesis in *Fusarium oxysporum* f.sp. *lycopersici* by sugars and its effect on symptom reduction in infected tomato plants. *Phytopathology*, **58**, 676–82.

PERL, M., COHEN, Y. & ROTEM, S. (1972). The effect of humidity during darkness on the transfer of assimilates from cucumber leaves to sporangia of *Pseudoperonospora cubensis*. *Physiological Plant Pathology*, **2**, 113–22.

PHILLIPS, D. A. & TORREY, J. G. (1972). Studies on cytokinin production by *Rhizobium*. *Plant Physiology*, **49**, 11–15.

QUISPEL, A. (1942). The lichenisation of aerophilic algae. *Proceedings. Koniklijke Nederlandse akademie van Wetenschappen* C **45**, 276–82.

QUISPEL, A. (1943). The mutual relations between algae and fungi in lichens. *Recueil des travaux Botaniques Néerlandais*, **40**, 413–541.

READ, C. P. (1970). *Parasitism and Symbiology*. New York: Ronald Press.

RODRIGUES PEREIRA, A. S., HOUWEN, P. J. W., DEURENBERG-VOS, H. W. J. & PEY, E. B. F. (1972). Cytokinins and the bacterial symbiosis of *Ardisia* species. *Zeitschrift für Pflanzenphysiologie*, **68**, 170–77.

SARGENT, J. A., TOMMERUP, I. C. & INGRAM, D. S. (1973). The penetration of a susceptible lettuce variety by the downy mildew fungus, *Bremia lactucae* Regel. *Physiological Plant Pathology*, **3**, 231–40.

SCOTT, G. D. (1969). *Plant Symbiosis*. London: Edward Arnold.

SCOTT, K. J. (1972). Obligate parasitism by phytopathogenic fungi. *Biological Reviews*, **47**, 537–72.

SEQUEIRA, L. (1963). Growth regulators and plant disease. *Annual Review of Phytopathology*, **1**, 1–30.

SHAW, M. & SAMBORSKI, D. J. (1956). The physiology of host–parasite relations. I. The accumulation of radioactive substances at infections of facultative and obligate parasites including tobacco mosaic virus. *Canadian Journal of Botany*, **34**, 389–405.

SMALL, J. G. L. & LEONARD, D. A. (1969). Translocation of C^{14}-labelled photosynthate in nodulated legumes as influenced by nitrate nitrogen. *American Journal of Botany*, **56**, 187–94.

SMITH, D. C. (1962). The biology of lichen thalli. *Biological Reviews*, **37**, 537–70.

SMITH, D. C. (1963). Experimental studies of lichen physiology. In *Symbiotic Associations, Symposia of the Society for General Microbiology*, **13**, 31–50. Ed. P. S. Nutman & B. Mosse. London: Cambridge University Press.

SMITH, D. C. (1974). Transport from symbiotic algae and symbiotic chloroplasts to host cells. In *Transport at the Cellular Level, Symposia of the Society for Experimental Biology*, **28**, in press. Ed. D. H. Jennings & M. A. Sleigh. London: Cambridge University Press.

SMITH, D. C., MUSCATINE, L. & LEWIS, D. (1969). Carbohydrate movement from autotrophs to heterotrophs in parasitic and mutualistic symbiosis. *Biological Reviews*, **44**, 17–90.

STANIER, R. Y., DOUDOROFF, M. & ADELBERG, E. A. (1971). *General Microbiology*, 3rd edition. London: Macmillan.

STAPP, C. (1961). *Bacterial Plant Pathogens*. Oxford: Clarendon Press.

STARR, M. P. (1959). Bacteria as plant pathogens. *Annual Review of Microbiology*, **13**, 211–38.

STARR, M. P., CHATTERJEE, A. K. (1972). The genus *Erwinia*: Enterobacteria pathogenic to plants and animals. *Annual Review of Microbiology*, **26**, 389–426.

STARR, M. P. & HUANG, J. C. C. (1972). Physiology of Bdellovibrios. *Advances in Microbial Physiology*, **8**, 215–61.

STEWART, W. D. P. (1966). *Nitrogen Fixation in Plants*. London: Athlone Press.

THIMANN, K. V. & SACHS, T. (1966). The role of cytokinins in the 'fasciation' disease caused by *Corynebacterium fascians*. *American Journal of Botany*, **53**, 731–9.

THORNTON, H. G. (1930). The influence of the host plant in inducing parasitism in lucerne and clover nodules. *Proceedings of the Royal Society*, B, **106**, 110–122.

THROWER, L. B. (1965). Host physiology and obligate fungal parasites. *Phytopathologische Zeitschrift*, **52**, 319–34.

ULRICH, J. M. (1960). Auxin production by mycorrhizal fungi. *Physiologia Plantarum*, **13**, 429–443.

VAN SUMERE, C. F., VAN SUMERE-DE PRETER, C. & LEDINGHAM, G. A. (1957). Cell-wall splitting enzymes of *Puccinia graminis* var. *tritici*. *Canadian Journal of Microbiology*, **3**, 761–70.

WAKSMAN, S. A. (1959). *The Actinomycetes*. London: Baillière, Tindall & Cox.

WAREING, P. F. & PHILLIPS, I. D. J. (1970). *The Control of Growth and Differentiation in Plants*. Oxford: Pergamon Press.

WEBSTER, J. (1970). *Introduction to Fungi*. London: Cambridge University Press.

WHEELER, C. T. (1969). The diurnal fluctuations in nitrogen fixation in the nodules of *Alnus glutinosa* and *Myrica gale*. *New Phytologist*, **68**, 675–82.

WHEELER, C. T. (1971). The causation of the diurnal changes in nitrogen fixation in the nodules of *Alnus glutinosa*. *New Phytologist*, **70**, 487–95.

WOOD, R. K. S. (1967). *Physiological Plant Pathology*. Oxford: Blackwell.

YARWOOD, C. E. (1956). Obligate parasitism. *Annual Review of Plant Physiology*, **7**, 115–42.

YARWOOD, C. E. (1967). Response to parasites. *Annual Review of Plant Physiology*, **18**, 419–38.

THE PRECELLULAR EVOLUTION AND ORGANIZATION OF MOLECULES

C. PONNAMPERUMA AND N. W. GABEL

Laboratory of Chemical Evolution,
University of Maryland, College Park,
Maryland 20742, *USA*

INTRODUCTION

According to the Oparin–Haldane hypothesis (Oparin, 1924; Haldane, 1928) of chemical evolution, the origin of life on earth requires the prior existence of chemicals, both organic and inorganic, which are constituent parts of living organisms. Before the emergence of these life forms, many chemical reactions must have taken place to produce the molecules which, as a result of their physical and chemical properties, were capable of self-organization and replication.

A question which may be asked about these chemical reactions is whether or not they are representative of evolution, in general, and in this way related to biological evolution. It is pertinent to recall Pirie's classic essay on 'The meaninglessness of the terms life and living' (Pirie, 1937). Any distinctions between life and non-life must of necessity be arbitrarily imposed by the observer. It is biologically observable that as each species or biochemical system becomes positioned by its ecological niche, other niches are created or nullified by interactions between functioning microenvironments. This phenomenon is paralleled by the interactions of electronic energy states on a molecular level. The plethora of experiments which have been carried out for the purpose of studying abiotic, chemical evolution indicates the necessity of microenvironments either of prior existence or whose formation occurs concomitantly with the formation of new chemical species (Ponnamperuma & Gabel, 1968; Lemmon 1970; Gabel & Ponnamperuma, 1972; Oro, Young & Ponnamperuma, to be published). Cloud (1968) has pointed out that the biological evolution of photosynthetic organisms also seems to have been dependent on the prior existence of ecological niches or microenvironments. Since life is considered to be a composite of non-living interrelationships, it would be incongruous within the analytic method of phenomenic observation to maintain that terminology describing the behavior of molecules and the behavior of complex living organisms are not related (Gabel, 1973).

SURFACE AND ATMOSPHERIC CONDITIONS OF PRIMITIVE EARTH-LIKE PLANETS

Composition

The age of the earth is believed to be about 4.5 billion years (Tilton & Steiger, 1965), that of the solar system approximately 5 billion years, and that of the universe 12 billion years (Dicke, 1962). Life on earth is considered to be more than 3 billion years old. The Onverwacht shales and the Fig Tree cherts from South Africa, dated at approximately 3.1–3.4 billion years, have evidence of some microfossils (Barghoorn & Schopf, 1966; Engel *et al.* 1968). Darwinian evolution would account for the developmental stages from the first organism to contemporary phyla; whereas all prior chemical transformations would fall into the realm of chemical evolution.

Natural scientists gave very little attention to the problem of the origin of life until Oparin (1924) and Haldane (1928) presented a documented, scientific hypothesis for the evolution of complex, biologically important molecules from the simple molecules of primitive atmospheres. The original atmosphere of the earth is believed to have consisted mainly of hydrogen and helium. Data from a number of reports on the rate of escape of these gases from the earth's gravitational field has been collated by Venkateswaran (1971). The present rarity of the noble gases in the earth's atmosphere in comparison to their distribution in the universe indicates that most of the original atmosphere of the earth was lost (Brown, 1952). Through the process of outgassing during planetary accretion a secondary primordial atmosphere consisting of hydrogen, methane, ammonia, and water would have appeared. Miller & Urey (1959) suggest the figure 1.5×10^{-3} atmospheres as a reasonable concentration for hydrogen in the primordial atmosphere. This would correspond to a pressure for methane of 4×10^{3} atmospheres. Miller & Urey (1959) concluded that, since ammonia is very rapidly converted to nitrogen and hydrogen by any energy source, a reducing atmosphere consisting of small amounts of hydrogen, ammonia, and water and a moderate pressure of methane and nitrogen would constitute a reasonable atmosphere for the primordial earth.

In one of the most exhaustive investigations on the products obtained by subjecting mixtures of gases to an energy source, Abelson (1956) studied twenty different combinations of hydrogen, methane, carbon monoxide, carbon dioxide, ammonia, nitrogen, water, and oxygen. Organic molecules, having two or more carbon atoms, formed only when the mixture was non-oxidizing with respect to methane. The

Table 1. *Energy sources available for organic synthesis on the primitive earth*

Source	Quantity (cal cm^{-2} yr^{-1})
Radiation from the sun (all wavelengths)	260000
Ultraviolet light	
$\lambda < 250$ nm	570
$\lambda < 200$ nm	85
$\lambda < 150$ nm	3.5
Electric discharges	4
Cosmic rays	0.0015
Radioactivity (to 1.0 km depth)	2.8
Heat from volcanoes	0.13
Meteorite impact	0.1

corollary implication of these experiments is that reducing conditions are necessary for the synthesis of organic molecules in significant quantities on primitive planetary objects. The escape velocity of hydrogen from any gravitational field is less than that of any other element. Therefore, during the process of chemical evolution, all astronomical systems are continually being depleted of hydrogen even when nuclear processes are not operative. Since the production of complex organic molecules from the primordial gases requires an increase in the oxidation state of some of the atoms of the synthesized molecules, it is just as important that the primordial atmosphere gradually becomes relatively more oxidizing as it is to begin with a reducing environment for the accumulation of diverse organic molecules. If the environment did not gradually become relatively more oxidizing, there would be no extensive chemical evolution. No system which is in equilibrium ever evolves.

Energy sources

The energy impinging upon the reducing atmosphere of the primitive earth which could initiate chemical reactions would be light from the sun, electric discharges, ionizing radiation from radioactive decay, and heat in the form of volcanoes and hot springs (Table 1).

Ultraviolet light is by far the most intense energy source available for the synthesis of more complex materials from the simple primordial molecules. The high intensities of the solar radiation and the long path length through the atmosphere may have resulted in significant amounts of photochemical reactions even if the absorption coefficients are very small (Fig. 1).

The intensity of the solar flux also decreases as the wavelength of the light decreases. It must be remembered, however, that the energy of

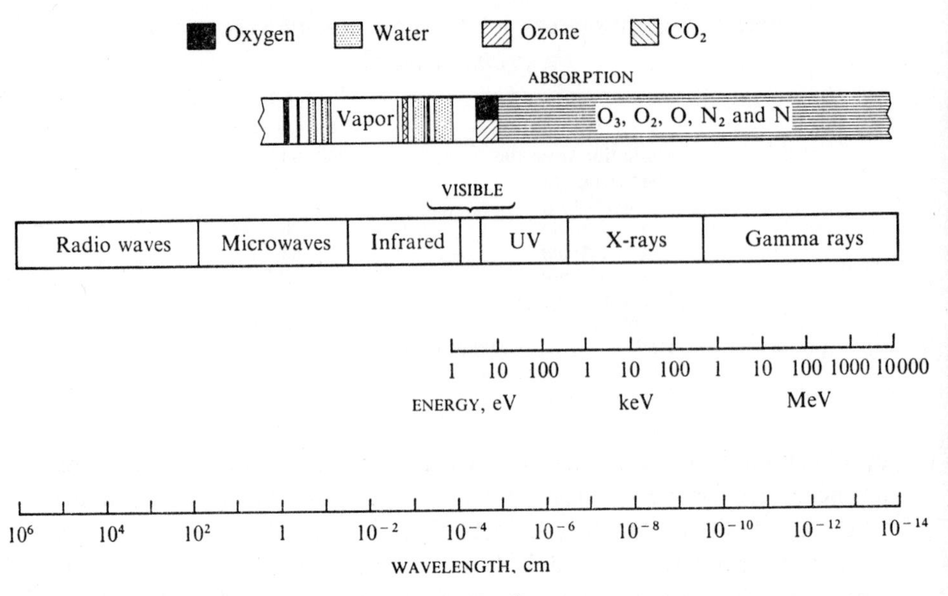

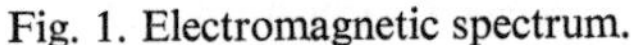

Fig. 1. Electromagnetic spectrum.

a photon of light increases with decreasing wavelength. To initiate bond-breaking of the simpler molecules and concomitant synthesis of more complex materials, the simpler molecules must be capable of absorbing the energy of a given wavelength of light which is impinging upon them. It would be instructive at this point to examine the absorption spectra of the gases of the primordial atmosphere. The absorption coefficients of a number of gases have been measured with argon in the reference cell (Watanabe, Zelikoff & Inn, 1953; Thompson, Harteck & Reeves, 1963). The two main constituents, methane and nitrogen, have absorption coefficients of less than 10^{-4} cm^{-1} between 185 and 200 nm. Fortunately, ammonia and water have an appreciable absorption in this region. Although the absorption of methane extends to 145 nm, it is highly probable that ammonia and water serve to a large extent both as initiators and reactants of any initial photochemical reactions.

THE SYNTHESIS OF COMPLEX ORGANIC MOLECULES

It is a tribute to Oparin (1924, 1964) and Haldane (1928) that, before the overwhelming abundance of hydrogen in the universe was determined astrophysically (Russell, 1935), they foresaw the need for the primordial atmosphere to be reducing in order for complex organic materials to form. One of the most common approaches to the prebiological chemistry of the earth involves the analysis of products from methane–ammonia–water mixtures subjected to one of several possible energy sources. These products are stable compounds which are probably derived from the reactions between chemical species initially produced from the gaseous mixture. Fig. 2 (from Yuasa & Imahori, 1971) is a generalized scheme of these reaction pathways.

By far the greatest amount of experimentation in the field of chemical evolution has been concerned with the origin of amino acids. The fact that amino acids are readily formed and detected is probably the main impetus for the avalanche of experiments performed on their prebiological synthesis. This subject has been extensively reviewed (Ponnamperuma & Gabel, 1968; Lemmon, 1970; Gabel & Ponnamperuma, 1972; Molton & Ponnamperuma, 1974). Miller (1957) has suggested that a Strecker synthesis is involved in the formation of amino acids by electric discharge in the heterogeneous phase-system of methane, ammonia, and water (Fig. 3).

The formation of so many amino acids under conditions which presumably simulate various environmental niches of a primordial planet does not mean that further research on the origin of amino acids is unnecessary. Many of the amino acids formed in these experiments do not occur naturally in biological material. Furthermore, a large number of the naturally occurring amino acids have not been found in these experiments. The preponderance of glycine and alanine amongst the products of these experiments is indicative of the high concentration of starting materials and relatively short reaction times compared to geophysical reality. It is very likely that high dilution, long reaction times, and a gradual removal of hydrogen would result in the detectable syntheses of many more amino acids. Since the products are the results of collisions of energetically excited particles, a high ratio of energy to particles is required with the further proviso that the relatively less thermodynamically stable aggregates (which in this case would be polyatomic carbon compounds), do not extensively collide with each other when energetically excited. It is significant to note that all of the non-protein amino acids formed in the aforementioned experiments have also been identified in the Murchison meteorite by Ponnamperuma and his co-workers (Kvenvolden, Lawless & Ponnamperuma, 1971).

The formation of some of the heterocyclic bases of nucleic acids has been shown to occur under primordial conditions (Figs. 4 and 5; Oro, 1961; Ponnamperuma, Sagan & Mariner, 1963; Sanchez, Ferris & Orgel, 1966; Yang & Oro, 1971). The sugar moietes of the nucleic acids are very likely the result of formaldehyde condensations (Fig. 6; Gabel

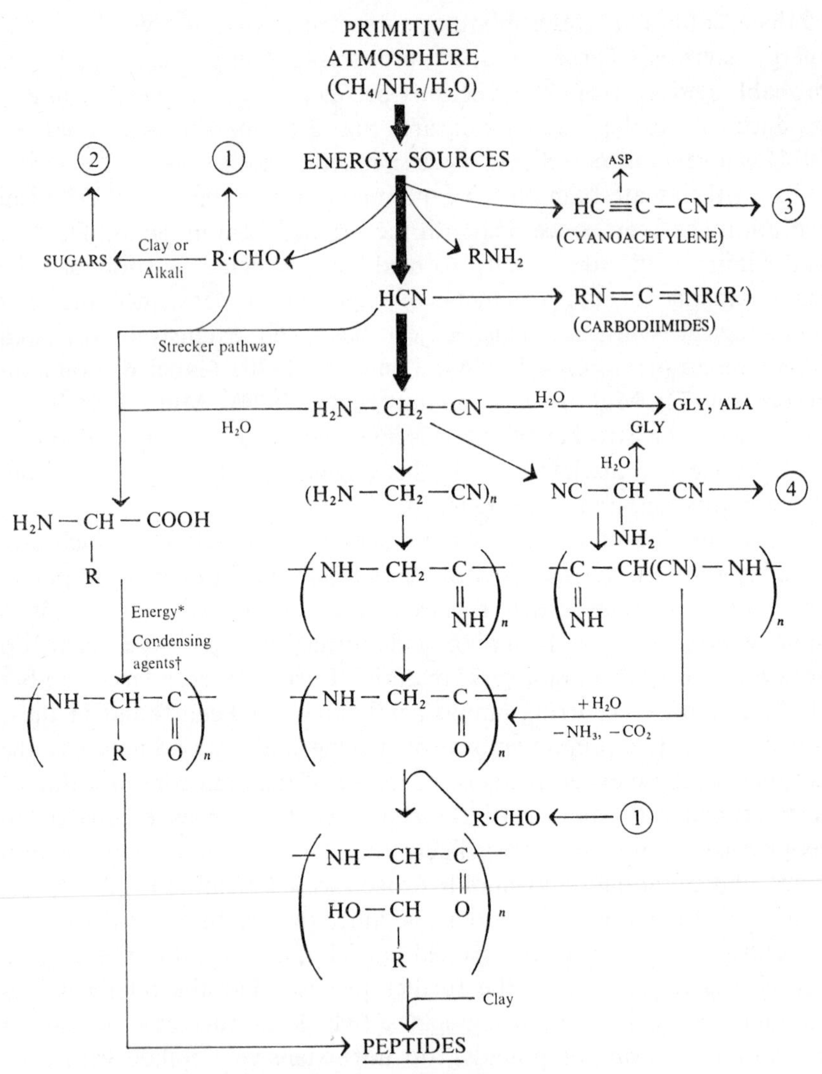

* UV radiation, heat, electric spark, etc.

† Carbodiimide, imidazole, diaminomaleonitrile, polyphosphate, etc.

Numbers in circles refer to contact points on other diagrams

Fig. 2 (part). *For legend see p. 400.*

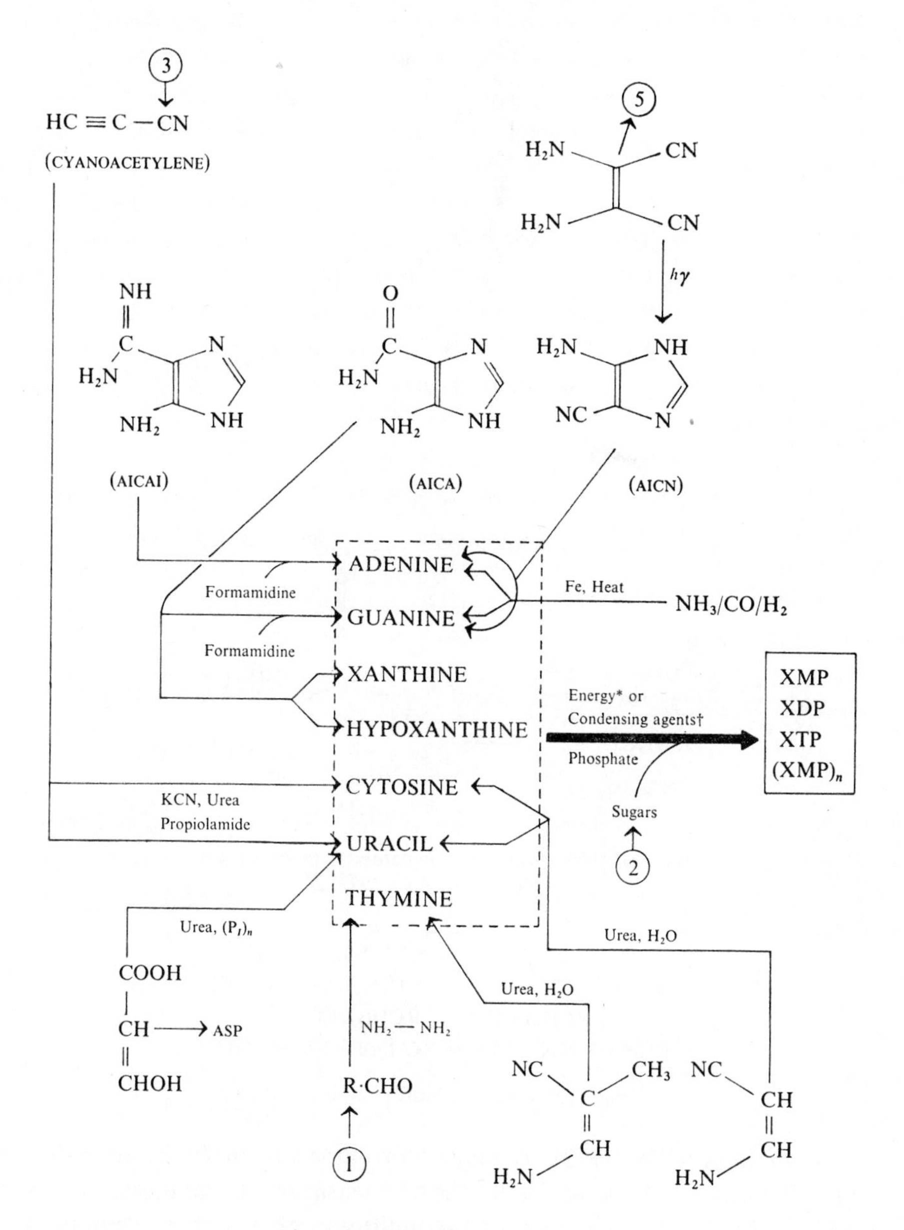

Fig. 2 (part). *For legend see p. 400.*

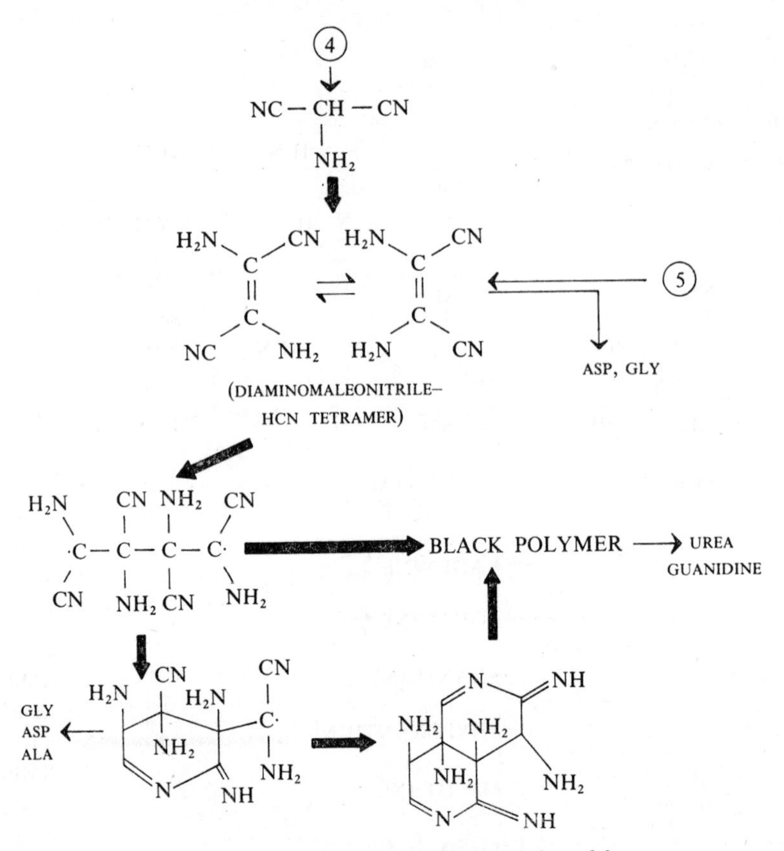

Fig. 2. Prebiological synthetic pathways (adapted from Yuasa & Imahori, 1971).

$$RCHO + NH_3 + HCN \rightleftharpoons RCH(NH_2)CN + H_2O$$
$$RCH(NH_2)CN + 2H_2O \rightarrow RCH(NH_2)COOH + NH_3$$
$$RCHO + HCN \rightleftharpoons RCH(OH)CN$$
$$RCH(OH)CN + 2H_2O \rightarrow RCH(OH)COOH + NH_3$$

Fig. 3. Strecker synthesis of amino acids.

& Ponnamperuma, 1972). A major current problem is to rationalize and demonstrate experimentally the condensation of the bases, sugars and orthophosphate under the conditions of suitable primordial microenvironments. When a dilute aqueous solution (10^{-3} M) of adenine, ribose and phosphate is exposed to ultraviolet light, adenosine has been reported to form (Ponnamperuma *et al.* 1963). Deoxyadenosine also appears to be synthesized in a similar manner by exposing adenine and deoxyribose to ultraviolet light in the presence of hydrogen cyanide.

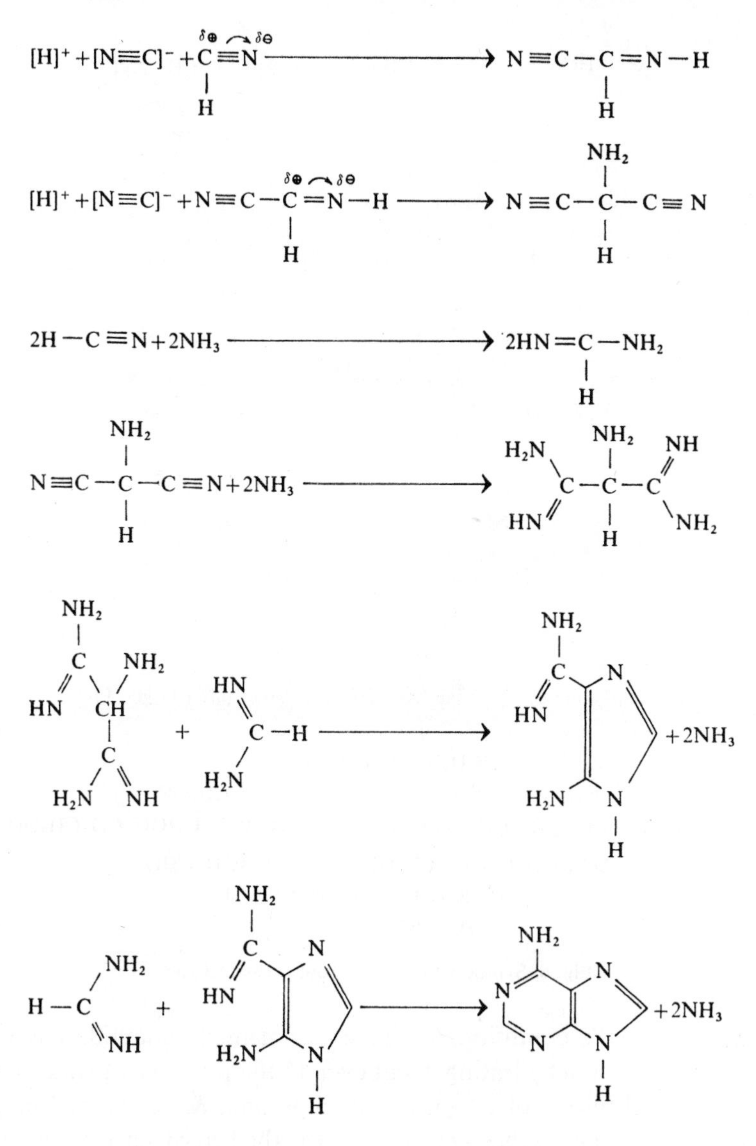

Fig. 4. Mechanism for formation of adenine from HCN.

Reid, Orgel & Ponnamperuma (1967) showed that several isomers are formed in this reaction. Most of them are hydrolytically unstable, with the exception of deoxyadenosine which yields an approximate conversion of not more than 0.3 %. The mechanism of the reaction is probably based upon the formation of cyanamide from the aqueous hydrogen

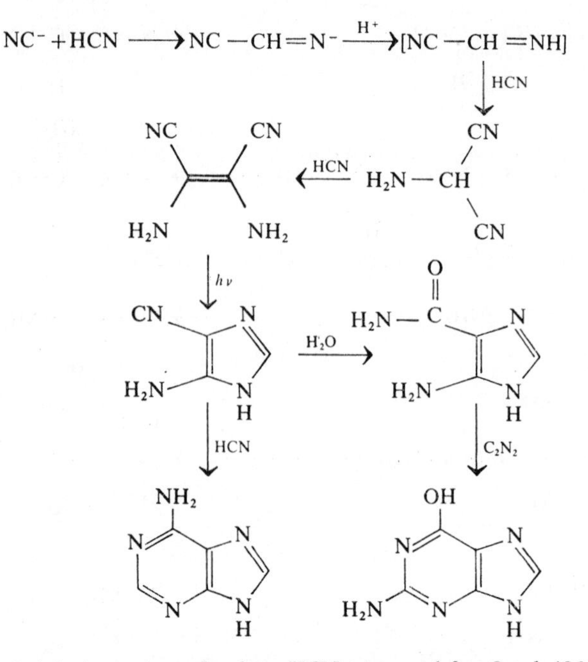

Fig. 5. Adenine and guanine from HCN tetramer (after Orgel, 1966).

$$2CH_2O \rightarrow CH_2OH.CHO$$
$$CH_2OH.CHO + CH_2O \rightarrow CH_2OH.CO.CH_2OH$$
$$CH_2OH.CHO + CH_2OH.CO.CH_2OH \rightarrow CH_2OH.CHOH.CHOH.CO.CH_2OH$$
$$2CH_2OH.CHO \rightarrow CH_2OH.CHOH.CHOH.CHO$$
$$\textit{or}\ CH_2OH.CO.CHOH.CH_2OH$$
$$2C_4 — [C_8]? \rightarrow C_3 + C_5$$

Fig. 6. Synthesis of sugars from formaldehyde.

cyanide under the conditions of the experiment. Cyanamide has been shown to act as a dehydrating agent even in the presence of water when irradiated with ultraviolet light (Ponnamperuma & Peterson, 1965).

In studies on the synthesis of nucleotides, the heterogeneous reactions that may have taken place in partially dried-up tidal areas has also been examined. In simulating these conditions, an intimate mixture of nucleosides was heated with an inorganic phosphate salt (Ponnamperuma & Mack, 1965). Mononucleotides were identified in the end products. Although the highest conversions were obtained at 160 °C, a small amount of the nucleotides was obtained at temperatures as low as 50 °C. Since water is not incompatible with these reactions and does not hinder them unless present in large excess, these conditions can be described as hypohydrous. Rabinowitz, Chang & Ponnamperuma (1968) have demonstrated that the phosphorylation which occurs in these reactions is the result of the prior thermal polymerization of orthophosphate to polyphosphate. Schwartz & Ponnamperuma (1968) phosphorylated adenosine with tripolyphosphate and higher polyphosphates in aqueous solutions which were *ca* 0.5 M in phosphorus. The 2′-, 3′- and 5′-isomers were obtained. These polyphosphates were found to be effective phosphorylating reagents for adenosine over a wide pH range. Thermal phosphorylation of nucleosides by inorganic phosphate salts produces some oligonucleotides (Ponnamperuma & Mack, 1965). The temperature required for this reaction is *ca* 150 °C and the yields are quite small. Just as in the synthesis of nucleotides via the same reaction conditions, Rabinowitz *et al.* (1968) have shown that the actual phosphorylating agent is a mixture of polyphosphate salts formed from the various orthophosphate salts used in the experiment.

The concept of the evolutionary relationship between proteins and polynucleotides has been reviewed by Jukes (1972) and Woese (1972). It is doubtful that polymerization of nucleotides occurred in a manner analogous to the polymerization of amino acids. For example, photodimerization occurs between the organic bases of nucleic acids (Wagner & Bucheck, 1970) and is responsible for the photodeactivation of DNA (Beukers & Berends, 1960; Setlow & Carrier, 1966). Radiolysis, however, promotes hydroxylation and could have been partly responsible for the formation of hydroxylated purines and pyrimidines.

The interaction between nucleotides and amino acids has been the subject of recent investigation (Saxinger, Ponnamperuma & Woese, 1970; Saxinger & Ponnamperuma, 1973). By means of a special chromatographic resin material containing amino acids covalently bound to the matrix, the relative affinities and relative selectivity coefficients of nucleotides for each of the amino acid resins were determined. There is a correlation between aromaticity and binding preference. Amino acids which have only electrostatic binding potential do not display this preference. The results seem to indicate that the major specificity of interaction in the polymer–monomer system is contributed by monomer–monomer interactions. There is thus evidence to assume that not only peptides and oligonucleotides, but also amino acids and mononucleotides, may have interacted in the 'primordial soup' giving rise to the earliest associations between nucleic acids and proteins.

Simulated primordial syntheses of polypeptides have received more attention than the syntheses of other biological polymers. This is probably due to the great emphasis that has been placed on the primordial origin of amino acids. Fox (1965) has long been an exponent of the anhydrous or hypohydrous thermal origin of all primordial organic compounds. Fox & Harada (1958) have shown that in the presence of a proportionally large amount of glutamic acid or aspartic acid, an intimate mixture of all eighteen amino acids normally present in proteins can be thermally polymerized at a temperature of 180–200 °C. These polymers have been described as proteinoids.

It has been argued that the energy available from volcanic activity is quite small compared to the other available sources of energy (Horowitz & Miller, 1962). Nevertheless, it cannot be denied that organic reactions, including polymerization of amino acids with the formation of peptide bonds, would occur under conditions simulating a volcano. A weakness in this type of pathway for the origin of polypeptides is the necessity of postulating highly concentrated, intimate mixtures of the amino acids which are to be polymerized.

The synthesis of peptides directly from dilute solutions of amino acids has also been accomplished through the use of such condensing agents as cyanamide and dicyanamide. Both compounds are known to be formed upon ultraviolet irradiation of aqueous solutions of hydrogen cyanide or through the irradiation of a mixture of methane, ammonia, and water.

Ponnamperuma & Peterson (1965) have reported that glycylleucine and leucylglycine were formed from a dilute aqueous solution of glycine and leucine subjected to ultraviolet irradiation in the presence of cyanamide at pH 5.

Rabinowitz & Ponnamperuma (1969) found that aqueous solutions which contained amino acids and polyphosphates (both 0.01 to 0.1 M) produced peptides at 70 °C or ambient room temperature at various pHs. The pH optimum is near neutrality (pH 7–8) and the reaction still proceeds in alkaline but not in acid medium. The condensation reaction occurs even with pyrophosphate and at room temperature, with small yields (0.4–0.5 %) (Rabinowitz, Flores, Krebsbach & Rogers, 1969). The yields increased with the length of the chain of the linear polyphosphate (up to 13.6 %). Ponnamperuma & Flores (1966) reported the formation of peptides from a methane, ammonia, and water mixture that had been subjected to an electric discharge. The peptides appeared to contain three to four amino acid residues. Upon hydrolysis, nine amino acids were identified. This condensation was probably mediated by the dimer or tetramer of hydrogen cyanide.

The amino acid adenylates have figured in some polymerization experiments (Banda & Ponnamperuma, 1971; Krampitz & Fox, 1969).

In their experiments Banda & Ponnamperuma solubilized a mixture of amino acid adenylates, including ^{14}C-alanine as a labelled marker, and allowed them to react at pH 8 for three hours at room temperature. The yield of polymeric material was below 5 %. Also significant was the fact that the initial amino acid components were all equimolar, whereas the hydrolyzed polymer was not, i.e. some selection had occurred. This experiment was modeled after the biological process of protein synthesis.

The Oparin–Haldane hypothesis assumes that the first organisms or proto-organisms were heterotrophs which utilized the organic material of their reducing environment. As the organic material became depleted and the environment gradually changed to an oxidizing one, an autotrophic organism would have had the best chance for survival at the earth's surface. What the latter statement implies is that some organisms had already incorporated porphyrins or proto-porphyrins within their metabolic systems before the reducing to oxidizing transition occurred. It is an open question whether the escape of hydrogen from the atmosphere or the evolvement of photosynthetic organisms had the greatest effect on the change from a reducing atmosphere to an oxidizing one. Nevertheless, photochemically oriented autotrophs must have existed during the transition. Cloud (1968) has documented well the geological evidence for this transition and presents a plausible argument for the evolutionary development of autotrophs.

Until very recently, all attempts to detect pyrroles or porphyrins in the end products of primitive atmospheres subjected to the various energy sources had been unsuccessful. Hodgson & Baker (1967) obtained evidence for the formation of porphyrins from a dilute aqueous solution of formaldehyde and pyrrole under conditions simulating geochemical abiogenesis. A more biochemically oriented approach used δ-aminolevulinic acid as the starting material. This amino acid is the biogenic precursor of pyrroles and it was found that ultraviolet irradiation could induce its condensation to porphyrin pigments (Szutka, 1966). Finally, Hodgson & Ponnamperuma (1968) successfully demonstrated that porphyrins are present in the residue formed by passing an electric discharge through methane, ammonia, and water vapor. The identification was made on the basis of absorption spectra, solubility, chromatographic behavior, metal complexing and fluorescence spectra.

PREBIOLOGICAL ENERGETICS

The first organisms or protolife forms are assumed by most investigators to have been obligate heterotrophs which used the materials of their environment for energy to reproduce and maintain their structure. Although it is conceivable that some simple chemicals of the environment could have been utilized directly, the universal occurrence of inorganic and organic phosphates in the metabolic pathways of every arbitrarily designated living thing indicates the fundamental importance of phosphates and especially the phosphoric anhydride bond for the maintenance of life processes.

Since the phosphoric anhydride bond is important as the repository of metabolic energy in biological systems, considerable attention has been focused by those interested in the origin of life on the abiotic formation of pyrophosphate from orthophosphate. Miller & Parris (1964) obtained a 1–2 % conversion of hydroxyapatite to calcium pyrophosphate in the presence of a dilute solution of cyanate salts.

There is ample reason for believing that cyanate ion may have been present in primordial waters. Urea, which is produced in abundance in experiments on so-called primordial gas mixtures, thermally decomposes in aqueous solution to ammonium cyanate. Orthophosphoric acid and cyanate salts in water readily form carbamyl phosphate, a mixed anhydride.

$$H_3PO_4 + NCO^{(-)} + H^{(+)}_{aq} \longrightarrow H_2N-\overset{\overset{\displaystyle O}{\|}}{C}-O-\underset{\underset{\displaystyle OH}{|}}{\overset{\overset{\displaystyle O}{\|}}{P}}-OH$$

It is probably through this intermediate that the pyrophosphate is formed.

The purine and pyrimidine nucleotide diphosphates and triphosphates, although universally occurring in all arbitrarily designated life forms, are not the only compounds with phosphoric anhydride bonds to be found in living tissue. Linear inorganic polyphosphate chains were first reported as a constituent of living cells by Wiame (1947*a*, *b*). Inorganic polyphosphates apparently occur throughout the biosphere (Harold, 1966; Gabel & Thomas, 1971) and have been proposed to have played a role in the structuralization (Gabel, 1965, 1973) and energetics of prebiological systems (Lipmann, 1965; Kulaev, 1971). Polyphosphates have been used as phosphorylating and condensation agents in aqueous solutions which attempt to simulate primordial environments

(Schwartz & Ponnamperuma, 1968; Rabinowitz *et al.* 1969). Their present-day regulatory function in biological transport has been discussed (Deierkauf & Booij, 1968; Gabel, 1972). It would appear from these reports that polyphosphates were not only important in the development of primordial phosphorus metabolism but are also functional in present-day biological systems.

PRECELLULAR ORGANIZATION

The formation of membranes which served to isolate microsystems from the environment probably took place at an early stage of chemical evolution (Gabel, 1965; Shah, 1972). Recent work by Folsome & Morowitz (1969) has demonstrated that complex structured products will form from alkanes which have been irradiated with ultraviolet light in the presence of aqueous phosphate and magnesium ions. Lasaga, Holland & Dwyer (1971) have suggested that the primitive earth was covered at one time with a layer of oil derived from methane photolysis.

If ultraviolet irradiation induced the formation of membranous structures, inorganic and organic molecules which would have been present in the environment would have been incorporated into these microsystems.

The two most extensively discussed models for precellular organization are the coacervates of Oparin (1964, 1965) and the microspheres of Fox (1965). Coacervation is a process in which a large part of the colloid of a sol separates as a second phase. The original phase system is depleted in colloidal material whereas the second phase is greatly enriched. If the new phase separates as amorphous droplets, the droplets are called coacervates. These droplets may or may not coalesce to a coacervate layer depending upon the amount of the second phase which is suspended in the original sol. Coacervation can be induced by various changes in the sol including hydrogen ion concentration and temperature.

Oparin has used natural and unnaturally occurring high molecular weight polymers and various mixtures of polymers to form coacervates. He has incorporated enzymes into the coacervate systems and has demonstrated an enhancement of activity in the coacervate drops above that which the enzyme would have in solution. Oparin does not in any way imply that the coacervates he and his associates have studied were the actual precursors of the protocell; but he does maintain that the process of coacervation, involving simpler biologically important materials, could have led to the formation of this protocell.

Fox & Yuyama (1963) have shown that when thermally produced polypeptides are rehydrated in warm water, cooling or salting out will produce a myriad of microspherules which, morphologically, closely resemble micrococci. These microspherules are called proteinoid microspheres. It is Fox's contention that biologically important polymers have a thermal origin and that the formation of proteinoid microspheres represents a plausible pathway by which the first protocells were produced.

Although there is some dispute over whether thermal polymerization of amino acids was the prime source of polypeptides, this should not detract from the usefulness of the proteinoid microspheres as a model for a protocell if microsphere formation were a general property of polypeptides. It has been argued that proteinoids are more stable than coacervates and represent, therefore, a more valid model of a protocell. Oparin, however, maintains that the instability of the coacervates is an advantage, since it makes them more responsive to their environment.

A radical departure from the aforementioned morphological concepts has been proposed by Gabel (1965). Based upon the idea that an excitable membrane preceded the formation of a protocell, this hypothesis traces the evolutionary development of a proto-life excitable membrane. From a critical examination of the properties of an excitable membrane (Abood & Gabel, 1965), it can be envisioned that a three-dimensional polyphosphate co-ordination complex with alkaline earth cations (Ca^{2+} and Mg^{2+}) as central atoms could serve as a primordial model for a proto-life membrane. The polyphosphates could be provided by leaching of primordial igneous phosphate deposits which would have separated from the cooling magma as polyphosphate salts. The positions in the complex to be occupied by alkali metal cations (Na^{+} and K^{+}) would depend on the ion-exchange characteristics of polyphosphates. The formation of this inorganic proto-life membrane would be in accord with the potential gradient of present-day biological membranes. It is further postulated that biologically important monomers and polymers, being polydentate ligands themselves, would participate in the formation of this macromolecular co-ordination complex and eventually assume structural and metabolic functions as the source of polyphosphates became depleted. Cogent evidence for this model has been provided by the universal occurrence of inorganic polyphosphates in living tissue (Gabel & Thomas, 1971). In addition, membrane transport phenomena appear to be dependent on the presence of inorganic polyphosphate chains, associated with the external membrane of yeast (Deierkauf & Booij, 1968).

CONCLUSION

One of the most overt but often ignored questions raised by the problem of the origin of life is: why did it all happen? Rosen (1973) maintains that any dynamical model of evolution which does not incorporate a principle of function cannot explain the *raison d'être* of evolution. Pattee (1973) has long argued that structure and function cannot be understood independently of each other and yet explain the origin of life.

One of the unresolved problems of morphological models for pre-cellular organization is the importance of response to environmental stimulus on evolution. Although Oparin's coacervate model would be responsive to its aqueous environment, its ability to maintain its structural integrity while being impinged by environmental information would be dependent on the composition (material and spatial) of the coacervate drop. Maintenance of structural integrity alone is not evolution. For evolution to occur, the morphological model must in some way utilize the impinging information to increase its responsiveness and impart stability to the *principles* which govern its structural *pattern*.

Gabel (1973) has extended the excitability hypothesis into a general theory in order to explain the origin of hierarchies and the presence of non-optimal species in evolution. Although formal and theoretical treatises on evolution run the risk of being expressions of Baconian optimism, they should certainly not be neglected for they may prove to have enduring value.

As an experimental pursuit, however, the origin of life must be treated from a mainly biochemical point of view. Current knowledge of biological self-replication would imply that a polynucleotide should have been the first form which most experimentalists would agree to designate as *living*.

The present work of our laboratory on the interaction between nucleotides and amino acids has given us some insight into the transcription of biological information between nucleic acids and proteins. We are hopeful that this would lead us to an understanding of the evolution of the genetic code, which is of paramount importance to the understanding of the origin of life.

REFERENCES

ABELSON, P. H. (1956). Amino acids formed in primitive atmospheres. *Science*, **124**, 935.

ABOOD, L. G. & GABEL, N. W. (1965). Relationship of calcium and phosphates to bioelectric phenomena in the excitatory membrane. *Perspectives in Biology and Medicine*, **9**, 1–12.

BANDA, P. W. & PONNAMPERUMA, C. (1971). Polypeptides from the condensation of amino acid adenylates. *Space Life Sciences*, **3**, 54–61.

BARGHOORN, E. S. & SCHOPF, J. W. (1966). Microorganisms three billion years old from the Precambrian of South Africa. *Science*, **152**, 758–63.

BEUKERS, R. & BERENDS, W. (1960). Isolation and identification of the irradiation product thymine. *Biochimica et Biophysica Acta*, **41**, 550–1.

BROWN, H. (1952). Rare gases and the formation of the earth's atmosphere. In *The Atmospheres of the Earth and Planets*, 2nd edition, ed. G. P. Kuiper, pp. 258–66. Chicago: University Press.

CLOUD, P. E., JR. (1968). Atmospheric and hydrospheric evolution on the primitive earth. *Science*, **160**, 729–36.

DEIERKAUF, F. A. & BOOIJ, H. L. (1968). Changes in the phosphatide pattern of yeast cells in relation to active carbohydrate transport. *Biochimica et Biophysica Acta*, **150**, 214–25.

DICKIE, R. H. (1962). Dating the galaxy by uranium decay. *Nature, London*, **194**, 329–30.

ENGEL, A. E., NAGY, B., NAGY, L. A., ENGEL, C. G., KREMP, C. W. W. & DREW, C. M. (1968). Alga-like forms in Onverwacht Series, South Africa: oldest recognized lifelike forms on earth. *Science*, **161**, 1005–8.

FOLSOME, C. E. & MOROWITZ, H. J. (1969). Prebiological membranes: synthesis and properties. *Space Life Sciences*, **1**, 538–44.

FOX, S. W. (1965). Simulated natural experiments in spontaneous organization of morphological units from proteinoid. In *The Origins of Prebiological Systems and of Their Molecular Matrices*, ed. S. W. Fox, pp. 361–82. New York: Academic Press.

FOX, S. W. & HARADA, K. (1958). Thermal copolymerization of amino acids to a product resembling protein. *Science*, **128**, 1214.

FOX, S. W. & YUYAMA, S. (1963). Abiotic production of primitive protein and formed microparticles. *Annals of the New York Academy of Sciences*, **108**, 487–94.

GABEL, N. W. (1965). Excitability and the origin of life: A hypothesis. *Life Sciences*, **4**, 2085.

GABEL, N. W. (1972). Could those rapidly exchangeable phosphoproteins be polyphosphate–protein complexes? *Perspectives in Biology and Medicine*, **15**, 640–3.

GABEL, N. W. (1973). Abiogenic aspects of biological excitability: A general theory for evolution. In *Biogenesis–Evolution–Homeostasis*, ed. A. Locker, pp. 85–91. Heidelberg & Berlin: Springer-Verlag.

GABEL, N. W. & PONNAMPERUMA, C. (1972). Primordial organic chemistry. In *Exobiology*, ed. C. Ponnamperuma, pp. 95–135. Amsterdam: North-Holland.

GABEL, N. W. & THOMAS, V. (1971). Evidence for the occurrence of inorganic polyphosphate in vertebrate tissue. *Journal of Neurochemistry*, **18**, 1229–42.

HALDANE, J. B. S. (1928). The origin of life. *Rationalist Annual*, **148**, 3–10.

HAROLD, F. M. (1966). Inorganic polyphosphates in biology: Structure, metabolism, and function. *Bacteriological Reviews*, **30**, 772–94.

Hodgson, G. W. & Baker, B. L. (1967). Porphyrin abiogenesis from pyrrole and formaldehyde under simulated geochemical conditions. *Nature, London*, **216**, 29–32.

Hodgson, G. W. & Ponnamperuma, C. (1968). Prebiotic porphyrin genesis: Porphyrins from electric discharge in methane, ammonia, and water vapor. *Proceedings of the National Academy of Sciences, USA*, **59**, 22–8.

Horowitz, N. H. & Miller, S. L. (1962). Current theories on the origin of life. *Fortschritte der Chemie organischer Naturstoffe*, **20**, 423–59.

Jukes, T. H. (1972). Recent advances in studies of evolutionary relationships between proteins and nucleic acids. *Space Life Sciences*, **1**, 469–90.

Krampitz, G. & Fox, S. W. (1969). A condensation of the adenylates of the amino acids common to protein. *Proceedings of the National Academy of Sciences USA*, **62**, 339–406.

Kulaev, I. S. (1971). Inorganic polyphosphates in evolution of phosphorus metabolism. In *Molecular Evolution*, vol. I, *Chemical Evolution and the Origin of Life*, ed. R. Buvet & C. Ponnamperuma, pp. 458–65. Amsterdam: North-Holland.

Kvenvolden, K., Lawless, J. & Ponnamperuma, C. (1971). Non-protein amino acids in the Murchison meteorite. *Proceedings of the National Academy of Sciences, USA*, **68**, 486–90.

Lasaga, A. C., Holland, H. D. & Dwyer, M. J. (1971). Primordial oil slick. *Science*, **174**, 53–5.

Lemmon, R. M. (1970). Chemical evolution. *Chemical Reviews*, **70**, 95–109.

Lipmann, F. (1965). Projecting backward from the present stage of evolution of biosynthesis. In *The Origins of Prebiological Systems and of Their Molecular Matrices*, ed. S. W. Fox, pp. 259–80. New York: Academic Press.

Miller, S. L. (1957). The mechanism of synthesis of amino acids by electric discharges. *Biochimica et Biophysica Acta*, **23**, 480–9.

Miller, S. L. & Parris, M. (1964). Synthesis of pyrophosphate under primitive earth conditions. *Nature, London*, **204**, 1248–50.

Miller, S. L. & Urey, H. C. (1959). Organic compound synthesis on the primitive earth. *Science, New York*, **130**, 245–51.

Molton, P. M. & Ponnamperuma, C. (1974). Chemical evolution in the universe. *Space Science Reviews*, in press.

Oparin, A. I. (1924). *Proischogdenie Zhizni.* Moscow: Moscovsky Robotchii.

Oparin, A. I. (1964). *The Chemical Origin of Life* (transl. by A. Synge). Springfield, Illinois: Charles C. Thomas.

Oparin, A. I. (1965). The pathways of the primary development of metabolism and artificial modeling of the development in coacervates drops. In *The Origins of Prebiological Systems and of Their Molecular Matrices*, ed. S. W. Fox, pp. 331–46. New York: Academic Press.

Oro, J. (1961). Mechanism for formation of adenine from hydrogen cyanide. *Federation Proceedings*, **20**, 352.

Pattee, H. H. (1973). Physical problems of the origin of natural controls. In *Biogenesis–Evolution–Homeostasis*, ed. A. Locker, pp. 41–9. Heidelberg and Berlin: Springer-Verlag.

Pirie, N. W. (1937). The meaninglessness of the terms life and living. In *Perspectives in Biochemistry*, ed. J. Needham & D. Green, pp. 11–22. New York: Macmillan.

Ponnamperuma, C. & Flores, J. (1966). Possible polymerization of amino acids in prebiotic synthesis. *Abstracts, 152nd National Meeting of the American Chemical Society*, New York, section C-33.

Ponnamperuma, C. & Gabel, N. W. (1968). Current status of chemical studies on the origin of life. *Space Life Sciences*, **1**, 64–96.

PONNAMPERUMA, C. & MACK, R. (1965). Formation of dinucleoside phosphates and dinucleotides in the study of chemical evolution. *Abstracts*, 150*th National Meeting of the American Chemical Society*, Atlantic City, New Jersey, section C-44.

PONNAMPERUMA, C. & PETERSON, E. (1965). Peptide synthesis from amino acids in aqueous solution. *Science*, **147**, 1572–4.

PONNAMPERUMA, C., SAGAN, C. & MARINER, R. (1963). Synthesis of adenosine triphosphate under possible primitive earth conditions. *Nature, London*, **199**, 222–6.

RABINOWITZ, J., CHANG, S. & PONNAMPERUMA, C. (1968). Phosphorylation by way of inorganic phosphate as a potential prebiotic process. *Nature, London*, **218**, 442–3.

RABINOWITZ, J., FLORES, J., KREBSBACH, R. & ROGERS, G. (1969). Peptide formation in the presence of linear or cyclic polyphosphates. *Nature, London*, **224**, 795–6.

RABINOWITZ, J. & PONNAMPERUMA, C. (1969). Peptide formation in aqueous solutions of amino acids in the presence of linear or cyclic polyphosphates: A possible prebiotic process. *Abstracts*, 158*th National Meeting of the American Chemical Society*, New York, Biol-212.

REID, C., ORGEL, L. E. & PONNAMPERUMA, C. (1967). Peptide synthesis under potentially prebiotic conditions. *Nature, London*, **216**, 936.

ROSEN, R. (1973). On the generation of metabolic novelties in evolution. In *Biogenesis–Evolution–Homeostasis*, ed. A. Locker, pp. 113–23. Heidelberg & Berlin: Springer-Verlag.

RUSSELL, H. N. (1935). *The Solar System and Its Origin*. New York: Macmillan.

SANCHEZ, R. A., FERRIS, J. P. & ORGEL, L. E. (1966). Conditions for purine synthesis: Did prebiotic synthesis occur at low temperatures? *Science*, **153**, 72–3.

SAXINGER, W. & PONNAMPERUMA, C. (1973). Interaction between amino acids and nucleotides in the prebiotic milieu. 4*th International Symposium on the Origin of Life*, Barcelona, Spain, 24–28 June.

SAXINGER, W., PONNAMPERUMA, C. & WOESE, C. (1970). Nucleic acid–amino acid interactions, nucleotide–amino amide binding studies. *Abstracts*, 160*th National Meeting of the American Chemical Society*, Chicago, Illinois, section 061.

SCHWARTZ, A. & PONNAMPERUMA, C. (1968). Phosphorylation of adenosine with linear polyphosphate salts in aqueous solution. *Nature, London*, **218**, 443.

SETLOW, R. B. & CARRIER, W. L. (1966). Pyrimidine dimers in ultraviolet-irradiated DNA's. *Journal of Molecular Biology*, **17**, 237–54.

SHAH, D. O. (1972). The origin of membranes and related surface phenomena. In *Exobiology*, ed. C. Ponnamperuma, chapter 7, pp. 235–65. Amsterdam: North-Holland.

SZUTKA, A. (1966). Formation of pyrrolic compounds by ultraviolet irradiation of δ-amino-levulinic acid. *Nature, London*, **212**, 401–2.

THOMPSON, B. A., HARTECK, P. & REEVES, R. R., JR. (1963). Ultraviolet absorption coefficients of CO_2, CO, O_2, H_2O, N_2O, NH_3, NO, SO_2 and CH_4 between 1850 and 4000 Å. *Journal of Geophysical Research*, **68**, 6431–6.

TILTON, G. R. & STEIGER, R. H. (1965). Lead isotopes and the age of the earth. *Science*, **150**, 1805–8.

VENKATESWARAN, V. (1971). Escape of hydrogen and helium from the earth's atmosphere. *Planetary and Space Science*, **19**, 275.

WAGNER, P. J. & BUCHECK, D. J. (1970). Photodimerization of thymine and uracil in acetonitrile. *Journal of the American Chemical Society*, **92**, 181–5.

WATANABE, K., ZELIKOFF, M. & INN, E. C. Y. (1953). Absorption coefficients of several atmospheric gases. *AFCRC Technical Report*, 53–23, Geophysical Research Paper, June 1953.

WIAME, J. M. (1947*a*). Etude d'une substance polyphosphores, basophile et metachromatique chez les levures. *Biochimica et Biophysica Acta*, **1**, 234–55.

WIAME, J. M. (1947*b*). The metachromatic reaction of hexametaphosphate. *Journal of the American Chemical Society*, **69**, 3146–7.

WOESE, C. (1972). The emergence of genetic organization. In *Exobiology*, ed. C. Ponnamperuma, pp. 301–41. Amsterdam: North-Holland.

YANG, C. C. & ORO, J. (1971). Synthesis of adenine, guanine, cytosine, and other nitrogen organic compounds by a Fischer–Tropsch-like process. In *Chemical Evolution and the Origin of Life*, ed. R. Buvet & C. Ponnamperuma, pp. 152–67. Amsterdam: North-Holland.

YUASA, S. & IMAHORI, K. (1971). Chemical evolution – Experimental approaches to the origin of life; formation of organic compounds on the primordial earth. *Science Reports*, Osaka University **20**, 5–41.

INDEX